Introduction to Quantum Nanotechnology

A Problem Focused Approach

Introduction to Quantum Nanotechnology

A Problem Focused Approach

Duncan G. Steel

Department of Electrical Engineering, Department of Physics,
The University of Michigan

OXFORD
UNIVERSITY PRESS

OXFORD
UNIVERSITY PRESS

Great Clarendon Street, Oxford, OX2 6DP,
United Kingdom

Oxford University Press is a department of the University of Oxford.
It furthers the University's objective of excellence in research, scholarship,
and education by publishing worldwide. Oxford is a registered trade mark of
Oxford University Press in the UK and in certain other countries

First Edition published in 2021
Impression: 2

Published in the United States of America by Oxford University Press
198 Madison Avenue, New York, NY 10016, United States of America

British Library Cataloguing in Publication Data
Data available

Library of Congress Control Number: 2020948796

ISBN 978–0–19–289507–3 (hbk.)
ISBN 978–0–19–289508–0 (pbk.)

DOI: 10.1093/oso/9780192895073.001.0001

Printed and bound by
CPI Group (UK) Ltd, Croydon, CR0 4YY

Cover image: Dr. Bo Sun, PhD

This book is dedicated to my family.

I shall pass this way but once; any good therefore that I can do or any kindness I can show to any human being, let me do it now. Let me not defer nor neglect it, for I shall not pass this way again.

Etienne de Grellet, Quaker Missionary

Preface

This book originated in lecture notes that were generated over a number of years working to develop an application-focused introduction to quantum theory. It isaimed at introducing upper level undergraduate engineers and applied physicists, as well as first year graduate students, to the emerging role of quantum physics in modern technology. Technology embraced concepts like Fermi–Dirac statistics, band structure, and energy quantization years ago, as semiconductor materials and devices rapidly developed. However, a revolution in thinking and development is occurring as Moore's law for exponential growth, based largely on classical physical behavior, is rapidly slowing down and new materials and nano-fabrication capabilities ramp up. Production of nano-structures and devices much smaller than current transistors are governed exclusively by quantummechanical behavior. Rules like Kirchhoff's laws, giving the current as a function of time, or Newton's laws, giving position as a function of time, are replaced by Schrödinger's equations and observables with their corresponding time independent operators. One of the most important results in quantum mechanics is the principle of superposition, where the state of the system can be both on and off, simultaneously. The principle of superposition can result in the production of quantum entangled states and used to teleport information, securely. The quantum vacuum that determines the performance of typical LEDs can now be engineered for specific performance objectives. Hence, the intent of this text is to create the beginning of a quantum toolbox for people interested in understanding and applying quantum mechanics to new ideas in technology or physics.

Pedagogically, the text assumes that students in engineering and applied physics are relatively comfortable with calculus, differential equations (reviewed in Appendix A) along with the first two years of introductory physics including mechanics and some electricity and magnetism. Hence, the technical discussion begins in Chapters 2–6 by simply stating Schrödinger's equation in differential form and examining problems in the context of devices such as nano-vibrators, tunnel junctions, and quantum dots. Solutions to most differential equations are simply stated, with their critical properties discussed. Most of the work is in one dimension so that the learning focuses on physical behavior important to engineering. While the focus of this book is on the relevance to emerging areas of technology, the style of presentation is very much based on the author's own experience in classes taught by Professor Eugene Merzbacher, at Chapel Hill, who focused on understanding the physics rather than complex calculations. His approach created the foundation for advanced understanding and work, and his focus was always on the student.

Starting in Chapter 7, the quantum postulates are presented as new design rules for quantum devices. The rules repeat, in a more general and abstract form, what the students used in chapters 2–6, but enable them to then see how problems like a quantum LC circuit (important for superconducting quantum computing) can be solved without solving a differential equation, laying the groundwork for discussing spin, which is anticipated in the analysis of a quantum gyroscope. Dirac notation and the Hilbert space emerge as key concepts for the remainder of the text.

Time dependence is included from the beginning, since Schrödinger's equation is time dependent. However, the discussion of dynamics becomes an important component of the discussion beginning in Chapter 9 and continuing, as well as understanding the language of pictures and representations. Semiclassical problems like a driven LC circuit or a driven two-level system (leading to Rabi oscillations and the quantum flip-flop) are analyzed, where the driver, such as a sinewave generator or an optical source, is treated classically.

The operator formulation of a quantum gyroscope and an introduction to spin is then presented in Chapter 10 and used to develop the basic ideas of quantum information. Following a discussion of approximation techniques in Chapter 11, Chapters 12 and 13 focus on fermions and bosons,quantum entanglement,and quantum measurement, including the Von Neumann hypothesis.

We then consider the problem of dissipation in Chapter 14, using the idea of a discrete state coupled to a continuum. We use this as a means to understand dissipation, where energy in one quantum system moves irreversibly to a second quantum system. Both systems are fully quantized. The math is nearly identical to that in the Weisskopf–Wigner theory of spontaneous emission. This problem is considered first in the Schrödinger picture and then in the Heisenberg picture where it becomes possible for students to analyze a driven RLC circuit in a fully quantum mechanical manner.

Chapter 14 prepares the student for the discussion in Chapter 15 of the quantized radiation field. The approach includes the development of the quantization concept, and then its application to spontaneous emission in the Weisskopf–Wigner approximation, and the effect of spontaneous emission on measurement.

Chapter 16 is an introduction to atomic operators, which is the name for the operators in a two-level system. The interaction with radiation is developed in the Heisenberg picture, along with the effects of spontaneous emission. This chapter prepares the student for more advanced discussions regarding field correlations and fluctuations. Chapter 17 introduces quantum electromagnetics, including a discussion of noise, the Hong–Ou–Mandel interferometer, and engineering the vacuum with a cavity. As interest increases in quantum sensing, quantum lidar, and quantum radar, this chapter is increasing in importance.

The final discussion in Chapter 18 is on the density matrix, a powerful formalism that is central to many discussions in quantum technology because it can simultaneously incorporate the effect of decay and decoherence into equations that deal explicitly only with the Hilbert sub-space of a specific quantum system.

The class at Michigan has received very strong reviews and has, in recent years, doubled in size every year. The students are primarily undergraduates, but some are first year graduate students who opt for this course rather than starting with the two-semester graduate sequence, the option allowing them to move more quickly to specialized classes in their area of interest.

It appeals to students in applied physics, electrical engineering, computer science, mechanical engineering, materials science, and nuclear engineering. Appendix A reviews the mathematics that they have covered earlier in calculus and summarizes the key results of linear algebra, though linear algebra is presented in the lectures as needed. Appendix A provides a convenient source to look up key relations and properties in linear algebra as well as reminding students how to multiply matrices and evaluate determinants. Problems are presented within the text, as new material is presented, rather than at the end of the chapter. The problems aim at facilitating understanding of the new material while often requiring the use and integration of previous material. The problems for the most part avoid questions that simply require finding the right formula somewhere, as well as problems that simply ask the students to fill in the details of some derivation, although some derivation type problems are useful for helping students to learn a new formalism such as with operators.

The text comprises material that, if taught completely, would take two semesters at Michigan. However, the book is structured so that it is possible to develop distinct one-semester courses that focus on different issues of interest. Several illustrative options are given below for one-semester courses. The discussions are cross-referenced to specific formulas or earlier discussions to facilitate learning and to enable ignoring some material while moving ahead to other material.

I would like to acknowledge the immense help provided by Prof. Paul Berman, over a period in excess of 40 years, in understanding advanced concepts in quantum theory, and the helpful conversations with him over theory and data that have help to inform me of deep underlying physics. I am also indebted to Prof. Herb Winful, who helped me to understand the approach I needed to communicate effectively with our undergraduate students to develop a successful course. I also want to acknowledge the important role of all the students in my courses who have provided valuable feedback on how to improve presentations and my communications.

Examples of different one-semester courses, all aimed at senior undergrads with no prior training in quantum physics, are as follows.

Course 1 The focus is on basic problems and a more problem-focused approach than the usual introduction to quantum mechanics. The first half of the semester is taken up with stationary state solutions to Schrödinger's equation in the coordinate representation, followed by a discussion of the postulates and then dynamics.

Course 2 The focus is on reducing the amount of material presented in the coordinate representation and increasing the emphasis on the use of operators and understanding the basic ideas of dissipation.

Course 3 This approach would work for students with an adequate math background. It starts with the postulates and continues through the density matrix, time permitting. This could be very useful for students who see themselves working on ideas where details such as calculating the matrix elements of an electric dipole operator are not important or are something they could teach themselves.

Course 1	Course 2	Course 3
Chapters 2–9. Chapter 4 on periodic potentials and the Bloch theorem could be skipped, though some discussion on the translation operator might be helpful when introducing the rotation operator in Chapter 10	Chapters 2 and 3, Chapters 7–10, Chapter 11 without second order terms and without degenerate state perturbation theory, Chapters 12–15 (skipping development of the Hamiltonian for the electromagnetic field in Chapter 15)	Chapters 7–11 (possibly skipping second order theory and degenerate time independent theory), Chapters 13–18
Chapters 10 and 11 if time allows (time independent perturbation theory with degenerate states could be skipped as well as second order terms in time independent and time dependent theory)	Chapter 17 on quantum electromagnetics, if there is time	

It is possible to skip Section II in Chapter 15, where the Hamiltonian for the electromagnetic field is derived.

Table of Contents

1 Introduction to Applied Quantum Mechanics: Why quantum behavior is impacting technology

There are many outstanding textbooks on quantum mechanics. Often, they take the historical approach of building up knowledge from simple systems and teaching the math and the new ways of thinking. Today, however, there is a need to move as effectively as possible from basic science understanding to sufficient understanding to facilitate transitioning of problem-solving skills to important problems that have an immediate impact on modern challenges in technology. Indeed, future engineers working to design new devices will need a skill set that is considerably broadened to include the behavior of materials and devices when they become sufficiently small. This text *seeks to provide a beginning for the creation of a quantum toolbox for those interested in these problems.* Using these tools requires a different way of thinking compared to the approach used for classical problems. And while the material works to give the reader the tools to solve problems, effective use of those tools *also requires a fundamental understanding of the basic underlying physics.*

Devices like transistors and quantum well lasers have already created a need to understand the impact of Fermi–Dirac statistics and energy quantization on devices. However, the emergent field of nano-technology is revealing that the concepts we have from our current scale devices are no longer adequate to predict correct device performance. Moreover, in this new regime, new physical properties such as quantum superposition and quantum entanglement are emerging that will revolutionize how we think about information, its storage, its transmission and processing or energy harvesting and transmission, sensing and metrology. *Quantum behavior can limit technology such as the growing problem of quantum tunneling in current computers or it can create new opportunities.*

In the typical approach to the beginning study of quantum mechanics, it is common to build up to Schrödinger's equation for a massive particle (e.g., an electron) written as a partial differential equation in space and time that reflects the well-known wave behavior of massive particles. Then the development progresses to various features of this equation in different systems and then examines analytical solutions in the simplest cases, usually the free particle, a particle experiencing some kind of 1-d potential (square box or square mountain), a particle sitting in a harmonic potential, a rigid rotator to introduce angular momentum, and finally the hydrogen atom. A few more subjects and the first semester is over with no more time for most students to learn more. The study of dynamics, where interesting things happen such as the fundamental physics of how light is emitted by an LED can be one to three semesters of post graduate studies away for the student, and quantum technology and information can be even further away.

This book takes a different approach. At the expense of understanding some of the most fundamental quantum systems in detail, the focus is on presenting the material in a way that is fully consistent with the rules of physics but in a way that enables the examination of features

Introduction to Quantum Nanotechnology: A Problem Focused Approach. Duncan Steel, Oxford University Press (2021).
 DOI: 10.1093/oso/9780192895073.003.0001

that are emerging in modern technology. We will start with some of the postulates of quantum mechanics in a limited way, such as postulating Schrödinger's equation as a partial differential equation, much as, say, mechanics starts with Newton's laws. We will then examine several systems of importance to technology from nano-vibrators to electronic structure, tunneling, and more. We will then write out the new design rules for quantum systems. Specifically, these rules are the formal postulates. *The only proof for physical postulates is that they lead to predictions in agreement with measurement.* You will need to temporarily suspend your usual class room expectation of learning by starting with a physical picture such as Kirchhoff's laws for circuits or Maxwell's equations, and open your mind to an entirely new approach to understanding and doing calculations for a physical system. For example, observables like position or voltage as a function of time do not exist in a quantum system or in your measurement.

We will be able to examine the features of a quantum LC circuit with the new approach, the basis of a leading approach to quantum computing. We will also introduce Dirac notation. Dirac introduced this notation in part to make the different "pictures" of quantum mechanics more similar, but also to deal with new physical observables, such as spin, that have no analogy in the classical system.

Schrödinger's equation, as a partial differential equation, follows from the postulates, but we will learn how to solve this equation using a more general method applied first to a nano-mechanical oscillator, nano-electronics, and other problems.

Math is the language of scientists and engineers: quantum is based on the mathematical manifestation of the postulates. The math is fairly simple, though it is likely that you have not seen some of this math presented in the way it will be used here. So, again, you need to be open to another transition. It may seem abstract or esoteric at first, but you will eventually see how it enables even more powerful thinking on even more usual problems than you have currently experienced. The basic math behind your experience with Fourier series and Fourier transforms is identical to what we will be doing but with functions (or math objects) different from sines, cosines, and exponentials. Appendix A provides a useful review of math that you have undoubtedly used in other classes.

This course will take you into understanding physical behavior when our physical systems become very small and sometimes very cold. If you think about CMOS technology, for example, you are likely familiar with Moore's law, where the number of transistors per unit area is growing exponentially with time. If you realize that atoms on a surface are typically separated by say 0.2–0.3 nm and then use typical plots of Moore's law that show that by 2030 there will be about 10^{10} transistors/mm^2, you will see that this corresponds to about 1000 atoms in a single planar transistor made of a monolayer of atoms. Based on the tools that you learn in this book, you will be able to show that transistors will begin to experience problems because of a failure of larger scale behavior (this explains why a crystal, like silicon, is different from just a bunch of closely packed silicon atoms). Even now, quantum tunneling is interfering with performance, as mentioned above. And of course, thermal loading (the result of dissipation, another complex concept in quantum that is discussed in this book) is causing problems with performance, not to mention huge demands on cooling resources. Effective use of quantum behavior could reduce energy consumption by orders of magnitude and create length scales that are at the limit of a few atoms.

The idea is that, as devices get smaller, quantum physics becomes increasingly important. But quantum can be important even in large systems. You might ask why quantum physics would impact something as crude (by today's standard) as the fluorescent lights overhead. There is a gas discharge and plasma inside the tube. This can be pretty classical (i.e., not really needing quantum physics to understand it) but the light is coming from individual atoms that are very small (about 0.1 nm in diameter). Why does an atom emit light? This is quantum in nature, and requires us to understand something called the vacuum radiation field (with many similarities to an LC circuit, it turns out). So, quantum is important again. At the same time, in the nano-world, it is possible to imagine the equivalent of a fluorescent light inside an opening in a small metal cavity. If the cavity is small enough, the atoms will not emit light... at all! The vacuum radiation field could also become important in nano-mechanical switches, as their separation becomes smaller because the quantum vacuum (what the radiation field is called even in the absence of light, not what you get from a vacuum to clean the house) might cause the switch to short circuit by forcing it closed.

Quantum behavior also opens up new opportunities because of quantum super position. You may recall that at one point, you might have learned how to mathematically write a pulse in time as a Fourier transform or Fourier sum of harmonic functions like sines and cosines or exponentials of time (a phasor, using Euler's formula). This was because an electromagnetic field (or a sound wave, or any other kind of function of time or space) could be written as a linear combination of these functions (either a sum or an integral). You learned that you could see the function of interest as a linear superposition of sines etc. In quantum mechanics, a quantum system can be in a linear superposition of quantum states (called eigenstates). Suppose that you have two states – just label the states with a (0) and a (1). In quantum, we will write this as $|0\rangle$ and $|1\rangle$ and the system will be described by a state vector $|\psi\rangle = \cos\theta|0\rangle + \sin\theta|1\rangle$. The states could be a switch in the off and on position or a bit in a computer representing a 0 and 1. In the case of the classical bit, you can write a (0) OR (1), but not both. An electric switch can be either on or off. In a quantum system, it can be in both states, simultaneously. This is the principle of superposition in quantum mechanics. Now, if we had three classical bits, we could write $2^3 = 8$ numbers, but only one at a time. In a quantum system, each quantum bit (qubit) would be described by its own $|\psi\rangle$. For three separate systems, we would have $|\psi\rangle_1$, $|\psi\rangle_2$, and $|\psi\rangle_3$, respectively. If we considered the entire system, we would write $|\psi\rangle = |\psi\rangle_1 |\psi\rangle_2 |\psi\rangle_3$. If this were classical, $|\psi\rangle_i$ would be a place holder for a 0 or 1 in base 2 for computing. But in the quantum world, each $|\psi\rangle_i$ itself can be in a superposition of a 0 and a 1. This means that you can write all eight numbers simultaneously. It looks like you can get remarkable parallelism. Things are more complicated, but perhaps your imagination is already beginning to see possibilities. An example of a $|\psi\rangle$ that is even more complicated than $|\psi\rangle = |\psi\rangle_1|\psi\rangle_2|\psi\rangle_3$ is deceptively simple looking. It is $|\psi\rangle = \frac{1}{\sqrt{2}}(|0\rangle_1|0\rangle_2|0\rangle_3 + |1\rangle_1|1\rangle_2|1\rangle_3)$. This $|\psi\rangle$ cannot be written as $|\psi\rangle = |\psi\rangle_1|\psi\rangle_2|\psi\rangle_3$, but it can still be created. It is an example of a GHZ (Greenberger–Horne–Zeilinger) state. This state is said to be quantum mechanically entangled. This kind of state can be used to teleport information from one location to another (using photons) with no chance of eavesdropping.

This course will look into these concepts. The primary goal is to familiarize you with some of the basic ideas that are emerging in technology and that are the result of engineering quantum concepts. The math is different, like a new language, and will take some time to get used to. The only way to really learn is to speak the language, so problems will stress the math and the technical applications and implications through problem solving.

2 Nano-Mechanical Oscillator and Basic Dynamics: Part I

2.1 Introduction

Rather than build up from things you might know, we are going to start right off with an important problem. Just as you *assumed* $F = ma$ in your studies of basic physics, we are going to assume that the new design rules are governed by Schrödinger's equation. For $F = ma$, you had to *assume* something for F, realize that the acceleration is defined by $a \equiv \frac{d^2x}{dt^2}$ and then solve a second order differential equation for $x(t)$. In that problem, Newton's laws are a set of postulates and the use of Newton's laws, like $F = ma$, for a specific problem requires the user to guess (postulate) a form for the force, like gravitation. The proof that the whole approach works is *only by measurements (i.e., experiments) verifying the theory.*

The problem we consider is shown in Fig. 2.1. This may seem like an idealized problem, but in nano-mechanical systems, this simple system has all the critical physics. Furthermore, related structures are being investigated for inertial guidance systems. Later, we will see why this basic system and slight variations are the simplest description of a major proportion of many of the problems you will encounter.

A set of unified postulates is presented in Chapter 7. At this stage, we will simply apply some of the consequences of the rules that you will learn shortly.

2.2 Classical Physics Approach: Finding *x(t)*

Classically, you would find the dependent variable, $x(t)$ the position (an observable) as a function of time, by solving for $x(t)$ from Newton's law:

$$F = ma = m\frac{d^2x}{dt^2} \tag{2.1}$$

where, from empirical evidence, the force from the spring is determined to be

$$F = -\kappa x \tag{2.2}$$

and κ is the spring constant. A second order differential equation in time is subject to two initial conditions.

Introduction to Quantum Nanotechnology: A Problem Focused Approach. Duncan Steel, Oxford University Press (2021).
 DOI: 10.1093/oso/9780192895073.003.0002

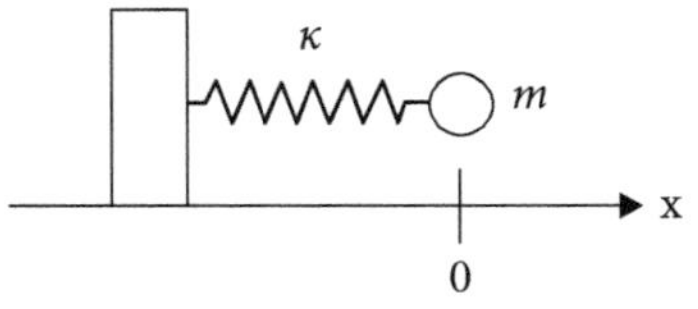

Fig. 2.1 The particle of mass m on a spring with spring constant κ. In the absence of any forces, the particle is at rest at the origin. This is the model for a nano-vibrator.

The *equation of motion* is then

$$\frac{d^2x}{dt^2} + \frac{\kappa}{m}x = 0 \tag{2.3}$$

with the solution given by

$$x(t) = A cos\omega t + B sin\omega t \tag{2.4}$$

A and B are determined by the initial conditions on $x\,(t = 0)$ and $\dot{x}\,(t = 0)$, where $\dot{x} = \frac{dx}{dt}$, and the natural frequency of the oscillator is given by

$$\omega = \sqrt{\frac{\kappa}{m}}. \tag{2.5}$$

Problem 2.1 *Provide the details of the math leading to the solution (2.4) from Eq. 2.3. Evaluate A and B, subject to $x\,(t = 0) = x_0$ and $\dot{x}\,(t = 0) = 0$. Find the average kinetic energy of the particle given by $\left\langle \frac{1}{2}m(\dot{x})^2 \right\rangle$ where the average is defined as the integration over one cycle (round trip) divided by the time period of the cycle.*

Another approach to this classical problem is provided by Hamilton rather than Newton. His postulate is based on two coupled first order differential equations called **Hamilton's equations**:

$$\frac{dp}{dt} = -\frac{\partial H}{\partial x} \text{ and } \frac{dx}{dt} = +\frac{\partial H}{\partial p} \tag{2.6}$$

where x and p are the **canonical coordinate** (meaning no constraints) and the corresponding **conjugate momentum**. H is the **Hamiltonian**. The Hamiltonian represents the total energy of the system, written in terms of two independent variables: x and p. In a simple problem like this, x and p are the same x and p that you are used to, namely $p = m\dot{x}$. The total energy is the sum of the **kinetic energy**, T, which is given by

$$T = \frac{1}{2}mv^2 \equiv \frac{p^2}{2m} \tag{2.7}$$

and the **potential energy**, which is related to the force by

$$F = -\nabla V(x) \tag{2.8}$$

giving

$$V(x) = \frac{1}{2}kx^2 \tag{2.9}$$

We get for the Hamiltonian:

$$H = T + V = \frac{p^2}{2m} + \frac{1}{2}kx^2 \tag{2.10}$$

Inserting this into Hamilton's equations we get:

$$\frac{dp}{dt} = -\frac{\partial H}{\partial x} = -\frac{\partial}{\partial x}\left(\frac{p^2}{2m} + \frac{1}{2}kx^2\right) = -kx \tag{2.11}$$

and

$$\frac{dx}{dt} = +\frac{\partial H}{\partial p} = \frac{\partial}{\partial p}\left(\frac{p^2}{2m} + \frac{1}{2}kx^2\right) = \frac{p}{m} \tag{2.12}$$

Taking the second derivative in Eq. 2.12 and inserting Eq. 2.11:

$$\frac{d^2x}{dt^2} = \frac{1}{m}\frac{dp}{dt} = -\frac{k}{m}x \tag{2.13}$$

or

$$\frac{d^2x}{dt^2} + \frac{k}{m}x = 0 \tag{2.14}$$

This is identical to Eq. 2.3 and hence gives the same result as Newton's law. However, Hamilton's equations describe the time evolution of other systems beside particles with mass moving in space. Later, you will use these equations to solve the classical LC circuit. It is only necessary to identify the canonical coordinate and conjugate momentum (that may not have units of position or linear momentum) for the problem, and write down the Hamiltonian.

The concept of the Hamiltonian is the foundation of quantum mechanics.

2.3 The Quantum Approach: Finding $\psi(x, t)$

Having understood the usual classical solution, it will now be possible to understand the profound change in the behavior when the system behaves quantum mechanically. The starting point for a quantum solution to this problem is the Hamiltonian for the oscillator, which we developed above.

In quantum mechanics, there is a postulate to be presented in Chapter 7 about converting the Hamiltonian into a **mathematical operator**. An operator is a math instruction that tells the reader what it does to math objects. An operator could be a derivative, a matrix, a variable, etc. We will do this more carefully in Chapter 7, but for now we will just give the appropriate operator. Specifically, the two observables (the canonical coordinate, $x(t)$ and the conjugate momentum, $p(t)$ become *time independent* operators, $\hat{x}$ and $\hat{p}$, respectively, such that in a coordinate basis

$$\begin{aligned} x(t) &\to \hat{x} = x \\ p(t) &\to \hat{p} = -i\hbar\frac{\partial}{\partial x} \\ H &\to \widehat{H} \end{aligned}$$

Notice that the operators and the corresponding functional form are time independent. This is quite general in quantum physics, and applies to any problem describing a point particle with mass m moving in space and can be generalized to two and three dimensions using the appropriate coordinate system. The corresponding form of the Hamiltonian is

$$\widehat{H} = \frac{\hat{p}^2}{2m} + \frac{1}{2}\kappa\hat{x}^2 = -\frac{\hbar^2}{2m}\frac{d^2}{dx^2} + \frac{1}{2}\kappa x^2 \tag{2.15}$$

The symbol $\hbar = \frac{h}{2\pi}$ and **Planck's constant** $h = 6.626 \times 10^{-34}$ joule-second, a very small number.

Problem 2.2 *Show that since we know that* $\hat{p} = -i\hbar\frac{d}{dx}$ *has units of linear momentum (the operator must have the same units as the corresponding variable), the units of Plank's constant must be joule-second.*

As discussed in classical physics, the governing equation for the problem is a time dependent equation, referred to as the *equation of motion*. An example is in Eqs 2.1 and 2.3, which defines the classical equation of motion for the variable, $x(t)$.

For quantum systems, the governing equation or equation of motion is given by **Schrödinger's equation**, shown here in operator form:

$$\widehat{H}\psi(x,t) = i\hbar\frac{d}{dt}\psi(x,t) \tag{2.16}$$

where

$$i \equiv \sqrt{-1} \tag{2.17}$$

Using the representation above, for $\hat{p}$ and $\hat{x}$, we see that we have a partial differential equation in with two independent variable, x and t.

$$\left(\frac{\hat{p}^2}{2m} + \frac{1}{2}\kappa\hat{x}^2\right)\psi(x,t) = \left(-\frac{\hbar^2}{2m}\frac{d^2}{dx^2} + \frac{1}{2}\kappa x^2\right)\psi(x,t) = i\hbar\frac{d}{dt}\psi(x,t) \tag{2.18}$$

Schrödinger's equation is the equation of motion, but not for an observable like x but rather for the **wave function**, $\psi(x,t)$. And $\psi(x,t)$ *contains all the information that can be known about the system.*

This is a very different looking equation from Newton's equations. First, we solved Eq. 2.3, a second order differential equation in time, for the position as a function of time ($x(t)$). In Eq. 2.16, we are solving a first order equation in time and a second order equation in space for a function $\psi(x,t)$ that has no physical meaning at this point, and in fact $\psi(x,t)$ is often complex (i.e., not real). What is even more unsettling to many is that we have no physical sense of any of the variables. Even the position has become independent of time, and there is no way to relate this to something that we have a physical sense for in the classical world.

Quantum analysis will never provide a form for $x(t)$. The most that can be known at this point about the change in x with time is the average value of repeated measurements (after the system has been recreated each time after the measurement) represented by the symbol $\langle x\rangle(t)$ called the time dependent **expectation value** where

$$\langle x\rangle(t) \equiv \langle \hat{x}\rangle(t) \equiv \int_{-\infty}^{\infty} dx\, \psi^*(x,t)\,\hat{x}\,\psi(x,t) = \int_{-8}^{\infty} dx\, \psi^*(x,t)\,x\,\psi(x,t) = \int_{-\infty}^{\infty} dx\, x\left|\psi(x,t)\right|^2 \tag{2.19}$$

Later, we will show why Eq. 2.19 is justified, though it is often included as a postulate in quantum (there will be more discussion in Chapter 7).

With Eq. 2.10 representing an average of the observable x, the function $|\psi(x,t)|^2 \equiv \psi^*(x,t)\,\psi(x,t)$ can be interpreted as a **probability density function** *for finding the system between x and x + dx (or in dx about x) at time t.* As a probability density function, we require:

$$\int_{-\infty}^{\infty} dx\, |\psi(x,t)|^2 = 1 \tag{2.20}$$

meaning that $\psi(x,t)$ must be **normalized** before it can qualify as a wave function. We will discuss this in more detail below.

We return now to Eq. 2.18 to solve a partial differential equation in x and t for the time dependent wave function, $\psi(x,t)$. Since the Hamiltonian, $\hat{H}$, is independent of time, we can use the **method of separation of variables** to simplify the equation (see Appendix A). Specifically, we can write $\psi(x,t) = \psi(x)g(t)$ and substitute into Eq. 2.16 to get

$$\psi(x,t) = \psi(x)g(t) \tag{2.21}$$

which gives

$$\hat{H}\psi(x)g(t) = i\hbar\frac{d}{dt}\psi(x)g(t) \tag{2.22}$$

Since x and t are independent variables and $\hat{H}$ is independent of time and $\left(i\hbar\frac{d}{dt}\right)$ is independent of space, it must be the case that we can rewrite Eq. 2.22 as

$$\frac{1}{\psi(x)}\hat{H}\psi(x) = i\hbar\frac{1}{g(t)}\frac{d}{dt}g(t) \tag{2.23}$$

This is the essence of the method of separation of variables described in texts on partial differential equations. Since x can vary on the left-hand side with no change in the right-hand side and t can vary on the right-hand side with no change on the left-hand side, it must be the case that both sides are constants; i.e.,

$$\frac{1}{\psi(x)}\hat{H}\psi(x) = i\hbar\frac{1}{g(t)}\frac{d}{dt}g(t) = E \tag{2.24}$$

where E is a constant and clearly has units of energy (the units of $\widehat{H}$). Then we have two separate equations:

$$\widehat{H}\psi(x) = E\psi(x) \tag{2.25}$$

called the **time independent Schrödinger equation**, and

$$i\hbar\frac{d}{dt}g(t) = Eg(t) \tag{2.26}$$

The solution to Eq. 2.26, which you can confirm by substitution, is

$$g(t) = e^{-iEt/\hbar} \tag{2.27}$$

This leaves us to solve for the spatial partial part of the solution: $\psi(x)$. For the nano-vibrator Hamiltonian, we have, for the time independent Schrödinger equation,

$$\left(-\frac{\hbar^2}{2m}\frac{d^2}{dx^2} + \frac{1}{2}\kappa x^2\right)\psi(x) = E\psi(x) \tag{2.28}$$

The form of Eqs 2.26 and 2.28 is that of an **eigenvalue equation**, since the result of the right-hand side (RHS) is the original function times a constant. The operator (on the left) transforms the function, $\psi(x)$, into itself, differing at most by a constant. Then $\psi(x)$ is called an **eigenfunction** and E is an **eigenvalue**. Eigenvalue problems, where the operator on the LHS of Eq. 2.28 changes, are frequently found in all areas of engineering and scientific problems. For people with experience in systems and linear algebra, the corresponding problem is the **eigenvector equation**. (See Appendix A.) More about eigenvectors and the corresponding state vector will be presented in later chapters.

If at all possible, differential equations should be solved analytically, in order to provide a means of quickly developing an intuitive understanding. This particular equation has been solved before, so we can use the standard approach. This book is *not* about solving differential equations. Closed form (i.e., analytical) solutions to differential equations are, in general, guessed where sometimes you go through some preliminary efforts to put it into a form that reveals a known solution. There are some rules and techniques for doing this. We will use a few, but in the end, we are simply going to write down the answer.

First, we must write this as a math equation, not a physics equation. One way to do this is to develop a dimensionless form of the coordinate, κ, which has units of length. This is done by combining fundamental constants from the equation, $\hbar$, m, and κ into a combination that has the unit of length. The approach, called **dimensional analysis**, is unique (within a multiplicative dimensionless constant) and gives:

$$\ell = \left(\frac{\hbar^2}{m\kappa}\right)^{\frac{1}{4}} \tag{2.29}$$

Problem 2.3 *Show that the units on the RHS of Eq. 2.29 reduce to a unit of length. Hint: the units for $\hbar$ are given above. The units for the force constant are immediately seen by recognizing* *that κx has units of force. You will need to work in the basic MKSA units which are length (meter), mass (kilogram), time (second), charge (coulomb). So, while a newton is the MKSA unit of force, it is not in the basic form.*

We can then define a new variable,

$$\xi = \frac{x}{\ell} \tag{2.30}$$

We define two more constants, namely a natural frequency given by

$$\omega_0 = \sqrt{\frac{\kappa}{m}} \tag{2.31}$$

and another dimensionless parameter,

$$\lambda = \frac{2E}{\hbar\omega_0} \tag{2.32}$$

Note that while $\frac{E}{\hbar\omega_0}$ is obviously dimensionless, the need for the factor of 2 is not at all obvious. The reason it is required is that in order to get the desired form of the equation that is tabulated, the factor of 2 is required. Using this result, Eq. 2.28 can be rewritten as

$$\frac{d^2}{d\xi^2}\psi(\xi) + (\lambda - \xi^2)\,\psi(\xi) = 0 \tag{2.33}$$

Problem 2.4 *Show that the units on the RHS of Eqs 2.31 and 2.32 are inverse time and dimensionless, respectively. Develop 2.33 from 2.28. It helps to remember that if x is an independent variable, then for an arbitrary constant, a, $\frac{d}{dax} = \frac{1}{a}\frac{d}{dx}$.*

This is a much simpler looking linear second order homogeneous differential equation. If ξ is large compared to λ, we can find a solution that eliminates the quadratic. So, without loss of any information or generality, we can assume

$$\psi(\xi) = H(\xi)\,e^{-\frac{1}{2}\xi^2} \tag{2.34}$$

If this is a bad guess, the equation will get uglier and we will have to start over. Substituting, we get a new differential equation:

$$\frac{d^2}{d\xi^2}H(\xi) - 2\xi\frac{d}{d\xi}H(\xi) + (\lambda - 1)\,H(\xi) = 0 \tag{2.35}$$

Problem 2.5 *Using Eq. 2.34, show that Eq. 2.35 now follows from Eq. 2.33.*

With functions like sines, cosines, and exponentials, their form seems well-known to us but, in fact, they are defined by a series solution to a differential equation. The same is done here. The approach learned in an introductory course on differential equations is the series method, where convergence is assured by arranging for the series to terminate. In general, this approach works, but it typically requires much more work than illustrated in such courses to fully understand the solution. In this case, Eq. 2.35 was studied in detail by Hermite. The solutions of interest here are **Hermite's polynomials**, where convergence is assured by requiring

$$\lambda = 2n + 1; n = 0, 1, 2, \ldots \tag{2.36}$$

Problem 2.6 *Assume a series solution to Eq. 2.35 of the form* $H(\xi) = \sum_{n=0}^{n=\infty} c_n \xi^n$ *and show that termination of the series requires the quantization condition given by Eq. 2.36.*

Equation 2.36 is a remarkable result. It says that there are an infinite number of solutions to Eq. 2.28 and that the allowed solutions only occur for specific values of the energy. Each solution corresponds to a well-defined value of the **eigenenergy**, now designated E_n. The allowed values of the energy are not continuous but discrete, given by:

$$E_n = \hbar\omega_0 \left(n + \frac{1}{2}\right); n = 0, 1, 2, \ldots \tag{2.37}$$

The energy is quantized (discrete). The quantization arises from a mathematical requirement that could be boundary conditions or convergence of a solution, for example. Physically, the quantization arises because the state is localized in space by the form of the potential and by the requirement that the solution vanish at $\pm\infty$. The **spectrum of eigenvalues** *is said to be discrete* and n is called a ***quantum number***. Notice that even for $n = 0$, the energy is not zero, it is $\frac{1}{2}\hbar\omega$. Sometimes this is the called the **zero-point energy**, the difference between the lowest allowed quantum energy and the lowest allowed classical energy. Note that the energy difference between successive levels is the same, and this means the energy spacing is **harmonic**. The quadratic potential is the only physical potential that gives harmonic spacing. All other potentials give a spacing that varies with the energy level quantum number, n, and this is called **anharmonic**.

Problem 2.7 *Show that 2.37 follows from 2.36.*

The first few *Hermite polynomials*, eigenfunctions, and eigenvalues are given in Table 2.1, where the ground state (lowest energy eigenstate) is a Gaussian, and all the remaining eigenfunctions are a Gaussian multiple by a polynomial.

Table 2.1 Analytical forms for the different states of the harmonic oscillator and eigenvalues.

n	$H_n(\xi)$	$\psi_n(\xi)$	E_n
$n=0$	$H_0(\xi) = 1$	$\psi_0(\xi) = D_0 e^{-\frac{1}{2}\xi^2}$	$E_0 = \frac{1}{2}\hbar\omega_0$
$n=1$	$H_1(\xi) = 2\xi$	$\psi_1(\xi) = D_1 2\xi e^{-\frac{1}{2}\xi^2}$	$E_1 = \frac{3}{2}\hbar\omega_0$
$n=2$	$H_2(\xi) = 4\xi^2 - 2$	$\psi_2(\xi) = D_2(4\xi^2 - 2) e^{-\frac{1}{2}\xi^2}$	$E_2 = \frac{5}{2}\hbar\omega_0$
$n=3$	$H_3(\xi) = 8\xi^3 - 12\xi$	$\psi_3(\xi) = D_3(8\xi^3 - 12\xi) e^{-\frac{1}{2}\xi^2}$	$E_3 = \frac{7}{2}\hbar\omega_0$
$n=4$	$H_4(\xi) = 16\xi^4 - 48\xi^2 + 12$	$\psi_4(\xi) = D_4(16\xi^4 - 48\xi^2 + 12) e^{-\frac{1}{2}\xi^2}$	$E_4 = \frac{9}{2}\hbar\omega_0$
$n=5$	$H_5(\xi) = 32\xi^5 - 16\xi^3 + 120\xi$	$\psi_5(\xi) = D_5(32\xi^5 - 16\xi^3 + 120\xi) e^{-\frac{1}{2}\xi^2}$	$E_5 = \frac{11}{2}\hbar\omega_0$

The constant, D_n, is given by

$$D_n = \left(2^n n! \sqrt{\pi}\right)^{-1/2} \tag{2.38}$$

and is determined such that $\psi_n(\xi)$ is normalized, i.e.,

$$\int_{-\infty}^{\infty} d\xi\, \psi_n^*(\xi)\, \psi_n(\xi) = 1 \tag{2.39}$$

Note: If we write $\psi_n(x)$ and substitute $\xi = \frac{x}{\ell}$ *in the table above we get, for the normalization coefficient,*

$$D_n = \left(2^n n! \ell \sqrt{\pi}\right)^{-1/2} \tag{2.40}$$

Hence, the normalization coefficient changes, depending on whether we integrate over dx (with units of length) or the dimensionless $d\xi$.

We will show later that the functions are also **orthogonal**, meaning that for a discrete spectrum of eigenvalues:

$$\int_{-\infty}^{\infty} d\xi\, \psi_n^*(\xi)\, \psi_{n'}(\xi) = \delta_{nn'} = \begin{bmatrix} 1 \text{ for } n = n' \\ 0 \text{ for } n \neq n' \end{bmatrix} \tag{2.41}$$

where $\delta_{nn'}$ is the **Kronecker delta function**. When a function is both *normalized* and *orthogonal* to the other functions in the set $\{\psi_n\}$, the function and set are said to be **orthonormal**. Like the functions sine and cosine, these functions are **complete**, which means that any function, say an arbitrary function, $\psi(x)$, can be expanded in terms of these functions such that

$$\psi(x) = \sum_{n=0}^{\infty} a_n \psi_n(x) \tag{2.42}$$

To find the expansion coefficient we multiply by $\psi_{n'}^*(x)$ and integrate, using Eq. 2.41

$$\int dx\, \psi_{n'}^*(x)\, \psi(x) = \sum_{n=0}^{\infty} a_n \int dx\, \psi_{n'}^*(x) \psi_n(x) = \sum_{n=0}^{\infty} a_n\, \delta_{nn'} = a_{n'} \tag{2.43}$$

i.e.,

$$a_n = \int dx\, \psi_n^*(x)\, \psi(x) \tag{2.44}$$

We note that Schrödinger's equation (Eq. 2.16) is first order in time. Hence, only the initial condition, $\psi(x, t = 0)$, needs to be specified. The general solution to Eq. 2.16, if the Hamiltonian is time independent, has the form $\psi(x)g(t)$. Both functions depend on E_n and hence $\psi(x)g(t) \rightarrow \psi_n(x)g_n(t)$ and then, using Eq. 2.27,

$$\psi(x,t) = \sum_{n=0}^{\infty} a_n\, \psi_n(x) g_n(t) = \sum_{n=0}^{\infty} a_n e^{-iE_n t/\hbar}\, \psi_n(x) \tag{2.45}$$

where the expansion coefficients, a_n, are given by 2.44 and $\psi(x) = \psi(x, t = 0)$.

2.4 Is It Classical or Quantum?

The classical solution of the oscillator or vibrator has done extremely well in understanding, predicting, and designing numerous things in our world. Hence, it is important to develop an intuitive feel for when this robust analysis might fail. If you strike a vibrator (a bell, tuning fork, cantilever, etc.), it will vibrate at the natural frequency, as in Eq. 2.31: $\omega_0 = \sqrt{\frac{\kappa}{m}}$. Hence, this frequency decreases with increasing mass (think of the low pitch of a big bell). Likewise, the frequency increases as the spring constant increases (think of a tighter spring which has a bigger spring constant). Consider now typical units in MKS: force is newtons, so the spring constant κ will have units of newtons/meter. Mass is a kilogram. If κ and m each have a numerical value of 1, the frequency is 1 radian/second (or divide by 2π to get hertz). You can imagine making masses as small or smaller than a picogram and maybe the force constant could be increased by many orders of magnitude. This could get the frequency up to a GHz or higher. For a GHz (multiply by 2π to convert to radians), the quantum energy corresponds to $E = \hbar\omega \sim 6\times10^{-25}$ joules. At T = 300K, $E = k_B T = 4 \times 10^{-21}$ joules.[1] So, $\hbar\omega \ll k_B T$; i.e., the energy of a typical vibration in the classical world is much smaller than typical thermal energy. Thermal energy in solids is manifested by vibrations in the solid. In ways that we will examine later, the thermal energy destroys quantum features, and most things under typical conditions behave classically. However, in modern technology, these kinds of problems are now being addressed to create new kinds of vibrators for new kinds of applications such as sensors. An example is shown at the end of the discussion in Fig. 2.4.

We have now seen that the energy is discrete in a harmonic oscillator, that the position as a function of time is *not knowable*, and that the energy associated with quantum vibration is often pretty small compared to other energies, meaning that to see these effects, the system has to be cooled down closer to 0K. But there are more strange things. In the classical oscillator, the particle moves back and forth between *two* **classical turning points** (CTP). The classical turning point occurs when all the energy is held as potential energy, and the particle is not moving. This would be when, in the oscillation for a given range of motion, the spring is maximally compressed or maximally extended (in the context of the vibration, not the mechanical spring limits); i.e., when $x = x_{CTP}$, and the kinetic energy is zero (the particle stops moving) and all the energy is in the potential:

$$E = \frac{1}{2}\kappa x_{CTP}^2 \equiv \frac{1}{2} m\omega^2 x_{CTP}^2 \tag{2.46}$$

[1] Recall that a system in thermal equilibrium is characterized by a temperature, T, and that the probability of the system being in a state with energy E is proportional to the Boltzmann factor given by $e^{-\frac{E}{k_B T}}$. In the limit where $E \ll k_B T$, the factor is 1 and that means that the state with energy E is naturally thermally excited. To exhibit quantum effects, you need these states to be empty so that deliberate excitation to the state can be observed.

or

$$x_{CTP} = \sqrt{\frac{2E}{\kappa}} \tag{2.47}$$

For the quantum oscillator, since $\ell = \left(\frac{\hbar^2}{m\kappa}\right)^{1/4}$ (Eq. 2.29)

$$\xi_{CTP} = \frac{x_{CTP}}{\ell} = \left(\frac{4E^2}{\kappa}\frac{m}{\hbar^2}\right)^{1/4} \tag{2.48}$$

The region beyond $|\ \xi\ | > \xi_{CTP}$ is called the **classically forbidden region** because a particle behaving classically cannot go beyond these points. Taking E as the lowest energy state in the quantum oscillator $E_0 = \frac{1}{2}\hbar\omega$, we get

$$\xi_{CTP} = (1)^{1/4} = 1$$

Looking now at the graph in Fig. 2.2 (the horizontal axis is the independent variable corresponding to the dimensionless parameter, ξ), we see that the probability density—the modulus squared of what is plotted in Fig. 2.2, which is $\psi_n(x)$—of finding the particle in the classically forbidden region is non-zero. The particle has **tunneled** into this region. This is distinctly a feature of the wave-like nature of Schrödinger's equation. The solution to Maxwell's equations shows similar behavior for electromagnetic waves under the right conditions and features

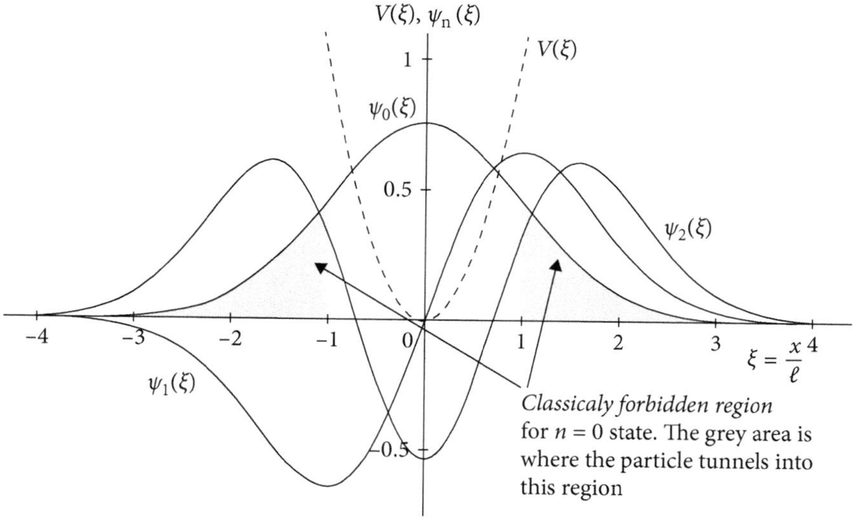

Fig. 2.2 The parabolic potential is shown (dashed) along with the form of the first three (un-normalized) eigenfunctions for the harmonic oscillator. Note that the x-axis is in units of ξ. The units of the potential energy are $\frac{1}{2}\hbar\omega$ so that for $E_0 = \frac{1}{2}\hbar\omega$, $V(\xi = 1) = 1$, etc. The classically forbidden region for the ground state $n = 0$ is for $|\ \xi\ | > 1$, as shown by the shaded region.

prominently in many important devices. The same is true for some quantum devices as shown below.

Problem 2.8 *A simple harmonic oscillator in siliconmight have a thin strip of material suspended in the vacuum between two supporting pads. The oscillation frequency (in hertz) for out of plane vibration of the thin strip is* $\frac{\omega}{2\pi} \sim \frac{t}{l^2}\sqrt{\frac{E}{\rho}}$. *E is Young's modulus, which is the ratio of applied pressure to the induced strain. For silicon,* $E = 160\ Gpa = 1.6 \times 10^{11}\ newtons/m^2$. *The density* $\rho = 2000\ kg/m^3$. *Take the dimensions to be* $t = 50$ *nm thickness and 1 micron long.*

a) *This is a manifestation of the nano-vibrator or harmonic oscillator. Evaluate the above expression for the oscillation frequency of the vibrator.*
b) *In materials, natural vibration frequencies (**phonons**) occur over a range set by the temperature of the lattice. If such vibrationsare strong enough at the operating frequency of the device, these phonons will likely destroy your effort to build the quantum effect you are working towards. So, vibration strength tends to go with the Boltzmann factor.*[2] *So, at room temperature (*$T \sim 300$ *K), determine whether you expect to see quantum behavior. Of course, you have to justify your answer with the numbers. Keep in mind that the energy of a vibration is* $E_{vibration} = \hbar\omega$. *It seems hopeless to build a quantum oscillator at room temperature in silicon. In this problem, use the numbers to show why. This is a simple type calculation to make an argument.*
c) *Now consider a carbon nanotube (CNT). Young's modulus is about* $5 \times 10^{12}\ newtons/m^2$. *The areal density for graphene (determined by the wall) for a CNT, is* $\sim 10^{-6}\ kg/m^2$. *For a 1 nm diameter tube, the mass density per unit length is* $\varrho_l = 2.5 \times 10^{-15}\ kg/m$. *Assume a length of 10nm and a tube thickness of* 0.1 *nm (a Bohr diameter). The fundamental frequency for a thin-walled pipe is* $\frac{\omega}{2\pi} \sim \frac{\pi}{2l^2}\sqrt{\frac{EI}{\rho_l}}$ *where I is the area moment of inertia* $I = \frac{\pi}{64}R^3 t \sim 5 \times 10^{-40}$ *where R is the pipe (CNT) radius and t is the wall/tube thickness. Consider both room temperature and low temperature (*$T = 1$ *K).*

Since answers to our usual classical questions are not appropriate for quantum systems and since everything that can be known about the system is contained in the wave function, the question now is what can we know? In fact, many people would ask how is this theory useful? The answer will emerge as we look more deeply into the various problems, but we begin developing the foundations of the answer here.

2.5 **What is Knowable in a Quantum System?**

We begin by using the mechanical oscillator/vibrator as an example of a much broader set of quantum devices and systems with observables more diverse than just the momentum and position.

[2] Recall that the Boltzmann factor $e^{\frac{-(E_j-E_k)}{k_B T}}$ is the relative probability of finding a particle in state E_j relative to state E_k.

To think about this, note that the information we are used to acquiring in the case of a vibrator is position, momentum (the quantum variable that would be typically velocity in a classical system) and energy. Often we want this information as a function of time. However, in a quantum system of this type, we know the wave function $\psi(x,t)$ for a problem (described in this case in Eq. 2.45), the solutions to the time independent Schrödinger equation, and the operators $\hat{p} \equiv -i\hbar \frac{d}{dx}$ and $\hat{x} \equiv x$, corresponding to the observables p and x. We also know the operator form of the total the energy, i.e., the Hamiltonian, Eq. 2.15, and the operator form for the kinetic energy $\left(\widehat{T} = \frac{\hat{p}^2}{2m} = -\frac{\hbar^2}{2m}\frac{d^2}{dx^2}\right)$ and potential energy, in this chapter for the nano-vibrator $\left(\widehat{V} = \frac{1}{2}kx^2\right)$. We have also found that the average value of an observable (x, p, or E) as a function of time is the expectation value of that observable given by Eq. 2.19.

To be clear, at present we do not know how the system came to be described by $\psi(x,t)$. In other words, we do not know what physical event took place to *create* the wave function in this problem. This will be discussed later in the text.

We will have to assume the initial state $\psi(x, t=0)$ in order to make progress. This is the initial condition we need, in order to find the expansion coefficients using Eq. 2.44, i.e.,

$$a_n = \int dx\, \psi_n^*(x)\, \psi(x, t=0) \tag{2.49}$$

where now $\psi_n(x)$ is an eigenfunction of some operator of interest (e.g., the energy $\widehat{H}$ or maybe the kinetic energy $\widehat{T} = \frac{\hat{p}^2}{2m}$. To understand the physical meaning of a_n, we first note that in Eq. 2.20 the physical interpretation of $|\psi(x,t)|^2$ also required normalization. That is, we require:

$$\int_{-\infty}^{\infty} dx\, \psi^*(x)\psi(x) = \int_{-\infty}^{\infty} dx\, \psi^*(x,t)\,\psi(x,t) = 1 \tag{2.50}$$

We also require $|\psi_n(x)|^2$ to be normalized (Eq. 2.39). This turns out to be critical for developing a physical meaning for a_n.

Since the expansion coefficients are given in Eq. 2.44 and we know that Eq. 2.45 gives us

$$\psi(x,t) = \sum_{n=0}^{\infty} a_n e^{-\frac{iE_n t}{\hbar}} \psi_n(x) \tag{2.51}$$

we see that the expansion coefficients give us information about the contribution of a given eigenstate, $\psi_n(x)$, to the entire wave function. From the normalization condition Eq. 2.50, and using Eq. 2.51 and orthonormality (Eq. 2.41) we get

$$\begin{aligned}\int_{-\infty}^{\infty} dx\, \psi^*(x,t)\,\psi(x,t) &= \int_{-\infty}^{\infty} dx \sum_{m=0}^{\infty} a_m^* e^{\frac{iE_m t}{\hbar}} \psi_m^*(x) \sum_{n=0}^{\infty} a_n e^{-\frac{iE_n t}{\hbar}} \psi_n(x) \\ &= \sum_{n=0}^{\infty}\sum_{m=0}^{\infty} a_m^* a_n e^{-\frac{i(E_n-E_m)t}{\hbar}} \int_{-\infty}^{\infty} dx\, \psi_n(x)\psi_m^*(x) \\ &= \sum_{n=0}^{\infty}\sum_{m=0}^{\infty} a_m^* a_n e^{-\frac{i(E_n-E_m)t}{\hbar}} \delta_{nm} = \sum_{n=0}^{\infty} |a_n|^2 = 1\end{aligned} \tag{2.52}$$

using the orthonormality of $\psi_n(x)$. Hence we see that $|a_n|^2$ is the fraction of the wave function described by $\psi_n(x)$ and represents the probability that a measurement of the system will give the eigenvalue corresponding to the eigenfunction, ψ_n; in this case we are talking about the energy, E_n. This is very similar to the ideas behind spectral analysis of a function or determining the power spectrum of a system. Since one of the rules in quantum mechanics is that the only thing allowed in a measurement of an observable, say of the energy, is an eigenvalue of the corresponding operator, $\hat{H}$, it is very important to know that the probability of measuring a specific eigenvalue, E_n is given by $|a_n|^2$. While $|a_n|^2$ is the probability of making such a measurement, a_n is the **probability amplitude**. Furthermore, note that while the phase of the probability amplitude is changing in time, there is no change as a function of time of being in a given state. Yet there remains a sense of classical behavior, as is discussed further below.

Consider now the interpretation of the following:

$$\langle\hat{H}\rangle(t) \equiv \int_{-\infty}^{\infty} dx\, \psi^*(x,t)\, \hat{H}\psi(x,t) \tag{2.53}$$

We have already explained that this represents the average value that one would get if repeatedly creating ψ and measuring the energy. The rules of quantum say that we will only measure eigenvalues of the operator corresponding to the observable in the measurement. If we use Eq. 2.51 and follow the steps in Eq. 2.52, recalling that $\hat{H}\psi_n(x) = E_n\psi_n(x)$, then

$$\langle\hat{H}\rangle(t) = \int_{-\infty}^{\infty} dx \sum_{m=0}^{\infty} a_m^* e^{\frac{iE_m t}{\hbar}} \psi_m^*(x)\hat{H} \sum_{n=0}^{\infty} a_n e^{\frac{-iE_n t}{\hbar}} \psi_n(x) = \sum_{n=0}^{\infty} E_n |a_n|^2 \tag{2.54}$$

We see that, as anticipated, $|a_n|^2$ represents the probability of measuring a given eigenvalue and that, again as expected, there is no time dependence in the average value of the energy. Equation 2.54 then means that $\langle\hat{H}\rangle$ is the average value of the energy in the system described by $\psi(x,t)$. This is a general result, namely that if an observable is described by an operator, $\hat{A}$, then the average value of $\hat{A}$ in a system described by the wave function $\psi(x)$ is

$$\langle\hat{A}\rangle \equiv \int_{-\infty}^{\infty} dx\, \psi^*(x)\hat{A}\psi(x) \tag{2.55}$$

This can be immediately seen in the same way by expanding the wave function in terms of the eigenstates of the operator $\langle\hat{A}\rangle$. Clearly, to evaluate $\langle\hat{A}\rangle$, it is not necessary to expand out $\psi(x)$. But if we want to know the probability of measuring an eigenvalue of the observable (recall that the only value that will be allowed is an eigenvalue) then we must solve the eigenvalue equation $\hat{A}u_{A_n} = A_n u_{A_n}$ and expand $\psi(x)$ such that $\psi(x,t) = \sum_{A_n} b_{A_n} u_{A_n}$ and that therefore the probability of measuring A_n is $|b_{A_n}|^2$. The pattern here is always of expanding a function in terms of the eigenfunctions of interest. *Depending on the problem, you may or may not need to do the expansion.*

Equation 2.55 is an example where expanding does not necessarily help, though it can sometimes reduce the algebra. Expanding also provides physical understanding of the basis for the result. The same is true in a Fourier decomposition, assuming that the basis functions of the Fourier decomposition have physical meaning in the problem, such as a power spectrum of, say, a pulse of RF radiation.

For the case of the harmonic oscillator above, we can now ask some interesting questions. Consider the ground state $n = 0$

$$\psi_0(x) = \frac{1}{(\pi \ell^2)^{1/4}} e^{-\frac{1}{2}\left(\frac{x}{\ell}\right)^2} \tag{2.56}$$

where again $\xi = \frac{x}{\ell}; \ell = \left(\frac{\hbar^2}{m\kappa}\right)^{\frac{1}{4}}$. This is the wave function for the particle that classically is bouncing back and forth between the classical turning points. We want to measure the momentum, p. To do this, we need to expand the ground state eigenfunction above in terms of eigenfunctions of the momentum operator.

The momentum operator was defined above to be:

$$\hat{p} = -i\hbar \frac{d}{dx} \tag{2.57}$$

The corresponding eigenvalue equation is

$$\hat{p} u_p(x) = p u_p(x) \tag{2.58}$$

or

$$\frac{d}{dx} u_p(x) = i\frac{p}{\hbar} u_p(x) \tag{2.59}$$

with solution:

$$u_p(x) = Ce^{i\frac{p}{\hbar}x} \equiv Ce^{ikx} \tag{2.60}$$

where the wave vector, k, is defined by the relation

$$p = \hbar k \tag{2.61}$$

k is the inverse of a length, and we define that length by the relation

$$k = \frac{2\pi}{\lambda_{DB}} \text{ or } \lambda_{DB} = \frac{2\pi\hbar}{p} \tag{2.62}$$

λ_{DB} is called the ***de Broglie wavelength***. More about this important concept later.

We notice that, from Eq. 2.61, the *eigenfunction of a particle with momentum p is a plane wave*. There are no constraints on the eigenvalue, p, so p can take on any value; i.e., there is a *continuous spectrum of eigenvalues*. By convention, when the eigenvalue is continuous and not characterized by an integer quantum number (the oscillator above has an integer quantum number, n), we usually write the eigenfunction a little differently:

$$u_p(x) \equiv u(p, x) \tag{2.63}$$

Since there are no constraints or boundary conditions, the plane wave state exists from $x = -\infty$ to $x = \infty$. This creates a problem when we try to find a value of C such that the function is normalized according to 2.20 since $u^*(p, x)\, u(p, x) = 1$ and then the integral in 2.20 is undefined (infinite, i.e., not square integrable). However, the formalism being developed works out well and is consistent with Fourier theory by introducing a new kind of function, first developed by Dirac, called the ***Dirac delta function***. Specifically,

$$\int_{-\infty}^{\infty} dx\; e^{ikx} = 2\pi\delta(k) \tag{2.64}$$

where the Dirac delta function is defined under an integral with the relation

$$\int_{-\infty}^{\infty} dk\; \delta(k)\, f(k) = f(0) \tag{2.65}$$

or more generally

$$\int_{k_0-\epsilon}^{k_0+\epsilon} dk\; \delta(k - k_0)\; f(k_0) = f(k_0) \tag{2.66}$$

where ϵ is real and $\epsilon > 0$. An important functional form for $\delta(k - k_0)$ is the Gaussian distribution function:

$$\delta(k - k_0) = \lim_{a\to 0} \frac{1}{a\sqrt{\pi}} e^{-\left(\frac{k-k_0}{a}\right)^2}. \tag{2.67}$$

In addition to the integral relation in 2.65, looking at the Gaussian form in 2.67, it appears that it is divergent (goes to infinity) when the argument is zero and is zero when the argument is non-zero. The integral remains unity, independent of the parameter a. Appendix C has other examples that can be useful in various problems as well as important properties. The Dirac delta function is a part of many mathematical problems in science and engineering, outside of quantum mechanics.

We then normalize the **momentum eigenfunction** to the Dirac delta function, setting $C = \frac{1}{\sqrt{2\pi\hbar}}$ giving the momentum eigenfunction as

$$u_p(x) = \frac{1}{\sqrt{2\pi\hbar}}\, e^{i\frac{p}{\hbar}x} \tag{2.68}$$

Alternatively, with $p = \hbar k$ and the identity (see Appendix C) $\delta(ax) = \frac{1}{|a|}\delta(x)$, we get $\delta(k - k') = \delta\left(\frac{p}{\hbar} - \frac{p'}{\hbar}\right) = \hbar\delta(p - p')$. Hence, using k, the normalized eigenfunction is

$$u_k(x) = \frac{1}{\sqrt{2\pi}}\, e^{ikx} \tag{2.69}$$

Eqs 2.68 and 2.69 are orthonormal according to

$$\int_{-\infty}^{\infty} dx\ u_{p'}^{*}(x) u_p(x) = \delta\left(p - p'\right) \text{ or}$$
$$\int_{-\infty}^{\infty} dx\ u_{k'}^{*}(x) u_k(x) = \delta\left(k - k'\right) \tag{2.70}$$

Sometimes, with continuous eigenvalues, the eigenfunctions are said to be normalized to the **Dirac delta function.**[3]

With this new knowledge, what will be the probability of measuring a momentum p for a system initially in the ground state of the harmonic oscillator? To answer this, we expand the initial wave function in the eigenstates of the momentum operator and look at the expansion coefficients. That is,

$$\psi(x) = \frac{1}{\left(\pi \ell^2\right)^{1/4}} e^{-\frac{x^2}{2\ell^2}} = \sum_{p=-\infty}^{\infty} C_p u\left(p, x\right) = \int_{-\infty}^{\infty} dp\ C(p) \frac{1}{\sqrt{2\pi\hbar}}\ e^{i\frac{p}{\hbar}x} \tag{2.71}$$

Note that the sum went over to an integral because the discrete spectrum of eigenvalues went over to a continuous spectrum of eigenvalues for the eigenstates of the momentum operator. This means the Kronecker delta function also goes over Dirac delta function for orthonormalization. To find the expansion coefficient, we multiply both sides by $u_{p'}^{*}(x)$ and use the result of Eq. 2.70 and get

$$C(p) = \int_{-\infty}^{\infty} dx\ \frac{1}{\sqrt{2\pi\hbar}}\ e^{-i\frac{p}{\hbar}x}\ \frac{1}{\left(\pi \ell^2\right)^{1/4}} e^{-\frac{x^2}{2\ell^2}} \tag{2.72}$$

where we have changed the symbol p' back to p. This integral is the Fourier transform of harmonic oscillator ground state, which is a Gaussian. To do this integral, it helps to know that

[3] These results are seen from Fourier theory. Using the definition of the Dirac δ functionin Eq. 2.67, we recall from Fourier theory, the Fourier transform of a function $f(x)$ is $F(y)$, given by

$$F(y) = \int_{-\infty}^{\infty} dx\ e^{-i2\pi yx} f(x)$$

and the inverse Fourier transform $F(y)$ gives $f(x)$:

$$f(x) = \int_{-\infty}^{\infty} dy\ e^{i2\pi yx}\ F(y)$$

Substituting for $F(y)$ using the first equation, we get

$$f(x) = \int_{-\infty}^{\infty} dy\ e^{i2\pi yx} \int_{-\infty}^{\infty} dx' e^{-i2\pi yx} f(x') = \int_{-\infty}^{\infty} dx' f(x') \int_{-\infty}^{\infty} dy\ e^{-i2\pi y(x-x')}$$

Since the right side must equal the left side, this means based on the definition in Eq. 2.67:

$$\int_{-\infty}^{\infty} dy\ e^{-i2\pi y(x-x')} = \delta\left(x - x'\right)$$

$$\int_{-\infty}^{\infty} dx\, e^{-ax^2+bx} = \sqrt{\frac{\pi}{a}}\, e^{\frac{b^2}{4a}} \textit{ for } Re\ a > 0 \tag{2.73}$$

Therefore,

$$C(p) = \left(\frac{\ell^2}{\pi\hbar^2}\right)^{1/4} e^{-\left(\frac{p}{\hbar}\right)^2 \frac{\ell^2}{2}} \tag{2.74}$$

So the probability of measuring a given p is given by $|C(p)|^2$. It is a Gaussian because the ground state of the harmonic oscillator is a Gaussian, and the Fourier transform (note that Eq. 2.72 is indeed a Fourier transform because of the form of the momentum eigenfunction) of a Gaussian is a Gaussian.

Since the momentum eigenvalues are continuous, we say that $|C(p)|^2$ is a probability density of finding the particle with momentum between p and $p + dp$. Since p and x are conjugate variables, $C(p)$ corresponds to p the same way $\psi(x)$ corresponds to x. $C(p)$ is often referred to as the **momentum representation** of the wave function and written as $\phi(p) \equiv C(p)$.

Problem 2.9 *Now consider a harmonic oscillator state with $n = 1$. See Table 2.1 for the form of the eigenfunction. Repeat the calculation above to find the probability amplitude density, $C(p)$.*

Problem 2.10 *Suppose now that a harmonic oscillator is in the ground state of a system with spring constant κ. The spring constant "instantly" changes to κ'. "Instantly" here means fast relative to all the time scales in the problem. This concept needs more discussion later. But for now, find the probability of finding the system in the ground state after the change in the spring constant. While it was unity before the change, it will now be less than unity after the change. Be sure to keep track of the changes throughout the problem such as the scale length parameter ξ.*

Finally, in this section, we recall that all the information that can be known is contained in the solution to Schrödinger's equation $\psi(x, t)$. Since Schrödinger's equation is a wave equation, the description of a massive particle is that of a wave. Nothing in $\psi(x, t)$ seems to show any behavior that is classical at this point. The position of the particle can be measured but it cannot be found as a function of time. This is the first of many times in your study of quantum that there is a sense of ambiguity between the intuitive idea of the particle and the reality that in quantum it is wave. More insight on this will emerge in later discussions.

2.6 **Coherent Superposition States and Coherent Dynamics**

So far, there has been no evidence of motion or behavior like we expect from a classical system. This section shows how something resembling classical behavior emerges by introducing one of the key features of quantum systems of critical importance to quantum technology, the **principle of superposition**. As in the case of electromagnetic waves, it is possible to have a system in more than one state. This follows from the fact that Schrödinger's equation is linear, meaning that if ψ_1 and ψ_2 are a solution, so is $\psi_1 + \psi_2$. The sum of two or more states is called a **coherent superposition state**. In the era of modern quantum technology, the coherent superposition state is perhaps the most important feature. Indeed, the example above shows how a harmonic oscillator in the ground state is in a coherent superposition of a continuum of momentum states as seen in Eq. 2.71. A harmonic oscillator could also be in a coherent superposition of harmonic oscillator states. For example, we could imagine a superposition of the two eigenstates:

$$\psi(\xi) = \cos\theta\ \psi_0(\xi) + \sin\theta\psi_1(\xi) \tag{2.75}$$

Even when the time dependence is added to Eq. 2.75, the relative phase is fully deterministic as a function of time and there are no random events changing the relative phase. This is what is meant by coherent superposition. The language of coherence reflects the fact that the phase difference between the two states is known for all time. Loss of coherence means that the phase relationship is lost by some process. Furthermore, as we examined earlier, if we included the time dependent phase factors to make it $\psi(\xi, t)$, *the probability of being in either state does not change with time.* A decrease in the probability amplitude of one or both states means that the total probability becomes less than 1. This is called decay, and represents dissipation. There will be much more discussion on this later.

The relative phase difference between the two states impacts where the particle is likely to be found since the states in Eq. 2.75 can constructively or destructively interfere at different points in space. Looking at Fig. 2.3a, which plots Eq. 2.75 for $\cos\theta = \sin\theta = \frac{1}{\sqrt{2}}$, you see that the wave function is localized more on one side of the origin than the other. The case of $\cos\theta = -\sin\theta = \frac{1}{\sqrt{2}}$ is also plotted and you see that the probability distribution of finding the particle has shifted to the other side. Using the sin and cos ensures that the wave function remains normalized for all values of θ. The probability of finding the system in the ground state and the first excited state remains

$$P(n=0) = \cos^2\theta \text{ and } P(n=1) = \sin^2\theta \tag{2.76}$$

Problem 2.11 *In the discussion of solving Schrödinger's equation, we used the method of separation of variables where we had*

$$\Psi_n(x,t) = \psi_n(x)e^{-\frac{iE_nt}{\hbar}} \tag{2.77}$$

Show now that if the state in 2.75 is $\psi(\xi, t=0)$ where $\xi = \frac{x}{\ell}$, then $\langle x(t)\rangle$ oscillates in time, by finding $\langle x\rangle(t)$. What is the oscillation frequency?

Note: If the particle had a charge, it would radiate. It will help in this problem if you use the spatial symmetry of the different terms in the integral to quickly establish which terms are zero. This will leave you with only one term plus its complex conjugate to evaluate.

Hint: Since at $t = 0$,

$$\psi(\xi) = \cos\theta\ \psi_0(\xi) + \sin\theta\ \psi_1(\xi) \tag{2.78}$$

we therefore know that

$$\psi(\xi, t) = \cos\theta\ \psi_0(\xi)\, e^{\frac{-iE_0t}{\hbar}} + \sin\theta\ \psi_1(\xi)\, e^{\frac{-iE_1t}{\hbar}} \tag{2.79}$$

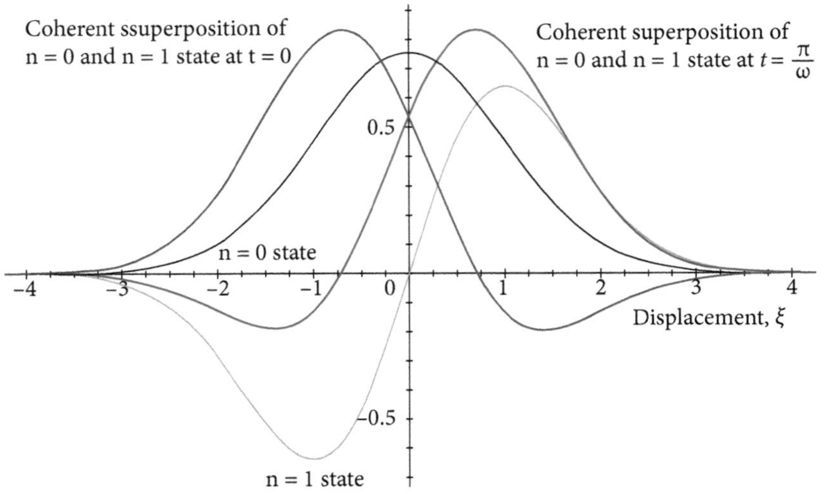

Fig. 2.3a The probability amplitude associated with $\psi(\xi, t)$ in Eq. 2.75 is shown here as a function of the normalized displacement, ξ. $\psi(\xi, t)$ is an equal linear superposition of $n = 0$ state (black curve) and $n = 1$ state (green curve) for a nano-vibrator. Because of the temporal dependence of the phase between the two components being different, the result of addition of the two amplitudes together changes in time. For example, at $t = 0$, constructive interference on the right and destructive interference on the left results in a shift to the right of the probability density of finding the particle (red curve). At $t = \frac{\pi}{\omega}$, the result of destructive and constructive interference has changed and there is now a shift to the left of the probability density of finding the particle.

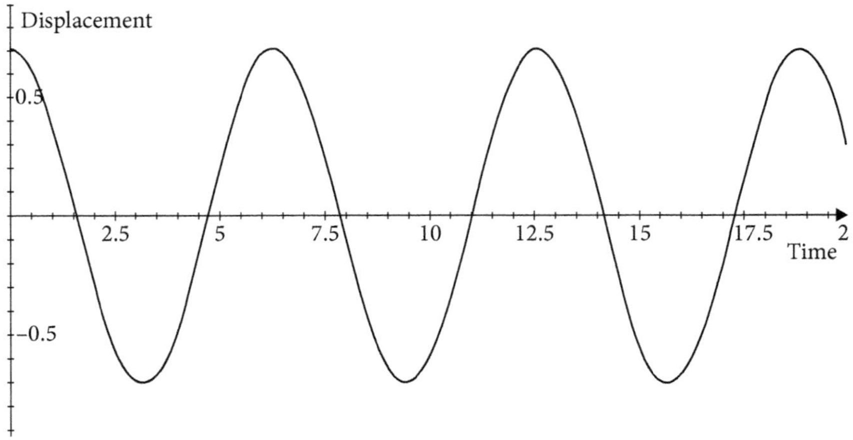

Fig. 2.3b Here, the expectation value of normalized position, $\xi(t)$, is shown as a function of time in units of $\frac{\pi}{\omega}$.

This problem reveals how the harmonic motion is recovered, which we associate with the classical system. In the quantum system, it results from the principle of superposition.

The state developed in Eq. 2.79 represents the time evolution of a coherent superposition state involving just two states. At any point in time, the two states $\psi_0(\xi)$ and $\psi_1(\xi)$ add together, resulting in constructive and destructive interference at different points in space, as indicated above. In constructive interference, the functions add to increase the magnitude of $\psi(\xi, t)$ and in destructive interference, they add to reduce the magnitude of $\psi(\xi, t)$. This behavior is a key property of waves. This can result in the probability amplitude oscillating in time at a given point in space or even the entire probability amplitude moving from one region of space to another. What is causing this change is the time evolution of the phase. Notice that even though the phase is changing, the probabilities of being in state 0 and state 1 remain constant in time as $cos^2\theta$ and $sin^2\theta$, respectively. Hence, as you show in Problem 2.11, *the harmonic motion of the classical system is recovered through the principle of superposition.* The oscillation in time is shown in Fig. 2.3b.

2.7 The Particle and the Wave

Our classical picture of massive particles being a point localized in space and described as a function of time is not a picture that works in a quantum system. Schrödinger's equation is a wave equation for the function that is clearly a wave. We saw above and will see in more detail in the next chapter, that the energy eigenfunction for a **free particle** (the potential is zero throughout all space) is simply $\frac{1}{\sqrt{2\pi}}e^{ikx-i\frac{Et}{\hbar}}$ and is clearly a plane wave. The above discussion shows that even in a well-defined energy eigenstate for the nano-vibrator, the system shows no behavior like oscillation that we associated with a classical vibrator with a given energy. The requirement to observe the expected oscillation is that the system be in a coherent superposition of two states. The ability to exist in a superposition is a feature of waves that gives rise to interference and diffraction. So the electron, a massive particle, is sometimes described well as a point particle, but in many cases this will fail, and the wave nature of the electron must be considered in order to get the correct description of behavior. This has historically been referred to a **wave–particle duality**. More important than the language, however, is understanding that in quantum-based technology, the electron or any other particle must be treated as a wave, using Schrödinger's equation.

An example of such a vibrator is shown in Fig. 2.4.

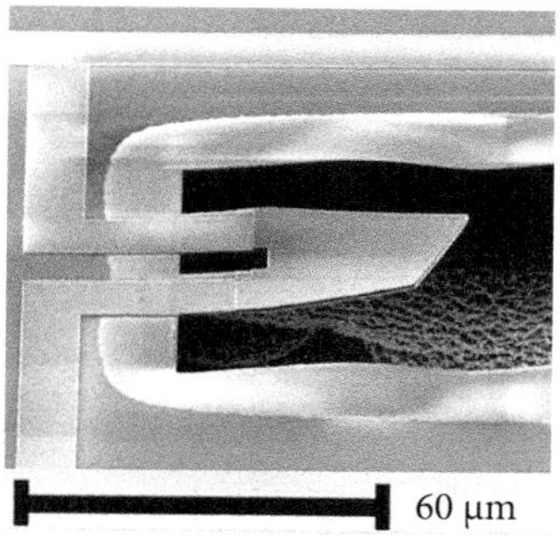

Fig. 2.4 A scanning electron micrograph of a nano-vibrator. The device, called a **quantum machine** by inventors, is operated at 25 mK to reduce the effects of thermal vibrations (**phonons**) and was developed by Aaron D. O'Connell et al. at UC-Santa Barbara. The image is taken from https://en.wikipedia.org/wiki/Quantum_machine.

2.8 Summary

In the classical world, dynamics and behavior are determined by equations of motion. Newton's law and Kirchhoff's laws are two examples. They are postulated (guessed, though both can be derived from a broader set of postulates). The equations of motion are second order in time requiring two initial conditions. In the physical description of the world, the rules are postulated in mathematical terms, not derived. The only proof of the appropriateness of these rules is their ability to correctly predict experimental observations.

In the quantum world, the equation of motion is Schrödinger's equation:

$$\hat{H}\psi = i\hbar\frac{\partial\psi}{\partial t}$$

$\hat{H}$ is an operator (a math instruction) that represents the total energy in the system. For a massive particle,

$$\hat{H} = \frac{\hat{p}^2}{2m} + \hat{V}$$

where $\hat{p}$ is the momentum operator given in the coordinate representation by

$$\hat{p} = -i\hbar\frac{\partial}{\partial x}$$

In general, the Hamiltonian is written in terms of two operators. The first is associated with the canonical coordinate $\hat{x}$, and the second is associated with its conjugate momentum, $\hat{p}$. In developing this, we are seeing that when a system is quantum mechanical, the observables—meaning the things you can measure—become operators. In this case, the three operators are $\hat{x}$, $\hat{p}$, and $\hat{H}$, corresponding to position, momentum, and energy, respectively.

You have also learned that if $\hat{H}$ does not depend explicitly on time (meaning there is no $f(t)$ in $\hat{H}$), then the equation of motion can be solved by the method of separation of variables; i.e.,

$$\psi(x,t) = \psi(x)g(t)$$

We found that $\psi(x)$ must satisfy the eigenvalue equation, called the time independent Schrödinger equation:

$$H\psi(x) = E\psi(x)$$

and that there are (usually) an infinite number of solutions. The spectrum of eigenvalues could be discrete, represented by E_n, or continuous, represented by E, and the corresponding eigenfunctions are $\psi_n(x)$ or $\psi(E,x)$. The set of all eigenfunctions is complete and meaning that any function in that space can be written as a linear combination of the eigenfunctions. For a discrete and continuous spectrum, this corresponds to:

$$\psi(x) = \sum_n C_n\psi_n(x) \text{ where } C_n = \int_{-\infty}^{\infty} dx\ \psi_n^*(x)\psi(x)$$

or

$$\psi(x) = \int_{-\infty}^{\infty} \mathrm{dE}\ C(E)\psi(E,x) \text{ where } C(E) = \int_{-\infty}^{\infty} dx\ \psi^*(E,x)\psi(x)$$

respectively. Then $\psi(x)$ is normalized meaning

$$\int_{-\infty}^{\infty} dx\ \psi^*(x,t)\psi(x,t) = 1$$

The eigenfunctions are orthonormal, where for a discrete spectrum of eigenvalues,

$$\int_{-\infty}^{\infty} dx\ \psi_n^*(x)\psi_m(x) = \delta_{nm}$$

and for a continuous spectrum of eigenvalues,

$$\int_{-\infty}^{\infty} dx\ \psi^*(E',x)\psi(E,x) = \delta(E'-E)$$

$\delta(E'-E)$ is a Dirac delta function, which is defined only under an integral such that, in general,

$$\int_{-\infty}^{\infty} dx\ \psi(x)\delta(x-x_0) = \int_{x_0-\epsilon}^{x_0+\epsilon} dx\ \psi(x)\delta(x-x_0)\psi(x_0) = \psi(x_0)$$

Beyond knowing $\psi(x,t)$, the only thing that is knowable is that for some observable like $x.p, or\ E$, you can know the expectation value as a function of time, where if we represent the corresponding operator for that observable as $\hat{A}$, the expectation value as a function of time is given by

$$\langle\hat{A}\rangle(t) = \int_{-\infty}^{\infty} dx\ \psi^*(x,t)\hat{A}\psi(x,t)$$

In general, the operator $\hat{A}$ has eigenfunctions and eigenvalues which we assume (for convenience) to be discrete, such that

$$\hat{A}u_n(x) = a_n u_n(x)$$

Then, because the set of all u_n forms a complete set and we can expand $\psi(x,t)$ in terms of these eigenfunctions such that, for the case of operator $\hat{A}$,

$$\psi(x) = \sum_n c_n u_n(x)$$

showing that (we consider the case of $t = 0$)

$$\langle \hat{A} \rangle = \int_{-\infty}^{\infty} dx \left(\sum_n c_n u_n(x) \right)^* \hat{A} \left(\sum_n c_n u_n(x) \right)$$

$$= \int_{-\infty}^{\infty} dx \left(\sum_n c_n u_n(x) \right)^* \left(\sum_n c_n a_n u_n(x) \right) = \sum_n (c_n)^2 a_n$$

Since $\psi(x)$ is normalized, this means that $(c_c)^2$ is the probability of finding the system in state u_n, with a value of a_n in a single measurement of $\hat{A}$. We note that evaluation of the expectation value does not require transforming the eigenfunction basis of the corresponding operator. But additional information is available if it is of interest to the design or measurement.

You have noticed that in the case of a free particle the wave function is clearly a wave, and that, even for the harmonic oscillator, the eigenfunctions have properties like standing waves. An important aspect that goes with a wave is the possibility that the system is described by more than one wave (or state) in the same function. This is the principle of superposition, which is a result of the linearity of Schrödinger's equation. In the case of the nano-oscillator, we had assumed a form of $\psi\,(x, t)$ involving only two states:

$$\psi\,(x, t) = a_n \psi_n e^{-i\frac{E_n t}{\hbar}} + a_{n'} \psi_{n'} e^{-i\frac{E_{n'} t}{\hbar}}$$

for

$$n \neq n'$$

where, since this state is taken to be normalized and the basis states are orthonormal, we require

$$|a_n|^2 + |a_{n'}|^2 = 1$$

More generally an arbitrary coherent superposition state

$$\psi\,(x, t) = \sum_n a_n \psi_n e^{-i\frac{E_n t}{\hbar}}$$

where

$$\psi\,(x, t) = \sum_n |a_n|^2 = 1$$

In a coherent superposition state, measurements of an observable, such as the position x, may show a time dependence if the states have different energies. For example, a measurement of x for the first coherent superposition state above gives an expectation value that oscillates in time:

$$\langle x \rangle(t) = a_n a_{n'}^* \left(\psi_n | x \psi_{n'} \right) e^{-i\frac{E_{n'} - E_n t}{\hbar}} + a_n^* a_{n'} \left(\psi_{n'} | x \psi_n \right) e^{i\frac{E_{n'} - E_n t}{\hbar}}$$

where, for shorthand, we introduce the notation:

$$(\psi_n|x\psi_{n'}) \equiv \int_{-\infty}^{\infty} dx\ \psi_n^*(x)x\psi_{n'}(x)$$

Notice that the expectations value results in a sum of bilinear products of probability amplitudes modulated in time by an oscillation at a frequency determined by their energy difference. In some areas now of quantum technology, it is this quantum coherence (the defined phase relationship between the two or more eigenstates) that is most exciting.

For particles with mass, such as the electron, you have seen the idea of the classical turning point and quantum tunneling. When a particle has a probability of being in a region of energy that is classically forbidden (e.g., where the kinetic energy, T, is less than the potential energy, V), the particle is said to have tunneled into this region. As it moves from a region where $T > V$ (classically allowed) to $T < V$, it passes through a classical turning point where $T = V$ or where V is discontinuous and on one side $T < V$ and on the other $T > V$.

The expansion of the wave function in terms of the basis states of the Hamiltonian or any other complete basis results in a wave function being represented by a coherent superposition of the basis states. The addition of these states can result in constructive and destructive interference, dramatically changing the spatial dependence of the probability amplitude compared to the dependence seen in any single eigenstate. The inclusion of time then results in a spatial motion of this probability amplitude. If the particle is charged, it can appear to be oscillating in space and that could result in a radiated electromagnetic field like an antenna.

Finally, in this section you studied the case of a nano-oscillator. The oscillator is characterized by a quadratic potential, which you see is the simplest potential that leads to a stable behavior, i.e., the particle remains in the vicinity of the origin and does not leave the immediate region. This potential is the basis of many important physical systems and, as you will eventually see, is even the potential that beautifully describes the quantization of the electromagnetic field and the quantum vacuum.

(i) Summary of rules

The main points to take away from Chapter 2 are not so much about the nano-vibrator, although that promises to be part of important devices in the future. The main points are imbedded in the discussion and are the key rules for working with quantum problems, using a real space (x-axis) representation. These points will eventually be formalized in the chapter on new design rules. Hence, in order to facilitate your work in the problems at this point, you might find the following summary of the equations useful.

1. Everything you can know about a system is contained in the wave function, $\psi(x,t)$.
2. $\psi(x,t)$ must be a solution to Schrödinger's equation: $\widehat{H}\psi(x,t) = i\hbar\frac{\partial}{\partial t}\psi(x,t)$ where $\widehat{H}$ is the operator form of the Hamiltonian, given by $\widehat{H} = \frac{\hat{p}^2}{2m} + \widehat{V}(\hat{x})$. Also, $\widehat{H}$ is the operator associated with the observable total energy, $\hat{p}$ is the operator associated with the observable momentum, and $\hat{x}$ is the operator associated with the observable position. In the coordinate representation, the form of the operators $\hat{x}$ and $\hat{p}$ is $\hat{x} = x$ and $\hat{p} = -i\hbar\frac{\partial}{\partial x}$.

3. Without further analyzing $\psi(x,t)$, you can learn two things: (1) $|\psi(x,t)|^2$ is the probability density of finding the particle between x and $x+dx$, and hence $\int_{-\infty}^{\infty} dx\, |\psi(x,t)|^2 = 1$; and (2) If a measurement of an observable "a" is made on a system prepared in state $\psi(x,t)$ and then the process of preparing the system in state $\psi(x,t)$ and measurement of "a" is repeated over and over again (assuming the system is reinitialized before each measurement), the average value "a" as a function of time is given by the expectation value of the associated operator $\hat{A}$ defined by $\hat{A}(t) = \int_{-\infty}^{\infty} dx \;\; \psi^*(x,t)\hat{A}\psi(x,t)$. (The order in the integrand is important since operators like this operate to the right.)
4. When making a measurement of the observable "a", the only values of "a" that are allowed in the measurement are those appearing in the solution of the eigenvalue equation, $\hat{A}u_n(x) = au_n(x)$. Furthermore, as a result of the requirement that the eigenvalues are real (because they can be measured in the lab), the eigenfunctions are orthonormal, meaning $\int_{-\infty}^{\infty} dx\, u_m^*(x)u_n(x) = \delta_{mn}$. The symbol $u_n(x)$ (or sometimes $\psi_n(x)$) is reserved for eigenfunctions with discrete eigenvalues, independent of the operator. It could be $\hat{H}$ or it could be something else. If the spectrum of eigenvalues is continuous, like momentum, then $\hat{P}u(p,x) = pu(p,x)$, and orthonormality is given by $\int_{-\infty}^{\infty} dx\, u^*(p,x)\, u(p',x) = \delta(p-p')$
5. To learn more, it is necessary to identify the observable of interest (e.g., momentum, energy, etc.) and expand the $\psi(x,t)$ in terms of the eigenfunctions of the operator associated with the observable of interest. For $\psi(x,t)$ it is typically only done for the total energy associated with the Hamiltonian and only if $\hat{H}$ is time independent. We then find the eigenfunctions of $\hat{H}$ by solving $\hat{H}u_n(x) = E_n u_n(x)$ and writing $\psi(x,t)$ as $\psi(x,t) = \sum_n a_n e^{-iE_n t/\hbar} u_n(x)$ where at $t = 0$, $\psi(x,t=0) = \sum_n a_n u_n(x)$ and $a_n = \int_{-\infty}^{\infty} dx\, u_n^*(x)\psi(x,t=0)$. Then $|a_n|^2$ is the probability of making a measurement of energy and getting E_n and finding the system in state $u_n(x)$. In the case where the eigenfunctions of the associated operator are not eigenfunctions of the Hamiltonian, we usually do not expand $\psi(x,t)$, since the time dependence in this case is more complicated. Instead, we ask the question of the state $\psi(x)$, which often represents some initial state, $\psi(x,t=0)$. So for, say, the momentum, we get $\psi(x,t=0)$ or $\psi(x,) = \int_{-\infty}^{\infty} dx\; a(p)u(p,x)$ where $a(p) = \int_{-\infty}^{\infty} dx\; u^*(p,x)\;\psi(x)$.

Vocabulary (page) and Important Concepts

- Hamilton's equations 6
- canonical coordinate 6
- conjugate momentum 6
- Hamiltonian 6
- kinetic energy 6
- potential energy 6
- mathematical operator 7
- Planck's constant 8
- Schrödinger's equation 8
- wave function 8

- expectation value 9
- probability density function 9
- normalization 9
- method of separation of variables 9
- time independent Schrödinger equation 10
- eigenvalue equation 10
- eigenfunction 10
- eigenvalue 10
- eigenvector equation 10
- dimensional analysis 10
- Hermite's polynomials 12
- eigenenergy 12
- spectrum of eigenvalues 12
- quantum number 12
- zero-point energy 12
- harmonic 12
- anharmonic 12
- orthogonal 13
- Kronecker delta function 13
- orthonormal 13
- complete 13
- classical turning points 14
- classically forbidden region 15
- tunneled 15
- phonons 16
- probability amplitude 18
- de Broglie wavelength 19
- momentum eigenfunction 20
- Dirac delta function 21
- momentum representation 22
- principle of superposition 22
- coherent superposition state 22
- free particle 25
- wave–particle duality 25
- quantum machine 25

3 Free Particle, Wave Packet and Dynamics, Quantum Dots, Defects and Traps

3.1 Introduction

In this chapter, we consider other potentials beyond the nano-vibrator (harmonic oscillator) that are important in modern technology. The quantum form of the Hamiltonian is the same as in Chapter 2, except we give the potential as an arbitrary function of x:

$$\hat{H} = \frac{\hat{p}^2}{2m} + V(x) = -\frac{\hbar^2}{2m}\frac{d^2}{dx^2} + V(x) \tag{3.1}$$

These structures are produced by design through growth and/or fabrication and also by defects in semiconductors that are exploited for application in devices. Again, we focus on one dimension to minimize the distraction from more complex math, but also because many devices today are adequately described by working in one dimension. For example, this is the same form of the equation that develops in a semiconductor like silicon or GaAs in the usual case of the effective mass approximation in a quantum well or a one-dimensional conducting wire. We can assume that m is the appropriate mass for an electron or a hole, for example. The mass in a vacuum for an electron is, of course, different than the effective mass in a semiconductor. So the fundamental behavior is really quite general. This same equation also shows the origin of band structure as we will show later—and we will also consider higher dimensions.

3.2 The Free Particle

In Chapter 2, we briefly considered the case when the spring constant is zero ($\kappa = 0$), in order to appreciate the fact that, mathematically, Schrödinger's equation is a kind of wave equation. We discuss this again in a little more detail. When $V = 0$, we have the **free particle Hamiltonian** in operator form:

$$\hat{H} = \frac{\hat{p}^2}{2m} = -\frac{\hbar^2}{2m}\frac{d^2}{dx^2} \tag{3.2}$$

Schrödinger's equation is

$$-\frac{\hbar^2}{2m}\frac{\partial^2}{\partial x^2}\psi(x,t) = i\hbar\frac{\partial}{\partial t}\psi(x,t) \tag{3.3}$$

Introduction to Quantum Nanotechnology: A Problem Focused Approach. Duncan Steel, Oxford University Press (2021).
 DOI: 10.1093/oso/9780192895073.003.0003

Since the Hamiltonian is again time independent, the form of the solution remains

$$\psi(x,t) = \psi(x)g(t) \tag{3.4}$$

and

$$g(t) = e^{-i\frac{Et}{\hbar}} \tag{3.5}$$

as before.

This leaves us to again solve a second order partial differential eigenvalue equation for $\psi(x)$ and E, namely the time independent Schrödinger equation (the energy eigenvalue equation),

$$\hat{H}\psi(x) = E\psi(x) \tag{3.6}$$

or substituting for the Hamiltonian

$$-\frac{\hbar^2}{2m}\frac{d^2}{dx^2}\psi(x) = E\psi(x) \tag{3.7}$$

Taking the form of the solution to be

$$\psi(x) = Ce^{ikx} \tag{3.8}$$

and substituting into Eq. 3.7, we find

$$\frac{\hbar^2 k^2}{2m} = E \tag{3.9}$$

or

$$k(E) = \pm\sqrt{\frac{2mE}{\hbar^2}} \tag{3.10}$$

Notice that this time, the solution to the time independent Schrödinger equation has no additional requirements (such as termination of a series expansion or boundary conditions) and hence there are no constraints on the eigenenergy, so E can take on any value. The *spectrum of eigenvalues is continuous* in this case (i.e., not discrete as in the case of the nano-vibrator in Chapter 2). Because there is no confining potential, the integration is over all space, so just as in Eq. 2.70 these functions are orthogonal for different E (and k) and are normalized to a Dirac delta-function, because the eigenvalues are continuous. The corresponding orthonormal energy eigenfunctions are given by

$$\psi(E,x) = \frac{1}{\sqrt{2\pi}}e^{\pm i\sqrt{\frac{2mE}{\hbar^2}}x} = \frac{1}{\sqrt{2\pi}}e^{\pm ikx} \tag{3.11}$$

where it is helpful to recall that a *common and important* representation of the Dirac delta-function introduced in Chapter 2 (see Appendix C) is given by

$$\int_{-\infty}^{\infty} dx\, e^{\pm i(k-k')x} = 2\pi\delta(k-k') \tag{3.12}$$

Hence, the normalization factor in Eq. 3.11 has to be $\frac{1}{\sqrt{2\pi}}$. Compared to 2.71, where we chose to use p as the eigenvalue rather than k, there is no $\frac{1}{\sqrt{\hbar}}$ in Eq. 3.11. Of course, it is clear that $p = \hbar k$. The discussion is very similar to that in the development of 2.71, except here the operator of interest is the Hamiltonian for a free particle, $\frac{\hat{p}^2}{2m}$, and there the operator of interest was for the observable $\hat{p}$. Because the functions in Eq. 3.11 (for any value of E) form a complete set (same as in Chapter 2 for the nano-vibrator), we obtain the full solution by following Eq. 2.42, except the sum over the eigenstates is converted to an integral, because the spectrum of eigenvalues here is continuous. The full time dependent solution for an arbitrary wave function is then:

$$\psi(x,t) = \int_{-\infty}^{\infty} dk\ C(k)\frac{1}{\sqrt{2\pi}}e^{ikx-i\frac{E(k)t}{\hbar}} = \int_{-\infty}^{\infty} dk\ C(k)\frac{1}{\sqrt{2\pi}}e^{ikx-i\frac{\hbar k^2 t}{2m}} \tag{3.13}$$

$$C(k) = \int_{-\infty}^{\infty} dx\frac{1}{\sqrt{2\pi}}e^{-ikx}\ \psi(x, t=0) \tag{3.14}$$

where, as discussed in the previous section, the expansion constant (i.e., expansion coefficient $C(k)$ also called the probability amplitude) is determined by the initial state of the system in Eq. 3.14. In the case of the harmonic oscillator, the individual expansion coefficients included a subscript (an integer) reflecting that it was associated with a specific energy eigenvalue. Here, the corresponding index is the value of k (or E) since each k corresponds to a specific energy eigenvalue. By common convention, since k is continuous, we include it as an argument of the expansion coefficient rather than a subscript, as discussed in Chapter 2. This set of eigenfunctions is also complete, meaning that any function defined as a function of x can be expanded in terms of these functions. Observe, too, that for a given value of E there are two eigenfunctions corresponding $\pm k$. This means that energy eigenvalues are doubly **degenerate** (the same). Noting that p $=$ $\hbar$k, $\pm k$ corresponds to a particle going toward $+\infty$ and $-\infty$, respectively.

Note that the time independent Schrödinger equation has solutions that have both continuous and discrete spectra of eigenvalues and eigenfunctions that are extended (in space) and confined. To be clear, a **bound state** means a **localized state** in some region of space and the wave function goes to zero at $\pm\infty$. Mathematically, localized states mean that the wave function is **square integrable**; i.e., $\int_{-\infty}^{\infty} |\psi(x)|^2 dx$ has a well-defined value. **Extended states** or **delocalized states**, like $\psi(x) = e^{\pm ikx}$, exist everywhere in space. The expression $\int_{-\infty}^{\infty} |\psi(x)|^2 dx$ has no value in the case where the wave function is extended, hence functions like this are not **square integrable. Bound states** have a discrete spectrum of eigenvalues and extended states have a continuous spectrum. Bound states occur when their energy is not sufficient to allow them to escape to $\pm\infty$. Even escaping to just $+\infty$ or $-\infty$ results in a continuous spectrum. Some forms of the time independent Schrödinger equation (depending on the form of the potential), such as the finite square potential below (Fig. 3.3) have a spectrum of eigenvalues containing both discrete and continuous values. For every form of the Hamiltonian, $\widehat{H}$, the set of all solutions (including both bound and extended) form the complete set of eigenfunctions of that Hamiltonian. Changing the form of the potential will change the details of the eigenfunctions of the complete basis, but the basis remains complete.

We will not discuss this now, but if you have thought about light or X-rays diffracting from a crystal, it will not surprise you to learn that electrons will also diffract from a crystal because of its wave-like nature. *In the quantum description of the electron, it cannot be considered to be some kind of massiveobject at some point in space.*

3.3 Localized State in Free Space: The Wave Packet

The eigenfunctions of the free particle are extended, *meaning that the probability is uniform over all space* (the x-axis in this case). So, if a particle is somehow localized in free space, it is not in an eigenstate of the free particle Hamiltonian. However, since the free particle states are complete, we can expand such a localized state in terms of the free particle states. A typical math form for a localized state that turns out to have a very important quantum significance is a **Gaussian wave packet**. As you will see, it is a Gaussian distribution of plane wave states. The normalized Gaussian wave function for a particle localized in space is given by:

$$\psi(x) = \frac{1}{(\pi\ell^2)^{1/4}} e^{-\frac{x^2}{2\ell^2}} \tag{3.15}$$

where ℓ is some scale length, not related to ℓ defined for the nano-vibrator. In this case, ℓ will be determined by the Hamiltonian (not discussed) used to localize the particle. The full width at half maximum for the probability density, $FWHM = 2\sqrt{\ln 2}\,\ell$.

The free particle eigenfunction is from Eq. 3.11, $\frac{1}{\sqrt{2\pi}}e^{\pm ikx}$ or $\frac{1}{\sqrt{2\pi\hbar}}e^{\pm i\frac{p}{\hbar}x}$. In this way, we showed in Chapter 2 (Eq. 2.71) that we could rewrite Eq. 3.15 as a coherent superposition of free particle eigenstates given by

$$\psi(x) = \frac{1}{(\pi\ell^2)^{1/4}} e^{-\frac{x^2}{2\ell^2}} = \int_{-\infty}^{\infty} dp\; C(p)\frac{1}{\sqrt{2\pi\hbar}}\, e^{i\frac{p}{\hbar}x} \tag{3.16}$$

and then by projecting onto a specific momentum eigenstate we found for $C(p)$

$$C(p) = \int_{-\infty}^{\infty} dx\; \frac{1}{\sqrt{2\pi\hbar}}\, e^{-i\frac{p}{\hbar}x}\, \frac{1}{(\pi\ell^2)^{1/4}} e^{-\frac{x^2}{2\ell^2}} = \left(\frac{\ell^2}{\pi\hbar^2}\right)^{1/4} e^{-\left(\frac{p}{\hbar}\right)^2\frac{\ell^2}{2}} \tag{3.17}$$

The evaluation of the integral in 3.17 uses the general result for the Gaussian integral in Eq. 2.73. Recall that this is also called the momentum representation. We can insert Eq. 3.17 back into 3.16:

$$\psi(x) = \int_{-\infty}^{\infty} dp\; C(p)\frac{1}{\sqrt{2\pi\hbar}}\, e^{i\frac{p}{\hbar}x} = \int_{-\infty}^{\infty} dp\left(\frac{\ell^2}{\pi\hbar^2}\right)^{1/4} e^{-\left(\frac{p}{\hbar}\right)^2\frac{\ell^2}{2}}\frac{1}{\sqrt{2\pi\hbar}}\, e^{i\frac{p}{\hbar}x} \tag{3.18}$$

If we carried out the integral over p, we would recover the original Gaussian wave function in Eq. 3.15.

However, we can now use this result to determine the time evolution or dynamics of the Gaussian wave packet, $\psi(x, t)$, because $\psi(x)$ is expanded in Eq. 3.18 in terms of the energy eigenfunctions of the free particle Hamiltonian. The time dependence for a given E is given by Eq. 3.5, namely

$$g(t) = e^{-i\frac{Et}{\hbar}} = e^{-i\frac{p^2 t}{2m\hbar}} \tag{3.19}$$

Then the eigenfunction as a function of time is

$$\frac{1}{\sqrt{2\pi\hbar}} e^{i\frac{p}{\hbar}x} e^{-iEt/\hbar} = \frac{1}{\sqrt{2\pi\hbar}} e^{i\left(\frac{p}{\hbar}x - \frac{p^2 t}{2m\hbar}\right)} \tag{3.20}$$

The time dependent form of the wave packet follows by inserting 3.20 into Eq. 3.18:

$$\begin{aligned}
\psi(x,t) &= \frac{1}{\sqrt{2\pi\hbar}}\left(\frac{\ell^2}{\pi\hbar^2}\right)^{\frac{1}{4}} \int_{-\infty}^{\infty} dp\; e^{-\left(\frac{p}{\hbar}\right)^2 \frac{\ell^2}{2}} e^{i\left(\frac{p}{\hbar}x - \frac{p^2 t}{2m\hbar}\right)} \\
&= \frac{1}{\sqrt{2\pi\hbar}}\left(\frac{\ell^2}{\pi\hbar^2}\right)^{\frac{1}{4}} \int_{-\infty}^{\infty} dp\; e^{-\left(\frac{\ell^2}{2\hbar^2} + i\frac{t}{2m\hbar}\right)p^2 - i\frac{p}{\hbar}x} \\
&= \frac{1}{\sqrt{2\pi\hbar}}\left(\frac{\ell^2}{\pi\hbar^2}\right)^{\frac{1}{4}} \int_{-\infty}^{\infty} dp\; e^{-\left(\frac{\ell^2}{2\hbar^2} + i\frac{t}{2m\hbar}\right)p^2 - i\frac{p}{\hbar}x} \\
&= \frac{1}{\sqrt{2\hbar}}\left(\frac{\ell^2}{\pi\hbar^2}\right)^{\frac{1}{4}} \frac{1}{\sqrt{\frac{\ell^2}{2\hbar^2} + i\frac{t}{2m\hbar}}} e^{\frac{-x^2}{4\hbar^2\left(\frac{\ell^2}{2\hbar^2} + i\frac{t}{2m\hbar}\right)}}
\end{aligned} \tag{3.21}$$

or

$$\psi(x,t) = \frac{1}{\pi^{\frac{1}{4}}\hbar}\left(\frac{\ell}{2}\right)^{\frac{1}{2}} \frac{1}{\sqrt{\frac{\ell^2}{2\hbar^2} + i\frac{t}{2m\hbar}}} e^{\frac{-x^2}{2\ell^2\left(1 + i\frac{\hbar t}{\ell^2 m}\right)}} \tag{3.22}$$

This is a remarkable result. It says that if you localize a particle in region with size of order ℓ it will expand or delocalize in time. The width of the probability density, $|\psi(x,t)|^2$, varies with time and is given by

$$FWHM = 2\sqrt{ln\,2}\,\ell\left(1 + \left(\frac{\hbar t}{\ell^2 m}\right)^2\right) \tag{3.23}$$

A question arises now as to, physically, why this happens. For the delocalized free particle, $\psi(x,t) = \frac{1}{\sqrt{2\pi\hbar}} e^{i\frac{p}{\hbar}x - i\frac{p^2 t}{2m\hbar}}$ and therefore the probability density of finding a particle at some point x at time t is $|\psi(x,t)|^2 = \frac{1}{2\pi\hbar}$. That is to say, it is a constant throughout space and time and very different from that predicted by taking $|\psi(x,t)|^2$ of Eq. 3.22 which shows a Gaussian with a width that broadens with time. The difference is seen in comparing the integrands in 3.18 and 3.21. In 3.18, we see that the **localized wave function**, the Gaussian, is created by *a coherent superposition of an infinite number of plane wave states*, described in the integrand by the eigenfunction of the free particle, $\frac{1}{\sqrt{2\pi\hbar}}e^{i\frac{p}{\hbar}x}$. It is important to note that the phase difference as a function p is well-defined in this system. It is this linear dependence of the phase on p that results in reproducing the localized state. In Eq. 3.21, we add the time dependence. We do this by inserting the time dependence for each free particle state seen in the time dependence for the corresponding eigenfunction given by $\psi(x,t) = \frac{1}{\sqrt{2\pi\hbar}}e^{i\frac{p}{\hbar}x - i\frac{p^2 t}{2m\hbar}}$. At $t = 0$, we recover the localized result of Eq. 3.18. However, for $t > 0$, we see that there is a quadratic term, p^2, in the phase shift. This quadratic term destroys the original phase relationship between all the different free particle states. The result is that there is no longer the necessary relative phase relationship needed to create the **constructive interference** that was the original state. Hence, the physical origin of the localized state is the **constructive interference** in a specific region of space of the superposition of an infinite number of waves and the **destructive interference** of the waves at large distances going to $\pm\infty$. The spreading in time is due to the loss of the constructive interference.

Problem 3.1

a) The initial wave function at $t = 0$ is given by

$$\psi(x) = \frac{1}{(\pi\ell^2)^{1/4}}e^{-\frac{x^2}{2\ell^2}}$$

Derive the FWHM of the probability densityand show that it is FWHM $= 2\sqrt{\ln 2}\,\ell$

b) Find the FWHM at time t.

c) Find the velocity as a function of time at which the FWHM is expanding. This is not about the velocity of the particle.

3.4 Nano-Heterostructures: Quantum Dots and Deep Traps

In modern semiconductor structures that are advancing the state of the art in nano-science, engineers and scientists are using various crystal growth methods and fabrication techniques to build structures that are planar and very thin, or thick but filled with small spherical-like structures (marbles) only a few tens of nanometers in diameter, or possibly making structures that look like a microwave resonator (microwave oven) but for electrons. Remarkable new ideas are emerging all the time.

Quantum dots are used in applications ranging from sensors in biomedical applications to the foundational structure for quantum information processing and transmission. The hard work is (a) in the laboratory, figuring out how to make these materials and (b) in the theory of how to use current technology to realize these new ideas. The behavior and rules for designing devices based on these structures are contained in Schrödinger's equation.

In solutions of partial differential equations of the type in this chapter, we need to consider the nature of the partial differential equation and the requirement for a solution to be well behaved. Singularities, where the value of a function that is associated with a measurement in the laboratory goes to $\pm\infty$, are certainly unphysical. The time independent Schrödinger equation must remain finite (i.e., not infinite) everywhere. In addition, since it is a second order differential equation, *the solution and its derivative must remain continuous* except as noted below. With this understanding, we begin by considering potentials described as **piecewise-constant potentials**.

Mathematically, the simplest concept is a potential that is an infinite square well, where the potential is zero at the bottom but goes to infinity at points to the left and right, as shown in Fig. 3.1. This is the simplest model, which turns out to be quite important for predicting and understanding quantum dots in the laboratory. This is a one-dimensional model, but the basic physics remains the same for higher dimensions (see Chapter 6).

With a point particle of mass m in the potential, this is the problem of the ***particle in a box***. It can be an excellent description for a particle in deep traps in solids. Finite potentials are considered below, but this potential reveals the essential physics and math.

In keeping with the effort to build up an intuitive understanding of quantum mechanics, what can we say about the problem? We know the massive particle will be localized in space between $\pm\frac{L}{2}$ and the confinement will lead to quantization of the allowed energies for the particle, i.e., the spectrum of eigenvalues with be discrete. Also, since the Hamiltonian on the left in Fig. 3.1 is even on reflection through the origin, the eigenfunctions will be even or odd. For the Hamiltonian, we again have

$$\hat{H} = \frac{\hat{p}^2}{2m} + V(x) = -\frac{\hbar^2}{2m}\frac{d^2}{dx^2} + V(x) \tag{3.24}$$

Based on the diagram on the left of Fig. 3.1 we have

$$V(x) = \begin{cases} 0 \; for \mid x \mid < \frac{L}{2} \\ \infty \; for \mid x \mid > \frac{L}{2} \end{cases} \tag{3.25}$$

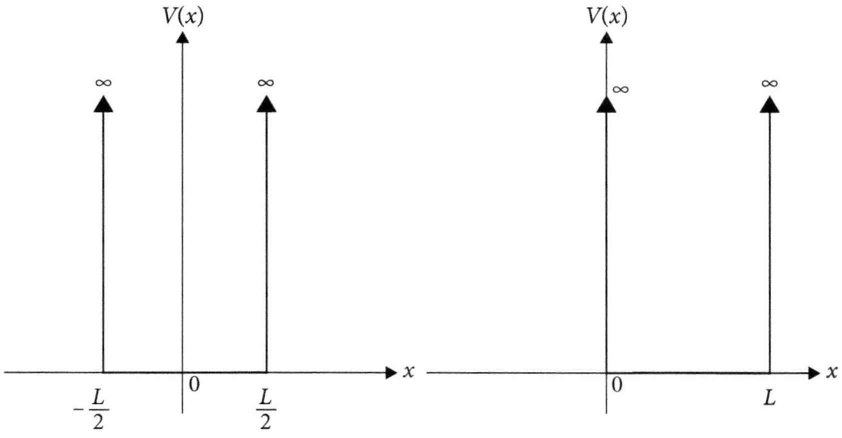

Fig. 3.1 A particle in a box has a constant potential at the bottom, usually taken to be zero and has a width of L. The identical box is shown in two different coordinate systems to illustrate a point about spatial symmetry described in the text. There is no difference in the physics between the two cases. Such a model is very effective at describing the basic quantum features of quantum dots and the effects of defects or traps on particles.

The time independent Schrödinger equation for the energy eigenvalues and eigenfunctions remains

$$\widehat{H}\psi(x) = \left(\frac{\hat{p}^2}{2m} + V(x)\right)\psi(x) = E\psi(x) \tag{3.26}$$

or

$$-\frac{\hbar^2}{2m}\frac{d^2}{dx^2}\psi(x) = E\psi(x) \text{ for } |x| \leq \frac{L}{2} \tag{3.27}$$

This equation appears identical to Eq. 3.7 for a free particle. However, Eq. 3.7 applies to all of space. Here in Eq. 3.27, it applies only to the region bounded by $\pm\frac{L}{2}$. Hence, the boundary conditions change the solution by placing constraints on where the function is non-zero.

For x outside the box, the potential goes to infinity, and that means the only solution outside the box is

$$\psi\left(|x| > \frac{L}{2}\right) \equiv 0 \tag{3.28}$$

This is also seen by considering the case where the potential is finite at $|x| = \frac{L}{2}$, namely $V\left(|x| = \frac{L}{2}\right) = V_0$ (discussed more completely below). In this case, for $|x| \geq \frac{L}{2}$, the solution to Eq. 3.26 has the form $\psi(x) \sim e^{-\kappa x}$ for $x > \frac{L}{2}$ and $e^{+\kappa x}$ for $x < -\frac{L}{2}$ where $\kappa^2 = \frac{2m(V_0 - E)}{\hbar^2}$. We see that in the limit that $V_0 \rightarrow \infty$, $\psi\left(x > \frac{L}{2}\right) \rightarrow 0$. Since ψ must be continuous everywhere, we see that $\psi\left(|x| \geq \frac{L}{2}\right) \equiv 0$.

The calculation of the energy eigenfunctions and eigenenergies will continue below, but we take advantage of this simple system to illustrate another very important property of this Hamiltonian that can be exploited to improve the understanding and reduce the amount of math needed to solve this problem. Fig. 3.1 contains two different displays for the calculation. On the left, the potential well is centered at the origin. On the right, the potential well begins at the origin and exists only for positive x. Notice that, for the case on the left, if you imagined a mirror perpendicular to the x-axis and located at the origin, the image in the mirror (aimed to the right) would be a reflection that looks exactly like the potential for negative x in the case on the left. This is not true for the picture on the right with the mirror remaining at the origin. So, for the picture on the left, we see that

$$V(x) = V(-x) \tag{3.29}$$

For a potential like this, the origin is located at the position of ***reflection symmetry*** and because of Eq. 3.29, $V(x)$ is an even function. The potential for the nano-vibrator in Chapter 2 had the same property, assuming that the potential was centered at the origin (as illustrated in Fig. 2.2).

Mathematically, the potential then in Eq. 3.29 describing the picture on the left is said to be *even on reflection through the origin*. Furthermore, the operator, $\frac{d^2}{dx^2}$, is even since it is the same for x as $-x$. This is not the same for say $\frac{d}{dx}$, which is $-\frac{d}{dx}$ for $x \rightarrow -x$. Hence, the entire operator for the Hamiltonian is even if we use the coordinate system on the left in the figure. Clearly, solving a differential equation using the picture on the right or the left will yield the same physics. But, if we can take advantage of *symmetry* (in this case, reflection symmetry), the solutions may be easier to obtain and easier to understand. We will use this simple math problem to illustrate the power of using symmetry (spatial or otherwise) as well as introducing an operator that represents the symmetry.

We define a new operator, $\hat{P}$, called a ***parity operator***, such that

$$\hat{P}f(x) \equiv f(-x) \tag{3.30}$$

We are then interested in finding the eigenvalues and eigenfunctions of $\hat{P}$. This sounds very abstract, but the result is important. We wish to find

$$\hat{P}f(x) = bf(x) \tag{3.31}$$

It is clear that

$$\hat{P}^2 f(x) = f(x) \tag{3.32}$$

Therefore

$$\hat{P}^2 f(x) = \hat{P}bf(x) = b\hat{P}f(x) = b^2 f(x) = f(x) \tag{3.33}$$

Based on the last equality, this says that the eigenvalues of the parity operator are

$$b = \pm 1 \tag{3.34}$$

or even ($+1$, **even parity**) and odd (-1, **odd parity**). Thus if $b = +1$, we have

$$\hat{P}f(x) = f(-x) = f(x) \tag{3.35}$$

So, $f(x)$ is an **even** function, such as $\cos(x)$ or x^2. Likewise, if $b = -1$, then $f(-x) = -f(x)$ and we say $f(x)$ is an odd function, such as $\sin(x)$ or x. *If the Hamiltonian is even, then the parity, b, is a good quantum number.*

To see the importance of this result, consider again the Hamiltonian written for the system in the left panel of Fig. 3.1. Since sending x to $-x$ does not change the Hamiltonian, this means that $\hat{P}\hat{H} = \hat{H}\hat{P}$ or

$$\hat{P}\hat{H} - \hat{H}\hat{P} \equiv \left[\hat{H}, \hat{P}\right] = 0 \tag{3.36}$$

where we call $\left[\hat{H}, \hat{P}\right]$ the **commutator** of $\hat{H}$ and $\hat{P}$. Because $\left[\hat{H}, \hat{P}\right] = 0$, we say that $\hat{H}$ and $\hat{P}$ commute. Note that as operators, it is important to recall that they operate on a function to the right, and hence it would *not be correct to say that $\hat{P}\hat{H}$ is the same as $\hat{H}$* since that ignores the presence of a function to the right of $\hat{H}$ when it is being used as an operator. This is obvious if you recall that an operator is a mathematical instruction. To say that $\hat{P}\hat{H} = \hat{H}\hat{P}$ means that the order in which the instructions are implemented does not matter. The importance of the commutator is associated with various kinds of symmetries and will emerge in the course of the discussion below and in future chapters.

Starting with Eq. 3.26, if $\psi(x)$ is an eigenfunction, then

$$\hat{P}\hat{H}\psi(x) = \hat{H}\hat{P}\,\psi(x) = E\hat{P}\,\psi(x), \tag{3.37}$$

showing that therefore $\hat{P}\,\psi(x)$ is also an eigenfunction. *This means we can choose solutions to* 3.26 *that are even or odd.* This is not possible for the figure on the right of Fig. 3.1. The use of parity enables us then to halve the amount of algebra. This is because if we choose the form of solutions that are even or odd, then when we satisfy the boundary condition at one of the boundaries, in the case of the figure on the left side of Fig. 3.1, it will be guaranteed that the function satisfies the boundary condition at the other boundary.

We write Eq. 3.27 as

$$\frac{d^2}{dx^2}\psi(x) = -k^2\psi(x) \text{ for } |x| \leq \frac{L}{2} \tag{3.38}$$

where again

$$k = \sqrt{\frac{2mE}{\hbar^2}}. \tag{3.39}$$

The solutions to this simple equation (3.38) are in terms of $e^{\pm ikx}$ or $\cos kx$ and $\sin kx$.[1] To solve 3.28, in general we create a linear combination of the two solutions to satisfy the boundary conditions. Recall now that the boundary conditions are that the eigenfunction must go to zero at the boundaries, i.e., for the figure on the left:

$$\psi\left(x = \pm\frac{L}{2}\right) = 0 \tag{3.40}$$

If we choose the exponential form, then $\psi(x) = Ae^{+ikx} + Be^{-ikx}$ and to find A and B, we solve the resulting set of two coupled homogeneous equations at the two boundary points. However, we know from the work on symmetry that the cosine is an even function and the sine is an odd function about the origin. For either option, we just have to arrange for the functions to vanish at the boundaries.

Hence, referring to Fig. 3.1 on the left, for **even eigenfunctions**,

$$\psi_{even}(x) = A\cos kx \tag{3.41}$$

and from the boundary conditions we require

$$\cos k\frac{L}{2} = \cos k\frac{-L}{2} = 0$$

[1] Note that, in general, the form of the second order differential equation that is encountered in this section is $\frac{d^2y}{dx^2} \pm k^2y = 0$. For the (+) sign, the solution is $e^{\pm ikx}$ or $\sin kx$ or $\cos kx$. For the (−) sign, the solution is $e^{\pm kx}$ or $\sinh kx$ or $\cosh kx$. If $y_1(x)$ and $y_2(x)$ represent the two linearly independent solutions, then the general solution is always of the form $y(x) = Ay_1(x) + By_2(x)$, where the constants are chosen to satisfy two boundary conditions. See Appendix A.

or

$$k_{n_{even}} = \left(n + \frac{1}{2}\right)\frac{2\pi}{L} \text{ for } n = 0, 1, 2, \ldots \tag{3.42}$$

Hence, the energy is quantized, and we require for the energy, E

$$E_{n_{even}} = \frac{\hbar^2 k_{n_{even}}^2}{2m} = \frac{\hbar^2}{2m}\left(\frac{\left(n+\frac{1}{2}\right)2\pi}{L}\right)^2 = E_{box}\left(n+\frac{1}{2}\right)^2 \text{ and } n = 0, 1, 2, \ldots E_{box} = \frac{\hbar^2}{2m}\left(\frac{2\pi}{L}\right)^2 \tag{3.43}$$

The normalized even eigenfunction is then:

$$\psi_{even}(x) = \sqrt{\frac{2}{L}}\cos\left(\left(n+\frac{1}{2}\right)\frac{2\pi x}{L}\right) \quad n = 0, 1, 2, \ldots \tag{3.43}$$

Referring to Fig. 3.1 on the left, for ***odd eigenfunctions***,

$$\psi_{odd}(x) = B\sin kx \tag{3.44}$$

we require

$$\sin k\frac{L}{2} = 0$$

or

$$k_{n_{odd}} = n\frac{2\pi}{L} \text{ for } n = 1, 2, \ldots \tag{3.45}$$

The eigenenergy is given by

$$E_{n_{odd}} = \frac{\hbar^2 k_{n_{odd}}^2}{2m} = \frac{\hbar^2}{2m}\left(\frac{2n\pi}{L}\right)^2 = E_{box}\, n^2 \; n = 1, 2, \ldots \tag{3.46}$$

The normalized odd eigenfunction is then

$$\psi_{odd}(x) = \sqrt{\frac{2}{L}}\sin\left(n\frac{2\pi x}{L}\right) \quad n = 1, 2, \ldots$$

The allowed energy eigenvalues are shown in the plot in Fig. 3.2a and illustrated in the potential well in Fig. 3.2b.

In nano-technology, quantum boxes are typically three-dimensional and sometimes rectangular so that the above solution applies for each of the three axes, x, y, and z.

Problem 3.2 *Normalize the even and odd eigenfunctions for an arbitrary n for the particle in the infinite potential well and plot the probability density of finding the particle between x and $x + dx$ for the lowest energy even and odd eigenfunction.*

Problem 3.3 *Find E_{box} for an electron in a box with a width of 10 nm. Is that energy much greater than or less than room temperature ($T = 300$ K). Since thermal vibrations often destroy quantum*

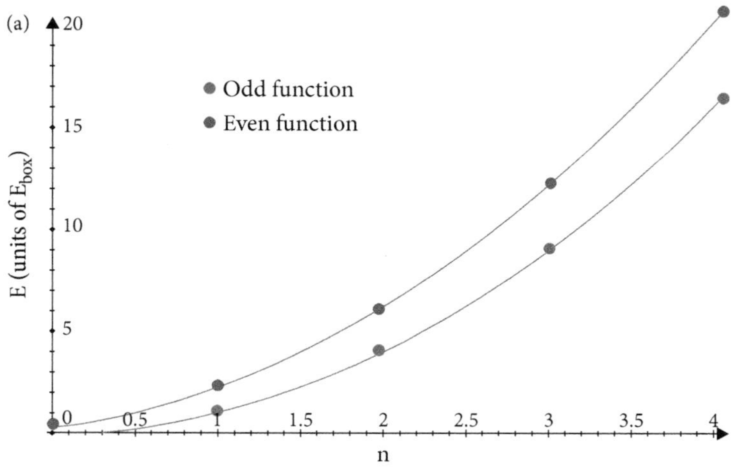

Fig. 3.2a Eigenstates and energy eigenvalues for the particle in a box. Notice that the lowest eigenenergy is not zero. Why? It is not obvious, but is quite profound. The reason

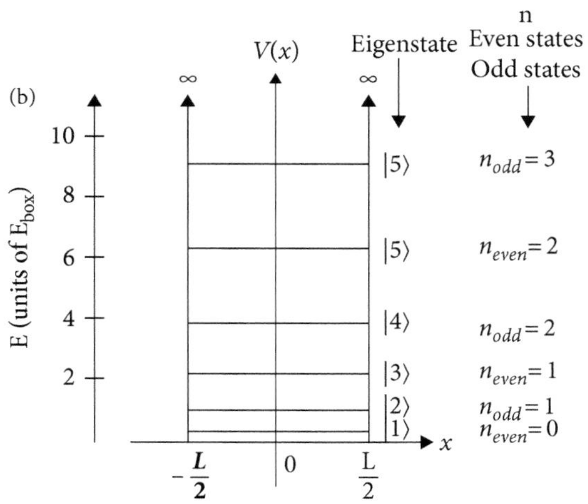

Fig. 3.2b Energy eigenstates in ascending order. The n-value corresponding to each state, depending on whether it is even or odd is shown on the right. Energy units in E_{box}

behavior, does your result preclude or predict possibly quantum behavior? To answer this question, the average energy of thermal motion must be much less than the energy of the quantum particle. In this case, that is E_{box}.

Problem 3.4 *Find the expectation value for $\hat{P}$ and $\hat{P}^2$ where $\hat{P} = -i\hbar\frac{d}{dx}$ for the lowest energy eigenstate.*

Problem 3.5 *Find the normalized energy eigenfunctions for the case on the right in Fig. 3.1 and show that the energy eigenvalues are exactly the same.*

Problem 3.6 *Using the approach in Chapter 2 (e.g., Eq. 2.79) to find the time dependence for the potential on the left of Fig. 3.1*

a) Find the $\psi(x,t)$ *if* $\psi(x, t = 0) = A_0 \cos k_{0_{even}} x + A_1 \sin k_{1_{odd}} x$

b) Find $\langle x \rangle (t)$

HINT: Use the spatial symmetry of the problem to show two of the four terms are zero, noting that x is an odd function. To evaluate the integral, use the trigonometric identity:

$$\cos u \cos v = \frac{1}{2}\left[\cos(u - v) + \cos(u + v)\right]$$

In comparing the results between the potential in Chapter 2 and the potential above, the math is different, but only quantitatively. Both solutions are characterized by being even or odd because the potentials are symmetric. The even functions have an even number of zero-crossings and the odd functions have an odd number of zero-crossings. Both potentials go to infinity as x goes to $\pm\infty$. The quadratic potential gives energy level spacings that are identical, harmonic, as we discussed in Chapter 2 but here the square potential gives spacings that monotonically increase with increasing n and are anharmonic. As in the case of the nano-vibrator, the minimum eigenenergy is above the minimum potential.

3.5 A Particle Trapped in a Shallow Defect

A similar problem, which is common in electronic devices but requires a slightly more complicated analysis, is considered in Fig. 3.3. This is a simple model for a shallow trap or defect in a semiconductor, for example, or a more realistic model for a quantum dot. It is also an accurate description for a thin layer of semiconductor placed between two layers of a higher bandgap material. Defects and traps are often quantum states where the depth of the potential is sufficiently small that the idea of the infinite potential above is no longer adequate because in the infinite potential, the particle can never escape.

Compared to Fig. 3.1, this system is like the box but the potential at the walls does not go to infinity but rather to V_0. The zero in the potential energy is taken to be the bottom of the trap. However, that is arbitrary: *one can always add a constant to the Hamiltonian without changing anything.* A shift in the energy means nothing unless it is relative to something else.

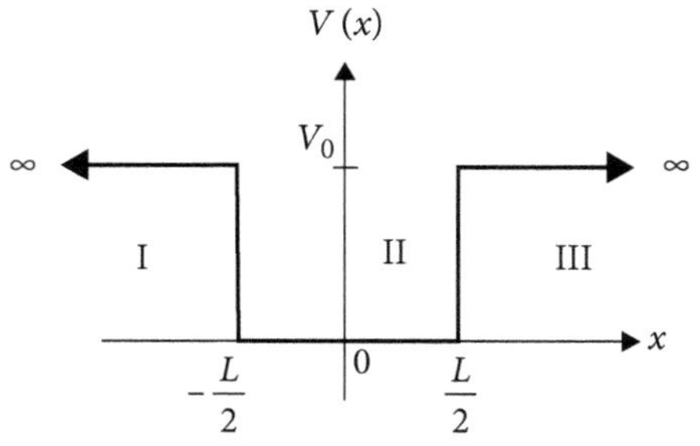

Fig. 3.3 A potential well is an excellent model for a defect, a quantum well, or even a quantum dot if generalized to three dimensions.

For piecewise constant potentials like this, it is common to divide the space into three separate regions where the potential is constant in each. The time independent Schrödinger equation is solved in each region and then boundary conditions for $\psi(x)$ and $\frac{d\psi(x)}{dx}$ (they are continuous at discontinuities in the potential, otherwise the second derivative would be singular and that is unphysical) are used to join the solutions. We are interested in the trap states, i.e., states where the particle is **localized** in the trap. Hence this requires $E < V_0$ where E is the energy of the particle. This region is identified as II in Fig. 3.3.

For regions I, II, and III respectively:

$$V = \begin{cases} V_o \text{ for } x \leq -\frac{L}{2} \\ 0 \text{ for } -\frac{L}{2} \leq x \leq \frac{L}{2} \\ V_o \text{ for } x \geq \frac{L}{2} \end{cases} \tag{3.47}$$

The symmetry of the Hamiltonian is *even* as it was for the left picture in Fig. 3.1. *Hence, parity remains a good quantum number and the solutions must be overall even or odd.*

In the region $-\frac{L}{2} \leq x \leq \frac{L}{2}$, where $V = 0$, the time independent Schrödinger equation is again Eq. 3.38:

$$\frac{d^2}{dx^2}\psi(x) = -\frac{2mE}{\hbar^2}\psi(x) = -k^2\psi(x)$$

In region II, where $V = 0$, the corresponding solutions will again be $\cos kx$ and $\sin kx$, as above, where $k^2 = \frac{2mE}{\hbar^2}$.

For the case $|x| \geq \frac{L}{2}$, *the solution is not zero (unlike the problem of the infinite potential) since the potential is finite*. The time independent Schrödinger equation becomes

$$\frac{d^2}{dx^2}\psi(x) = \frac{2m}{\hbar^2}(V_0 - E)\psi(x) = \kappa^2\psi(x) \text{ for } |x| \geq \frac{L}{2} \tag{3.48}$$

where now

$$\kappa^2 = \frac{2m}{\hbar^2}(V_0 - E) > 0 \tag{3.49}$$

The solutions now are in terms of exponential functions $e^{\pm\kappa x}$. Since $e^{-\kappa x} \to \infty$ for $x \to -\infty$ and $e^{\kappa x} \to \infty$ for $x \to \infty$, these solutions are divergent since they are singular at $x = \pm\infty$, respectively, and are unphysical since they place the greatest probability of finding the particle at infinity rather than in the trap. Hence, we set the coefficient in front of them (i.e., their amplitude) to 0. The general solution is then given in each region.

For the even solution:

$$\psi(x) = \begin{cases} \psi_I(x) = A\, e^{\kappa x} & x \leq -\frac{L}{2} \\ \psi_{II}(x) = B\, \cos kx & -\frac{L}{2} \leq x \leq \frac{L}{2} \\ \psi_{III}(x) = C\, e^{-\kappa x} & \frac{L}{2} \leq x \end{cases} \tag{3.50}$$

where the Roman numerals correspond to those shown in Fig. 3.3.

For the odd solution

$$\psi(x) = \begin{cases} \psi_I(x) = A'\ e^{\kappa x} & x \leq -\frac{L}{2} \\ \psi_{II}(x) = B'\ \sin kx & -\frac{L}{2} \leq x \leq \frac{L}{2} \\ \psi_{III}(x) = C'\ e^{-\kappa x} & \frac{L}{2} \leq x \end{cases} \tag{3.51}$$

The prime on the constant in Eq. 3.51 is to distinguish those letters from the similar constant in Eq. 3.50. For $|x| \geq \frac{L}{2}$ we require for the even solution $C = A$. For the odd solution, we require $C' = -A'$ so that the overall $\psi(x)$ is even or odd.

The boundary conditions are that both $\psi(x)$ and $\frac{d\psi(x)}{dx} \equiv \psi'(x)$ are continuous at points of discontinuity of the boundary, or the second derivative will be divergent. Since we have chosen the coefficients and functions to assure that the overall result is even or odd, we have only to match boundary conditions at one of the boundaries.

For the even functions, we require at $x = \pm\frac{L}{2}$ for continuity of $\psi(x)$ and $\psi'(x)$ meaning

$$\psi_I\left(x = -\frac{L}{2}\right) = \psi_{II}\left(x = -\frac{L}{2}\right) \tag{3.52}$$

$$\psi_{II}\left(x = +\frac{L}{2}\right) = \psi_{III}\left(x = +\frac{L}{2}\right) \tag{3.53}$$

$$\left.\frac{d\psi_I(x)}{dx}\right|_{x=-\frac{L}{2}} = \left.\frac{d\psi_{II}(x)}{dx}\right|_{x=-\frac{L}{2}} \tag{3.54}$$

$$\left.\frac{d\psi_{II}(x)}{dx}\right|_{x=+\frac{L}{2}} = \left.\frac{d\psi_{III}(x)}{dx}\right|_{x=+\frac{L}{2}} \tag{3.55}$$

Where the notation like $\left.\frac{d\psi_I(x)}{dx}\right|_{x=-\frac{L}{2}}$ means taking the derivative of the function $\psi_I(x)$ and evaluating at $x = -\frac{L}{2}$. A more compact notation is $\psi'_I\left(-\frac{L}{2}\right)$. Because of symmetry, if we satisfy these conditions at say $x = +\frac{L}{2}$, then we are guaranteed to satisfy them at $x = -\frac{L}{2}$. So we arbitrarily choose to work at just $x = +\frac{L}{2}$. Hence using 3.53 and 3.55 we get

$$B\cos\frac{kL}{2} = Ce^{-\frac{\kappa L}{2}} \tag{3.56}$$

$$B\sin\frac{kL}{2} = C\frac{\kappa}{k}e^{-\frac{\kappa L}{2}} \tag{3.57}$$

Rewriting these two equations as

$$B\cos\frac{kL}{2} - Ce^{-\frac{\kappa L}{2}} = 0 \tag{3.58}$$

and

$$B\sin\frac{kL}{2} - C\frac{\kappa}{k}e^{-\frac{\kappa L}{2}} = 0 \tag{3.59}$$

We see that with B and C being the unknowns, this is a set of two homogeneous equations in two unknowns where homogeneous means that the right-hand side of both of these equations is 0 (see Appendix A). Solutions to these equations exist only if the determinant of the coefficients is zero, i.e., for *even solution*

$$\frac{k}{\kappa} = \cot\frac{kL}{2} \tag{3.60}$$

For the *odd solution*, we repeat the process and find:

$$-\frac{k}{\kappa} = \tan\frac{kL}{2} \tag{3.61}$$

Recall that both $\kappa^2 = \frac{2m}{\hbar^2}(V_0 - E)$ and $k^2 = \frac{2m}{\hbar^2}E$ are functions of E; it is not possible to reduce these solutions to an expression for E. The equations are said to be transcendental and must be solved numerically or graphically. Figure 3.4 is an example of a graphical solution. Both sides of the function are plotted. Where they overlap is a solution.

These equations show several remarkable features. As expected now, two classical turning points lead to quantization of the energy if, past the turning point, the eigenfunction is forced to go to zero as the distance goes to $\pm\infty$. However, unlike the first example above, the particle can be found on the other side of the classical turning point. Again, as in the case of the harmonic oscillator, the particle is seen to **tunnel** into the barrier (i.e., the classically forbidden region). We will discuss the case when the energy is above the potential barrier energy in the

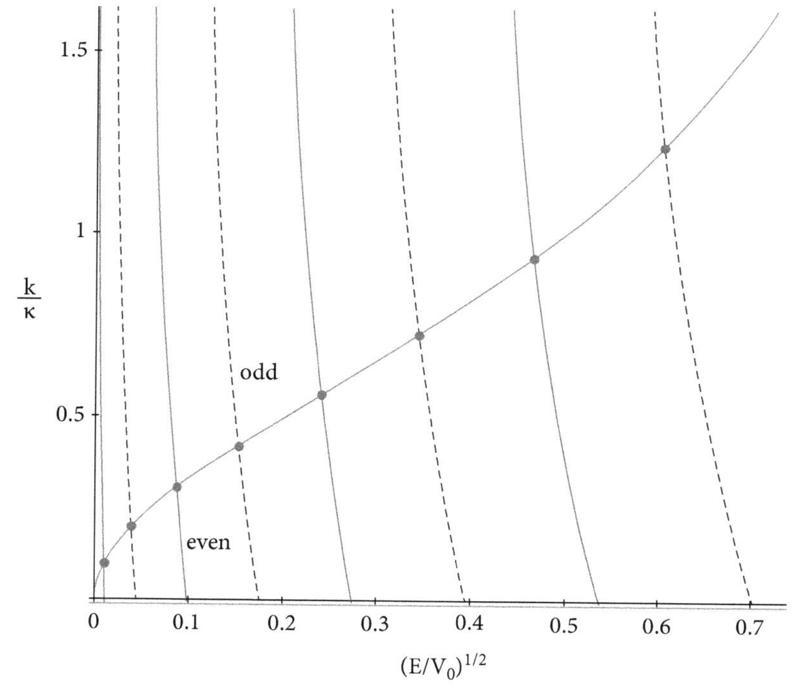

Fig. 3.4 A graphical solution to the transcendental equations for a particle in a well for a value of $\sqrt{\frac{2mV_0}{\hbar^2}}L = 30$.

next chapter, but in practice, a particle moving along at higher energy often encounters an interaction with another particle and loses energy in the process and falls into this kind of trap.

Problem 3.7 *You know from the last chapter that the probability density of finding the particle between the x and x+ dx is given by* $[\psi(x)]^2$. *Find the total probably (by integrating) of finding the particle in the forbidden region for the lowest energy eigenfunction. To do that, follow the steps:*

a) *Begin first by normalizing the entire eigenfunction. Since you had two homogeneous equations in two unknowns (Eqs 3.58 and 3.59) use one to find C in terms of B. Symmetry gave you A (above). To normalize it now, evaluate B such that* $\int dx\, [\psi(x)]^2 = 1$.

b) *The classically forbidden region is outside the well in Fig. 3.3, so integrate over the entire forbidden space.*

3.6 A Particle Trapped in a Point Defect Represented by a Dirac Delta-Function Potential

A point defect that can bind a particle like the above example is often represented by a Dirac delta-function as the potential (see Appendix A), so that the time independent Schrödinger equation becomes:

$$\left(-\frac{\hbar^2}{2m}\frac{d^2}{dx^2} - \widetilde{V}_0\delta(x)\right)\psi(x) = -\mid E\mid \psi(x) \tag{3.62}$$

Again, bound states exist only where the particle cannot be found at $\pm\infty$, which means for this problem $E < 0$. For $E > 0$, the particle is not bound. To avoid confusion, we write $-\mid E\mid$ since we are working with eigenvalues less than 0. We will examine how a physically reasonable potential can be represented by a Dirac delta-function below. However, the reason to do this is because once you are comfortable with this function, it can considerably simplify the algebra.

The problem is illustrated in Fig. 3.5a. A minus sign is placed in front of the $\mid E\mid$ and $\widetilde{V}_0$ to account for the fact that the potential is negative and the energy for a bound state is necessarily less than zero while we want the value for $\widetilde{V}_0$ and $\mid E\mid$ to be a positive number. There is a tilde mark on the potential symbol because the entire term has units of energy, and since the unit of the delta-function has to be the inverse of the unit of the argument (length in this case), $\widetilde{V}_0$ has units of energy-length. This is a form of a limiting case of the infinite potential box or dot above. We consider this potential for two different cases shown in Fig. 3.5 that enable new physical insight with reduced algebra compared to using piecewise constant potentials.

For the **Dirac potential** in Fig. 3.5a, there are only two regions, I and II, rather than three. Following the approach above, the solutions in the two regions for a localized state are given by:

$$\begin{aligned} I\ \psi(x) &= Ae^{\kappa x}\ \ x \leq 0 \\ II\ \psi(x) &= Be^{-\kappa x}\ \ x \geq 0 \end{aligned} \tag{3.63}$$

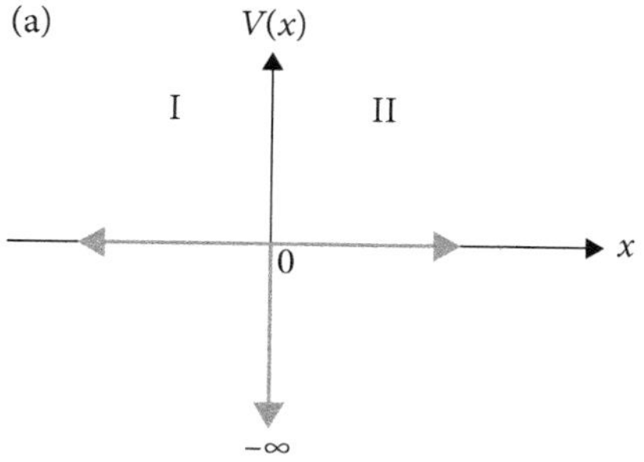

Fig. 3.5a Negative Dirac delta-function potential for representing a point defect. We are interested in finding a bound state and hence we consider the case where the energy eigenvalue is less than 0.

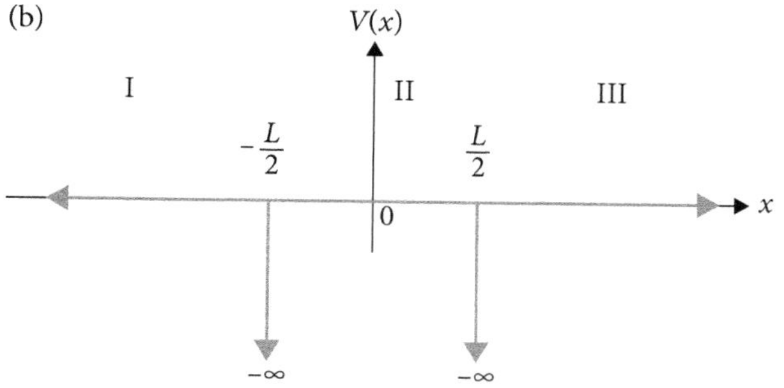

Fig. 3.5b Two Dirac-delta-function potentials show how the degeneracy (i.e., if the separation was large, the eigenenergies for each state would be the same) is lifted in the presence of coupling (i.e., when the states are close enough for some eigenfunction overlap).

where as usual

$$\kappa = \sqrt{\frac{2mE}{\hbar^2}} \quad E > 0 \tag{3.64}$$

We require continuity again of the function at $x = 0$ and so for the left figure (3.5a) we have

$$A = B \tag{3.65}$$

Interestingly, this condition eliminates the possibility of an odd solution.

For the boundary condition on the derivative, we must consider the delta-function. We do this by integrating the time independent Schrödinger equation over the location of the delta-function. The result is called the ***jump condition***. We have then:

$$\lim_{\epsilon\to 0}\int_{0-\epsilon}^{0+\epsilon} dx \left(-\frac{\hbar^2}{2m}\frac{d^2}{dx^2} - \widetilde{V}_0\delta(x)\right)\psi(x) = -\mid E \mid \lim_{\epsilon\to 0}\int_{0-\epsilon}^{0+\epsilon} dx\, \psi(x) \tag{3.66}$$

The RHS side is zero since the continuity of the function at the origin assures that the values at the two limits are the same as $\epsilon \to 0$, leaving us with the **jump condition for a negative going potential**:

$$\psi_{II}'(0+\epsilon) - \psi_I'(0-\epsilon) = -\frac{2m}{\hbar^2}\widetilde{V}_0\psi(0) \equiv -\sigma\, \psi(0) \tag{3.67}$$

where

$$\sigma = \frac{2m}{\hbar^2}\widetilde{V}_0 \tag{3.68}$$

Note that for a positive going potential the jump condition is

$$\psi'_{II}(0+\epsilon) - \psi'_I(0-\epsilon) = \frac{2m}{\hbar^2}\widetilde{V}_0\psi(0) \equiv \sigma\,\psi(0) \tag{3.69}$$

ψ' means the derivative. Using the form of the wave function in Eq. 3.63 we get for Eq. 3.67

$$B + A = \frac{\sigma}{\kappa}B \tag{3.70}$$

Combining this with Eq 3.65, we find

$$\kappa = \frac{\sigma}{2} \tag{3.71}$$

This then places a constraint on the value of E, namely since $\kappa^2 = \frac{2mE}{\hbar^2}$ we get that

$$E = \frac{2m}{\hbar^2}\widetilde{V}_0^2 \tag{3.72}$$

Recall from Eq. 3.66 that the energy eigenvalue is $-E$ to ensure that the value of E is a positive number. *Remarkably, there is only one bound state in the negative going* ***Dirac delta-function potential.***

Problem 3.8 *The kind of structures in Figs 3.1 and 3.2 are very useful models for calculating the behavior of semiconductor heterostructures like quantum wells. Imagine a structure that has an extremely high and narrow barrier right in the middle of 3.1 as shown in Fig. 3.6. The potential for the time independent Schrödinger equation is then*

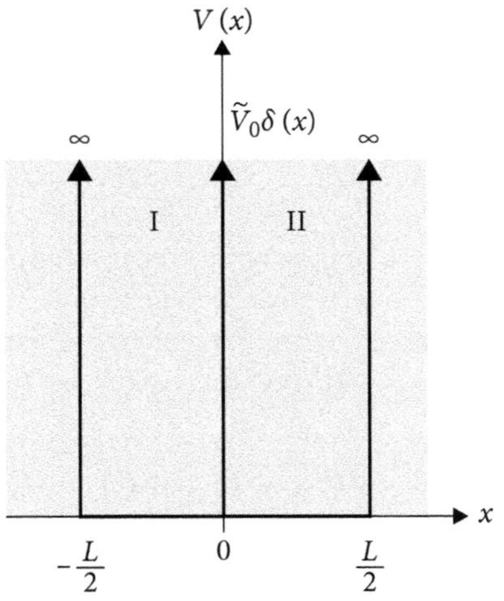

Fig. 3.6 An infinite quantum well potential with a barrier in the middle.

$$V(x) = \begin{matrix} \infty & x \leq -\frac{L}{2} \\ 0 & -\frac{L}{2} \leq x \leq 0 \\ \widetilde{V}_0 \delta(x) & x = 0 \\ 0 & 0 \leq x \leq \frac{L}{2} \\ \infty & x \geq \frac{L}{2} \end{matrix}$$

Find the unnormalized eigenfunctions and eigenvalues for even and odd solutions.

We now consider the case shown in Fig. 3.5b, which has two negative delta-function potentials. The math and physics we have encountered is now applied to a problem that is of considerable practical importance. This problem will set the stage for the problem we consider in Chapter 4, namely the origin of band structure in crystals.

Individually, they each have one bound state, but if they are sufficiently close, each potential will sense the other. What does it mean to be close enough? How is eigenfunction associated with one potential impacted by the other potential. We solve the problem exactly to find out. Now we have three regions. The origin has been placed at the center to enable the use of symmetry. In accordance with the above discussion of reflection symmetry, since $V(x) = V(-x)$, the parity operator leaves the Hamiltonian unchanged and hence the solutions to this problem are again either even or odd. Parity remains a good quantum number.

For the **even solution**

$$\psi(x) = \begin{cases} \psi_I(x) = A\ e^{\kappa x} & x \leq -\frac{L}{2} \\ \psi_{II}(x) = B \cosh \kappa x & -\frac{L}{2} \leq x \leq \frac{L}{2} \\ \psi_{III}(x) = A\ e^{-\kappa x} & \frac{L}{2} \leq x \end{cases} \tag{3.73}$$

For the **odd solution**

$$\psi(x) = \begin{cases} \psi_I(x) = A'\ e^{\kappa x} & x \leq -\frac{L}{2} \\ \psi_{II}(x) = B' \sinh \kappa x & -\frac{L}{2} \leq x \leq \frac{L}{2} \\ \psi_{III}(x) = -A'\ e^{-\kappa x} & \frac{L}{2} \leq x \end{cases} \tag{3.74}$$

where the symmetry requirements for even and odd have already been implemented.

(a) Even solution

We begin by applying the continuity at $x = \frac{L}{2}$

$$Ae^{-\kappa\frac{L}{2}} = B \cosh \kappa \frac{L}{2} \tag{3.75}$$

and then the jump condition

$$A\kappa e^{-\kappa\frac{L}{2}} + B\kappa \sinh \kappa \frac{L}{2} = \sigma A e^{-\kappa\frac{L}{2}} \tag{3.76}$$

or

$$(\sigma - \kappa)\, Ae^{-\kappa\frac{L}{2}} = B\kappa \sinh \kappa\frac{L}{2} \tag{3.77}$$

where again, $\sigma = \frac{2m}{\hbar^2}\widetilde{V}_0$.

Combining Eq. 3.75 with Eq. 3.77, we find a condition on E for a bound state to exist, namely

$$\frac{\sigma - \kappa}{\kappa} = \tanh \kappa\frac{L}{2} \tag{3.78}$$

(b) **Odd solution**

We again apply the continuity and jump condition at $x = \frac{L}{2}$

$$-A'e^{-\kappa\frac{L}{2}} = B' \sinh \kappa\frac{L}{2} \tag{3.79}$$

$$-A'\kappa e^{-\kappa\frac{L}{2}} + B\kappa \cosh \kappa\frac{L}{2} = -\sigma A'e^{-\kappa\frac{L}{2}} \tag{3.80}$$

or

$$(\kappa - \sigma)\, A'^{e^{-\kappa\frac{L}{2}}} = B\kappa \cosh \kappa\frac{L}{2} \tag{3.81}$$

Again combining Eq. 3.79 with Eq. 3.81, we find a condition on $E\left(\text{recall } \kappa^2 = \frac{2mE}{\hbar^2}\right)$ for a bound state to exist, namely

$$\frac{\sigma - \kappa}{\kappa} = \coth \kappa\frac{L}{2} \tag{3.82}$$

In problems like this it is helpful to find dimensionless quantities, which allows you to change the independent and, if appropriate, the dependent variable into quantities that take on values of unity. So, for example, we know that $\kappa\frac{L}{2}$ is dimensionless, and that $\kappa^2 = \frac{2mE}{\hbar^2}$ where E is the independent variable. So, we convert $\kappa\frac{L}{2} \to f\left(\frac{\text{E}}{\text{E}'}\right)$ where E′ is a characteristic energy in the system that follows from writing $\kappa\frac{L}{2}$ as $f\left(\frac{\text{E}}{\text{E}'}\right)$. The following illustrates the method for E_L:

$$\kappa\frac{L}{2} = \sqrt{\frac{2mL^2E}{4\hbar^2}} = \sqrt{\frac{E}{E_L}};\text{where } E_L = \frac{2\hbar^2}{mL^2} \tag{3.83}$$

In the transcendental equation, we also need to express $\frac{\sigma-\kappa}{\kappa}$ then as a function $g\left(\frac{E}{E_L}\right)$

$$\frac{\sigma - \kappa}{\kappa} = \frac{\frac{2m}{\hbar^2}\widetilde{V}_0\frac{L}{2} - \sqrt{\frac{E}{E_L}}}{\sqrt{\frac{E}{E_L}}} = \frac{2\frac{\frac{\widetilde{V}_0}{L}}{E_L} - \sqrt{\frac{E}{E_L}}}{\sqrt{\frac{E}{E_L}}} \tag{3.84}$$

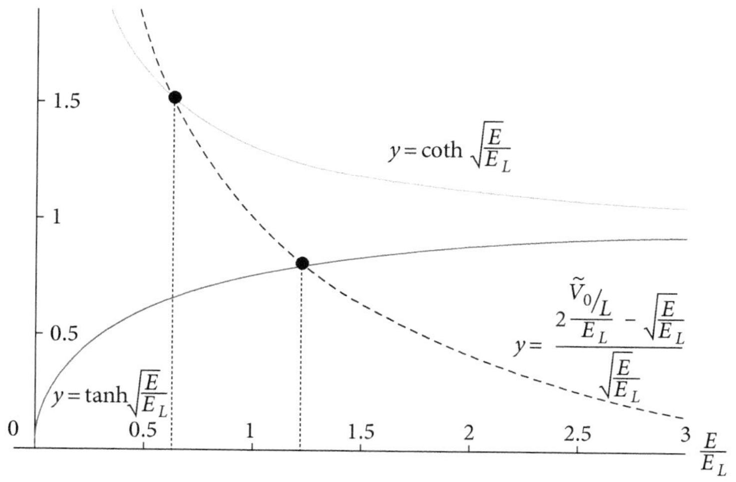

Fig. 3.7 A graph showing the conditions for the two energy eigenvalues for the even and odd solution. The higher energy eigenvalue corresponds to the even solution and the lower energy eigenvalue corresponds to the odd solution. For this case, $2\frac{\tilde{V}_0/L}{E_L} = 2$.

In this case then, we can easily find the graphical behavior of the transcendental equations for the even and old solutions. For the case $2\frac{\tilde{V}_0/L}{E_L} = 2$, shown in Fig. 3.7.

What does this mean? There are two negative going potentials. Each potential was described by a single eigenstate with eigenvalue $\kappa = \frac{\sigma}{2}$. Each potential supported one state, and since the potentials have the same magnitude, the energy eigenvalues are **degenerate** and the eigenfunctions are identical. But when the two potentials are close to each other (L is small enough), then the new states become a linear combination of the two identical old states; one combination is even and the other is odd. Furthermore, since the two states have the same energy when the wells were isolated, it means that the system is doubly degenerate. As the wells approached each other, each is mathematically impacted by the adjacent potential. The eigenfunctions of each one overlap in space with a significant non-zero value. As a result—the degeneracy is lifted. The coupled system still now has two different states and the energy of each state is different. As you add more trapping potentials, you form more states. With a large number, you form a band, as is found in a solid crystal such as silicon. Chapter 4 will discuss this in more detail, but the physics is readily apparent in this simple problem.

3.7 Physical Interpretation of the Dirac Delta-Function Potential

Physically, in the laboratory, how do we think about a Dirac function representing something physical like a point defect, which is not really a point? Clearly, this seems unreasonable. An infinite value over an infinitely small distance. However, there are many mathematical ways to represent $\delta(x)$. A common way is to use a Gaussian as in Eq. 2.67 in the following form:

$$\delta(x) = \lim_{\epsilon \to 0} \frac{1}{\epsilon\sqrt{\pi}} e^{-\left(\frac{x}{\epsilon}\right)^2} \tag{3.85}$$

Suppose the real potential is similar to

$$V(x) = \pm V_0 e^{-\left(\frac{x}{a}\right)^2} = \left(V_0 a\sqrt{\pi}\right) \frac{\pm 1}{a\sqrt{\pi}} e^{-\left(\frac{x}{a}\right)^2} \tag{3.86}$$

over some region of space that is comparable to a. We take V_0 to be positive for clarity in the discussion. While $+V_0$ $(-V_0)$ is the maximum (minimum) value for the real potential, the area under the curve is given by $\pm V_0 a\sqrt{\pi}$. The value a is the width associated with the Gaussian. So if we write

$$\widetilde{V}_0 = V_0 a\sqrt{\pi} \tag{3.87}$$

and then let the remaining symbol denoted by $a \to \epsilon$ and $\epsilon \to 0$, we have recovered the form in the above Hamiltonian

$$V(x) = \pm\widetilde{V}_0 \delta(x) \tag{3.88}$$

What is required for this to be a reasonable idea (i.e., an approximation representing the complicated real system by something that is more mathematically tractable without recourse to numerical analysis)? Typically, you let V_0 represent the actual depth or height of the potential and a is the spatial extent of the potential. But it is clear that this approximation has quantitative flaws, since it has only one bound state for a negative potential and a simple piecewise constant potential may have more. Nevertheless, this approximation is important because it shows *qualitative* features like lifting of the degeneracy between two adjacent potential wells. Indeed, in Chapter 4, we will analyze a periodic array of such potentials. The result will be band of states, identical to that found in any periodic crystal.

3.8 **Summary**

In this section you learned about the free particle state described by the wave function

$$\psi(x,t) = \frac{1}{\sqrt{2\pi}} e^{\pm ikx}$$

which is an eigenfunction of the free particle Hamiltonian

$$\widehat{H} = \frac{\widehat{p}^2}{2m}$$

with eigenenergy

$$E = \frac{\hbar^2 k^2}{2m}$$

A free particle can be localized in space through a superposition of plane wave states, an example being the Gaussian wave packet:

$$\psi(x) = \frac{1}{(\pi \ell^2)^{1/4}} e^{-\frac{x^2}{2\ell^2}} = \int_{-\infty}^{\infty} dp \left(\frac{\ell^2}{\pi \hbar^2} \right)^{1/4} e^{-\left(\frac{p}{\hbar}\right)^2 \frac{\ell^2}{2}} \frac{1}{\sqrt{2\pi\hbar}} e^{i\frac{p}{\hbar}x}$$

Where $e^{i\frac{p}{\hbar}x} = e^{ikx}$ is the plane wave state and $p = \hbar k$. Such a state, once created (the mechanism for creation is not described here), immediately begins to delocalize in space as a function of time.

Basic Hamiltonians, where the potential is piecewise continuous, are very effective for describing the basic quantum features in many materials, such as quantum dots. The quantum well in the model in Fig. 3.3 has been around since the beginning of quantum mechanics. But it was not until the second half of the 20th century that this model became the basis for many modern devices in semiconductors, based on multiple quantum wells. It was not until then that they learned how to build such structures. In semiconductor physics, under the effective mass approximation mentioned in the next chapter, the problem reduces to exactly this problem in the Schrödinger picture. These systems above, as crude as they might seem, produce adequate results when correctly applied, to represent many physical problems of interest without recourse to more detailed numerical analysis. The physical intuition that results from their application advances the rates of design innovation enormously. *Without physical intuition, it is difficult to think creatively about new possibilities.*

In the process of studying these systems, the concept of symmetries was introduced. Here, simple spatial symmetry based on reflection was used to simplify the algebra by defining an operator, $\hat{P}$, that represents that symmetry and by finding the eigenvalues of that operator. We found that such an operator commutes with the Hamiltonian that is also characterized by that symmetry, i.e.,

$$\left[\hat{H}, \hat{P}\right] = 0,$$

Beyond piecewise continuous potentials, Dirac delta-functions are also extremely useful in providing important understanding of quantum phenomena, and feature regularly in quantum studies because of their relatively simple mathematical properties. Here the potential was used to see the behavior of deep traps and also the beginning concept that leads to band theory in solids, namely the coupling between adjacent identical states leads to a lifting of the energy degeneracy and the formation of two non-degenerate states.

Vocabulary (page) and Important Concepts

- extended state 34
- localized state 34
- delocalization state 34
- Gaussian wave packet 35
- localized wave function 36
- constructive interference 37
- destructive interference 37
- piecewise-constant potential 38
- particle in a box 38
- reflection symmetry 39
- parity operator 40
- even parity 40
- odd parity 40
- even eigenfunctions 41
- odd eigenfunctions 42
- commutator 40
- tunneling 47
- Dirac delta-function potential 50
- jump condition 49
- degenerate eigenvalues 53

4 Periodic Hamiltonians and the Emergence of Band Structure: The Bloch Theorem and the Dirac Kronig–Penney Model

4.1 Introduction

Periodicity characterizes all the Hamiltonians that are used to understand crystalline solids like silicon and gallium arsenide and nitrides. The challenge is always to take some standard potential, like the hydrogenic Coulomb potential (Chapter 6) that applies to the electrons at each lattice site and then adapt the solution to the problem of when there is a periodic array of this potential. Here we are not interested in considering the details of the Coulomb coupling. Rather, the interest is to prepare you for this more challenging problem by introducing you to the general method of diagonalizing the Hamiltonian containing a periodic potential, i.e., solving the time independent Schrödinger equation and finding the eigenfunctions and energy levels.

It is likely obvious that if the wave function of an electron confined by a binding (e.g., Coulomb, square well, etc.) potential at a specific lattice site does not have some spatial overlap with the binding potential at an adjacent lattice site, then there is nothing gained by considering the periodic nature of the potential. The crystal is probably below solid density, and amounts to an array of individual atoms. However, if the wave function of the electron does spatially overlap the corresponding wave function from the adjacent potentials, then we have a new kind of structure and we have to consider the periodic nature of the Hamiltonian. What you will see is that while the electron is localized at the lattice site, its wave function extends spatially over the entire crystal. The language is that the wave function is **delocalized** over the crystal (as opposed to **localized** in some region of space like for a particle in a square well potential). To see this, we will again exploit the symmetry of the problem, namely the periodicity.

4.2 The Translation Operator

It is often the case in quantum mechanics that when you can identify a symmetry associated with the problem—such as we did in Chapters 2 and 3 for Hamiltonians that were invariant on reflection through the origin in one dimension—identifying an operator that represents that symmetry and commutes with the Hamiltonian (commuting with the Hamiltonian means that the Hamiltonian is not affected by the operator) often provides insight into how to obtain a solution to Schrödinger's equation. In the case of crystals, they are characterized by periodic potentials along specific axes of the crystal. Hence, along such an axis, the Hamiltonian is invariant under translation by the spatial period of the potential, and so a **translation operator** that translates a function by that amount commutes with the Hamiltonian. In this section,

Introduction to Quantum Nanotechnology: A Problem Focused Approach. Duncan Steel, Oxford University Press (2021).
 DOI: 10.1093/oso/9780192895073.003.0004

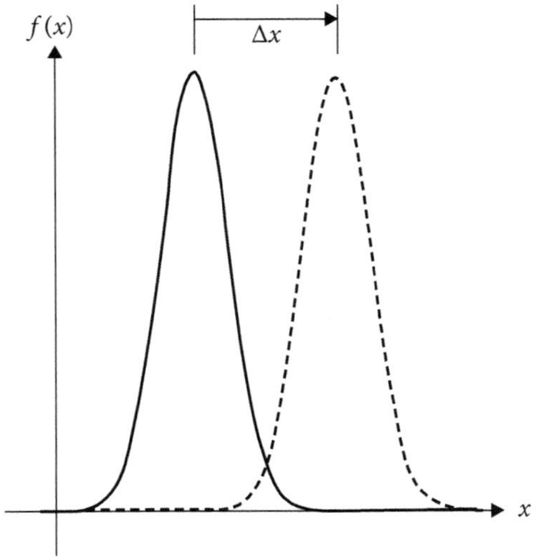

Fig. 4.1 An example of a wave function (solid line) for a particle that has been translated (dashed line) by an amount Δx.

we develop the translation operator. We will use the same method in Chapter 10 when we encounter spherical symmetry and angular momentum operators.

We begin by considering an operator that describes the translation of a particle described by a wave function, $f(x)$, in this case a one-dimensional system, as shown in Fig. 4.1. If the particle is now moved to the right (positive direction) an amount, Δx, we then have that

$$f(x) \to f(x - \Delta x) \tag{4.1}$$

where $f(x) \to f(x - \Delta x)$ is shown by the dashed line in Fig. 4.1.

Assuming now that Δx is *infinitesimally small*, we can use the definition of the derivative[1] to relate $f(x - \Delta x)$ and $f(x)$, where $f(x - \Delta x)$ is in the translated system and $f(x)$ is in the untranslated system, we get

$$f(x - \Delta x) = f(x) - \Delta x \frac{d}{dx} f(x) = \left(1 - \Delta x \frac{d}{dx}\right) f(x) \tag{4.3}$$

where we have expanded $f(x - \Delta x)$ in a Taylor series, keeping just the first two terms since in the limit $\Delta x \to 0 + \epsilon$ where ϵ is vanishing small but non-zero, the next higher order terms go like ϵ^2. We now define a translation operator, $\widehat{T}_{\Delta x}$ **a generator of infinitesimal translations**, in terms of the momentum operator, $\hat{p} = -i\hbar \frac{d}{dx}$:

$$\widehat{T}_{\Delta x} \equiv \left(1 - \Delta x \frac{d}{dx}\right) = \left(1 - \frac{i}{\hbar} \Delta x \hat{p}\right) \tag{4.4}$$

[1] Recall that the derivative is defined by:

$$\frac{df(x)}{dx} = \lim_{\epsilon \to 0} \frac{f(x + \epsilon) - f(x)}{\epsilon}$$

and rewrite Eq. 4.3

$$f(x - \Delta x) = \widehat{T}_{\Delta x} f(x) \tag{4.5}$$

If the operator $\widehat{T}_{\Delta x}$ is applied to a function an infinite number of times, the translation can become finite. To see this, we write Δx as

$$\Delta x = \lim_{n \to \infty} \frac{\xi}{n} \tag{4.6}$$

where ξ would be a finite translation. To apply this an infinite number of times, we write

$$f(x - \xi) = \lim_{n \to \infty} \left(1 - \frac{i}{\hbar} \frac{\xi}{n} \hat{p}\right)^n f(x) \tag{4.7}$$

The translation operator for this translation becomes

$$\widehat{T}_\xi = \lim_{n \to \infty} \left(1 - \frac{i}{\hbar} \frac{\xi}{n} \hat{p}\right)^n = 1 - \frac{n!}{1!\,(n-1)!} \frac{i}{\hbar} \frac{\xi}{n} \hat{p} + \frac{n!}{2!\,(n-2)!} \left(\frac{i}{\hbar} \frac{\xi}{n} \hat{p}\right)^2$$
$$- \frac{n!}{3!\,(n-3)!} \left(\frac{i}{\hbar} \frac{\xi}{n} \hat{p}\right)^3 0 \cdots = 1 - \frac{1}{1!} \xi \frac{i}{\hbar} \hat{p} + \frac{1}{2!} \left(\frac{i}{\hbar} \xi \hat{p}\right)^2 - \frac{1}{3!} \left(\frac{i}{\hbar} \xi \hat{p}\right)^3 \cdots = e^{-\frac{i\hat{p}}{\hbar}\xi} \tag{4.8}$$

where the expansion was done using the binomial theorem. Hence, the **translation operator** is

$$\widehat{T}(\xi) = e^{-\frac{i\hat{p}}{\hbar}\xi} \tag{4.9}$$

and

$$\widehat{T}_\xi f(x) = f(x - \xi) \tag{4.10}$$

To complete the discussion, we now just expand the exponential in a power series (see Appendix B):

$$f(x - \xi) = \widehat{T}_\xi f(x) = e^{-i\frac{\hat{p}\xi}{\hbar}} f(x) = \left(1 - i\frac{\hat{p}\xi}{\hbar} + \frac{1}{2!}\left(-i\frac{\hat{p}\xi}{\hbar}\right)^2 - \ldots\right) f(x)$$
$$= f(x) - \xi f'(x) + \frac{\xi^2}{2!} f''(x) + \ldots \tag{4.11}$$

and compare the result on the far right in the above equation Eq. (4.8) to the usual form of the Taylor series expansion of $f(x)$ about a:

$$f(x) = f(a) + f'(a)\,(x - a) + \frac{1}{2!} f''(a)(x - a)^2 + \ldots. \tag{4.12}$$

To see the identity, we write x as $x - \xi$, and a as x

$$
\begin{aligned}
f(x-\xi) &= f(x) + f'(x)(x-\xi-x) + \frac{1}{2!}f''(x)(x-\xi-x)^2 + \ldots \\
&= f(x) - \xi'(x) + \frac{1}{2!}\,\xi^2 f''(x) + \ldots
\end{aligned}
\tag{4.13}
$$

We see that the development of the translation operator based on the generator of infinitesimal translations gives the same result as if we had started by guessing the generator and proving it using the Taylor series expansion.

4.3 Crystals and Periodic Potentials: The Bloch Theorem and the Dirac Kronig–Penney Model

Atoms that form crystals, such as silicon, diamond, gallium arsenide and many others, are central to modern technology. They are formed by a periodic array of atoms. The period of the structure may depend on the direction one is measuring in a crystal and much is to be learned by understanding the various periodic patterns in these materials. In this chapter, we are going to just slightly extend what we have done in Chapter 3 by considering a potential that is periodic in space. In the final example of the previous chapter, we considered the case of two negative Dirac delta-function potentials compared to a single Dirac potential. We found that the two degenerate states became two non-degenerate states with two distinct eigenfunctions. So, we might imagine if we had three such identical potentials, we would get three different eigenenergies and three different corresponding eigenfunctions. This would be correct. As we get n such potentials where n is very large, the separation between the different eigenenergies becomes increasingly smaller as n increases, eventually giving rise to a "band" of energies that form a continuum for an infinite number of such potentials. Crystals do not have an infinite number, but effectively the spacings are so close that, for all practical purposes, they form a continuous band. The energy level structure as a function of the energy in the band is called the band structure. In this chapter, we will see how this comes about.

As always, the Hamiltonian for the time independent Schrödinger equation is of the form

$$
\hat{H} = \frac{\hat{p}^2}{2m} + V(x) \tag{4.14}
$$

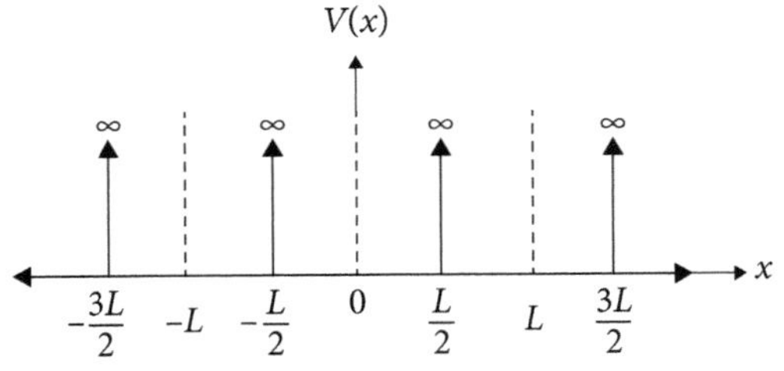

Fig. 4.2 A periodic potential represented by Dirac delta-functions (solid vertical lines with arrow) at $x = \left(n + \frac{1}{2}\right)L$ separated by L, called the **lattice spacing**. Each delta-function is in the middle of an imaginary cell, bounded by dashed lines, called **the unit cell**. The **lattice site** is the center of the unit cell.

The difference for the case of an infinite one-dimensional crystal is that the Hamiltonian is periodic, as shown in Fig. 4.2, with a spatial period of L. That is, we take $V(x)$ such that

$$V(x+L) = V(x) \tag{4.15}$$

An example of a **periodic potential** is shown in Fig. 4.2. To show the physics and reduce some of the mathematical complexity, we approximate the real potential using Dirac delta-function potentials, as we did in Chapter 3. The **lattice spacing** is L and represents the distance between the nuclei. The nucleus may move slightly in response to acoustic waves in the crystal called phonons. Each **lattice site** is centered in the **unit cell**. $L \sim 0.1 - 0.5$ nm in solids.

The explicit form of the periodic potential is

$$V(x) = \widetilde{V}_0 \sum_{n=-\infty}^{n=\infty} \delta\left(x - \left(n + \frac{1}{2}\right) L\right) \tag{4.16}$$

The details of the form of $V(x)$ are not important for the main result. Changing Eq. 4.16 to a more realistic Coulomb coupling will change the details to an extent but not the basic physics that will emerge in this section. It is Eq. 4.15 that makes all the difference, along with the idea that '$|n|$' in Eq. 4.16 ranges from zero to a large number. For a crystal of macroscopic extent, the periodic potential in the middle appears to extend (effectively) to infinity (a typical crystal might have $\sim 5 \times 10^7$ atoms/cm along a crystal axis). The physics at the boundary is not expected to affect the middle of the crystal.

We now note that, given the above Hamiltonian is periodic in x, then

$$\widehat{T}_L \widehat{H} = \widehat{H} \widehat{T}_L \tag{4.17}$$

or

$$\widehat{T}_L \widehat{H} - \widehat{H} \widehat{T}_L \equiv \left[\widehat{T}_L, \widehat{H}\right] = 0 \tag{4.18}$$

That is to say, the commutator of $\widehat{T}_L$ and $\widehat{H}$ is zero, or $\widehat{T}_L$ and $\widehat{H}$ commute. Recall that the implication of commuting is that the order in which the operators operate on a function does not matter. Hence, we know that the eigenfunctions of $\widehat{H}$ can be eigenfunctions of $\widehat{T}_L$. We follow the same procedure, as we did with the parity operator, and find the eigenfunctions and eigenvalues of $\widehat{T}_L$:

$$\widehat{T}_L f(x) = \alpha_L f(x) \tag{4.19}$$

It is clear that since

$$\widehat{T}_L f(x) = f(x+L) \tag{4.20}$$

then

$$\widehat{T}_{nL} f(x) = f(x+nL) \tag{4.21}$$

and

$$\hat{T}_{mL}\hat{T}_{nL}f(x) = \hat{T}_{(m+n)L}f(x) \tag{4.22}$$

Furthermore, given $\hat{T}_L f(x) = \alpha_L f(x)$ (Eq. 4.19), we then have with Eq. 4.22

$$\alpha_{nL}\alpha_{mL} = \alpha_{(m+n)L} \tag{4.23}$$

With these properties in mind, the solution still remains a *guess*. Nevertheless, knowing that the argument of products of exponentials is the exponential of the sum of the arguments (the arguments being numbers), we guess that

$$\alpha_L = e^{ikL} \tag{4.24}$$

This then implies that the form of the eigenfunction, $f(x)$, must be

$$f(x) = e^{ikx}u(x) \tag{4.25}$$

where $u(x)$ is periodic in L:

$$u(x+L) = u(x) \tag{4.26}$$

To see that $f(x)$ is an eigenfunction then of $\hat{T}_L$, we apply $\hat{T}_L$ to $f(x)$:

$$\hat{T}_L f(x) = \hat{T}_L e^{ikx}u(x) = e^{ik(x+L)}u(x+L) = e^{ikL}e^{ikx}u(x) = e^{ikL}f(x) \tag{4.27}$$

Hence, $f(x) = e^{ikx}u(x)$ is an eigenfunction of $\hat{T}_L$ and the corresponding eigenvalue is e^{ikL}.

From this result, we see that the form of the solution to the time independent Schrödinger equation with a periodic potential is given by the form $e^{ikx}u(x)$ where $u(x+L) = u(x)$. This is the **Bloch theorem.** At this point, k has no apparent physical meaning other than representing an overall phase. The e^{ikx} term is called the **envelope function** while $u(x)$ is the **unit cell function**. We note that the eigenvalue, e^{ikL}, is periodic in k; i.e.,

$$e^{ikL} = e^{i\left(k+\frac{2\pi}{L}\right)L} \tag{4.28}$$

and so, k is restricted to the values

$$-\frac{\pi}{L} \le k \le \frac{\pi}{L} \tag{4.29}$$

In condensed matter physics, this is called the **reduced zone representation** or **the first Brillouin zone**. *Notice that while the unit cell function is a function associated with each lattice site, the envelope function results in an extended state of the wave function that is* ***delocalized*** *throughout the crystal, not localized to a single unit cell.* In other words, the eigenfunctions are delocalized in space, in contrast to the cases in Chapters 2 and 3, where the eigenfunctions are localized in space.

In the discussion so far, we have considered the Hamiltonian with a periodic potential such as might occur in a crystal. Such a structure has boundaries. The argument above concluded that boundaries were too far away to be important. In nano-structures, the crystal may be quite small. However, most crystals are large enough (much larger than even 10^6 lattice sites in all directions) that whatever is going on in the middle of the crystal is not expected to be impacted by these boundaries. In this case, it was proposed to assume that the system has periodic boundary conditions. Physically, it is as though the crystal was deformed into a ring where the opposing boundaries are in contact with each other. If we have N atoms in length, then we have

$$e^{ikNL} = 1 \tag{4.30}$$

Equation 4.30 is called the **Born–von-Karman boundary condition**. This means then that $kNL = 2m\pi$ or $k = \frac{2m\pi}{NL}, m = 1, 2, \ldots$ and that therefore k has a period of $\frac{2\pi}{NL}$. As N is very large, we see that effectively, k becomes continuous. For sufficiently small crystals, the allowed values of k will be discrete.

With the lattice sites located at $x = \left(n + \frac{1}{2}\right) L$ in Fig. 4.2, the unit cell boundaries are located at $x = nL$ The solution to the time independent Schrödinger equation for $E > 0$ for the space *between* the lattice sites then has the form

$$\psi_n(x) = A_n e^{ik'x} + B_n e^{-ik'x} \tag{4.31}$$

where $k' = \sqrt{\frac{2mE}{\hbar^2}}$ is a real number and $\left(n - \frac{1}{2}\right) L \leq x \leq \left(n + \frac{1}{2}\right) L.$

We then use ψ_0 and ψ_1 corresponding to the region to left and right, respectively, of the delta-function at $x = +\frac{L}{2}$ in the picture, without loss of generality, and the continuity and the jump conditions (Eq. 3.69) are applied to $\psi\left(x = +\frac{L}{2}\right)$ and give for continuity at $x = +\frac{L}{2}$ for the first boundary condition

$$\left(A_0 e^{ik'\frac{L}{2}} + B_0 e^{-ik'\frac{L}{2}}\right) = \left(A_1 e^{ik'\frac{L}{2}} + B_1 e^{-ik'\frac{L}{2}}\right) \tag{4.32}$$

And for the jump condition across the delta-function for the *second boundary conditions*

$$\left(A_1 e^{ik'\frac{L}{2}} - B_1 e^{-ik'\frac{L}{2}}\right) - \left(A_0 e^{ik'\frac{L}{2}} - B_0 e^{-ik'\frac{L}{2}}\right) = \frac{2m}{k'\hbar^2} \tilde{V}_0 \psi\left(\frac{L}{2}\right) \equiv \frac{\sigma}{k'}\left(A_1 e^{ik'\frac{L}{2}} + B_1 e^{-ik'\frac{L}{2}}\right) \tag{4.33}$$

where $\sigma = \frac{2m}{\hbar^2}\tilde{V}_0$. We also require that u and u' be continuous everywhere with $u(x + L) = u(x)$. This means then that u and u' must be continuous across this unit cell boundary. In the limit $\varepsilon \to 0$, this means that

$$u'(0 + \varepsilon) = u'(L - \varepsilon) \tag{4.34}$$

and

$$u'(0 + \varepsilon) = u'(L - \varepsilon) \tag{4.35}$$

From Bloch theorem, $\psi(x) = e^{ikx}u(x)$, we then find $u(x)$ in terms of $\psi(x)$ as

$$u(x) = e^{-ikx}\psi(x) \tag{4.36}$$

We get for the condition in Eq. 4.34

$$\psi_0(0) = e^{-ikL}\psi_1(L) \tag{4.37}$$

After inserting for regions 0 and 1 in Eq. 4.37 we then get

$$(A_0 + B_0) = e^{-ikL}\left(A_1 e^{ik'L} + B_1 e^{-ik'L}\right) \tag{4.38}$$

For u' we note

$$u'(x) = -ike^{-ikx}\psi(x) + e^{-ikx}\psi'(x) \tag{4.39}$$

Then for the conditions in Eq. 4.35, we have

$$-iku(\varepsilon) + e^{-ik\varepsilon}\psi'(\varepsilon) = -iku(L-\varepsilon) + e^{-ik(L-\varepsilon)}\psi'(L-\varepsilon) \tag{4.40}$$

But since $\varepsilon \to 0$, we have in Eq. 4.40 that $u(0) = u(L)$, a requirement from the Bloch theorem, giving for the final result for the requirement of the continuity of the derivative that

$$\psi'(0) = +e^{-ikL}\psi'(L) \tag{4.41}$$

Hence,

$$(A_0 - B_0) = e^{-ikL}\left(A_1 e^{ik'L} - B_1 e^{-ik'L}\right) \tag{4.42}$$

We rewrite the four equations (Eqs 4.32, 4.33, 4.38, and 4.42) to make them a little more consolidated in the same order as above and let $\alpha^{\pm} \equiv e^{\pm ik'\frac{L}{2}}$ and $\beta^{\pm} \equiv e^{\pm ik'L}$:

$$\alpha^+ A_0 + \alpha^- B_0 - \alpha^+ A_1 - \alpha^- B_1 = 0 \tag{4.43}$$

$$\alpha^+ A_1\left(1 - \frac{\sigma}{ik'}\right) - \alpha^- B_1\left(1 + \frac{\sigma}{ik'}\right) - \alpha^+ A_0 + \alpha^- B_0 = 0 \tag{4.44}$$

$$A_0 + B_0 - \beta^+ e^{-ikL} A_1 - \beta^- e^{-ikL} B_1 = 0 \tag{4.45}$$

$$A_0 - B_0 - \beta^+ e^{-ikL} A_1 + \beta^- e^{-ikL} B_1 = 0 \tag{4.46}$$

If we first add 4.45 and 4.46 and then subtract 4.46 from 4.45, we get:

$$A_1 = \beta^- e^{+ikL} A_0 \tag{4.47}$$

$$B_1 = \beta^+ e^{+ikL} B_0 \tag{4.48}$$

Substituting into 4.43 and 4.44

$$\left(1-\beta^{-}e^{+ikL}\right)\alpha^{+}A_0+\left(1-\beta^{+}e^{+ikL}\right)\alpha^{-}B_0=0 \tag{4.49}$$

$$\left(1-\left(1-\frac{\sigma}{ik'}\right)\beta^{-}e^{+ikL}\right)\alpha^{+}A_0-\left(1-\left(1+\frac{\sigma}{ik'}\right)\beta^{+}e^{+ikL}\right)\alpha^{-}B_0=0 \tag{4.50}$$

These are two homogeneous equations in two unknowns (see Appendix A). For a solution to exist, we require the determinant of the coefficients to be zero:

$$\begin{aligned}&\left(1-\beta^{-}e^{+ikL}\right)\left(1-\left(1+\frac{\sigma}{ik'}\right)\beta^{+}e^{+ikL}\right)\alpha^{-}\alpha^{+}\\&+\left(1-\beta^{+}e^{+ikL}\right)\left(1-\left(1-\frac{\sigma}{ik'}\right)\beta^{-}e^{+ikL}\right)\alpha^{-}\alpha^{+}=0\end{aligned} \tag{4.51}$$

A few of the key points in the reduction are shown below:

$$\begin{aligned}&1-\beta^{-}e^{+ikL}-\left(1+\frac{\sigma}{ik'}\right)\beta^{+}e^{+ikL}+\left(1+\frac{\sigma}{ik'}\right)\beta^{+}e^{+ikL}\beta^{-}e^{+ikL}\\&\quad+1-\beta^{+}e^{+ikL}-\left(1-\frac{\sigma}{ik'}\right)\beta^{-}e^{+ikL}+\beta^{+}e^{+ikL}\left(1-\frac{\sigma}{ik'}\right)\beta^{-}e^{+ikL}=0\end{aligned}$$

$$2e^{-ikL}-\left(\beta^{-}+\beta^{+}\right)-\left(1+\frac{\sigma}{ik'}\right)\beta^{+}+\left(1+\frac{\sigma}{ik'}\right)e^{+ikL}-\left(1-\frac{\sigma}{ik'}\right)\beta^{-}+e^{+ikL}\left(1-\frac{\sigma}{ik'}\right)=0$$

$$2e^{-ikL}-\left(\beta^{-}+\beta^{+}\right)-\left(1+\frac{\sigma}{ik'}\right)\beta^{+}+2e^{+ikL}-\left(1-\frac{\sigma}{ik'}\right)\beta^{-}=0$$

$$4\cos kl-\left(\beta^{-}+\beta^{+}\right)-\left(1+\frac{\sigma}{ik'}\right)\beta^{+}-\left(1-\frac{\sigma}{ik'}\right)\beta^{-}=0$$

$$4\cos kl-2\left(\beta^{-}+\beta^{+}\right)-\frac{\sigma}{ik'}\beta^{+}+\frac{\sigma}{ik'}\beta^{-}=0$$

$$4\cos kl-4\cos k'l-\frac{\sigma}{ik'}\left(\beta^{+}-\beta^{-}\right)=0$$

$$\cos kl=\cos k'l+\frac{\sigma}{2k'}\sin k'l$$

Resulting in a transcendental equation for the *eigenenergy E as a function of k, known as the* ***dispersion relation.***

$$\cos kL=\cos k'L+\frac{\sigma}{2k'}\sin \kappa k'L \tag{4.52}$$

where again $k'=\sqrt{\frac{2mE}{\hbar^2}}$ and $\sigma=\frac{2m\widetilde{V_0}}{\hbar^2}$.

The k, from the Bloch theorem, appears only on the left-hand side of equation 4.52 and ranges from $-\frac{\pi}{L}$ to $\frac{\pi}{L}$. Because $\cos kL$ lies between -1 and $+1$, all values of $E=\frac{\hbar^2 k'^2}{2m}$ such that

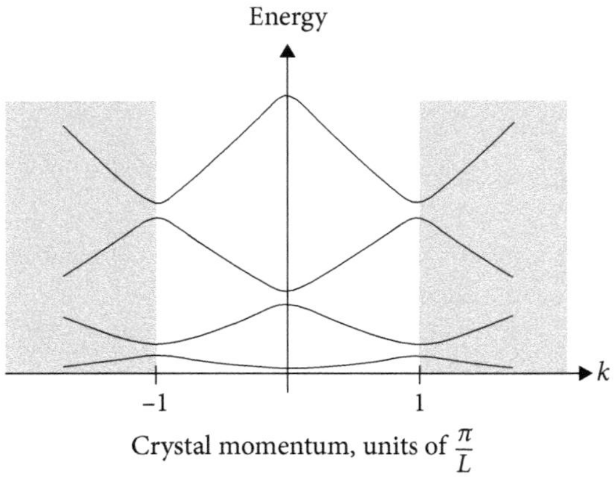

Fig. 4.3 This is the result of the solution to the transcendental equation showing the solution for the energy as a function of *k*. The light region represents the first Brillouin zone. The lowest lying state has a minimum at zone center and hence is the location of the electrons in this state. The next band has a maximum at zone center and hence pushes the electrons to the edges of the band.

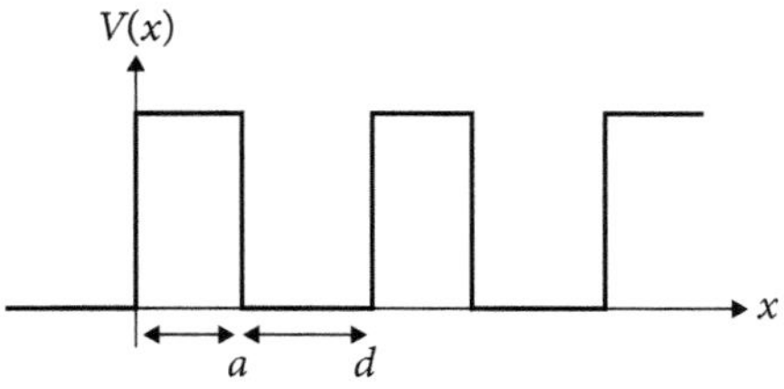

Fig. 4.4 Periodic square well potential.

$-1 \leq \cos k'L + \frac{\sigma}{2k} \sin k'L \leq 1$ are allowed and hence form a band. The results for this calculation are shown in Fig. 4.3. k is called the **crystal momentum** and is restricted to the first Brillouin zone- $\frac{\pi}{L} \leq k \leq \frac{\pi}{L}$, as discussed above. k is a quantum number in that it determines the energy, same as say the k for a free particle. Interestingly, the free particle is quadratic in k (as we expect since it is quadratic in p and $p = \hbar k$) and the band structure is often quadratic, at least over some regions. Further discussion and the transition to the effective mass approximation typical of semiconductors is the field of condensed matter and/or semiconductor physics.

Finally, it might be more intuitive to use a negative going Dirac delta-function rather than a positive function. A negative potential would certainly more resemble say a Coulomb potential of an electron in the presence of a nucleus. However, in Chapter 3, we showed that there is only one bound state and so a periodic array of that potential would show only one band associated with that state. A different approach that is often followed to introduce band structure is to assume a periodic potential of the type in Fig. 4.4.

Problem 4.1 *Find the transcendental equation that describes the band structure for a periodic piecewise constant potential shown in Fig. 4.4.*

4.4 Summary

In this section, you learned more about the importance of symmetry in quantum mechanics and in math, and the power of developing operators that reflect this symmetry. The specific problem was that of periodic potentials, where the Hamiltonian has the form:

$$\hat{H} = \frac{\hat{p}^2}{2m} + V(x)\text{: where } V(x) = V(x+L)$$

When $V(x) = V(x+L)$, the potential has translation symmetry, reflecting the *periodic* modulation of the potential. Through translation symmetry resulting in a periodic structure, the translation operator was developed where

$$\hat{D}_\xi f(x) = f(x+\xi)$$

and

$$\hat{D}_\xi = e^{i\frac{\hat{p}\xi}{\hbar}}$$

and applied to the development of the Bloch theorem, namely a solution to the time independent Schrödinger equation:

$$\hat{H}\psi(\mathrm{x}) = E\psi(\mathrm{x})$$

This must have the form

$$\psi(x) = e^{ikx}u(x) \text{ where } e^{ikL} = e^{i\left(k+\frac{2\pi}{L}\right)L}$$

The Bloch theorem is the basis of developing band structure in crystals, but this theorem applies to the study of the eigenmodes of any periodic system, such as is encountered in electromagnetic systems (e.g., photonic crystals). The e^{ikx} is called the envelope function and $u(x)$ the unit cell function. The bands result from the continuum of states created with a large number of identical systems with degenerate eigenstates that interact, lifting the degeneracy, and ultimately forming the band. The crystal momentum, k, that results from the application of the Bloch theorem is a new quantum number associated with the energy in the band. k is restricted to the values

$$-\frac{\pi}{L} \le k \le \frac{\pi}{L}$$

called the reduced zone representation, or the first Brillouin zone. The finite size of any real crystal, when sufficiently large, should not impact the behavior away from the boundaries, so the boundary conditions at the edges are called the Born–von-Karman boundary condition corresponding to a long scale length periodicity of NL where N is the number of lattice sites and L is the periodicity (size of the unit cell).

In a periodic structure, the electron, and the resulting positively charged hole (a concept from semiconductor physics), take on an effective mass because of their interaction with the periodic lattice. This was not discussed here, but in this case, the behavior of things like traps in Chapter 3 and other structures discussed in later chapters apply direction in these materials but the mass of the electron or hole is replaced by an effective mass. In that case, the Schrödinger equation remains the same, except for the mass, a very important result.

Vocabulary (page) and Important Concepts

- delocalized 57
- localized 57
- translation operator 57
- generator of infinitesimal translations 58
- periodic potential 60, 61
- unit cell 61
- lattice site 61
- lattice spacing 61
- Bloch theorem 62
- envelope function 62
- unit cell function 62
- reduced zone representation 62
- first Brillouin zone 62
- Born–von-Karman boundary condition 63
- dispersion relation 65
- crystal momentum 66

5 Scattering, Quantum Current, and Resonant Tunneling

5.1 Introduction

Free particles were discussed in Section 3.2. They have a Hamiltonian given by

$$H = \frac{p^2}{2m} + V$$

where V is a constant. Since a constant can always be added to the Hamiltonian with no impact on the results (either classically or quantum mechanically), the V is usually ignored. The wave function of the particle extends to infinity and hence the particle is not localized in space, unlike the bound discrete states that we studied in the previous chapters. The eigenvalues of the Hamiltonian form a continuum spectrum, rather than being discrete as seen for bound states.

5.2 Scattering

However, when there is a disruption in the potential, for example $V = 0$ everywhere, and $E > 0$, except in one location where V can go positive or negative, the particle will scatter. It is scattering, for example, that limits the conductivity of an electron in a metal or carrier mobility in a semiconductor. Scattering can also be used as a means of controlling particles in a quantum device.

In one dimension, scattering means that the particle continues in its original direction or is forward scattered (transmitted) or is back scattered (reflected). We consider the simple case in Fig. 3.3 when $E > V_0$ (bound states for $E < V_0$) or for $E > 0$ in Fig. 5.1. In both cases, there are no bound states because there are no potential energy boundaries to confine the particles when the energies satisfy the corresponding inequality above. Because a particle with energy E experiences the potential even in the case in Fig. 3.3 (classically the particle speeds up over the well), it will undergo **elastic scattering**, meaning that it can *change the direction of* k *and the probability amplitude in space*; but because it is elastic, it means that the kinetic energy $T = \frac{\hbar^2 k^2}{2m}$ is not changed after scattering and thus the magnitude of $k = \frac{p}{\hbar}$ does not change.

Since you are seeing that even massive particles behave like waves and you know that waves like sound or light scatter when the propagation medium changes, for example light moving from air to glass, the same thing will happen because the particle interacts with the potential. Scattering occurs when something in the medium of propagation changes the propagation

Introduction to Quantum Nanotechnology: A Problem Focused Approach. Duncan Steel, Oxford University Press (2021).
 DOI: 10.1093/oso/9780192895073.003.0005

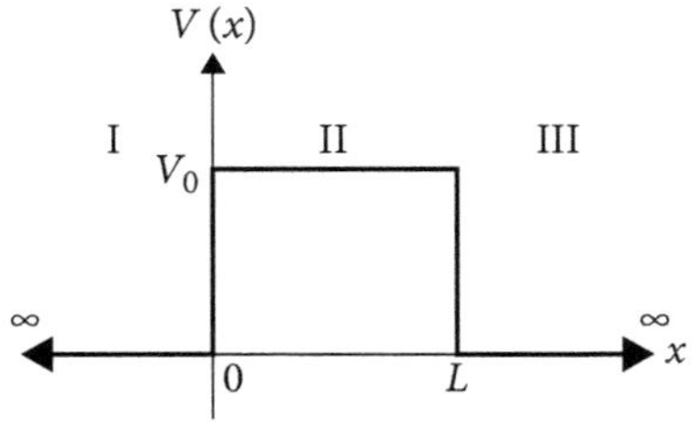

Fig. 5.1 This is a potential barrier. Because you are now aware of tunneling, you can imagine that a particle incident from the left with $E < V_0$ will encounter a classical turning point but tunnel into the barrier with some chance of escaping out of the barrier at position L. The particle does not lose energy in this process.

medium in a local area. In the case of Fig. 3.3, those particles with $E \geq V_0$ are described by **scattering states** and in the case of Fig. 5.1, all particles with $E > 0$ are described by scattering states. The spectrum of energy eigenvalues is continuous (E can take on any value) and the eigenfunctions will exist throughout space. In thinking of scattering in a quantum system, your intuitive thinking about waves will likely be more helpful than imagining scattering by say marbles.

To illustrate this problem, we will consider just the **potential barrier** in Fig. 5.1 to make it different from the earlier potentials. The math is nearly identical. We have conditions on the energy. The energy of the particle, E, is either $E \geq V_0$ or $0 \leq E \leq V_0$. We will consider here just the case for $0 \leq E \leq V_0$.

Again, there are three regions in the figure, as in Fig. 3.3, but we are interested in what happens if the particle is characterized by a wave that propagates to the right in the absence of the potential. This also means that we do not consider particles coming from the right in the absence of the potential. These assumptions *break the spatial symmetry* of the problem and there is no longer any value to work with even and odd solutions. Physically, this corresponds to there being a source of particles at $-\infty$. The source is not part of the Hamiltonian.[1] Because of this **symmetry breaking**, the earlier problems where the parity was ± 1 no longer hold, and hence parity is not a good quantum number. You can still use trigonometric and hyperbolic functions, but the algebra is a little easier using exponentials with real and imaginary arguments. Remember that you *do not "take the real part"* in quantum. Often the functions are fully complex. To make the algebra a little less cumbersome, the origin has been shifted. Since symmetry is no longer an issue, there is no reason to maintain the origin in the middle.

The time independent Schrödinger equation is again:

$$\left(-\frac{\hbar^2}{2m}\frac{d^2}{dx^2} + V(x)\right)\psi(x) = E\,\psi(x) \tag{5.1}$$

Defining $k = \sqrt{\frac{2mE}{\hbar^2}}$ and $\kappa = \sqrt{\frac{2m(V_0 - E)}{\hbar^2}}$ in the usual way, the form of the solution for $E < V_0$ is

[1] Any time the Hamiltonian is not complete, it is possible that eventually a paradox or some other problem will arise in the analysis. The problem may be related to this approximation. This will not be the case here.

$$\psi(x) = \begin{cases} \psi_I(x) &= e^{ikx} + re^{-ikx} & x \leq 0 \\ \psi_{II}(x) &= Ae^{\kappa x} + Be^{-\kappa x} & 0 \leq x \leq L \\ \psi_{III}(x) &= te^{ikx} & L \leq x \end{cases} \tag{5.2}$$

As this is a scattering problem, if you think classically, you imagine something like say a meteor approaching the earth, feeling the gravitational field, picking up speed and then, depending on the approach and speed, either scattering off in a new direction or perhaps going into orbit around the earth. The problem is solved using Newton's law and integrating over time, as usual. Here, there is no time dependence and once the four unknown coefficients become known, this will be the solution. This is called the **method of stationary states**. The same technique is also used in other wave problems such as to solve the problem of scattering of electromagnetic waves.

Notice first that we have taken the amplitude of the first term in region I to be 1 since this is the solution we restricted ourselves to in the absence of the potential. It has a positive eigenvalue for the momentum operator and hence is going from left to right, if you recall that the eigenvalue of the operators, $\hat{p}$, correspond to the value of the observable, adding the appropriate time dependence (giving $e^{i(kx-\omega t)}$ where $k = \sqrt{\frac{2mE}{\hbar^2}}$ and $\hbar\omega = E$). Taking the amplitude of this term to be unity means that *the solutions will then not be normalized.* For our purposes, this is OK, since we will only be interested in ratios of say reflected or transmitted wave amplitudes to the amplitude of the incident wave.[2]

Inside the potential barrier, the eigenfunction is described by growing and damped exponentials. In region III, the term varying like e^{-ikx} is suppressed because this would correspond to a particle coming in from $+\infty$ and that would not happen if we assume particles are only coming from a source at $-\infty$. Boundary conditions are the same—continuity of the function $\psi(x)$ and its derivative $\psi'(x)$—except that we must now match them at both boundaries.

At $x = 0$ we have

$$1 + r = A + B \tag{5.3}$$

and for continuity of the derivative

$$1 - r = \frac{\kappa}{ik}(A - B) \tag{5.4}$$

at $x = L$

$$Ae^{\kappa L} + Be^{-\kappa L} = te^{ikL} \tag{5.5}$$

and for continuity of the derivative

$$Ae^{\kappa L} - Be^{-\kappa L} = t\,\frac{ik}{\kappa}e^{ikL} \tag{5.6}$$

[2] Normalization of scattering states in piecewise constant potentials is discussed in Paul R. Berman's *Introductory Quantum Mechanics*, Section 6.7.1, Springer (2018).

This is a set of four inhomogeneous equations. Solving for t, we get:

$$t = \frac{4e^{-ikL}}{4\cosh \varkappa L + i2\left(\frac{\varkappa}{k} - \frac{k}{\varkappa}\right)\sinh \varkappa L} \tag{5.7}$$

And for the **transmission**, meaning the probability of finding the particle in Region III

$$T = |t|^2 = \frac{1}{1 + \frac{\sinh^2 \varkappa L}{\frac{4E}{V_0}\left(1 - \frac{E}{V_0}\right)}} \tag{5.8}$$

where again $\varkappa = \sqrt{\frac{2m(V_0 - E)}{\hbar^2}}$ with $V_0 \geq E$.

The **reflectance**, being the probability amplitude of the particle being in region I and moving toward increasingly negative x, is given by r and the **reflection** is

$$R = |r|^2 = 1 - T \tag{5.9}$$

The most important piece of insight to get from this is that even for $E \leq V_0$ the transmission is greater than 0. This is an example of the particle tunneling entirely through the barrier. One of the more important devices in electrical engineering is called a tunnel junction. Tunneling is a routine technique used in the design of modern electronics and optoelectronics. It is also becoming a limiting problem in classical electronics as dimensions become very small.

Problem 5.1 *Find the transmission for $E > V_0$ for Fig. 5.1 and plot as a function of E. For the plot, you could take the barrier height V_0 to be say 1 eV and assume an electron moving at energies between say >1 and <5 eV. What is a likely value of L? Try taking $L = 10\lambda_{DB}$ where we will take λ_{DB} to be De Broglie wavelength at V_0. Since $k = \sqrt{\frac{2mE}{\hbar^2}}$, let $k' = \sqrt{\frac{2m(E-V_0)}{\hbar^2}}$.*

In the case of these problems, it is clear that there is a source of particles at $x = -\infty$. In the absence of a scattering potential, it is not hard to imagine that with a source to the left, there

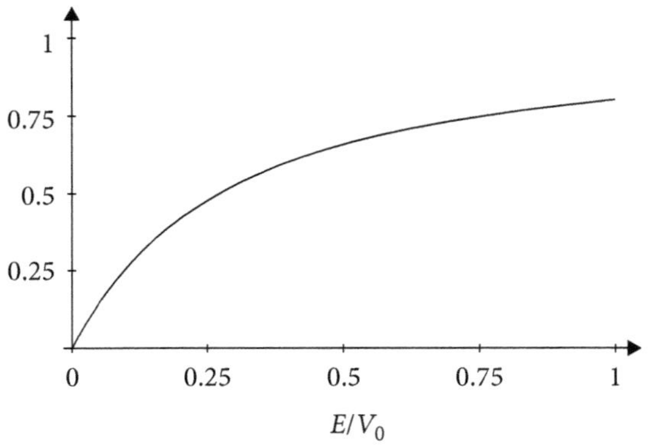

Fig. 5.2 Transmission as a function of incident energy. $\sqrt{\frac{2mV_0}{\hbar^2}}L = 1$

must be a current of particles going to the right. The concept of a probability current is not hard to understand. Suppose a particle is known to be in a volume, V. Then the probability of finding the particle at some point $\boldsymbol{r}$ at some time t in the volume is

$$P(\boldsymbol{r}, t) \equiv \left|\psi\left(\boldsymbol{r}, t\right)\right|^2 \tag{5.10}$$

and because of normalization:

$$\int_{Volume} d^3r P(\boldsymbol{r}, t) \equiv \int_{Volume} d^3r \left|\psi\left(\boldsymbol{r}, t\right)\right|^2 = 1 \tag{5.11}$$

Since the particle is confined to the volume, there is no loss, and so as a function of time:

$$\frac{d}{dt}\int_{Volume} d^3r P(\boldsymbol{r}, \boldsymbol{t}) \equiv \int_{Volume} d^3r \frac{d}{dt}\left|\psi\left(\boldsymbol{r}, \boldsymbol{t}\right)\right|^2 = 0 \tag{5.12}$$

If $P(\boldsymbol{r}, \boldsymbol{t})$ represented a classical density, then we would write:

$$\begin{aligned} \frac{d}{dt} P(\boldsymbol{r}, \boldsymbol{t}) &= \left(\frac{\partial P(\boldsymbol{r}, \boldsymbol{t})}{\partial t} + \frac{\partial \boldsymbol{r}}{\partial t}\frac{\partial P(\boldsymbol{r}, \boldsymbol{t})}{\partial \boldsymbol{r}}\right) = \left(\frac{\partial}{\partial t} + \boldsymbol{v}\cdot\nabla_r\right) P(\boldsymbol{r}, \boldsymbol{t}) \\ &= \frac{\partial P(\boldsymbol{r}, \boldsymbol{t})}{\partial t} + \nabla_r \cdot \boldsymbol{v} P(\boldsymbol{r}, \boldsymbol{t}) = \frac{\partial P(\boldsymbol{r}, \boldsymbol{t})}{\partial t} + \nabla_r \cdot \boldsymbol{J} \end{aligned} \tag{5.13}$$

where

$$\boldsymbol{J} = \boldsymbol{v} P(\boldsymbol{r}, \boldsymbol{t}) \tag{5.14}$$

is the classical probability current and $\boldsymbol{v}$ is the velocity of the particles.

And since

$$\frac{d}{dt}\int_{Volume} d^3r P(\boldsymbol{r}, \boldsymbol{t}) = 0 \tag{5.15}$$

this requires

$$\frac{\partial P(\boldsymbol{r}, \boldsymbol{t})}{\partial t} + \nabla_r \cdot \boldsymbol{J} = 0 \tag{5.16}$$

This is sometimes called a continuity equation (hydrodynamics) or conservation equation (e.g., charge conservation).

In the quantum form we have:

$$\frac{\partial\left|\psi(\boldsymbol{r}, t)\right|^2}{\partial t} + \nabla_r \cdot \boldsymbol{J} = 0 \tag{5.17}$$

where $\boldsymbol{J}$ is now the **quantum probability current** defined by

$$\boldsymbol{J} = -\frac{i\hbar}{2m}\left(\psi^*(\boldsymbol{r}, t)\,\nabla\psi(\boldsymbol{r}, t) - \psi(\boldsymbol{r}, t)\,\nabla\psi^*(\boldsymbol{r}, t)\right) \tag{5.18}$$

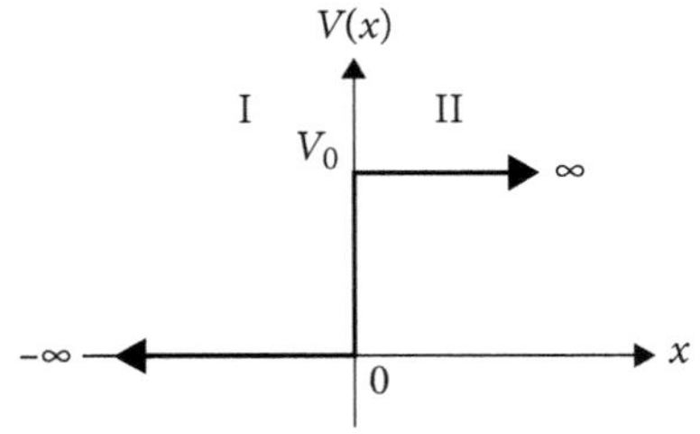

Fig. 5.3 Step potential of the type that can represent the interface between two semiconductors with different bandgaps or the interface between, say, a conductor like copper and an insulator.

Problem 5.2 *Use the full time dependent Schrödinger equation to show that, given Eq. 5.18, you recover Eq. 5.17.*

Start by writing down the full Schrödinger equation and the complex conjugate, remembering that the Hamiltonian, given by $\hat{H} = \frac{\hat{p}^2}{2m} + V(x)$, *is fully real, meaning* $\hat{H}^* = \hat{H}$. *Then multiply the Schrödinger equation by* $\psi^*(\mathbf{r}, t)$ *and multiply the complex conjugate of the Schrödinger equation by* $\psi(\mathbf{r}, t)$ *and subtract the two equations.*

Problem 5.3 *For the* ***step potential*** *in Fig. 5.3, find the solution to the time independent Schrödinger equation in regions I and II, for r and t, respectively, assuming the source of particles is on the left with* $E > V_0$.

Find the ratio to the incident particles of the probabilities for the reflection and transmission. For the case of Fig. 5.1, you can show that if the incident wave has amplitude of unity, then $R + T = 1$, *implying a conservation of probability. For Fig. 5.3, show that* $R + T \neq 1$. *The problem is that in region II, the particles are moving more slowly and so the amplitude of the wave is bigger. Hence, probability is not always the right quantity to be conserved. This is more appropriately addressed by considering the quantum probability current associated with the incoming, reflected, and transmitted waves.*

Find the quantum current for the incoming wave, the reflected wave, and transmitted wave. Show that $|J_{inc}| = |J_{ref}| + |J_{trans}|$. *The magnitude of the quantum current is used since the quantum current is a vector.*

Problem 5.4 *Solve the problem again but for* $E < V_0$ *in Fig. 5.3. Show that there is no quantum current in region II, even though the probability is non-zero, i.e., the particle has tunneled into the barrier. Physically, why is there no current here. Find the total current in region I. Explain your result, physically.*

5.3 Tunneling Through a Repulsive Point Defect Represented by a Dirac Delta-Function Potential

A point defect can lead to scattering of charges in a current and can be represented by a positive or negative potential, depending on the signs of the relative charges. A point defect can be represented by Dirac **delta-function barrier** in the potential, so that the time independent Schrödinger equation becomes

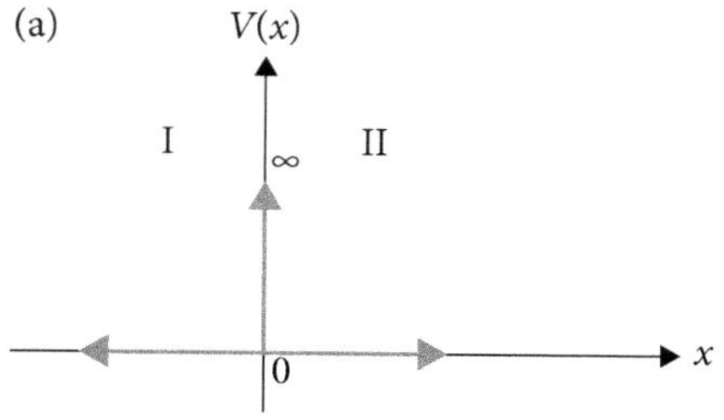

Fig. 5.4a A repulsive potential or barrier represented by a Dirac function located at the origin.

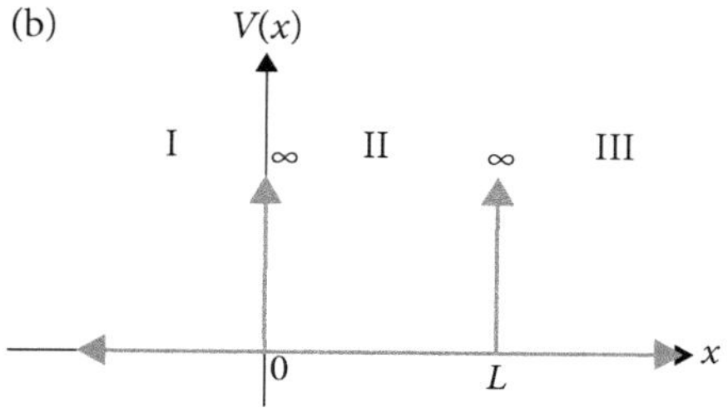

Fig. 5.4b A double barrier problem represented by two positive Dirac potentials.

$$\left(-\frac{\hbar^2}{2m}\frac{d^2}{dx^2} + \widetilde{V}_0\delta(x)\right)\psi(x) = E\,\psi(x) \tag{5.19}$$

where $\widetilde{V}_0 = V_0 L$ where L is the effective size of the trap.

Under what conditions is this likely to be a reasonable model? The question must be answered by comparing length scales. The one length scale is the effective size of the defect. Where is the other length scale? Recall that to the left of the defect is the incoming particle (wave) and the reflected wave, given below as $\psi_I(x) = e^{ikx} + re^{-ikx}$. The other length scale is the de Broglie wavelength $\lambda_D = \frac{2\pi}{k}$. Hence, the particle needs a de Broglie wavelength that is large compared to L: $\lambda_D \gg L$. This means that the scattering of the particle is not sensitive to details in the trap.

The problem is laid out in Fig. 5.4a. We solve this problem in the usual way, based on boundary conditions, but since the potential is infinitely narrow, there are only two regions: left of the origin (I) and right of the origin (II). Since $E > 0$ the wave function has the form:

$$\begin{cases} \psi_I(x) = e^{ikx} + re^{-ikx} & x \le 0 \\ \psi_{II}(x) = \quad t\,e^{ikx} & 0 \le x \end{cases} \tag{5.20}$$

Again, we must have continuity of the wave function at the origin:

$$1 + r = t \tag{5.21}$$

We then use the jump condition (Eq. 3.69) for connecting the first derivative of the wave function on both sides:

$$\psi'_{II}(0+\epsilon) - \psi'_{I}(0-\epsilon) = \frac{2m}{\hbar^2}\tilde{V}_0\psi(x) \equiv \sigma\,\psi(x) \tag{5.22}$$

where $\sigma = \frac{2m}{\hbar^2}\tilde{V}_0$ where ψ' means the derivative. Using the form of the wave function in Eq. 5.20 we get

$$t - (1 - r) = -i\frac{\sigma}{k}t \tag{5.23}$$

where we used the function to the right of the origin for the RHS of the jump condition. The two equations in two unknowns then become:

$$\begin{aligned} t - r &= 1 \\ t\left(1 + i\frac{\sigma}{k}\right) + r &= 1 \end{aligned} \tag{5.24}$$

or

$$t = \frac{1}{1 + i\frac{\sigma}{2k}} \tag{5.25}$$

and from Eq. 5.24

$$r = t - 1 = -\frac{i\frac{\sigma}{2k}}{1 + i\frac{\sigma}{2k}} \tag{5.26}$$

So the transmission is then

$$T = |t|^2 = \frac{1}{1 + \left(\frac{\sigma}{2k}\right)^2} \tag{5.27}$$

Looking more closely at the expressions for t and r (Eqs 5.25 and 5.26), we note that they are both complex. The scattering in this case results in a **scattering phase shift**. Looking at t as an example:

$$t = \frac{1}{1 + i\frac{\sigma}{2k}} = \frac{\left(1 - i\frac{\sigma}{2k}\right)}{1 + \left(\frac{\sigma}{2k}\right)^2} \tag{5.28}$$

The phase is given by

$$\phi = \arctan\left(-\frac{\sigma}{2k}\right) \tag{5.29}$$

Physically, it makes sense that as the strength of the scattering potential goes to zero, the phase shift also goes to zero as does the amplitude. The phase shift is a measure of the strength of the scattering. Phase will become increasingly important in our discussions in quantum devices. Phase here is a parameter that, in principle, can be controlled, which is not available for particles behaving classically.

We approach this now using a more powerful mathematical method based on matrices called the **transfer matrix** approach. We start again with

$$\begin{cases} \psi_I(x) &= Ie^{ikx} + re^{-ikx} \quad x \leq 0 \\ \psi_{II}(x) &= t\,e^{ikx} + Fe^{-ikx} \quad x \geq 0 \end{cases} \tag{5.30}$$

A letter I (representing incident) is inserted into the term coming from the left as a place holder for the amplitude of the incident wave. Previously, we had taken that to be unity, but for this, we want to keep track of the prefactor. Similarly, in region II, an incoming term is included though normally, $F = 0$ in this problem. It may not always be zero, in more complex problems as we see below.

We start again with the continuity and the jump condition:

$$I + r = t + F \tag{5.31}$$

$$(I - r) = (t - F) + i\frac{\sigma}{k}(t + F) \tag{5.32}$$

where again

$$\sigma = \frac{2m}{\hbar^2}\widetilde{V}_0 \tag{5.33}$$

We have written the equations so I and r (coefficients in region I) are on the left and t and F (coefficients in region II) are on the right. Had we evaluated the right-hand side of the jump condition using $\psi(x)$ from region I, the equation would have been different and the resulting matrix different but still correct. Putting the results in matrix form, we have

$$\begin{bmatrix} 1 & 1 \\ 1 & -1 \end{bmatrix}\begin{bmatrix} I \\ r \end{bmatrix} = \begin{bmatrix} 1 & 1 \\ 1 + i\frac{\sigma}{k} & -\left(1 - i\frac{\sigma}{k}\right) \end{bmatrix}\begin{bmatrix} t \\ F \end{bmatrix} \tag{5.34}$$

Carrying out the above multiplication results in a 2×1 vector with entries corresponding to Eqs 5.31 and 5.32. We now rewrite 5.34 by multiplying both sides by the inverse of the matrix (see Appendix A) on the left.

$$\begin{bmatrix} I \\ r \end{bmatrix} = \begin{bmatrix} 1 & 1 \\ 1 & -1 \end{bmatrix}^{-1}\begin{bmatrix} 1 & 1 \\ 1 + i\frac{\sigma}{k} & -\left(1 - i\frac{\sigma}{k}\right) \end{bmatrix}\begin{bmatrix} t \\ F \end{bmatrix} \tag{5.35}$$

$$\begin{bmatrix} 1 & 1 \\ 1 & -1 \end{bmatrix}^{-1} = \frac{1}{2}\begin{bmatrix} 1 & 1 \\ 1 & -1 \end{bmatrix} \tag{5.36}$$

$$\begin{bmatrix} I \\ r \end{bmatrix} = \frac{1}{2}\begin{bmatrix} 1 & 1 \\ 1 & -1 \end{bmatrix}\begin{bmatrix} 1 & 1 \\ 1 + i\frac{\sigma}{k} & -\left(1 - i\frac{\sigma}{k}\right) \end{bmatrix}\begin{bmatrix} t \\ F \end{bmatrix} = \begin{bmatrix} 1 + i\frac{\sigma}{2k} & i\frac{\sigma}{2k} \\ -i\frac{\sigma}{2k} & 1 - i\frac{\sigma}{2k} \end{bmatrix}\begin{bmatrix} t \\ F \end{bmatrix} \tag{5.37}$$

If $F = 0$ and $I = 1$ we recover the result from above:

$$1 = \left(1 + i\frac{\sigma}{2k}\right)t \tag{5.38}$$

$$t = \frac{1}{\left(1 + i\frac{\sigma}{2k}\right)}; T = \frac{1}{1 + \left(\frac{\sigma}{2k}\right)^2} \tag{5.39}$$

The transfer matrix, $\boldsymbol{M}$, is then defined as the matrix that relates $\begin{bmatrix} t \\ F \end{bmatrix}$ to $\begin{bmatrix} I \\ r \end{bmatrix}$ in Eq. 5.37:

$$\boldsymbol{M} = \begin{bmatrix} 1 + i\frac{\sigma}{2k} & i\frac{\sigma}{2k} \\ -i\frac{\sigma}{2k} & 1 - i\frac{\sigma}{2k} \end{bmatrix} \tag{5.40}$$

This will become quite useful below in a resonant tunneling problem.

Problem 5.5 *Show that the transfer matrix for Problem 5.3 is given by*

$$M = \frac{1}{2}\begin{bmatrix} 1 + \frac{k'}{k} & 1 - \frac{k'}{k} \\ 1 - \frac{k'}{k} & 1 + \frac{k'}{k} \end{bmatrix} \tag{5.41}$$

where again $k = \sqrt{\frac{2mE}{\hbar^2}}$, *let* $k' = \sqrt{\frac{2m(E-V_0)}{\hbar^2}}$

5.4 **Resonant Tunneling**

In the case of other kinds of waves (electromagnetic, acoustic, etc.), it is well-known that physical structures can result in resonances, where a resonance is one that shows an especially enhanced interaction with the wave at a specific energy. The behavior of a more complex structure can be more useful than a simpler structure. The enhanced interaction is due to constructive or destructive interference. In many cases, resonances result in enhanced scattering, in this case, meaning reflection or transmission or increasing the wave amplitude in some limited region of space.

In quantum devices, tunneling and **resonant tunneling** are used to create special devices. The list is quite long by now, but one of the most famous was a tunnel junction that led to high frequency performance, negative differential resistance, and the Nobel prize.

Using the transfer matrix above we can understand the physics of this system by considering the system with two infinitely narrow potential barriers, Fig. 5.4b. The barriers are represented by Dirac delta-functions that go to infinity in magnitude.

Once again, we begin with the one-dimensional time independent Schrödinger equation.

$$\left(-\frac{\hbar^2}{2m}\frac{d^2}{dx^2} + V_0(x)\right)\psi(x) = E\,\psi(x) \tag{5.42}$$

where now

$$V_0(x) = \widetilde{V}_0\delta(x) + \widetilde{V}_0\delta\,(x - L) \tag{5.43}$$

We take the same approach as above for writing the solutions down in three regions:

$$\psi(x) = \begin{cases} \psi_I(x) &= Ie^{ikx} + re^{-ikx} & x \leq 0 \\ \psi_{II}(x) &= Ae^{+ikx} + Be^{-ikx} & 0 \leq x \leq L \\ \psi_{III}(x) &= t\, e^{ikx} + F\, e^{-ikx} & L \leq x \end{cases} \tag{5.44}$$

Where again we insert I and F as place holders. We solve this problem using the transfer matrix approach, though the usual way would also work. For $x = 0$, we have for the two boundary conditions again,

$$I + r = A + B \tag{5.45}$$

$$(I - r) = (A - B) + i\frac{\sigma}{k}(A + B) \tag{5.46}$$

With the above, we can rewrite the equations in matrix form as

$$\begin{bmatrix} I \\ r \end{bmatrix} = \begin{bmatrix} 1 + i\frac{\sigma}{2k} & i\frac{\sigma}{2k} \\ -i\frac{\sigma}{2k} & 1 - i\frac{\sigma}{2k} \end{bmatrix} \begin{bmatrix} A \\ B \end{bmatrix} = M \begin{bmatrix} A \\ B \end{bmatrix} \tag{5.47}$$

We now apply the continuity and jump condition to $x = L$ between regions II and III:

$$Ae^{+ikL} + Be^{-ikL} = te^{+ikL} + Fe^{-ikL} \tag{5.48}$$

$$Ae^{+ikL} - Be^{-ikL} = te^{+ikL} - Fe^{-ikL} + i\frac{\sigma}{k}\left(te^{+ikL} + Fe^{-ikL}\right) \tag{5.49}$$

$$\begin{bmatrix} 1 & 1 \\ 1 & -1 \end{bmatrix} \begin{bmatrix} Ae^{+ikL} \\ Be^{-ikL} \end{bmatrix} = \begin{bmatrix} 1 & 1 \\ \left(1 + i\frac{\sigma}{k}\right) & -\left(1 - i\frac{\sigma}{k}\right) \end{bmatrix} \begin{bmatrix} te^{+ikL} \\ Fe^{-ikL} \end{bmatrix} \tag{5.50}$$

or

$$\begin{bmatrix} 1 & 1 \\ 1 & -1 \end{bmatrix} \begin{bmatrix} e^{+ikL} & 0 \\ 0 & e^{-ikL} \end{bmatrix} \begin{bmatrix} A \\ B \end{bmatrix} = \begin{bmatrix} 1 & 1 \\ \left(1 + i\frac{\sigma}{k}\right) & -\left(1 - i\frac{\sigma}{k}\right) \end{bmatrix} \begin{bmatrix} te^{+ikL} \\ Fe^{-ikL} \end{bmatrix} \tag{5.51}$$

or

$$\begin{bmatrix} A \\ B \end{bmatrix} = \begin{bmatrix} e^{-ikL} & 0 \\ 0 & e^{+ikL} \end{bmatrix} \frac{1}{2} \begin{bmatrix} 1 & 1 \\ 1 & -1 \end{bmatrix} \begin{bmatrix} 1 & 1 \\ \left(1 + i\frac{\sigma}{k}\right) & -\left(1 - i\frac{\sigma}{k}\right) \end{bmatrix} \begin{bmatrix} te^{+ikL} \\ Fe^{-ikL} \end{bmatrix} \tag{5.52}$$

$$\begin{bmatrix} A \\ B \end{bmatrix} = \begin{bmatrix} e^{-ikL} & 0 \\ 0 & e^{+ikL} \end{bmatrix} \begin{bmatrix} 1 + i\frac{\sigma}{2k} & i\frac{\sigma}{2k} \\ -i\frac{\sigma}{2k} & 1 - i\frac{\sigma}{2k} \end{bmatrix} \begin{bmatrix} te^{+ikL} \\ Fe^{-ikL} \end{bmatrix} \tag{5.53}$$

If we define the translation matrix as

$$T = \begin{bmatrix} e^{-ikL} & 0 \\ 0 & e^{+ikL} \end{bmatrix} \tag{5.54}$$

We can write the transfer matrix as

$$\begin{bmatrix} A \\ B \end{bmatrix} = TMT^{*} \begin{bmatrix} t \\ F \end{bmatrix} \tag{5.55}$$

Therefore

$$\begin{aligned} \begin{bmatrix} I \\ r \end{bmatrix} &= MTMT^{*} \begin{bmatrix} t \\ F \end{bmatrix} \\ &= \begin{bmatrix} 1 - i\frac{\sigma}{2k} & -i\frac{\sigma}{2k} \\ i\frac{\sigma}{2k} & 1 + i\frac{\sigma}{2k} \end{bmatrix} \begin{bmatrix} e^{-ikL} & 0 \\ 0 & e^{+ikL} \end{bmatrix} \begin{bmatrix} 1 - i\frac{\sigma}{2k} & -i\frac{\sigma}{2k} \\ i\frac{\sigma}{2k} & 1 + i\frac{\sigma}{2k} \end{bmatrix} \begin{bmatrix} e^{ikL} & 0 \\ 0 & e^{-ikL} \end{bmatrix} \begin{bmatrix} t \\ F \end{bmatrix} \end{aligned} \tag{5.56}$$

With $I = 1$ and $F = 0$, multiplication of the 2×2 matrices gives a final 2×2 matrix that leads immediately to an expression for the transmission. We get after the algebra:

$$t = \frac{1}{1 - \left(\frac{\sigma}{2k}\right)^2 (1 - \cos 2kL) - i\left(2\left(\frac{\sigma}{2k}\right) - \left(\frac{\sigma}{2k}\right)^2 \sin 2kL\right)} \tag{5.57}$$

$$T = |t|^2 = \frac{1}{\left[1 - \left(\frac{\sigma}{2k}\right)^2 (1 - \cos 2kL)\right]^2 + \left[\left(\frac{\sigma}{2k}\right)^2 \sin 2kL - 2\left(\frac{\sigma}{2k}\right)\right]^2} \tag{5.58}$$

The result as plotted in Fig. 5.5 is quite remarkable. For a single barrier (Eq. 5.27), the transmission is quite low and broad band (i.e., not strongly dependent on energy away from the origin. However, by designing in a second barrier, identical to the first, the transmission approaches unity (Eq. 5.58), albeit at a specific energy.

The ability to transmit only a selected energy gives the option for tuning the energy of the electrons that enter another device after the resonant tunnel junction. It can also be used to measure energy of the electrons or filter the electrons.

Problem 5.6 *Show that, for Fig. 5.1 and $E < V_0$, the matrices of the transfer matrix are*

$$\begin{bmatrix} I \\ r \end{bmatrix} = \frac{1}{2}\begin{bmatrix} 1 & 1 \\ 1 & -1 \end{bmatrix} \begin{bmatrix} 1 & 1 \\ \frac{\kappa}{ik} & -\frac{\kappa}{ik} \end{bmatrix} \begin{bmatrix} e^{-\kappa L} & 0 \\ 0 & e^{\kappa L} \end{bmatrix} \frac{1}{2}\begin{bmatrix} 1 & 1 \\ 1 & -1 \end{bmatrix} \begin{bmatrix} 1 & 1 \\ \frac{ik}{\kappa} & -\frac{ik}{\kappa} \end{bmatrix} \begin{bmatrix} e^{ikL} & 0 \\ 0 & e^{-ikL} \end{bmatrix} \begin{bmatrix} t \\ F \end{bmatrix}$$

5.5 Summary

This section has focused on the subject of scattering, the word assigned to describing the effect of a generally delocalized wave, incident on a potential that can be a barrier or even a potential well containing a bound state. In the latter case, the energy of the particle is greater than the potential energy creating bound states. In the case of barriers where the particle energy is below the maximum energy, the scattering includes tunneling through the barrier to the free particle region again. Two quantum wells led to interesting behavior when the states were degenerate in Chapter 3. Here, scattering from two barriers leads to resonance that can be used to achieve new device performance. This problem has illustrated the power of using the transfer or scattering matrix when considering multiple structures. The scattering matrix concept, like the Bloch

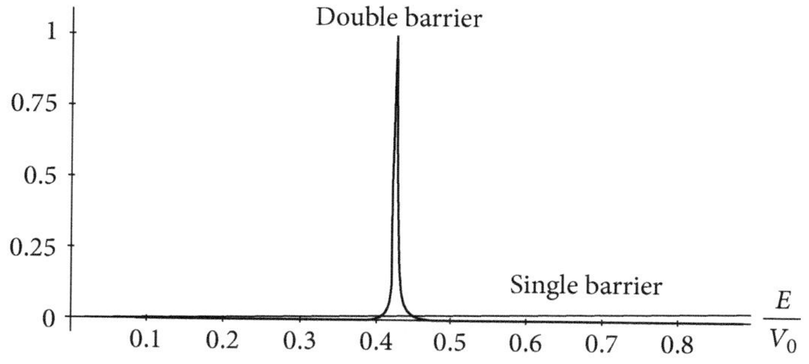

Fig. 5.5 As a function of E/V_0 (arbitrary units), the transmission for the single barrier and the double barrier. The double barrier reaches nearly 100% on resonance.

theorem in Chapter 4, is also a powerful mathematical tool in other fields, where the net effect of numerous structures (e.g., antennas, optical elements) must be considered.

In the classical thinking of particle scattering, the image of marbles scattering from each other gives the impression of a dynamical process, but you notice in this analysis that there was no mention of time. The solution here is independent of time and the approach is called the method of stationary states. It is a common technique for waves and is applied to other kinds of waves such as electromagnetic fields. Scattering is typically in three dimensions, but in one dimension, the scattering is either backward (reflection) or forward (transmission). Scattering leads to loss of a stream of particles such as electrons in a wire, so it is often viewed as degrading performance. However, it can be used to enhance performance, also, as in the case of resonant tunneling. Ideas like resonance and phase create new opportunities for the design of devices from sources to gates to information storage and metrology.

Vocabulary (page) and Important Concepts

- elastic scattering 69
- scattering states 70
- symmetry breaking 70
- potential barrier 70
- method of stationary states 71
- reflection 72
- transmission 72
- quantum probability current 73
- step potential 74
- delta-function barrier 74
- scattering phase shift 76
- transfer matrix 77
- resonant tunneling 78

6 Bound States in Three Dimensions: The Atom

6.1 Introduction

The main objective of this text is to introduce the reader to the important applications emerging in technology that require quantum mechanics to accurately describe their behavior. While mathematics is the language of physical science, the focus is on one-dimensional systems to reduce the extra burden of more complex math, while making the new physical behavior clear.

However, in some cases, a full three-dimensional solution of the time independent Schrödinger equation is needed in order to get some specific numbers or to deal with some important three-dimensional behavior. In this chapter, we briefly consider the basic problem and then solve the problem for the hydrogen atom.

If the potential can be written in the form $V(x, y, z) = V(x) + V(y) + V(z)$, then the time independent Schrödinger equations becomes

$$\left[-\frac{\hbar^2}{2m}\left(\hat{p}_x^2 + \hat{p}_y^2 + \hat{p}_z^2\right) + V(x) + V(y) + V(z)\right]\psi(x, y, z) = E\psi(x, y, z) \tag{6.1}$$

This is now broken down into three separate partial differential equations using the usual method of separation of variables. We did this in Chapters 2 and 3 to separate time and space when the Hamiltonian was time independent.

We start with

$$\psi(x, y, z) = \psi(x)\psi(y)\psi(z) \tag{6.2}$$

In this case, it is clear that the eigenvalue for the energy in the time independent Schrödinger equation is given by

$$E = E_x + E_y + E_z \tag{6.3}$$

and that

$$\hat{H}_q\psi(q) = E_q\psi(q) \tag{6.4}$$

where q = x, y, or z and $\hat{H}_q$ is that part of the Hamiltonian that depends only on q:

$$\hat{H}_q = \frac{\hat{p}_q^2}{2m} + \hat{V}_q = -\frac{\hbar^2}{2m}\frac{\partial^2}{\partial q^2} + V(q) \tag{6.5}$$

Introduction to Quantum Nanotechnology: A Problem Focused Approach. Duncan Steel, Oxford University Press (2021).
 DOI: 10.1093/oso/9780192895073.003.0006

Problem 6.1 *Find the energy eigenfunctions and eigenvalues for the infinite potential problem in Section 3.2 in three dimensions for a box centered at the origin with dimension* $\pm\frac{L_x}{2}, \pm\frac{L_y}{2}, \pm\frac{L_z}{2}$ *for the x, y, z dimensions, respectively. Be sure to normalize the function. Feel free to copy the solution from the earlier work in Chapter 3. Recall from the earlier Problem 3.2 that the normalization for the even and odd functions for the infinite potential is given by* $\sqrt{\frac{2}{L}}$.

Problem 6.2 *Find the energy eigenfunctions for a free particle in the three-dimensional meaning* $V(x, y, z) = 0$. *Take the momentum* $\mathbf{p} = p_x\hat{x} + p_y\hat{y} + p_z\hat{z}$ *and the corresponding k-vector as defined by* $\hbar\mathbf{k} = \mathbf{p}$. *Show that the normalized energy eigenfunction for the free particle is then a normalized plane wave given by*

$$\psi(x, y, z) = \frac{1}{(2\pi\hbar)^{3/2}} e^{i\frac{p\cdot r}{\hbar}} \text{ with energy } E = \frac{p^2}{2m} \tag{6.6}$$

In the above discussion and problems, it is important to notice that the problems have three **quantum numbers** associated with the three independent **degrees of freedom**, x, y, z. These are canonical coordinates that have no constraints. This does not seem surprising, but it is very important. Even if the coordinate system is changed, the number of degrees of freedom remains the same and there will be a **quantum number** for each degree of freedom.

Analysis of some systems of critical importance to nano-technology necessitates moving to higher-dimensional analysis in other coordinate systems. Indeed, the basis of any material is the atom and this is a three-dimensional object. Most of the above problems have even more features that are interesting for devices and technology if viewed from a three-dimensional analysis. Typically, the problems of interest at this level are solvable analytically after the corresponding differential equation is put in "standard form", so that the solution can be looked up in standard tables.

Whether the problem is something as simple as $f'' + k^2 f = 0$ $\left(\text{note } f = f(x) \text{ and } f'' \equiv \frac{d^2 f}{dx^2}\right)$ with solutions such as $\sin kx$ or something more complicated, the solutions are ultimately developed as a power series solution and then given a name such as the sine function, Bessel function, etc. Many of these differential equations result from writing a wave-like equation in three dimensions where the $\frac{d^2}{dx^2}$ is replaced by the equivalent two- or three-dimensional form ∇^2 where $\nabla^2 = \frac{\partial^2}{\partial x^2} + \frac{\partial^2}{\partial y^2} + \frac{\partial^2}{\partial z^2}$ in Cartesian coordinates or whatever coordinate system is appropriate for the problem. Recall that when dealing with functions of more than one independent variable, operators like $\frac{d}{dx}$ are replaced with $\frac{\partial}{\partial x}$. Depending on the spatial symmetry, this operator is rewritten in the coordinates associated with symmetry such as spherical, cylindrical, elliptical, etc. The resultant forms of the time independent Schrödinger equation can then be looked up in tables such as in Abramowitz and Stegun.[1] Many software packages that provide symbolic manipulation to solve various mathematical equations produce these solutions, also.

[1] Milton Abramowitz and Irene A. Stegun, *Handbook of Mathematical Functions*, National Bureau of Standards Applied Mathematics Series 55, US Government Printing Office (1964).

The approach in this book is not about getting good at that kind of math. However, it is important that you are aware of the basic quantum features of some three-dimensional systems such as the atom, and that you are not intimidated if faced with developing or understanding the solution in spherical coordinates to say the three-dimensional version of a spherically symmetric nano-vibrator in Chapter 2. The solutions to the equations are standard in math handbooks and we are going to exploit our ability to look up solutions rather than derive these solutions to make this as straightforward as possible.

The form of the Hamiltonian remains the same but, of course, the potential is now a three-dimensional potential as is the coordinate operator, $\hat{r}$, and the momentum operator, $\hat{p}$. Here, we will only consider problems such as the atom that have **spherical symmetry**, meaning that the potential $V(\mathbf{r})$ depends only on the magnitude of $\mathbf{r}$. Hence, $V(\mathbf{r}) = V(r)$.

$$\hat{H} = \frac{\hat{p}^2}{2m} + V(r) \tag{6.7}$$

where we assume that the mass of the nucleus is infinite compared to the electron, so we are only concerned about the kinetic energy of the electron. Otherwise, we would have to transform into the **center of mass** coordinate system which would still retain the form of the operator and not change in the essential results. The coordinate representation for the operator $\frac{\hat{p}^2}{2m}$ becomes (see Appendix D):

$$\frac{\hat{p}^2}{2m} = -\frac{\hbar^2}{2m}\nabla^2 = -\frac{\hbar^2}{2m}\left(\frac{1}{r^2}\frac{\partial}{\partial r}r^2\frac{\partial}{\partial r} + \frac{1}{r^2\sin\theta}\frac{\partial}{\partial\theta}\sin\theta\frac{\partial}{\partial\theta} + \frac{1}{r^2\sin^2\theta}\frac{\partial^2}{\partial\phi^2}\right) \tag{6.8}$$

The time independent Schrödinger equation then becomes

$$\left[-\frac{\hbar^2}{2m}\left(\frac{1}{r}\frac{\partial^2}{\partial r^2}r + \frac{1}{r^2}\left(\frac{1}{\sin\theta}\frac{\partial}{\partial\theta}\sin\theta\frac{\partial}{\partial\theta} + \frac{1}{\sin^2\theta}\frac{\partial^2}{\partial\phi^2}\right)\right) + V(r)\right]\psi(r,\theta,\phi) = E\psi(r,\theta,\phi) \tag{6.9}$$

where we have made a common substitute of for the radial operator: $\frac{1}{r^2}\frac{\partial}{\partial r}r^2\frac{\partial}{\partial r} = \frac{1}{r}\frac{\partial^2}{\partial r^2}r$

6.2 The Hydrogenic Atom

For a hydrogenic atom, we assume hydrogen (even in GaAs, hydrogen is the basis of the calculations) and so for the **Coulomb potential** between the two charges:

$$V(r) = -\frac{e^2}{4\pi\epsilon_0}\frac{1}{r} \tag{6.10}$$

where r is the distance from the nucleus to the electron. We assume a single charge on the nucleus. If you need to consider a charge, Z, on the nucleus, just replace e^2 with Ze^2. Note that, in fact, the atomic problem is a two-body problem: a nucleus and an electron. So there is a set of coordinates for both. As mentioned above, the problem is solved by moving to the center of mass coordinate system. In practice, because the nucleus is almost 2000 times as heavy as the electron, we assume that the mass of the nucleus is infinite and then the coordinate r represents a vector from the nucleus to the electron. In a semiconductor, an excited electron can go into orbit around the oppositely charged hole, resulting from the excitation forming a **quasiparticle** called an **exciton**. In this case, the two carriers have an effective mass that is not quite as different from the other as it is in the case of the electron and the nucleus and so calculations must be performed in the center of mass. Most importantly, once the transformation is made, the resulting equation is the same as in the atomic case but some numerical values change.

Finding the eigenvalues for the energy and the eigenfunctions for the electron is a major challenge in solving a three-dimensional partial differential equation, if you are the first one to do it. Fortunately, because of the spherical symmetry, the equation can be put into a form where it and its solution are contained in tables of standard differential equations. In addition, instead of going through the usual details, we will take a slightly different approach, beginning by using a result presented in the discussion of angular momentum in Chapter 10.

Specifically, to be introduced more formally in Chapter 10, we introduce the concept of **orbital angular momentum** and the quantum mechanical description. It is represented by the symbol $\boldsymbol{L}$, where $\boldsymbol{L}$ is defined in classical mechanics as

$$\boldsymbol{L} = \boldsymbol{r} \times \boldsymbol{p} \tag{6.11}$$

It is the important physical parameter associated with rotation, where $\boldsymbol{r}$ is the distance from the center of rotation and $\boldsymbol{p}$ is the momentum associated with the rotating mass. The vector $\boldsymbol{L}$ is parallel to the **axis of rotation** and perpendicular to the **plane of rotation**. With the right-hand rule, with counter-clockwise rotation in the plane, the angular momentum points upward.

We will not go through the math here, but if $\boldsymbol{r}$ and $\boldsymbol{p}$ are replaced by their equivalent three-dimensional quantum operators in Cartesian coordinates, they can then be converted to spherical coordinates, which is more natural for describing angular momentum. The discussion for finding the appropriate eigenvalues and eigenfunctions for angular momentum is more involved than is necessary for this discussion. However, such a discussion would focus on identifying what is called a **complete set of commuting operators** (CSCO) that describe the entire space defined by the observable. In the case of angular momentum and using spherical coordinates, it is a three-dimensional object, but the coordinate r is fixed. In describing the rotation then, θ and ϕ are the only coordinates that change. From an observable standpoint, we know that the total angular momentum is conserved, meaning that in the absence of dissipation it does not change. So, $|\,\boldsymbol{L}\,|$ is a constant of the motion in the absence of any external forces, as is the magnitude of linear momentum $|\,\hbar\boldsymbol{k}\,|$ for a free particle. Rather than converting this to an operator, it is standard to work with a scalar corresponding to $\boldsymbol{L} \cdot \boldsymbol{L} \equiv L^2$. The other observable that does not change is the projection of the angular momentum along one axis. Think of a gyroscope spinning around the z-axis that is, say, aligned with gravity. If you then tip the gyroscope at some angle to that z-axis, it precesses around that axis, but L^2 and $\boldsymbol{L} \cdot \hat{\boldsymbol{z}}$ ($\hat{\boldsymbol{z}}$ is the unit vector in the z-direction) remain constant. So L_z is another constant of the motion.

It makes sense to choose the z-axis, since the variables, θ and ϕ, are defined relative to the z-axis as shown in Fig. 6.1. Specifically, θ measures the angle between the z-axis and the angular momentum vector (ranges from 0 to π) while ϕ measures the angle between the x-axis and the projection of the angular momentum in the x-y plane (ranges from 0 to 2π). The z-axis is called the **axis of quantization**. If we convert L^2 and L_z to quantum operators, we can show we have a CSCO for angular momentum.

In spherical coordinates, any standard introductory text on quantum mechanics[2] will show that working from Cartesian coordinates and starting with Eq. 6.11 and the derivative form of the momentum operator, the coordinate representations for the operators $\hat{\boldsymbol{L}}^2$ and $\hat{L}_z$ in spherical coordinates are

$$\hat{\boldsymbol{L}}^2 = -\hbar^2 \left(\frac{1}{\sin\theta} \frac{\partial}{\partial\theta} \sin\theta \frac{\partial}{\partial\theta} + \frac{1}{\sin\theta^2} \frac{\partial^2}{\partial\phi^2} \right) \tag{6.12}$$

and

$$\hat{L}_z = -i\hbar \frac{\partial}{\partial\phi} \tag{6.13}$$

Note that Eq. 6.12 (without the $\hbar^2$) is identical to the angular part in Eq. 6.9. Hence, knowing the eigenfunctions and eigenvalues of $\hat{\boldsymbol{L}}^2$ and $\hat{L}_z$ is very important for solving Eq. 6.8.

The eigenfunctions for these two operators are ***spherical harmonics*** symbolized by $Y_{lm}(\theta, \phi)$ where

$$\hat{L}^2 Y_{lm}(\theta, \phi) = l(l+1)\hbar^2 Y_{lm}(\theta, \phi) \tag{6.14}$$

where l and m are integers, $l \geq 0$ and $-l \leq m \leq l$. The same function is also an eigenfunction for $\hat{L}_z$ and

$$\hat{L}_z Y_{lm}(\theta, \phi) = m\hbar Y_{lm}(\theta, \phi). \tag{6.15}$$

The z-axis is the axis of quantization and $\hat{L}_z$ is the projection of the angular momentum along the z-axis. The m is often called the **magnetic quantum number** for the magnetic substate with quantum number l and m. The reason for the nomenclature will be seen more clearly in Chapters 10 and 11. Some of the basic physics will be seen below in Problem 6.4.

These functions are also **orthonormal** for different quantum numbers l and m:

$$\int d\Omega Y^*_{l'm'}(\theta, \phi) Y_{lm}(\theta, \phi) = \int_0^{\pi} \sin\theta d\theta \int_0^{2\pi} d\phi Y^*_{l'm'}(\theta, \phi) Y_{lm}(\theta, \phi) = \delta_{l'l}\delta_{m'm} \tag{6.16}$$

We note that in considering angular momentum, we had two unconstrained (i.e., canonical) coordinates, θ and ϕ. This corresponds to two degrees of freedom, and then as expected, two quantum numbers, l and m, both having a discrete spectrum of eigenvalues. Alternatively, we had two operators, $\hat{L}^2$ and $\hat{L}_z$ that formed a complete set of commuting operators with eigenvalues $l(l+1)\hbar^2$ and $m\hbar$, respectively.

Aside from problems like the atom, you might imagine this is relatively specialized physics. You will see shortly that for some very profound reasons, the concepts and behavior of angular momentum are central to much of the newest thinking in advanced technology. It goes far beyond the idea of a rotating object.

[2] See for example Paul R. Berman, Introductory Quantum Mechanics, pp. 189–192, Springer International Publishing, 2018.

Table 6.1 The first few spherical harmonics in both spherical and Cartesian coordinates. From: https://en.wikipedia.org/wiki/Table_of_spherical_harmonics

***l* = 0**
$Y_{0\,0}(\theta,\varphi) = \frac{1}{2}\sqrt{\frac{1}{\pi}}$
***l* = 1**
$Y_{1-1}(\theta,\varphi) = \frac{1}{2}\sqrt{\frac{3}{2\pi}}e^{-i\varphi}\sin\theta$
$Y_{1\,0}(\theta,\varphi) = \frac{1}{2}\sqrt{\frac{3}{\pi}}\cos\theta$
$Y_{1\,1}(\theta,\varphi) = -\frac{1}{2}\sqrt{\frac{3}{2\pi}}e^{i\varphi}\sin\theta$
***l* = 2**
$Y_{2-2}(\theta,\varphi) = \frac{1}{4}\sqrt{\frac{15}{2\pi}}e^{-2i\varphi}\sin^2\theta$
$Y_{2-1}(\theta,\varphi) = \frac{1}{2}\sqrt{\frac{15}{2\pi}}e^{-i\varphi}\sin\theta\cos\theta$
$Y_{2\,0}(\theta,\varphi) = \frac{1}{4}\sqrt{\frac{5}{\pi}}(3\cos^2\theta - 1)$
$Y_{2\,1}(\theta,\varphi) = -\frac{1}{2}\sqrt{\frac{15}{2\pi}}e^{i\varphi}\sin\theta\cos\theta$
$Y_{2\,2}(\theta,\varphi) = \frac{1}{4}\sqrt{\frac{15}{2\pi}}e^{2i\varphi}\sin^2\theta$

The first few spherical harmonics are shown in Table 6.1.

With this information for the hydrogen atom, Schrödinger's equation can now be written in terms of the $\hat{L}^2$ operator, rather than the angle dependent differential operators:

$$\left(-\frac{\hbar^2}{2m}\left(\frac{1}{r}\frac{\partial^2}{\partial r^2}r - \frac{1}{\hbar^2}\frac{\hat{L}^2}{r^2}\right) - \frac{e^2}{4\pi\epsilon_0}\frac{1}{r}\right)\psi(r,\theta,\phi) = E\psi(r,\theta,\phi) \tag{6.17}$$

And since we know the eigenfunctions of $\hat{L}^2$, we can now write the solution for the eigenfunction for the hydrogen atom as

$$\psi_{nlm}(r,\theta,\phi) = R_{nl}(r)Y_{lm}(\theta,\phi) \tag{6.18}$$

Substituting Eq. 6.18 into Eq. 6.17, we get for the equation for the **radial function**

$$\left(-\frac{\hbar^2}{2m}\left(\frac{1}{r}\frac{\partial^2}{\partial r^2}r - \frac{l(l+1)}{r^2}\right) - \frac{e^2}{4\pi\epsilon_0}\frac{1}{r}\right)R_{nl}(r) = E_nR_{nl}(r) \tag{6.19}$$

The solution is developed in various forms and is often given the name of the person who analyzed the solution, such as the **Laguerre polynomials** in the case of the $1/r$ potential. As we

Table 6.2 The first few radial eigenfunction for hydrogenic atoms. Z is the charge on the nuclei. The text assumes $Z = 1$. See the text for the definition of the Bohr radius, a_0. https://quantummechanics.ucsd.edu/ph130a/130_notes/node233.htm

$R_{1\,0}(r) = 2\left(\frac{Z}{a_0}\right)^{\frac{3}{2}} e^{-\frac{Zr}{a_0}}$
$R_{2\,1}(r) = \frac{1}{\sqrt{3}}\left(\frac{Z}{2a_0}\right)^{\frac{3}{2}} \frac{Zr}{a_0} e^{-\frac{Zr}{2a_0}}$
$R_{2\,0}(r) = 2\left(\frac{Z}{2a_0}\right)^{\frac{3}{2}} \left(1 - \frac{Zr}{2a_0}\right) e^{-\frac{Zr}{2a_0}}$
$R_{3\,2}(r) = \frac{2\sqrt{2}}{27\sqrt{5}}\left(\frac{Z}{3a_0}\right)^{\frac{3}{2}} \left(\frac{Zr}{a_0}\right)^2 e^{-\frac{Zr}{3a_0}}$
$R_{3\,1}(r) = \frac{4\sqrt{2}}{3}\left(\frac{Z}{3a_0}\right)^{\frac{3}{2}} \left(\frac{Zr}{a_0}\right)\left(1 - \frac{Zr}{6a_0}\right) e^{-\frac{Zr}{3a_0}}$
$R_{3\,0}(r) = 2\left(\frac{Z}{3a_0}\right)^{\frac{3}{2}} \left(1 - \frac{2Zr}{3a_0} + \frac{2(Zr)^2}{27a_0^2}\right) e^{-\frac{Zr}{3a_0}}$

did for the spherical harmonics, we will just tabulate the solution in Table 6.2, and several of the lowest energy eigenfunctions are tabulated in Table 6.3 (combining the results of 6.2 and 6.1).

There is no dependence of the energy eigenvalues on l even though the radial function does depend on l. The energy eigenvalues are given by

$$E_n = -\frac{\hbar^2}{2ma_0^2}\frac{1}{n^2} \equiv -E_R\frac{1}{n^2} \quad n = 1, 2, 3, \ldots \tag{6.20}$$

where E_R is the ***Rydberg constant***:

$$E_R = \frac{\hbar^2}{2ma_0^2} = 2.18 \times 10^{-18} \text{ joules} = 13.6 \text{ eV} \tag{6.21}$$

and represents the energy of the lowest $n = 1$ level. Recall that with the potential varying as $(-1/r)$, discrete bound states, i.e., eigenstates, will exist only for $E < 0$.

$$a_0 = \frac{4\pi\epsilon_0\hbar^2}{me^2} = 0.0529 \text{ nm} \tag{6.22}$$

is the **Bohr radius** and is the distance between the electron and the nucleus with the highest probability density. Importantly, the solution to Eq. 6.19 requires

$$l \leq n - 1 \tag{6.23}$$

The normalized eigenfunction for hydrogen is then ψ_{nlm} where n, l, and m are the **quantum numbers** and n is called the ***principal quantum number***.

Table 6.3 Lowest energy hydrogen eigenfunctions.
https://nanohub.org/resources/5011/download/zeemansplitting.pdf

Lowest energy hydrogenic eigenfunctions for $\psi_{nlm}(r,\theta,\phi)$	Complete eigenfunction	Orbital
$\psi_{100}(r,\theta,\phi)$	$\frac{1}{\sqrt{\pi a_0^3}} e^{\frac{-r}{a_0}}$	s
$\psi_{200}(r,\theta,\phi)$	$\frac{1}{2\sqrt{2\pi a_0^3}}\left(1-\frac{r}{2a_0}\right)e^{\frac{-r}{2a_0}}$	s
$\psi_{210}(r,\theta,\phi)$	$\frac{1}{2\sqrt{2\pi a_0^3}}\frac{r}{a_0}e^{\frac{-r}{2a_0}}\cos\theta$	p
$\psi_{21\pm1}(r,\theta,\phi)$	$\frac{\mp 1}{2\sqrt{2\pi a_0^3}}\frac{r}{a_0}e^{\frac{-r}{2a_0}}\cos\theta\, e^{\pm i\phi}$	p
$\psi_{300}(r,\theta,\phi)$	$\frac{1}{3\sqrt{3\pi a_0^3}}\left(1-\frac{2}{3}\left(\frac{r}{a_0}\right)+\frac{2}{27}\left(\frac{r}{a_0}\right)^2\right)e^{\frac{-r}{3a_0}}$	s
$\psi_{310}(r,\theta,\phi)$	$\frac{2}{9}\sqrt{\frac{2}{\pi a_0^3}}\left(\frac{r}{a_0}\right)\left(1-\frac{1}{6}\left(\frac{r}{a_0}\right)\right)e^{\frac{-r}{3a_0}}\cos\theta$	p
$\psi_{320}(r,\theta,\phi)$	$\frac{1}{486}\sqrt{\frac{6}{\pi a_0^3}}\left(\frac{r}{a_0}\right)^2 e^{\frac{-r}{3a_0}}\left(3\cos^2\theta-1\right)$	d
$\psi_{32\pm1}(r,\theta,\phi)$	$\mp\frac{1}{81}\sqrt{\frac{1}{\pi a_0^3}}\left(\frac{r}{a_0}\right)^2 e^{\frac{-r}{3a_0}}\sin\theta\cos\theta\, e^{\pm i\phi}$	d
$\psi_{32\pm2}(r,\theta,\phi)$	$\frac{1}{162}\sqrt{\frac{1}{\pi a_0^3}}\left(\frac{r}{a_0}\right)^2 e^{\frac{-r}{3a_0}}\sin^2\theta\, e^{\pm 2i\phi}$	d

It is important to study the three-dimensional structure of the eigenfunctions, because this structure determines the important properties for materials. Figure 6.1 shows a modern rendering of the structure. The $n = 1$ state has both $l = 0$ and $m = 0$. Linear combinations of the degenerate states result in spatial shapes that lack spherical symmetry. These combinations then become essential in understanding molecular bonding, the basis of crystal formation and chemistry.

We note a tradition in labeling that started many decades ago before the quantum basis for the spectral feature was understood. However, it persists today. It is letter-based code for the l-value for any state. Specifically, the $l = 0$ states are all called **s-orbitals** (the s stands for "sharp"). The $l = 1$ state is called a **p-orbital** or p-state (p stands for principal), For $l = 2$, it is called a d orbital (**d-stands** for diffuse) and $l = 3$ is called an **f-orbital** (f stands for fine). After f, it is alphabetical (g, h, ... etc.) The p-states combine in different ways to form the basis for creation of important semiconductors such as GaAs. Table 6.3 shows the lettering convention.

Problem 6.3 *Using the normalized expression for the eigenfunction for the ground state ($n = 1$) of hydrogen, find $\langle r\rangle$. Show the maximum radial probability density $P(r)$ of finding the particle between rand $r + dr$ is a_0.*

Problem 6.4 *Magnetic properties of materials have been incredibly important for the rapid increase and the dramatic drop in cost of many forms of memory. Magnetic properties can be quite*

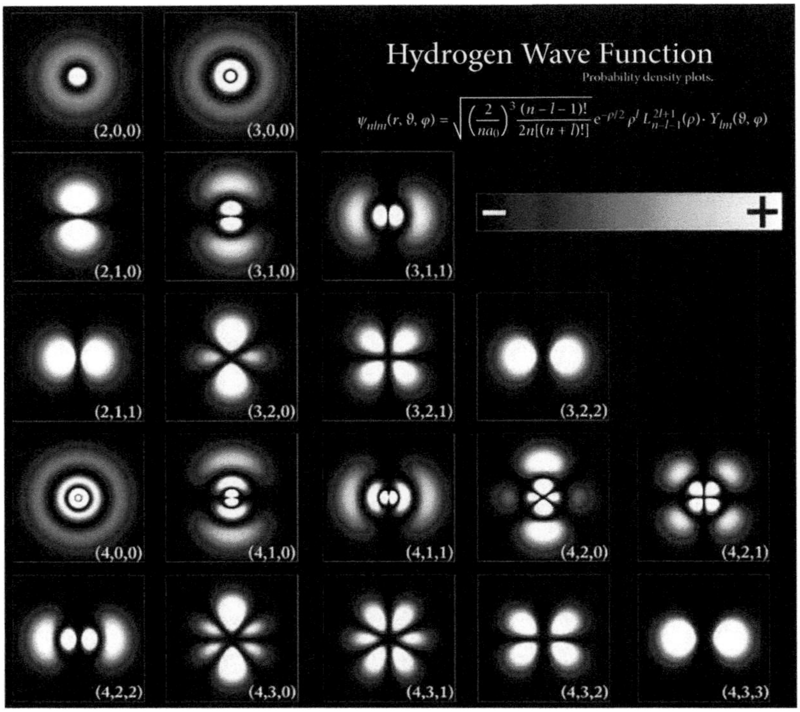

Fig. 6.1 Images of the 3-dimensional structure of various hydrogen eigenfunctions. The axis of quantization is always z and is taken in the vertical direction. In each box the numbers in the parentheses are. There is an expression of the eigenfunctions in the upper right corner. The symbols are defined in the text except for which is part of the solution for radial eigenfunction. It is a Laguerre polynomial and is also a tabulated function. To fully understand the images, you need to rotates in your mind around the vertical axis. This represents the rotational symmetry about the z-axis. https://en.wikipedia.org/wiki/Atomic_orbital#/media/File:Hydrogen_Density_Plots.png
Permission details: This work has been released into the public domain by its author, PoorLeno at English Wikipedia. This applies worldwide. In some countries this may not be legally possible; if so: PoorLeno grants anyone the right to use this work for any purpose, without any conditions, unless such conditions are required by law.

involved, but already you can see evidence of a simple magnetic memory. To understand this, you must understand that the ***magnetic moment*** *of an electron in orbit*[3] *is* $\boldsymbol{\mu}_L = -\frac{e}{2m}\mathbf{r}\times\mathbf{p} = -\frac{e}{2m}\mathbf{L}$. *The quantity* $\beta_o = \frac{\hbar e}{2m}$ *is called the* ***Bohr magneton*** *for the electron, and e is elementary charge (a positive number) and the charge on the electron (−1)e. This is the origin of the minus sign in* $\hat{\boldsymbol{\mu}}_L$. *If in a device, you can induce a magnetization then you have stored a bit*

[3] The magnetic moment of a circular loop of current has magnitude defined by $\mu_M = IA$ where I is the current and A is the area of the loop. The magnetic moment is a vector pointing in a direction perpendicular to the plane containing the current. If the current loop lies in the x-y plane, and the current moves at velocity v counter-clockwise (from intersecting the x-axis to then intersecting the y-axis (right-hand thumb rule), then the current is the number of electrons passing a point per unit time. If there are N electrons moving at velocity v moving counter-clockwise in a loop of radius r, then $I = -\frac{eNv}{2\pi r} = -\frac{eNp}{2m\pi r}$. The minus sign arises because electrons moving counter-clockwise give a current moving clockwise. Therefore $\boldsymbol{\mu}_M = -\frac{eNp}{2m\pi r}\pi r^2 = -\frac{eNp}{2m}r$. Since $\boldsymbol{\mu}_M$ is a vector pointing in the perpendicular direction to the surface, we can write it as $\boldsymbol{\mu}_M = -\frac{e}{2m}r\times\mathbf{p} = -\frac{e}{2m}\mathbf{L}$, where we assume $N = 1$ for an electron in orbit around a nucleus.

Table 6.4 Letter names assigned to various values of total orbital angular momentum. Starting with g, the naming continues alphabetically.

ℓ – *value*	**Letter name**
0	s
1	p
2	d
3	f
4	g

of information. In the quantum picture, $\boldsymbol{\mu}_L \to \hat{\boldsymbol{\mu}}_L = -\frac{e}{2m}\hat{\mathbf{L}}$*; i.e., the observable (the magnetic moment) becomes an operator. Consider a p-state (*$l = 1$*). There are three different values that m can take. Calculate* $\langle \hat{\boldsymbol{\mu}}_{L_z} \rangle$ *for each of the three possible eigenfunctions:* ψ_{n1m}*. For this problem, n does not matter. As you can see, you can store three different values.*

Problem 6.5 *As we did in Chapter 3, we looked at a square potential well that had boundaries that were infinitely high in potential energy (Fig. 3.1). This is an excellent model for understanding and estimating the confinement effect in a layered system. However, other structures may be three-dimensional. An easy system to examine to see the impact of spherical symmetry in this system is to just assume a spherical version of Fig. 3.1. Then Eq. 6.1 is*

$$\hat{H} = \frac{\hat{p}^2}{2m} + V(r), \text{ where } V(r) = \begin{bmatrix} 0 & 0 \le r \le a \\ \infty & a \le r \end{bmatrix}$$

And then using the three-dimensional version of the kinetic term in spherical coordinates from Eq. 6.17

$$\frac{\hat{p}^2}{2m} = -\frac{\hbar^2}{2m}\nabla^2 = -\frac{\hbar^2}{2m}\left(\frac{1}{r}\frac{\partial^2}{\partial r^2}r - \frac{\hat{L}^2}{\hbar^2 r^2}\right)$$

Now solve the time independent Schrödinger equation with angular momentum $l = 0$*. Show that the energy levels are quantized subject to*

$$E_n = \frac{1}{2m}\left(\frac{n\pi\hbar}{a}\right)^2, n = 1, 2, 3, \ldots$$

Assuming that $n = 1$ *and* $a = 1$ *nm, confirm that the energy difference between the* $n = 2$ *and* $n = 1$ *state is ~1.1 eV. Find the corresponding wavelength of light that could drive this transition, recalling that the energy of a photon is* $\hbar\omega$*,* $\omega = ck$*, k is the wave vector of the field where* $k = \frac{2\pi}{\lambda}$*. It is impressive to realize that a 1.1 eV (*1.76×10^{-19} *J) can change the voltage potential in the sphere by 1 volt. Orders of magnitude of energy less than required in a typical electronic device. How to exploit this kind of thinking is driving a lot of research now.*

6.3 Summary

In this chapter, a brief overview of three-dimensional quantum problems based on more complicated potentials was presented. Without going into great detail on the math, you have a basic understanding of the effects of Coulomb confinement of the electron around the nucleus and the basic idea of geometrical confinement by boundaries of an electron. Importantly, you also see that in spherically symmetric systems, the idea of rotation in a classical system is transformed into the angular distribution of the probability amplitude of the electron (i.e., the eigenfunctions).

Vocabulary (page) and Important Concepts

- degrees of freedom 83
- quantum number 83
- center of mass 84
- Coulomb potential 84
- spherical symmetry 84
- quasiparticle 85
- exciton 5
- angular momentum 85
- orbital angular momentum 85
- axis of rotation 85
- plane of rotation 85
- complete set of commuting operators 85
- axis of quantization 86
- spherical harmonics 86
- orthonormal 86
- magnetic quantum number 86
- radial function 87
- Laguerre polynomials 87
- Rydberg constant 88
- Bohr radius 88
- quantum numbers 88
- principal quantum number 88
- s, p, d, f orbital 89
- magnetic moment 90
- Bohr magneton 90

7 New Design Rules for Quantum: The Postulates

7.1 Introduction

Like everything else you have learned, mathematical descriptions of physical systems begin with the rules or more formally, the **postulates**, that govern that system. These are statements of the mathematical foundations for understanding the problem. Even in mechanics, when you learned $F = ma$ or in circuits when you learned about Kirchhoff's circuit laws, you were given the rules. $F = ma$ was not derived, nor is the form of F for a given force. Kirchhoff's law can be derived from Maxwell's equations, but Maxwell's equations were also postulated. All of the mathematical descriptions in the physical sciences are based on postulates. *The only proof that they are correct is that the predictions made, based on those postulates, agree with experiment.*

However, unlike the earlier experience where things like $F = ma$ were likely understandable in terms of your daily experience (forces, acceleration, energy, etc.), quantum mechanics does not begin this way. This is not surprising. The daily experience of things like a force is part of the classical world and classical physics. The quantum world is generally not evident in daily experience, which explains in part why it developed so much later than classical physics. To isolate quantum behavior in the laboratory, historically, it has taken some effort.

This section describes the rules that we followed in the earlier chapters, which form the basis of any kind of design that you want to develop for a quantum device. These rules are entirely different from other rules you have learned. The rules in this chapter are not complete, so we will extend them as we introduce new ideas in later chapters. Importantly, these rules are also not entirely unique. Different texts take different approaches, but the approach here is fairly standard. Regardless of the approach, the mathematical result is the same.

The approach below also introduces **Dirac notation** in parallel with the equivalent notation of functions that have been used up to this point. The full scope of Dirac notation will be developed and discussed throughout the remainder of the text. Understanding Dirac notation is very important in quantum since some physical behavior cannot be described with functions in real space. Where feasible, we will use Dirac notation with **bra**s and **ket**s, as well as the functional representation in parallel. However, they are not equal. One would never write a bra or ket equaling a function. The relationship between them will be discussed later in this chapter.

Note that, as discussed in the preface, it is possible to start the study of quantum mechanics in this chapter rather than at the beginning. Because of that, there is some redundancy in the discussion. There is also the need of making sure that the pedagogical mapping of the understanding developed in Chapters 2–6 carries forward with the introduction of Dirac notation.

Introduction to Quantum Nanotechnology: A Problem Focused Approach. Duncan Steel, Oxford University Press (2021).
 DOI: 10.1093/oso/9780192895073.003.0007

7.2 The Postulates of Quantum Mechanics

Postulate 1: Everything that can be known about a quantum system is contained in the mathematical object, $|\psi(t)\rangle$ called a **state vector**, or a **wave function,** designated as $\psi(\boldsymbol{r}, t)$, where $\boldsymbol{r}$ is a vector in real space and t is time.

Note that there is no $\boldsymbol{r}$ in $|\psi(t)\rangle$, and $\boldsymbol{r}$ itself could be replaced with another kind of degree of freedom that was not in real space. The symbol for a quantum vector is $|\ \rangle$ and is called a **ket.**

Math Note 1

To understand Dirac notation and the meaning of a state vector or eigenvector, we consider a vector represented by $\boldsymbol{A}$ in real space. $\boldsymbol{A}$ can be written as $\boldsymbol{A} = A_x\check{\boldsymbol{x}} + A_y\check{\boldsymbol{y}} + A_z\check{\boldsymbol{z}}$ where symbols like $\check{\boldsymbol{x}}$ represent the **unit vector** or **basis vector** along an axis (usually in real space, hence the three unit vectors). The unit vector can be represented as a **column vector**:

$$\check{\boldsymbol{x}} = \begin{bmatrix} 1 \\ 0 \\ 0 \end{bmatrix}, \check{\boldsymbol{y}} = \begin{bmatrix} 0 \\ 1 \\ 0 \end{bmatrix}, \check{\boldsymbol{z}} = \begin{bmatrix} 0 \\ 0 \\ 1 \end{bmatrix} \tag{7.1}$$

and the vector $\boldsymbol{A}$ would then be written as

$$\boldsymbol{A} = \sum_{i=x,y,z} A_i\check{i} = \begin{bmatrix} A_x \\ A_y \\ A_z \end{bmatrix} \tag{7.2}$$

The unit vectors are **orthonormal**, meaning that $\check{\boldsymbol{r}} \cdot \check{\boldsymbol{s}} = \delta_{\hat{r},\hat{s}}$, where $\check{\boldsymbol{r}}, \check{\boldsymbol{s}} \in \{\check{\boldsymbol{x}}, \check{\boldsymbol{y}}, \check{\boldsymbol{z}}\}$ and the **Kronecker delta** is defined by

$$\delta_{\check{r},\check{s}} = \begin{cases} 1 & \check{r} = \check{s} \\ 0 & \check{r} \neq \check{s} \end{cases} \tag{7.3}$$

where $\check{\boldsymbol{r}} \cdot \check{\boldsymbol{r}} = 1$, showing that the unit vectors are **normalized**, and 0 if $\check{\boldsymbol{r}} \neq \check{\boldsymbol{s}}$, meaning that the vectors are **orthogonal**. A **row vector** is a $n \times 1$ **matrix** where a matrix in general is an $m \times n$ rectangular array of numbers with m rows and n columns.

A quantum vector would be written as $|\psi\rangle$ and, like the case for the real space vector (Eq. 7.2), the quantum vector can be written in terms of an infinite set of orthonormal **basis vectors:**

$$|\psi\rangle = \sum_n c_n \,|\, n\rangle \tag{7.4}$$

The column vectors in the quantum case, designated $|\, n\rangle$, are also unit vectors. Typically there are an infinite number of basis vectors. The ket $|\, n\rangle$ then means an infinitely long column vector with a 1 in the nth position:

$$|n\rangle = \begin{bmatrix} 0 \\ 0 \\ \vdots \\ 0 \\ 1_n \\ 0 \\ \vdots \end{bmatrix} \tag{7.5}$$

Just as $\{\check{x}, \check{y}, \check{z}\}$ span (i.e., define) a three-dimensional real space, the set $\{|n\rangle\}$ define an **abstract vector space** called a **Hilbert space**. It is important to understand that it is not necessary to represent $|n\rangle$ as a column vector. It is sufficient to say $|n\rangle$ is a unit vector. The same is true of the unit vectors for real space. However, it is sometimes convenient for clarity in discussions to use an explicit form such as in Eq. 7.5.

The **inner product** of two vectors, $\boldsymbol{A}$ and $\boldsymbol{B}$, such as we described above in the case of real space, is $\boldsymbol{A} \cdot \boldsymbol{B} = a_j b_j + a_k b_k + a_l b_l$. In linear algebra, this is done using the concepts of matrices. The inner product becomes the product of two matrices, a row vector (a 1×3 matrix) times a **column vector** (a 3×1 matrix):

$$\boldsymbol{A} \cdot \boldsymbol{B} = \begin{bmatrix} a_x & a_y & a_z \end{bmatrix} \begin{bmatrix} b_x \\ b_y \\ b_z \end{bmatrix} = a_x b_x + a_y b_y + a_z b_z = \sum_{i=x,y,z.} a_i b_i \tag{7.6}$$

resulting in a **scalar** (a 1×1 matrix).

In quantum mechanics, the numbers and functions are generally complex because of the nature of Schrödinger's equation and the postulates. In linear algebra with complex numbers we define the **adjoint** of a matrix, $\hat{\mathbf{A}}$, designated $\hat{\mathbf{A}}^{\dagger}$. This is the **complex transpose** of the original matrix (e.g., a vector) identified above. For a matrix $\hat{\mathbf{A}}$ such as

$$\hat{\mathrm{A}} = \begin{bmatrix} a_{11} & a_{12} & a_{13} & \\ a_{21} & a_{22} & a_{23} & \cdots \\ a_{31} & a_{32} & a_{33} & \\ & \vdots & & \end{bmatrix} \tag{7.7}$$

Then the **complex transpose** is written as

$$\hat{\mathrm{A}}^{\dagger} = \begin{bmatrix} a_{11}^* & a_{21}^* & a_{31}^* & \\ a_{12}^* & a_{22}^* & a_{32}^* & \cdots \\ a_{13}^* & a_{23}^* & a_{33}^* & \\ & \vdots & & \end{bmatrix} \tag{7.8}$$

Then the inner product is written as $\hat{A}^{\dagger} \cdot \hat{B}$.

In the case of a quantum vector, the adjoint of the **ket** in Eq. 7.4 can be written in terms of the **bra**:

$$\langle \psi | = \sum_n c_n^* \langle n | \tag{7.9}$$

Where $|n\rangle$ in Eq. 7.5 is a column vector, $\langle n|$ is the corresponding row vector. In Dirac notation, the inner product is written as

$$\langle\psi|\psi\rangle = \sum_n c_n^*\langle n|\sum_{n'} c_{n'}|n'\rangle = \sum_{nn'} c_n^* c_{n'}\langle n|n'\rangle = \sum_n |c_n|^2 \tag{7.10}$$

We have used the fact stated above that the basis vectors are orthonormal: i.e., $\langle n|n'\rangle = \delta_{nn'}$. Notice that the adjoint of the ket of the vector is the bra of the vector:

$$\langle\psi| = |\psi\rangle^\dagger \tag{7.11}$$

Dirac notation is not used as a substitute for a function, like $\psi(x)$. While ψ could be written in a bra or ket, e.g., $|\psi\rangle$, one would not write $\psi(x)$ or any $f(x)$ inside such a symbol. Dirac notation is basis independent. More discussion on the comparison and use of functions and Dirac notation will be given below.

Postulate 2: Associated with every physical observable is a **Hermitian operator**, $\hat{Q}$. The only allowed values for the observable are **eigenvalues** of the operator. The **eigenvectors** or **eigenfunctions** of the operator form a complete orthonormal set, meaning that the set of eigenfunctions or eigenvectors of a Hermitian operator are orthonormal and that any arbitrary function or vector in the same space can be written as a linear combination of the eigenfunctions or eigenvectors, respectively.

Completeness can be proven but it is sometimes postulated in quantum for simplicity (or assumed to be understood). The reason the operator must be Hermitian is that Hermitian operators have real (vs complex) eigenvalues, as discussed in more detail below.

We note here that the language can get confusing because of using the concept of function and vector in the same discussion. As will be seen, eigenfunctions and wave functions, i.e., both functions in real space, may or may not be a part of a given calculation. Other times, only vectors will be used. It is also possible to go smoothly from function to vector representations but going from vector to function is not always possible or useful. The same point holds for operators.

For operators based on a canonical coordinate (x, in this case) like $\hat{\boldsymbol{p}} = -i\hbar\frac{d}{dx}$, an operator is Hermitian if for any *arbitrary* functions $f(x)$ and $g(x)$ which are well behaved (i.e., vanish at $\pm\infty$):

$$\int dx\, g^*(x)\hat{p}f(x) = \int dx(\hat{p}g(x))^* f(x) \tag{7.12}$$

The Hermiticity of an operator can depend on the boundary conditions, and hence the boundary conditions are sometimes a part of the definition of an operator.

In Dirac notation, the equivalent statement is:

Given

$$\hat{Q}|u\rangle = |v\rangle \text{ and } \langle u|\hat{Q}^\dagger = \langle v| \tag{7.13}$$

$\hat{Q}$ is Hermitian if

$$\hat{Q} = \hat{Q}^\dagger. \tag{7.14}$$

In the case of the matrix in Eq. 7.7, this means $\boldsymbol{A}$ is Hermitian if $a_{ij} = a_{ji}^*$.

Problem 7.1

a) Show that $\hat{p} = -i\hbar\frac{d}{dx}$ *is Hermitian.*

b) Show also that if

$\widehat{Q} = \begin{bmatrix} a & b \\ b^* & c \end{bmatrix}$ (i.e., $\widehat{Q}$ is Hermitian) where a and c are both real, then for an arbitrary

$$|u\rangle = \begin{bmatrix} \alpha \\ \beta \end{bmatrix}$$

$\widehat{Q}\,|\,u\rangle = |\,v\rangle$ and $\langle u\,|\,\widehat{Q}^{\dagger} = \langle v\,|$ or $\widehat{Q}\,|\,u\rangle = \left(\langle u\,|\,\widehat{Q}^{\dagger}\right)^{\dagger}$

Hint: For part (a) (proving Eq. 7.12), integrate by parts and assume that the functions are physical and hence vanish at $\pm\infty$.

Math Note 2

Operators are mathematical objects that perform an operation (formally, a mapping) on another mathematical object to turn it into another mathematical object. A matrix, $\hat{A}$, is an example of an operator. When it multiplies (operates on) an appropriate vector, it changes it into another vector. Another operator corresponds to the scalar variable for position. The **identity matrix** (including the scalar 1) is an operator. Likewise, the object $-i\hbar\frac{d}{dx}$ is also an operator, where again $\hbar$ is ***Planck's constant*** (h) divided by 2π). The symbol for an operator is usually not written in bold font unless it is referring to a matrix.

An operator like a matrix $\hat{A}$ or the operator $\hat{p} = -i\hbar\frac{d}{dx}$ has eigenvalues and corresponding eigenvectors or eigenfunctions for each eigenvalue. If the eigenvalue can take on discrete values, the *spectrum of eigenvalues is* ***discrete.*** If the eigenvalue can take on a continuum of values, the *spectrum of eigenvalues is* ***continuous.*** In some cases, an operator will have a combination of a continuous and a **discrete spectrum of eigenvalues**, such as a finite potential well with both discrete bound and continuous scattering states.

In the case of some square matrix, say $\hat{\mathbf{A}}$, the eigenvalues and eigenvectors of $\hat{\mathbf{A}}$ are defined by the eigenvector equation

$$\hat{A}\,|A\rangle = A\,|A\rangle \qquad (7.15)$$

Rearranging, Eq. 7.15 gives

$$(\hat{A} - A\hat{I})\,|A\rangle = 0 \qquad (7.16)$$

where $\hat{I}$ is the identity matrix for this problem (a square matrix of the same order, n, as $\hat{A}$ and with 0's everywhere except along the diagonal, where all the elements are 1. In other words, the matrix elements of $\hat{I}$ are $\hat{I}_{ij} = \delta_{ij}$.) The eigenvalues are determined by setting the determinant of the matrix $(\hat{A} - A\hat{I})$ in Eq. 7.16 to 0:

$$\det\left(\hat{A} - A\hat{I}\right) = 0 \tag{7.17}$$

We consider two simple examples based on a matrix operator $\hat{A}$ of order $n = 2$. In the first case, consider

$$\hat{A} = \hat{A}_D = \begin{bmatrix} A_{11} & 0 \\ 0 & A_{22} \end{bmatrix} \tag{7.18}$$

where $\hat{A}_D$ is a **diagonal matrix**. It is easy to then show that the eigenvalues are A_{11} and A_{22} and the normalized eigenvectors are

$$|A_{11}\rangle = \begin{bmatrix} 1 \\ 0 \end{bmatrix} \tag{7.19}$$

and

$$|A_{22}\rangle = \begin{bmatrix} 0 \\ 1 \end{bmatrix} \tag{7.20}$$

with

$$\hat{A}_D\,|\,A_{11}\,\rangle = A_{11}|\,A_{11}\rangle \tag{7.21}$$

and

$$\hat{A}_D\,|\,A_{22}\,\rangle = A_{22}|\,A_{22}\rangle \tag{7.22}$$

This is readily verified by substituting the matrix for $\hat{A}_D$ and the eigenvector for the corresponding eigenvalue and showing that when $\hat{A}_D$ operates on the eigenvector you get back the eigenvector and the corresponding eigenvalue. One can also show that the above is true by setting

$$|A\rangle = \begin{bmatrix} a \\ b \end{bmatrix}$$

where a and b are unknowns to be determined by first solving for the eigenvalues using Eq. 7.17 and then, for each eigenvalue, using Eq. 7.15 and **normalization** to find a and b.

Suppose now if $\hat{A}$ becomes non-diagonal, e.g.,

$$\hat{A} = \hat{A}_D + \begin{bmatrix} 0 & A_{12} \\ A_{21} & 0 \end{bmatrix} = \begin{bmatrix} A_{11} & A_{12} \\ A_{21} & A_{22} \end{bmatrix} \tag{7.23}$$

Then to find the new eigenvalues, we must go back and redo the calculation starting with Eq. 7.15. Let $|\,\alpha_i\rangle$ represent a new eigenvector with eigenvalue α_i. Then

$$\hat{A}|\,\alpha_i\,\rangle = \alpha_i|\,\alpha_i\rangle \tag{7.24}$$

or

$$\left[\hat{A} - \alpha_i\hat{I}\right]|\,\alpha_i\rangle = 0 \tag{7.25}$$

We expand $|\alpha_i\rangle$ in terms of the original eigenvectors:

$$|\alpha_i\rangle = a_i\,|A_{11}\rangle + b_i\,|A_{22}\rangle = a_i\begin{bmatrix}1\\0\end{bmatrix} + b_i\begin{bmatrix}0\\1\end{bmatrix} = \begin{bmatrix}a_i\\b_i\end{bmatrix} \tag{7.26}$$

This means that Eq. 7.16 now becomes

$$\begin{bmatrix}A_{11}-\alpha_i & A_{12}\\ A_{21} & A_{22}-\alpha_i\end{bmatrix}\begin{bmatrix}a_i\\b_i\end{bmatrix} = \begin{bmatrix}0\\0\end{bmatrix} \tag{7.27}$$

Equation 7.27 represents a set of two homogeneous equations (see Appendix A, then we use Eq. 7.17 to find the new eigenvalues:

$$\det\begin{bmatrix}A_{11}-\alpha_i & A_{12}\\ A_{21} & A_{22}-\alpha_i\end{bmatrix} \equiv \begin{vmatrix}A_{11}-\alpha_i & A_{12}\\ A_{21} & A_{22}-\alpha_i\end{vmatrix} = 0 \tag{7.28}$$

Giving

$$\begin{aligned}\alpha_i \to \alpha_\pm &= \frac{1}{2}\left((A_{11}+A_{22}) \pm \sqrt{(A_{11}+A_{22})^2 - 4(A_{11}A_{22}-A_{21}A_{12})}\right)\\ &= \frac{1}{2}\left((A_{11}+A_{22}) \pm \sqrt{(A_{11}-A_{22})^2 + 4A_{21}A_{12}}\right)\end{aligned} \tag{7.29}$$

There are two eigenvalues with corresponding eigenvectors

$$|\alpha_\pm\rangle = \begin{bmatrix}a_\pm\\b_\pm\end{bmatrix} \tag{7.30}$$

Without loss of generality, we assume $A_{22} > A_{11}$. Then in the limit, the off-diagonal terms, $A_{21} = A_{12}^* \to 0$, $\alpha_+ \to A_{22}$, and $\alpha_- \to A_{11}$.

As explained in Appendix A, two homogeneous equations in two unknowns where the determinant of the coefficients is 0 leads to a solution of one unknown in terms of the other. In other words, the unknowns are not uniquely determined. The remaining unknown is determined by the condition that the vector is normalized: i.e., $\langle\alpha_+|\alpha_+\rangle = \langle\alpha_-|\alpha_-\rangle = 1$. For the first equation, then, we return to Eq. 7.24 and use:

$$\begin{bmatrix}A_{11} & A_{12}\\ A_{21} & A_{22}\end{bmatrix}\begin{bmatrix}a_\pm\\b_\pm\end{bmatrix} = \alpha_\pm\begin{bmatrix}a_\pm\\b_\pm\end{bmatrix}$$

For reasons that will become evident below, while we only need one of the two equations, we will use both to make the physics clearer. For the first equation, we use the upper equality for the left and right side and determine the coefficients for α_-

$$(A_{11}-\alpha_-)\,a_- + A_{12}b_- = 0$$

$$b_- = -\frac{A_{11}-\alpha_-}{A_{12}}a_-$$

$$|\alpha_-\rangle = a_-\begin{bmatrix} 1 \\ -\frac{A_{11}-\alpha_-}{A_{12}} \end{bmatrix}$$

after normalization

$$|\alpha_-\rangle = \frac{1}{\sqrt{1+\left|\frac{A_{11}-\alpha_-}{A_{12}}\right|^2}}\begin{bmatrix} 1 \\ -\frac{A_{11}-\alpha_-}{A_{12}} \end{bmatrix} = \frac{1}{\sqrt{1+\left|\frac{A_{11}-\alpha_-}{A_{12}}\right|^2}}\left(|A_{11}\rangle - \frac{A_{11}-\alpha_-}{A_{12}}|A_{22}\rangle\right)$$

and for α_+

$$A_{21}a_+ + (A_{22}-\alpha_+)\,b_+ = 0$$

$$|\alpha_+\rangle = a_+\begin{bmatrix} -\frac{A_{22}-\alpha_+}{A_{21}} \\ 1 \end{bmatrix}$$

$$|\alpha_+\rangle = \frac{1}{\sqrt{1+\left|\frac{A_{22}-\alpha_+}{A_{21}}\right|^2}}\begin{bmatrix} -\frac{A_{22}-\alpha_+}{A_{21}} \\ 1 \end{bmatrix} = \frac{1}{\sqrt{1+\left|\frac{A_{22}-\alpha_+}{A_{21}}\right|^2}}\left(-\frac{A_{22}-\alpha_+}{A_{21}}|A_{11}\rangle + |A_{22}\rangle\right)$$

It is now easy to show that in the limit $A_{21} = A_{12}^* \to 0$, $|\,\alpha_+\rangle \to |\,A_{22}\rangle$, and $|\,\alpha_-\rangle \to |\,A_{11}\rangle$.

It is very important in terms of the basic behavior here to notice that, compared to the case of $\hat{\boldsymbol{A}}_D$ with eigenstates of $|\,A_{11}\rangle$ and $|\,A_{11}\rangle$, the solution here is a linear combination of $|\,A_{11}\rangle + |\,A_{22}\rangle$. This means that the effect of adding a matrix that is *non-diagonal* to a matrix that is diagonal leads to **state mixing**. The implication of this is that the properties of the new lower state $|\,\alpha_-\rangle$ and upper state $|\,\alpha_+\rangle$ now have some of the properties of the old upper state $|\,A_{22}\rangle$ and lower state $|\,A_{11}\rangle$, respectively. For example, if $|\,a\,\rangle$ and $|\,d\,\rangle$ are states with defined and opposite spatial symmetry (parity), such as the ground and first excited state of a nano-vibrator or a particle in a well, states $|\,\alpha_\pm\rangle$ in the presence of the non-diagonal coupling will not have such parity. The non-diagonal matrix represents the effect of adding a new term to the Hamiltonian. The non-diagonal elements in the Hamiltonian couple or mix the states resulting in state mixing. A specific pair of off-diagonal elements, e.g., a_{ij} and a_{ji}, mix states i and j.

The above discussion focused on using matrices as operators with corresponding vectors since any operator, $\hat{A}$, in a Hilbert space described by the set of complete basis vectors, e.g., the energy eigenstates, $|\,E_j\rangle$ described by some Hamiltonian, can be written as a matrix in that basis:

$$\hat{A} = \sum_{ij}\langle E_i|\hat{A}|E_j\rangle\,|\,E_i\,\rangle\langle\,E_j\,| = \begin{bmatrix} \langle E_1|\hat{A}|E_1\rangle & \langle E_1|\hat{A}|E_2\rangle & \cdots \\ \langle E_2|\hat{A}|E_1\rangle & \langle E_2|\hat{A}|E_2\rangle & \cdots \\ \vdots & \vdots & \ddots \end{bmatrix} \tag{7.31}$$

where $|\,E_i\rangle$ is the usual unit vector, given by a column of 0's with a 1 in the i-position. This is true even if one is working with eigenfunctions and differential equations.

Problem 7.2 *Find the eigenvalues and normalized eigenvectors for the operator* $\hat{A} = \begin{bmatrix} 1 & 1 \\ 1 & -1 \end{bmatrix}$.

Problem 7.3 *Assume an arbitrary non-diagonal matrix* $\hat{A}$, *like Eq. 7.7, where the problem of finding the eigenvalues and eigenvectors leading to* $\hat{A}\,|v_n\rangle = a_n\,|v_n\rangle$ *has been completed and*

$$|v_n\rangle = \begin{bmatrix} v_{n_1} \\ v_{n_2} \\ \vdots \\ v_i \\ \vdots \end{bmatrix} \tag{7.32}$$

We now form a matrix of these vectors:

$$\hat{S}^{\dagger} = [|v_1\rangle\ |v_2\rangle\ \cdots\ |v_i\rangle\ \cdots] = \begin{bmatrix} v_{1_1} & v_{2_1} & v_{3_1} & \cdots \\ v_{1_2} & v_{2_2} & v_{3_2} & \cdots \\ v_{1_3} & v_{2_3} & v_{3_3} & \cdots \\ \vdots & \vdots & \vdots & \ddots \end{bmatrix} \tag{7.33}$$

Show that $\hat{\boldsymbol{S}}\hat{\boldsymbol{A}}\hat{\boldsymbol{S}}^{\dagger}$ diagonalizes $\hat{\boldsymbol{A}}$. That is, show that

$$\hat{S}\hat{A}\hat{S}^{\dagger} = \begin{bmatrix} a_1 & 0 & 0 & 0 & 0 \\ 0 & a_2 & 0 & 0 & 0 \\ 0 & 0 & \ddots & 0 & 0 \\ 0 & 0 & 0 & a_i & 0 \\ 0 & 0 & 0 & 0 & \ddots \end{bmatrix} \tag{7.34}$$

The entire process of finding the new eigenvalues and eigenvectors or eigenfunctions is called **matrix diagonalization**.

Problem 7.4 *Show that* $u_p(x) = Ce^{\frac{ipx}{\hbar}}$ *is an eigenfunction of* $\hat{p}$ *with eigenvalue p. Note that at this point, the spectrum of eigenvalues is continuous. A common requirement that arises in the math of quantum physics is that a function is periodic such as in a crystal. If the periodicity of the system is L, we then require for any function, specifically* $u_p(x)$ *that* $u_p(x) = u_p(x+L)$.

a) Use the expression for $u_p(x) = Ce^{\frac{ipx}{\hbar}}$ *in this, and find the condition on p, i.e., find the allowed eigenvalues to satisfy the requirement that* $u_p(x) = u_p(x+L)$. *Recall that* $e^{i2n\pi} = 1$. *In this case the spectrum of eigenvalues is now discrete.*

b) For this case, normalization is set by the requirement that $\int_x^{x+L} dx\, u_p^*(x)u_p(x) = 1$. *Find C.*

c) Show that $\int_x^{x+L} dx\, u_p^*(x)u_{p'}(x) = 0$ *when the eigenvalues are not the same.*

Problem 7.5 *To see how to apply the above analysis on finding eigenvalues, consider any of the systems discussed in the earlier sections. Assume that the Hamiltonian for that problem has been diagonalized, meaning that the eigenvalues and eigenfunctions have been identified. We designate Hamiltonian* $\widehat{H}_0$ *and the corresponding eigenfunctions* $u_n(x)$. *Then we have* $\widehat{H}_0 u_n(x) = E_n u_n(x)$. *Now we turn on an interaction,* $\widehat{V} = V(x)$ *that couples two of the levels,* $n = j$ *and* $n = k$. *The total Hamiltonian is then* $\widehat{H} = \widehat{H}_0 + \widehat{V}$.

Use the above procedure to find the new eigenfunctions of the Hamiltonian, $\hat{H}$. There are only two levels that are impacted. Begin by expanding the new unknown eigenfunction in terms of the two original eigenfunctions; e.g., $u_\pm = a_\pm u_j + b_\pm u_k$. Then, for the two new levels, solve the time independent Schrödinger equation to find the new eigenvalues, $E_\pm$ and the unknown coefficients for $u_\pm$. You will need to use the fact that u_j and u_k *are orthonormal, meaning $\int dx\ u^*_{j(k)}(x)u_{k(j)}(x) = \delta_{jk}$. To reduce the clutter in the algebra, use the standard notation that $V_{jk(kj)} = \int dx\ u^*_{j(k)}(x)V(x)u_{k(j)}(x)$ and assume $V_{jj(kk)} = 0$.*

Postulate 2 states the set of eigenstates is complete. To understand the meaning of this statement, we use the real space analogy used above for a vector, $\boldsymbol{r}$. This vector can be written as a linear combination of its unit vectors as $\boldsymbol{r} = x\check{x}+y\check{y}+z\check{z}$ since the set of unit vectors $\{\check{x}, \check{y}, \check{z}\}$ forms a complete set of vectors in real space, meaning that any arbitrary vector $\boldsymbol{r}$ can be written as a linear combination of the set of unit vectors. This is a kind of **Hilbert** space, defined by the three unit vectors.

We see then that the 2×2 matrix above for $\hat{A} = \frac{1}{2}\begin{bmatrix} 1 & 0 \\ 0 & -1 \end{bmatrix}$ defines a complete set of vectors (not real space, but an abstract space) because any 2×1 vector can be expressed in terms of a linear combination eigenvectors of $\hat{A}$, namely $\begin{bmatrix} a \\ b \end{bmatrix} = a\begin{bmatrix} 1 \\ 0 \end{bmatrix} + b\begin{bmatrix} 0 \\ 1 \end{bmatrix}$. Similarly, with the appropriate eigenvectors found above, for $\hat{A} = \begin{bmatrix} 1 & 1 \\ 1 & -1 \end{bmatrix}$ are complete over the same space.

In the case of eigenfunctions, the eigenfunctions also form a complete set. For the case $u_p(x) = \alpha e^{\frac{ipx}{\hbar}}$, which is an eigenfunction of the momentum operator $\hat{p} = -i\hbar\frac{d}{dx}$, we can express any function, $f(x)$, as a linear combination of functions of $u_p(x) = \alpha e^{\frac{ipx}{\hbar}}$. Just like the case of vectors, we have to sum over the different functions, but in this case, there is a different function for each value of p. Since there are no boundary conditions on the function, p is continuous. The **continuous spectrum of eigenvalues** means that we must integrate (rather than having a discrete sum: i.e., $f(x) = \alpha\int_{-\infty}^{\infty} dpF(p)e^{\frac{ipx}{\hbar}}$. For the particular form of the operator above, $\hat{p} = -i\hbar\frac{d}{dx}$, we find that the function $f(x)$ is expressed as a Fourier transform of a function $F(p)$, a result that is a consequence of the form of the eigenfunctions for $\hat{p}$.

Math Note 3

Theorem: Eigenvalues of Hermitian operators are real and the eigenvectors (eigenfunctions) are orthogonal. Since measurements in the laboratory *always give real numbers* and only allowed eigenvalues of the operator associated with an observable are allowed (by Postulate 2), it is required that the operator yield only real eigenvalues. This is the reason that the postulate requires the operators to be Hermitian.

Proof: We will use Dirac notation but you can also do this for functions, with integration. We take $\hat{A}$ to be Hermitian, meaning $\hat{A} = \hat{A}^\dagger$

$\hat{A} \mid a_n \rangle = a_n \mid a_n\rangle$ and then by definition of a Hermitian operator $\langle a_n \mid \hat{A}^\dagger = a_n^*\langle a_n \mid$

Then

$$\langle a_m|\hat{A}|a_n\rangle = a_n \langle a_m|a_n\rangle \tag{7.35}$$

and

$$\langle a_m|\hat{A}|a_n\rangle = a_m^* \langle a_m|a_n\rangle \tag{7.36}$$

But since $\hat{A}$ is Hermitian, the left-hand sides of both equations are the same, so subtracting:

$$(a_n - a_m^*)\langle a_m|a_n\rangle = 0 \tag{7.37}$$

If $n \neq m$ and the eigenvalues are not the same (non-degenerate), then we require

$$\langle a_m|a_n\rangle = 0 \tag{7.38}$$

So the eigenvectors are orthogonal. If $n = m$, then

$$\langle a_n|a_n\rangle \neq 0 \tag{7.39}$$

requiring

$$(a_n - a_n^*) = 0 \tag{7.40}$$

meaning that a_n is real.

The physical significance is clear, namely that if the number we get for a specific measurement of an observable is limited to the eigenvalues of the corresponding operator, those numbers must be real since measurements give only real numbers. This was mentioned above in the context of Postulate 2.

Math Note 4

If a set of eigenfunctions (or eigenvectors) is complete, then any function (vector) in the same space (defined by the operator and boundary conditions) can be expanded in terms of those functions (vectors). In the case of a discrete spectrum of eigenvalues, this means that

$$\psi(x) = \sum_n c_n u_n(x) \tag{7.41}$$

Because the $\{u_n(x)\}$ are orthonormal

$$\int dx\ u_n^*(x) u_m(x) = \delta_{nm} \tag{7.42}$$

Then multiplying the expression above for $\psi(x)$ by $u_n^*(x)$, changing the subscript in the sum to m and integrating over all space, we get

$$\int dx\ u_n^*(x)\,\psi(x) = \sum_m c_m \int dx\ u_n^*(x) u_m(x) = \sum_m c_m \delta_{nm}. \tag{7.43}$$

or

$$c_n = \int dx\, u_n^*(x)\, \psi(x). \tag{7.44}$$

Substituting in the expansion for $\psi(x)$ we get

$$\psi(x) = \sum_n \int dx' u_n^*(x')\,\psi(x')\,u_n(x) = \int dx' \left(\sum_n u_n^*(x')\,u_n(x)\right)\psi(x') \tag{7.45}$$

Since the right-hand side must be $\psi(x)$, it is clear that

$$\sum_n u_n^*(x')\,u_n(x) = \delta(x - x') \tag{7.46}$$

This relation is called the **closure relation** for eigenfunctions with a discrete spectrum of eigenvalues and is a result of completeness and orthonormality of the eigenfunctions.

In the case of a continuous spectrum of eigenvalues, we get

$$\psi(x) = \int dkc(k)u(k,x) \tag{7.47}$$

where k is the continuous eigenvalue. The set $\{u(k,x)\}$ is also orthonormal, where for a continuous spectrum of eigenvalues this means that

$$\int dx\, u^*(k,x)\, u(k',x) = \delta(k - k') \tag{7.48}$$

Then multiplying the expression above for $\psi(x)$ by $u^*(k,x)$, changing the eigenvalue in the integral to k' and integrating over all space, we get

$$\begin{aligned}\int dx\, u^*(k,x)\,\psi(x) &= \int dx\, u^*(k,x) \int dk' c(k')\, u(k',x)\\ &= \int dk' c(k') \int dx\, u^*(k,x)\, u(k',x) = \int dk' c(k')\, \delta(k-k') = c(k)\end{aligned} \tag{7.49}$$

giving

$$c(k) = \int dxu^*(k,x)\,\psi(x) \tag{7.50}$$

Substituting back into the expression for $\psi(x)$ we get

$$\begin{aligned}\psi(x) &= \int dkc(k)u(k,x) = \int dk \int dx' u^*(k,x')\,\psi(x')\,u(k,x)\\ &= \int dx'\psi(x') \int dku^*(k,x')\,u(k,x)\end{aligned} \tag{7.51}$$

Showing that in order to get the equality, you again require

$$\int dk\, u^* (k, x')\, u(k, x) = \delta (x' - x) \tag{7.52}$$

This is the **closure relation** for eigenfunctions with a continuous spectrum of eigenvalues.

If we repeat the above discussion now in Dirac notation, the statement of closure looks slightly different. Hence, for a discrete spectrum of eigenvalues,

$$|\psi\rangle = \sum_n b_n \,|\, q_n\rangle \tag{7.53}$$

where again the set of all eigenvectors is complete and the eigenvectors are orthonormal. We evaluate b_n by projection on to a specific $\langle q_n|$ state to get

$$b_n = \langle u_n|\psi\rangle \tag{7.54}$$

then

$$|\psi\rangle = \sum_n \langle u_n|\psi\rangle \,|\, u_n\rangle = \sum_n |q_n\rangle \langle q_n|\psi\rangle \tag{7.55}$$

Since, as above using the functional form, the right-hand side must reduce to $|\psi\rangle$ so this means:

$$\sum_n |u_n\rangle\langle q_n| = \hat{I} \tag{7.56}$$

This relation is the statement of **closure relation** for eigenvectors with a discrete spectrum of eigenvalues and is a result of completeness and orthonormality of the eigenvectors.

Repeat this for a continuous spectrum of eigenvalues described by the eigenvectors $|k\rangle$ where k represents the eigenvalue that has a continuous range. Then by completeness

$$|\psi\rangle = \int dk\; c(k) \,|\, k\rangle. \tag{7.58}$$

Projecting onto a specific eigenvector, $\langle k|$ and using the orthonormality relation for eigenvectors with a continuous spectrum of eigenvalues:

$$\langle k|k'\rangle = \delta (k - k') \tag{7.59}$$

we get

$$c(k) = \langle k|\psi\rangle \tag{7.60}$$

Substituting into the expression for $|\psi\rangle$ we

$$|\psi\rangle = \int dk\, c(k)\,|k\rangle = \int dk \langle k|\psi\rangle \,|\, k\rangle = \int dk\, |k\rangle \langle k|\psi\rangle \tag{7.61}$$

again showing that we get the equality stating the closure relation:

$$\int dk\, |k\rangle\langle k| = \hat{I} \tag{7.62}$$

This relation is the statement of the **closure relation** for eigenvectors with a continuous spectrum of eigenvalues.

Math Note 5

A **Hilbert space** is an **abstract vector space** as we discussed earlier. In quantum mechanics, it is usually defined by a Hermitian operator like a Hamiltonian, $\hat{H}$. By analogy, in real space (**Euclidean space**) the three unit vectors, $\hat{\mathbf{x}}$, $\hat{\mathbf{y}}$, and $\hat{\mathbf{z}}$ form a complete orthonormal set (orthogonal and normalized meaning that the inner product for each is 1, and the inner product between any two different unit vectors is 0), and any vector in that space can be expanded in terms of a linear combination of these unit vectors. We say that the unit vectors span or define the Euclidean space. In quantum mechanics, the eigenfunctions or eigenvectors of Hermitian operators, such as the Hamiltonian, form a complete orthonormal set, as discussed above. As it is a complete orthonormal set, any function in that same space can be expanded in terms of those eigenfunctions. By analogy with the unit vectors in Euclidean space, we say that the eigenfunctions or eigenvectors are like the unit vectors and span the Hilbert space defined by the original operator, in this case, $\hat{H}$.

This actual language is not usually used, say in signal processing, but the Fourier series and Fourier transform are examples of using a complete orthonormal basis that defines a Hilbert space to expand functions in that space. Those functions used in the Fourier series or Fourier transform are also eigenfunctions of a wave equation with an associated operator.

If we take the eigenfunction, $\frac{1}{\sqrt{2\pi\hbar}}e^{\frac{ip_x x}{\hbar}}$, of the momentum operator in the x-direction, $\hat{p}_x$, any function, $f(x)$, can be expanded in terms of that eigenfunction. But clearly, $g(y)$ cannot be expanded in terms of eigenfunctions of $\hat{p}_x$ because it is not in the space defined by $\hat{p}_x$. For that, we would introduce the momentum operator in the y-direction, $\hat{p}_y$, and its eigenfunctions. The operator $\hat{P} = \hat{p}_x + \hat{p}_y$ with eigenfunctions $\left(\frac{1}{\sqrt{2\pi\hbar}}e^{\frac{ip_x x}{\hbar}}\right)\left(\frac{1}{\sqrt{2\pi\hbar}}e^{\frac{ip_x x}{\hbar}}\right) = \frac{1}{2\pi\hbar}e^{\frac{i(p_x x+p_y y)}{\hbar}}$ now defines a Hilbert space that can by used to expand both $f(x)$ and $g(y)$, or $f(x)g(y)$. The eigenfunctions of $\hat{P}$ are a product of the eigenfunctions of $\hat{p}_x$ and $\hat{p}_y$. This is a **product space**. In the language of linear algebra, the Hilbert space defined by $\hat{P}$ is a product space of the eigenfunctions of $\hat{p}_x$ and $\hat{p}_y$ represented by $\hat{p}_x \otimes \hat{p}_y$. If the eigenfunctions of $\hat{p}_x$ and $\hat{p}_y$ are described in Dirac notation by $|\hat{p}_x\rangle$ and $|\hat{p}_y\rangle$, then the basis set for the product space defined by $\hat{p}_x \otimes \hat{p}_y$ is given by $|\hat{p}_x\rangle\,|\hat{p}_y\rangle$ (or sometimes by $|\hat{p}_x\hat{p}_y\rangle$). In explicit vector form $|\hat{p}_x\rangle\,|\hat{p}_y\rangle$ is written as two adjacent column vectors. The user must know (by an appropriate notation) which column vector is associated with which sub-space. Note that once a product space is created, such as $\hat{p}_x \otimes \hat{p}_y$, the individual spaces defined by $|\hat{p}_x\rangle$ and $|\hat{p}_y\rangle$ are sub-spaces.

Usually, the language of a Hilbert space is discussed in the context of eigenvectors and Dirac notation. Consider two Hamiltonians, $\hat{H}_1$ and $\hat{H}_2$, where $\hat{H}_1|E_{1_n}\rangle = E_{1_n}|E_{1_n}\rangle$ and $\hat{H}_2|E_{2_n}\rangle = E_{2_n}|E_{2_n}\rangle$. Each Hamiltonian defines its own Hilbert space. It could be that they represent say an atom and a free electron. When dealing with the two systems (the atom and the electron) simultaneously, such as a problem where say we have another Hamiltonian defined by $\hat{H} = \hat{H}_1 + \hat{H}_2 + \hat{V}_{12}$, then the Hilbert space for $\hat{H}$ is the product space defined by $\hat{H}_1 \otimes \hat{H}_2$ and represented by $|E_{1_n}\rangle|E_{2_n}\rangle$. The product space is sometimes called a **tensor**

product. It is interesting to note that in the case of, say, the proton and electron, they are in the same Euclidean space but in different Hilbert spaces.

Finally, to fully define a Hilbert space in more complex problems such as, say, a square box potential in two or three dimensions, the hydrogen atom, or say a free electron in three-dimensional space with intrinsic angular moment, spin, it is necessary to have what was identified in Chapter 6 as a **complete set of commuting observables (CSCO),** and the responding operators. This is a set of operators whose eigenvalues completely specify the state of the system. It is easily shown then that such a set of operators have common eigenfunctions or eigenvectors that form a complete set.[1] Consider the problem of a free particle along the x-axis. The linear momentum is described by the operator $\hat{p}_x$. The eigenfunctions or eigenvectors of $\hat{p}_x$ define the entire space for this problem. Suppose now we include the y-dimension, the particle is free to move in the x-y plane. Now, it is evident that $\hat{p}_x$ is not adequate to define the entire Hilbert space of the problem. The problem is solved by including $\hat{p}_y$. Note that $\hat{p}_x$ and $\hat{p}_y$ commute, $\left[\hat{p}_x, \hat{p}_y\right] = 0$. It is evident that between these two operators, the entire Hilbert space is defined for the x-y plane and hence $\hat{p}_x$ and $\hat{p}_y$ form a **complete set of commuting operators**.

Postulate 3: Let $\widehat{Q}$ be an operator associated with an observable with eigenvalues q_n where $\widehat{Q}|v_n\rangle = q_n|v_n\rangle$. If a measurement of the observable is made multiple times on a quantum system, described by $|\psi\rangle$ and identically created before each measurement, the average value of the measurement is given by the expectation value:

$$\langle q\rangle = \langle\widehat{Q}\rangle \equiv \langle\psi|\widehat{Q}|\psi\rangle \tag{7.63}$$

Problem 7.6 *Using the property of orthonormality for eigenfunctions with a discrete spectrum of eigenvalues, show that if* $|\psi\rangle = \sum_n c_n |v_n\rangle$, *then* $\langle q\rangle = \sum_n |c_n|^2 q_n$ *where* $\widehat{Q}\,|\,v_n\rangle = q_n\,|v_n\rangle$ *This means that* $|c_n|^2$ *is the probability given the system is described by* $|\psi\rangle = \sum_n c_n |v_n\rangle$ *that a measurement of the system will give a value* q_n. *The c-number* (c_n, *in general complex) is called the probability amplitude. We will discuss later the implication for an operation with a continuous spectrum of eigenvalues.*

[1] *Proof: (1) For the case on non-degenerate eigenvalues*—Consider two commuting operators, $\hat{P}$ and $\widehat{Q}$, such that $\left[\hat{P}, \widehat{Q}\right] = 0$. In Dirac notation, the set of eigenkets, $\{|p_j\rangle\}$ of $\hat{P}$, form a complete set with *non-degenerate* eigenvalues given by $\hat{P}\,|\,p_j\rangle = p_j\,|\,p_j\rangle$. Since $\hat{P}\widehat{Q}\,|\,p_j\rangle = \widehat{Q}\hat{P}\,|\,p_j\rangle = p_j\widehat{Q}\,|\,p_j\rangle$, $\widehat{Q}\,|\,p_j\rangle$ is also an eigenket of P with eigenvalue p_j. Since the eigenvalues are non-degenerate, $\widehat{Q}\,|\,p_j\rangle$, can be different from $|\,p_j\rangle$ by at most a constant. *(2) For the case of degenerate eigenvalues*—In the event that an eigenket, $|p_j\rangle$, is m-fold degenerate, let the corresponding eigenkets be $|p_j^{(k)}\rangle$ where $k = 1, 2, \ldots m$ and $|p_j^{(k)}\rangle = p_j\,|\,p_j^{(k)}\rangle$. Because the operators commute, $\widehat{Q}\,|\,p_j^{(k)}\rangle$ is also an eigenket of $\hat{P}$ with eigenvalue p_j. We now expand the eigenket $\widehat{Q}\,|\,p_j^{(k)}\rangle$ in terms of the m eigenkets $|\,p_j^{(n)}\rangle$ with degenerate eigenvalues: $\widehat{Q}\,|p_j^{(k)}\rangle = \sum_{n=1}^{m} a_{k_n}\,|p_j^{(n)}\rangle$. If we now multiply both sides by the expansion coefficient b_k and sum over k, we have $\widehat{Q}\sum_{k=1}^{m} b_k\,|p_j^{(k)}\rangle = \sum_{k=1}^{m}\sum_{n=1}^{m} b_k a_{k_n}\,|p_j^{(n)}\rangle$. Then $\sum_{k=1}^{m} b_k\,|p_j^{(k)}\rangle$ will be an eigenket of $\widehat{Q}$ with eigenvalue q_j if $\sum_{k=1}^{m} b_k a_{k_n} = q_j b_n$. This gives m equations in $m+1$ unknowns, namely $\{b_k\}$ and q_j. Setting the determinant of the coefficients to 0 gives an equation with m roots for q_i. Solving for the coefficients of the expansion for each $q_{j,}$ we get the corresponding eigenket $|\,q_r\rangle = \sum_{k=1}^{m} b_k^{q_r}\,|p_j^{(k)}\rangle$.

Postulate 4: The equation of motion for the state of the system is given by Schrödinger's equation:

$$\hat{H}\,|\psi\rangle = i\hbar\frac{d\,|\psi\rangle}{dt} \tag{7.64}$$

Or with functions:

$$\hat{H}\psi\,(r,t) = i\hbar\frac{d\psi\,(r,t)}{dt} \tag{7.65}$$

The operator in this equation represented by $\hat{H}$ is the operator form of the **Hamiltonian.** Up to this point in the earlier chapters, we defined and used the Hamiltonian appropriately. However, here we discuss the meaning of the Hamiltonian and the requirements for writing a Hamiltonian before it becomes a quantum operator. These are part of the rules of quantum physics.

The Hamiltonian is a concept that emerged through the work of Lagrange and Hamilton, in the mid 1700s and 1800s, in math and physics. The work took the thinking of Newton and $F = ma$ in mechanics to a new level of sophistication which then impacted many areas of physics. The ideas and concepts in this work are a foundation for all of quantum physics. Fortunately, we can exploit the large body of knowledge and—with some simplifying assumptions that will be removed later and using our same approach to the rules of this formalism as we are using here (i.e., postulates)—we can completely construct the essence of what is currently needed.

To see this, we will go back to a very simple problem in Figure 2.1 with a mass on a spring in the nano-vibrator that we considered in Chapter 2. This problem will shortly become the basis for understanding such complex issues as a quantum electronic circuit as well as the vacuum radiation field and quantum electrodynamics (QED). As we said in Chapter 2, this simple problem is remarkably profound in science and technology. While $F = ma$ was *postulated* by Newton (and confirmed by experiment), Newton did not give us the prescription for the force, F. How do we know the mathematical form of the force? We learned it is Hooke's law ($F = -kx$), but why is that right? This is important: *it was a guess*. Based on experiment (before or afterward, it doesn't matter), you can prove that this is correct. But it was a guess. The solution was worked out in Chapter 2.

To understand the basic idea in the approach of **Lagrangian** and **Hamiltonian** mechanics, you recall in your earlier training in basic physical science that you learned about kinetic energy and potential energy. Kinetic energy you learned is given by

$$T = \frac{1}{2}mv^2 = \frac{1}{2}m\dot{x}^2 \tag{7.66}$$

The potential energy, $V(x)$, for this kind of problem is defined by the relation to the force $F = -\nabla V(x)$. (Note: This relationship came from understanding that the work done in a closed path, often in two or three dimensions, was zero in a so-called conservative force field like gravity or the harmonic oscillator. Since the force is a vector, this is shown to mean that the curl of the force is zero and hence the force must be proportional to the gradient of the scalar, from the rules of vector algebra.) Hence, we immediately see that for the harmonic oscillator:

$$V(x) = \frac{1}{2}kx^2 \tag{7.67}$$

Without going into too much detail here because it isn't necessary, Lagrange developed the calculus of variations (Newton developed calculus). Hamilton defined a function called the ***Lagrangian*** as

$$L = T - V \tag{7.68}$$

and showed that its integral was an extremum (calculus of variations) when it was integrated along the right path.

Hamilton took this further. He pointed out that if the coordinate was **canonical** (often shown as a q), i.e., meaning general, with no constraints (a swinging pendulum moves in x and y but x and y are constrained to the length of the pendulum and hence are not canonical, whereas θ is free of constraints), then it was possible to define a **conjugate momentum** by the relation:

$$p = \frac{\partial L}{\partial \dot{q}} \tag{7.69}$$

So, for the harmonic oscillator (where $q = x$):

$$L(x, \dot{x}, t) = \frac{1}{2}m\dot{x}^2 - \frac{1}{2}kx^2 \tag{7.70}$$

Then, using $p = \frac{\partial L}{\partial \dot{q}}$ in one dimension, we get that

$$p = m\dot{x} \tag{7.71}$$

This is no different from how you learned to define the linear momentum, but imagine what you would have if we considered the problem of pendulum where $q = \theta$, the units will not even be right for what you have historically learned is momentum.

The rule of Hamilton then said the Hamiltonian for a potential such as we have is the total energy written in terms of the **canonical coordinate** and its **conjugate momentum**. For the harmonic oscillator, the Hamiltonian is given by

$$H = \frac{p^2}{2\,m} + V(x) = \frac{p^2}{2\,m} + \frac{1}{2}kx^2 \tag{7.72}$$

Using the Hamiltonian, we recover the results from Newton's law using **Hamilton's equations**:

$$\dot{q} = \frac{\partial H}{\partial p} \tag{7.73}$$

and

$$\dot{p} = -\frac{\partial H}{\partial q} \tag{7.74}$$

Using Eqs 7.73 and 7.74 with Eq. 7.72, we get

$$\dot{x} = \frac{\partial H}{\partial p} = \frac{p}{m}$$

and

$$\dot{\mathrm{p}} = -\frac{\partial \mathrm{H}}{\partial \mathrm{x}} = -\mathrm{kx}$$

The first is correct because we know from mechanics that $mv = p$ and for the second, we know that since in general $\dot{p} = F$, we have for the spring $\dot{p} = F = -kx$. We will use this simple result to understand the quantum vacuum in a later chapter.

We will need and use the balance of this discussion later when we encounter new physics, but for now, Hamilton used this definition along with the two linear first-order coupled differential equations for p and q to develop new equations of motion (Newton and Lagrange give one linear second order differential equation). Of course, the Newton, Lagrangian, and Hamiltonian all give the same result.

It is by this discussion that we now understand the Postulate 4 when it presents the equation of motion for the quantum state where, for the nano-vibrator, the Hamiltonian is given by

$$\hat{H} = \frac{\hat{p}^2}{2\,m} + V(\hat{x}) = \frac{\hat{p}^2}{2\,m} + \frac{1}{2}k\hat{x}^2 \tag{7.75}$$

Then, according to Postulate 2, p and x, as observables, become Hermitian operators in quantum mechanics.

To make progress in doing real calculations in quantum mechanics, the remaining issue is to understand the form for the operators. Specifically, how do we understand what the operators for p and x do, explicitly. This is resolved completely, though not obviously, by the Postulate 5:

Postulate 5: The form of the operators for the canonical coordinate and the corresponding conjugate momentum is given by the requirement that

$$[\hat{x}, \hat{p}] = i\hbar \tag{7.76}$$

The square brackets here represent the ***commutator***, discussed earlier, defined as:

$$\left[\hat{A}, \hat{B}\right] \equiv \hat{A}\,\hat{B} - \hat{B}\hat{A} \tag{7.77}$$

If $\hat{A}$ and $\hat{B}$ are matrices, it is clear from your experience in linear algebra that, in general, the product of matrices is not the same if their order is changed. To understand and design many various quantum systems, it is frequently not necessary to be any more explicit about the form of these operators than we have described so far. However, sometimes, it is important to have a coordinate representation for the operators. We could not have solved for the eigenfunctions in Chapters 2 and 3 without this. In this case, you can show that if

$$\hat{x} \rightarrow x \tag{7.78}$$

and

$$\hat{p} \rightarrow -i\hbar \frac{d}{dx} \tag{7.79}$$

that the requirement of Postulate 5 is satisfied.

Problem 7.7

a) Show that making the above assignments for the canonical coordinate and conjugate momentum satisfies Postulate 5. To do this, recall that, as operators, it is assumed there is a mathematical object on the right of the product of operators on which they operate.

b) Show that alternatively,

$$\hat{p} \rightarrow p$$

and

$$\hat{x} \rightarrow i\hbar \frac{d}{dp}$$

also satisfy Postulate 5.

Hint: Recall that when using an operator, operators must be considered in the context of operating on their corresponding mathematical object (a function, a vector, etc.). Hence, when considering an operator like $\frac{d}{dx}x$, *then you must consider the equation like* $\frac{d}{dx}xf(x) = f(x) + x\frac{d}{dx}f(x) = \left(1 + x\frac{d}{dx}\right)f(x)$, *therefore if* $\frac{d}{dx}$ *is an operator and* x *is an operator, then* $\frac{d}{dx}x = \left(1 + x\frac{d}{dx}\right)$. *Without this understanding, you would say* $\frac{d}{dx}x = 1$, *which is wrong.*

7.3 The Heisenberg Uncertainty Principle: The Minimum Uncertainty State

A direct result of the non-commutation of the canonical coordinate and its conjugate momentum is that both observables cannot be simultaneously observed to arbitrary accuracy. This is called the **Heisenberg uncertainty principle**.

The **quantum variance** of a measurement is defined by the quantum analog of the classical variance, namely if $\hat{A}$ and $\hat{B}$ are two observables then the variance of each is defined by:

$$\Delta\hat{A}^2 \equiv \int dx\, \psi^*(x)\left(\hat{A}^2 - \langle\hat{A}\rangle^2\right)\psi(x) = \langle\hat{A}\rangle^2 - \langle\hat{A}\rangle^2 \tag{7.80}$$

$$\Delta\hat{B}^2 \equiv \int dx\, \psi^*(x)\left(\hat{B}^2 - \langle\hat{B}\rangle^2\right)\psi(x) = \langle\hat{B}\rangle^2 - \langle\hat{B}\rangle^2 \tag{7.81}$$

Physically, the variance is a measure of fluctuations or noise in the measurement. Classical variance is defined in much same way, but wave functions are replaced by the probability distribution. If a measurement of the position, x, is taken, the noise in the measurement is then given by the **quantum standard deviation**, which is $\sqrt{\Delta\hat{x}^2}$.

To prove the Heisenberg uncertainty principle, then, we have to show that if $[\hat{A}, \hat{B}] = i\hat{C}$

$$\Delta\hat{A}^2\Delta\hat{B}^2 \geq \left(\frac{1}{2}\langle\hat{C}\rangle\right)^2 > 0 \tag{7.82}$$

The proof begins by noting that the following inequality is obvious, because the operators associated with observables are Hermitian and hence the integrand is equal to or greater than zero over the range of x:

$$\int dx\left[(\hat{A} + i\lambda\hat{B})\,\psi(x)\right]^*\left[(\hat{A} + i\lambda\hat{B})\,\psi(x)\right] \geq 0 \tag{7.83}$$

where λ is a real number. Because the operators are Hermitian and using $[\hat{A}, \hat{B}] = i\hat{C}$, we can simplify the integrand in Eq. 7.83 to get:

$$\begin{aligned}\int dx\left[(\hat{A} + i\lambda\hat{B})\,\psi(x)\right]^*\left[(\hat{A} + i\lambda\hat{B})\,\psi(x)\right] &= \int dx\,\psi^*(x)\left[(\hat{A} - i\lambda\hat{B})\right]\left[(\hat{A} + i\lambda\hat{B})\,\psi(x)\right]\\ &= \int dx\,\psi^*(x)\left[\left(\hat{A}^2 + \lambda^2\hat{B}^2 + i\lambda\left(\hat{A}\hat{B} - \hat{B}\hat{A}\right)\right)\right]\psi(x)\\ &= \int dx\,\psi^*(x)\left[\left(\hat{A}^2 + \lambda^2\hat{B}^2 + i\lambda\left[\hat{A},\hat{B}\right]\right)\right]\psi(x) = \Delta\hat{A}^2 + \lambda^2\Delta\hat{B}^2 - \lambda\langle\hat{C}\rangle\end{aligned} \tag{7.84}$$

Allowing us to conclude from Eq. 7.83 that

$$\lambda^2\Delta\hat{B}^2 - \lambda\hat{C} + \Delta\hat{A}^2 \geq 0 \tag{7.85}$$

where

$$[\hat{A}, \hat{B}] = i\hat{C} \tag{7.86}$$

$$\langle\hat{C}\rangle = \int dx\,\psi^*(x)\hat{C}\,\psi(x) \tag{7.87}$$

We have assumed $\langle\hat{A}\rangle$ and $\langle\hat{B}\rangle$ are zero to make the proof simpler. The minimum value of the left-hand side of the inequality in Eq. 7.85 is had by taking the derivative with respect to λ and setting it to 0, namely:

$$2\lambda\Delta\hat{B}^2 - \langle\hat{C}\rangle = 0 \tag{7.88}$$

giving

$$\lambda = \frac{\langle\hat{C}\rangle}{2\Delta\hat{B}^2} \tag{7.89}$$

Inserting into Eq. 7.85 we find

$$\frac{\langle\hat{C}\rangle^2}{4\left(\Delta\hat{B}^2\right)^2}\Delta\hat{B}^2 - \frac{\langle\hat{C}\rangle^2}{2\Delta\hat{B}^2} + \Delta\hat{A}^2 \geq 0 \quad (7.90)$$

$$\Delta\hat{A}^2\Delta\hat{B}^2 \geq \frac{\langle\hat{C}\rangle^2}{4} \quad (7.91)$$

For the case of the postulate$[\hat{x},\hat{p}] = i\hbar$, where again $\hat{x}$ and $\hat{p}$ represent the canonical coordinate and conjugate momentum, respectively, we get

$$\Delta\hat{x}^2\Delta\hat{p}^2 \geq \frac{\hbar^2}{4} \quad (7.92)$$

This is a profound result. It says that when the operators associated with two observables do not commute, they cannot be measured to arbitrary accuracy, simultaneously. Furthermore, a system cannot simultaneously be in an eigenstate of both operators since that would mean measurements would lead to knowledge of both observables to arbitrary accuracy.

Returning to the original inequality in Eq. 7.83, we see that for the equality to be true, since the integrand is positive definite, we require:

$$(\hat{A} + i\lambda\hat{B})\,\psi(x) = 0 \quad (7.93)$$

In the case that $\hat{A} = \hat{x} = x$ and $\hat{B} = \hat{p} = -i\hbar\frac{d}{dx}$, we have

$$\lambda = \frac{\langle\hat{C}\rangle}{2\Delta\hat{B}^2} = \frac{\hbar}{2\Delta\hat{p}^2} \quad (7.94)$$

$$\left(x + \frac{\hbar}{2\Delta\hat{p}^2}\hbar\frac{d}{dx}\right)\psi(x) = 0 \quad (7.95)$$

$$\left(\frac{d}{dx} + \frac{2\Delta\hat{p}^2}{\hbar^2}x\right)\psi(x) = \left(\frac{d}{dx} + \frac{1}{2\Delta\hat{x}^2}x\right)\psi(x) = 0 \quad (7.96)$$

This is easily integrated to give, after normalization,

$$\psi(x) = \left(\frac{1}{2\pi\Delta\hat{x}^2}\right)^{1/4} e^{-\frac{x^2}{4\Delta\hat{x}^2}} \quad (7.97)$$

This is called a **minimum uncertainty state**.

This theorem does not prescribe how to generate the Gaussian form for a wave function, but in the event that the system is described by operators $\hat{x}$ and $\hat{p}$, the system must be in the state given by Eq. 7.97, in order to provide the minimum error for measurement. The full width at half-maximum of the probability density given by $|\psi(x)|^2$ is given by:

$$FWHM = 2\sqrt{ln\ 2\Delta\hat{x}^2} \quad (7.98)$$

Problem 7.8

a) *Find the probability density of finding the particle's momentum p between p and p + dp the minimum uncertainty state give in Eq. 7.97. Recall from Chapter 2 that this is equivalent to the momentum representation.*
b) *Find the variance of p and show that the equality in Eq. 7.92 is satisfied for this state.*

7.4 Interpreting the Expansion Coefficients: Relating Functional Form to Dirac Form

We have discussed the fact that because of completeness, any arbitrary state of the system can be expanded in terms of the eigenfunctions of any operator, A, where $\hat{A}u_n(x) = a_n u_n(x)$, or $\hat{A}|\mathrm{n}\rangle = a_n|\mathrm{n}\rangle$. So for an operator with a discrete spectrum of eigenvalues $\psi(x) = \sum_n c_n u_n(x)$ where $c_n = \int_{-\infty}^{\infty} dx\ u_n^*(x)\psi(x)$ or in Dirac notation $|\psi\rangle = \sum_n c_n |\mathrm{n}\rangle$ with $c_n = n\langle|\psi\rangle$. Because of normalization, $\int_{-\infty}^{\infty} dx\ \psi^*(x)\psi(x) = 1$. If we insert the expansion for $\psi(x)$ and use the fact that the eigenfunctions are orthonormal, we find $\sum_n |c_n|^2 = 1$. This means that $|c_n|^2$ is the **probability** of finding the system with eigenvalue a_n is $|c_n|^2$ and c_n is the **probability amplitude.** In Dirac notation $\langle\psi|\psi\rangle = 1$ and $c_n = \langle n|\psi\rangle$. In the language of Dirac notation taken from the understanding of the inner (dot) product for simple vectors, $\langle n|\psi\rangle$ is the projection of $|\ \psi\rangle$ onto the unit vector $|\ n\rangle$. For an observable with a continuous spectrum of eigenvalues $\psi(x) = \int dk c(k) u\,(k, x)$, $|c(k)|^2$ is then the **probability density** of finding the system with eigenvalue k between k and $k + dk$.

Now, we recall that the average value of an observable is given by the expectation value of the corresponding operator, namely, $\langle\hat{A}\rangle = \langle a\rangle = \int_{-\infty}^{\infty} dx \psi^*(x)\hat{A}\psi(x) = \sum_n a_n |c_n|^2$, then as mentioned above, $|c_n|^2$ is the probability of finding the system with eigenvalue a_n in a measurement of $\hat{A}$. This is sometimes called **Born's rule** or postulate. In case of a measurement of the position, x, associated with the operator $\hat{x}$, we have for the assignment $\hat{x} = x$, $\langle\hat{x}\rangle = \langle x\rangle = \int_{-\infty}^{\infty} dx \psi^*(x) x \psi(x) = \int_{-\infty}^{\infty} dx\ \ x|\psi(x)|^2$, so we see now that $|\psi(x)|^2$ is the probability density of finding the particle between x and $x + dx$. Following the discussion in Davydov[2] and recalling above the discussion of the meaning of $|\langle n|\psi\rangle|^2$, the equivalent statement in Dirac notation is that since we have that $|\langle x|\psi\rangle|^2$ is the probability density of finding the particle between x and $x + dx$ then $|\psi(x)|^2 \equiv |\ \langle x|\psi\rangle|^2$ or $\psi(x) = \langle x|\psi\rangle$. It is possible to put this mapping on far more rigorous grounds[3], but the result is the same, namely that $\langle x|\psi\rangle$ is the projection of $|\ \psi\rangle$ onto real space. Similarly, the projection of $|n\rangle$ onto real space is $\langle x|n\rangle = u_n(x)$. Another way to see this is to consider $|x\rangle$ to be an eigenvector of the operator $\hat{x}$ meaning: $\hat{x}\ |x\rangle = x\ |x\rangle$. By completeness, $\int_{-\infty}^{\infty} dx |x\rangle\langle x| = \hat{I}$, the identity operator. For the expectation value $\langle\hat{x}\rangle = \langle x\rangle = \langle\psi|\hat{x}|\psi\rangle$. By completeness, $\langle x\rangle = \langle\psi|\hat{x}|\psi\rangle = \int_{-\infty}^{\infty} dx \int_{-\infty}^{\infty} dx'$ $\langle\psi|x'\rangle\langle x'|\hat{x}|x\rangle\langle x|\psi\rangle = \int_{-\infty}^{\infty} dx \int_{-\infty}^{\infty} dx' \langle\psi|x'\rangle x\delta\,(x - x')\langle x|\psi\rangle = \int_{-\infty}^{\infty} dx\ \ x\langle\psi|x\rangle\langle x|\psi\rangle$. Comparing this to $\langle x\rangle = \int_{-\infty}^{\infty} dx\ \ x|\psi(x)|^2$, we see again that $\langle x|\psi\rangle = \psi(x)$. Whether you

[2] A.S. Davydov, "Quantum Mechanics" Pergamon Press, Oxford, 1965.

[3] P.R. Berman, "Introduction to Quantum Mechanics" Springer, Cham Switzerland, 2018.

use the more mathematical approach in note 3 or the above approach from note 2, you see that it is now possible to go seamlessly from functional form to Dirac form.

7.5 Summary

The above provides the basic postulates of quantum mechanics. There are additional postulates that we will add later to deal with Fermions, Bosons, and spin, plus another postulate regarding measurement (the von Neumann postulate). There is a very important shift in thinking that must occur in this chapter. First, the previous rules you have learned for physics have now been replaced by a set of rules that also apply to small systems. In the process, things you have taken for granted and as obvious—such as always calculating your observable (position, voltage, energy, etc.) as a function of time—are now no longer an option. In the past, you had an equation of motion for your observable. In quantum, *there is no such thing*. The only equation of motion you have from the postulates is that given in Postulate 4: Schrödinger's equation. We will eventually develop equations of motion for the operators and their expectation values. Remarkably, but in some sense an obvious requirement is that these will often look very similar to the classical equations of motions for those observables, since it eventually becomes necessary to recover classical behavior if these rules are to be universal. This will not be extensively discussed in this text. Finally, the postulates in this chapter represent the core postulates of non-relativistic quantum mechanics. Three additional postulates will be introduced later.

Summary of Postulates

Postulate 1: Everything that can be known about a quantum system is contained in the mathematical object, $|\,\psi(t)\rangle$ called a ***state vector***, or a ***wave function*** designated as $\psi\,(\boldsymbol{r},t)$, where $\boldsymbol{r}$ is a vector in real space and t is time.

Postulate 2: Associated with every physical observable is a **Hermitian *operator***, $\widehat{Q}$, where Hermitian means $\widehat{Q} = \widehat{Q}^{\dagger}$. The only allowed values for the observable are the eigenvalues of the operator. The eigenvectors (or eigenfunctions) of the operator form a complete set.

Postulate 3: Let $\widehat{Q}$ be an operator associated with an observable with eigenvalues q_n where $\widehat{Q}\mid v_n\rangle = q_n\mid v_n\rangle$. If a measurement of the observable is made multiple times on a quantum system, described by $|\,\psi\rangle$ and identically created before each measurement, the average value of the measurement is given by the expectation value: $\langle q\rangle = \langle\hat{Q}\rangle \equiv \left\langle\psi|\widehat{Q}|\psi\right\rangle$

Postulate 4: The equation of motion for the state of the system is given by Schrödinger's equation:

$$\hat{H}\,|\,\psi\rangle = i\hbar\frac{d\,|\,\psi\rangle}{dt}$$

or with functions:

$$\hat{H}\psi(r,t) = i\hbar\frac{d\psi(r,t)}{dt}$$

Postulate 5: The form of the operators for the canonical coordinate and the corresponding conjugate momentum is given by the requirement that

$$[\hat{x},\hat{p}] = i\hbar$$

Table 7.1 Various common mathematical forms and operations as done when the system is described by a state function on the left and a state vector on the right.

$\psi(x), \psi(\mathbf{r}), \psi(\mathbf{r},t)$	$\lvert\psi\rangle, \lvert\psi(t)\rangle$
$\hat{A}\psi(x) = a\psi(x)$ $\hat{A}u_n(x) = a_n\psi(x)$	$\hat{A}\lvert\psi\rangle = a\lvert\psi\rangle$ $\hat{A}\lvert n\rangle = a_n\lvert n\rangle$
$\psi(x) = \sum_n c_n u_n(x);\quad c_n = \int dx\, u_n^*(x)\,\psi(x)$	$\lvert\psi\rangle = \sum_n c_n\lvert u_n\rangle;\ c_n = \langle u_n\lvert\psi\rangle$
Expectation value	Expectation value
$\langle\hat{A}\rangle = \int dx\psi^*(x)\,\hat{A}\psi(x)$	$\langle\hat{A}\rangle = \langle\psi\lvert\hat{A}\rvert\psi\rangle$
Closure relation	Closure relation
$\sum_n u_n^*(x')\,u_n(x) = \delta(x-x')$ $\int dk\, u^*(k,x')\; u(k,x) = \delta(x'-x)$	$\sum_n \lvert u_n\rangle\langle u_n\rvert = \hat{I}$ $\int dk\,\lvert k\rangle\langle k\rvert = \hat{I}$
Schrödinger's equation	Schrödinger's equation
$\hat{H}\psi(x) = i\hbar\frac{d\psi(x)}{dt}$	$\hat{H}\lvert\psi\rangle = i\hbar\frac{d\lvert\psi\rangle}{dt}$

Vocabulary (page) and Important Concepts

- postulate 93
- Dirac notation 93
- state vector 94
- wave function 94
- bra 93, 95
- ket 93, 94
- unit vector 94
- basis vector 94
- column vector 94

- Kronecker delta 94
- normalization 94
- orthogonal 94
- orthonormal 94
- row vector 94
- matrix 94
- inner product 95
- scalar 95
- adjoint 95
- abstract vector space 95
- Hilbert space 95
- complex transpose 95
- Hermitian operator 96
- eigenfunctions 96
- eigenvectors 96
- eigenvalues 96
- operator 96
- discrete eigenvalues 97
- continuous eigenvalues 97
- completeness 96
- identity matrix 97
- discrete spectrum of eigenvalues 97
- diagonal matrix 98
- state mixing 100
- matrix diagonalization 101
- continuous spectrum of eigenvalues 102
- Born's rule 114
- closure 114
- product space 106
- complete set of commuting operators (CSCO) 107
- Hamiltonian 108
- Lagrangian 108
- canonical coordinate 109
- conjugate momentum 109
- Hamilton's equations 109

- commutator 110
- Heisenberg uncertainty principle 111
- quantum variance 111
- quantum standard deviation 112
- minimum uncertainty state 113
- probability 114
- probability amplitude 114
- probability density 114

8 Heisenberg Operator Approach: Nano-Mechanical Oscillator (Part 2) and the Quantum LC Circuit

8.1 Introduction

In Chapter 7, the formal postulates of quantum mechanics were introduced. There we saw that indeed, as we did in the earlier chapters, we were able to write Schrödinger's equation in a coordinate representation by setting the momentum operator $\hat{p} = -i\hbar\frac{d}{dx}$ and the position operator $\hat{x} = x$. Doing this ensured that the two operators satisfied the commutator relation in Postulate 5. We then solved the resulting equation for the time independent form of the Schrödinger equation for the energy eigenfunctions and eigenvalues and ultimately the wave function.

Here, we examine two problems: the nano-mechanical vibrator and the quantum LC circuit. Unlike with the earlier approach for the nano-vibrator, we use only operators and postulates. Sometimes, this is called the ***Heisenberg approach*** or ***operator approach***. We do this because there are many problems that are approached using this kind of analysis and that have an important impact on nano-technology. You will also see in the case of the LC circuit that the wave function of the form $\psi(x)$ does not even exist. This is an extremely powerful approach. In Chapter 9 we will see how we can incorporate time dependence in the operators and eliminate the need for Schrödinger's equation in some cases. In Chapter 10, we will see how the behavior of the operators suggests the existence (confirmed experimentally) of spin. In Chapter 16 we will see how operators become a powerful means to understand the dynamical behavior of a complex system.

8.2 Heisenberg or Operator Approach to Solving the Time Independent Schrödinger Equation

Figure 2.1 showing the potential for the nano-vibrator remains central to this discussion. Our starting point is the Hamiltonian for the nano-vibrator, which was introduced in Eq. 2.15:

$$\widehat{H} = \frac{\hat{p}^2}{2m} + V(\hat{x}) = \frac{\hat{p}^2}{2m} + \frac{1}{2}k\hat{x}^2 \tag{8.1}$$

The full time dependent Schrödinger's equation:

$$\widehat{H}\left|\psi(t)\right\rangle = i\hbar\frac{d\left|\psi(t)\right\rangle}{dt} \tag{8.2}$$

Introduction to Quantum Nanotechnology: A Problem Focused Approach. Duncan Steel, Oxford University Press (2021).
© Duncan Steel. DOI: 10.1093/oso/9780192895073.003.0008

or

$$\left(\frac{\hat{p}^2}{2m} + \frac{1}{2}k\hat{x}^2\right)|\psi(t)\rangle = i\hbar\frac{d\,|\psi(t)\rangle}{dt} \tag{8.3}$$

In this problem, the operator equation depends on the operators for x and p and also on time, t. The Hamiltonian is time independent. This allows us to use the method of separation of variables as we did before. The temporal part remains a differential equation. Again,

$$|\psi(t)\rangle = |\psi\rangle\, g(t) \tag{8.4}$$

allowing us to rewrite the operator equation as

$$\frac{1}{|\psi\rangle}\left(\frac{\hat{p}^2}{2m} + \frac{1}{2}k\hat{x}^2\right)|\psi\rangle = i\hbar\frac{1}{g(t)}\frac{dg(t)}{dt} = E \tag{8.5}$$

When written like this, the two sides are independent of each other and so they both must be equal to the same constant, allowing us to break up Schrödinger's equation as two separate equations giving us the Dirac form of the equations in Eqs 2.25 and 2.26:

$$\left(\frac{\hat{p}^2}{2m} + \frac{1}{2}k\hat{x}^2\right)|\psi\rangle = E\,|\psi\rangle \tag{8.6}$$

$$i\hbar\frac{df(t)}{dt} = Eg(t) \tag{8.7}$$

As before, E is the energy eigenvalue, and the solution to 8.7 is given as

$$g(t) = e^{-\frac{iEt}{\hbar}} \tag{8.8}$$

This leaves us with solving the eigenvector Eq. 8.6, the time independent Schrödinger equation. The discussion is nearly identical to that in Chapter 2, except that here we have maintained the operators without inserting their equivalent functional form.

We begin by recognizing that since the Hamiltonian has units of energy, if we factor out a quantity with the units of energy, the terms inside the bracket will then be overall dimensionless. We know that $\hbar$ has units of energy-time ($ML^2\ T^{-1}$). The natural frequency in the system is $\omega = \sqrt{\frac{k}{m}}$, which has units of T^{-1}. Hence, it is natural to take as a unit of energy, $\hbar\omega$. We rewrite the Hamiltonian for 8.6 as

$$\hat{H} = \hbar\omega\left(\frac{\hat{p}^2}{2m\hbar\omega} + \frac{1}{2}\frac{k}{\hbar\omega}\hat{x}^2\right) \tag{8.9}$$

where now each of the two terms in the brackets is dimensionless.

Equation 8.9 has the form of $\left(\hat{A}^2 + \hat{B}^2\right)$, where both operators are **Hermitian**. This can be rewritten in terms of $(\hat{A} + i\hat{B})(\hat{A} - i\hat{B})$. Since, in general, $\hat{A}$ and $\hat{B}$ do not commute, $\left(\hat{A}^2 + \hat{B}^2\right)$ and $(\hat{A} + i\hat{B})(\hat{A} - i\hat{B})$ are not equal. Using the commutator of $\hat{x}$ and $\hat{p}$, we rewrite 8.9 as

$$\hat{H} = \hbar\omega\left(\frac{\hat{p}^2}{2m\hbar\omega} + \frac{1}{2}\frac{k}{\hbar\omega}\hat{x}^2\right)$$
$$= \hbar\omega\left(\sqrt{\frac{k}{2\hbar\omega}}\hat{x} + i\frac{\hat{p}}{\sqrt{2m\hbar\omega}}\right)\left(\sqrt{\frac{k}{2\hbar\omega}}\hat{x} - i\frac{\hat{p}}{\sqrt{2m\hbar\omega}}\right) - \frac{\hbar\omega}{2}$$
$$= \hbar\omega\left(\hat{a}\hat{a}^{\dagger} - \frac{1}{2}\right) \tag{8.10}$$

where we define

$$\hat{a} = \left(\sqrt{\frac{k}{2\hbar\omega}}\hat{x} + i\frac{\hat{p}}{\sqrt{2m\hbar\omega}}\right) \tag{8.11}$$

and the **adjoint**

$$\hat{a}^{\dagger} = \left(\sqrt{\frac{k}{2\hbar\omega}}\hat{x} - i\frac{\hat{p}}{\sqrt{2m\hbar\omega}}\right). \tag{8.12}$$

Problem 8.1 *Show the steps needed to get the final result in 8.10 using 8.11 and 8.12 along with the commutator relation from the postulates:*

$$[\hat{x}, \hat{p}] = i\hbar \tag{8.13}$$

Do not convert these operators in their functional form (meaning, do not send $\hat{x} \rightarrow x$ and $\hat{p} \rightarrow -i\hbar\frac{d}{dx}$).

We have switched the order of the operators for $\hat{p}$ and $\hat{x}$ in Eqs 8.11 and 8.12 to be consistent with other presentations commonly found in discussions of the Heisenberg's approach.

From the commutator relations for $\hat{x}$ and $\hat{p}$ given above in Eq. 8.13, it follows that

$$\left[\hat{a}, \hat{a}^{\dagger}\right] = 1 \tag{8.14}$$

Allowing us (for motivation to be seen shortly) to rewrite 8.10 as

$$\hat{H} = \hbar\omega\left(\hat{a}\hat{a}^{\dagger} - \frac{1}{2}\right) = \hbar\omega\left(\hat{a}^{\dagger}\hat{a} + \frac{1}{2}\right) \tag{8.15}$$

Problem 8.2 *Show* Eq. 8.14 *follows from the primary postulate stated in 8.13 using the definitions for $\hat{a}^{\dagger}$ and $\hat{a}$ and then show 8.15 results from 8.10.*

To be clear, no progress has been made in finding the eigenvectors and the eigenvalues for the time independent Schrödinger equation at this point, nor is anything known now about the features of the solution.

We adopt a fairly flexible but useful form based on Dirac notation from Chapter 7. The energy eigenstate is designated $|E_n\rangle$ with eigenvalue E_n so that

$$\hat{H}|E_n\rangle = E_n|E_n\rangle \tag{8.16}$$

The subscript n is intended to denote different eigenvalues. The symbol n is chosen because the spectrum of energy eigenvalues will be shown to be discrete (ignoring our earlier demonstration of this in Chapter 2 for now). We use Eqs 8.14–8.16 as three statements of what is given (based on our previous definitions and postulates) to be true. If Eq. 8.16 were a differential equation, it would be clear what to do.

Since $|E_n\rangle$ is a solution of Eq. 8.16, we consider the vector created by $\hat{a}\,|E_n\rangle$ and look to find the impact of $\hat{a}$ on $|E_n\rangle$. Then $\hat{a}\,|E_n\rangle$ is another ket (vector) but we have no information on its properties. To determine what this is, we operate with $\widehat{H}$:

$$\widehat{H}\hat{a}\,|E_n\rangle = \hbar\omega\left(\hat{a}^\dagger\hat{a} + \frac{1}{2}\right)\hat{a}\,|E_n\rangle \tag{8.17}$$

Now, using Eq. 8.14, the commutator of $\hat{a}^\dagger\hat{a}$ and $\hat{a}$ is:

$$\left[\hat{a}^\dagger\hat{a}, \hat{a}\right] = \hat{a}^\dagger\hat{a}\hat{a} - \hat{a}\hat{a}^\dagger\hat{a} = -\left[\hat{a}, \hat{a}^\dagger\right]\hat{a} = -\hat{a} \tag{8.18}$$

that is

$$\hat{a}^\dagger\hat{a}\hat{a} = \hat{a}\hat{a}^\dagger\hat{a} - \hat{a} \tag{8.19}$$

For Eq. 8.17 we can now rewrite it as

$$\begin{aligned}\widehat{H}\hat{a}\,|E_n\rangle &= \hbar\omega\left(\hat{a}^\dagger\hat{a}\hat{a} + \frac{1}{2}\hat{a}\right)|E_n\rangle = \hbar\omega\left(\hat{a}\hat{a}^\dagger\hat{a} - \hat{a} + \frac{1}{2}\hat{a}\right)|E_n\rangle = \hbar\omega\hat{a}\left(\hat{a}^\dagger\hat{a} - \frac{1}{2}\right)|E_n\rangle \\ &= \hbar\omega\hat{a}\left(\hat{a}^\dagger\hat{a} + \frac{1}{2} - 1\right)|E_n\rangle = \left(E_n - \hbar\omega\right)\hat{a}\,|E_n\rangle\end{aligned} \tag{8.20}$$

This is a rather remarkable result. It says that if $|E_n\rangle$ is an eigenvector with eigenvalue E_n, then $\hat{a}\,|E_n\rangle$ is an eigenvector with eigenvalue $E_n - \hbar\omega$. Using proof by induction, you can easily show (it is likely intuitively obvious to you) that $\hat{a}^m\,|E_n\rangle$ has eigenvalue $E_n - m\hbar\omega$ where m is an integer. This demonstrates that the spectrum of eigenvalues is discrete.

Extending this approach, it can be shown that

$$\widehat{H}\hat{a}^m\,|E_n\rangle = \left(E_n - m\hbar\omega\right)\hat{a}^m\,|E_n\rangle \tag{8.21}$$

$$\widehat{H}\hat{a}^{\dagger m}\,|E_n\rangle = \left(E_n + m\hbar\omega\right)\hat{a}^{\dagger m}\,|E_n\rangle \tag{8.22}$$

And therefore

$$\hat{a}^m\,|E_n\rangle = C\,|E_n - m\hbar\omega\rangle \tag{8.23}$$

$$\hat{a}^{\dagger m}\,|E_n\rangle = D\,|E_n + m\hbar\omega\rangle \tag{8.24}$$

where C and D are constants to be determined.

Problem 8.3 *Use the principle of mathematical induction to show that* $\widehat{H}\hat{a}^m\,|E_n\rangle = \left(E_n - m\hbar\omega\right)$ $\hat{a}^m\,|E_n\rangle$ *has eigenvalue* $E_n - m\hbar\omega$*. Hint: We have shown above that this is true for* $m = 1$*. In the* *proof by mathematical induction (or just proof by induction), you now* ***assume*** *that it is true for some m (other than 1) and then prove that it is true for m + 1.*

Problem 8.4 *Show now that if* $|E_n\rangle$ *is an eigenvector with eigenvalue* E_n*, then* $\hat{a}^\dagger |E_n\rangle$ *is an eigenvector with eigenvalue* $E_n + \hbar\omega$*. Using proof by induction, then show that* $\hat{a}^{\dagger m} |E_n\rangle$ *has eigenvalue* $E_n + m\hbar\omega$*.*

Now we work to determine the explicit eigenvalues for a given value of n. To get this, we consider the potential energy term in the Hamiltonian: $V(x) = \frac{1}{2}kx^2$. Since $k > 0$ and $x^2 \geq 0$, the potential energy is always greater than or equal to zero. A similar argument applies to the kinetic energy, $T = \frac{p^2}{2m}$. So, the total energy must be $E \geq 0$. But for m large enough, $E_n - m\hbar\omega < 0$. Hence, there is a maximum value of m such that $E_n - m\hbar\omega \geq 0$. Let m be that value. Then we have that $\widehat{H}\hat{a}^m |E_n\rangle \neq 0$, but we must require that $\widehat{H}\hat{a}^{m+1} |E_n\rangle = 0$. If not, then for the maximum value of m where $E_n - m\hbar\omega \geq 0$, we would get a negative energy eigenvalue for $E_n - (m+1)\hbar\omega < 0$.

Let us take $|E_0\rangle$ to be the state with the lowest allowed energy eigenvalue. Then

$$\widehat{H} |E_0\rangle = E_0 |E_0\rangle \tag{8.25}$$

Since $|E_0\rangle$ is the lowest eigenvalue then, to avoid a negative energy, we require

$$\hat{a} |E_0\rangle = 0 \tag{8.26}$$

and therefore

$$\hat{a}^\dagger \hat{a} |E_0\rangle = 0 \tag{8.27}$$

or because $\widehat{H} = \hbar\omega\left(\hat{a}^\dagger\hat{a} + \frac{1}{2}\right)$

$$\hbar\omega\hat{a}^\dagger\hat{a} = \widehat{H} - \frac{1}{2}\hbar\omega \tag{8.28}$$

and therefore

$$\left(\widehat{H} - \frac{\hbar\omega}{2}\right)|E_0\rangle = \left(E_0 - \frac{\hbar\omega}{2}\right)|E_0\rangle = 0 \tag{8.29}$$

or

$$E_0 = \frac{\hbar\omega}{2} \tag{8.30}$$

and

$$\widehat{H} |E_0\rangle = \frac{\hbar\omega}{2} |E_0\rangle \tag{8.31}$$

This is a remarkable result. By just using the commutation relation, Schrödinger's equation, and the Hamiltonian, we have found the lowest energy eigenvalue for the harmonic oscillator.

Now, using the result above in Eqs 8.20, 8.21,

$$|E_0 + m\hbar\omega\rangle \equiv |E_m\rangle \tag{8.32}$$

and

$$\widehat{H}\,|E_m\rangle = \hbar\omega\left(m + \frac{1}{2}\right)|E_m\rangle \tag{8.33}$$

Problem 8.5 *Show that* Eq. 8.34, *8.35, and* 8.36 *are true:*

$$\hat{a}^\dagger\hat{a}\ |E_m\rangle = m\ |E_m\rangle \tag{8.34}$$

$$\hat{a}\ |E_m\rangle = \sqrt{m}\ |E_m - \hbar\omega\rangle \tag{8.35}$$

$$\hat{a}^\dagger\ |E_m\rangle = \sqrt{m+1}\ |E_m + \hbar\omega\rangle \tag{8.36}$$

Problem 8.6 *Show that if we require* $\langle E_m|E_m\rangle = 1$, *then for* $|E_m\rangle = C_m(\hat{a}^\dagger)^m\,|E_0\rangle$ *we get that*

$$C_m = \sqrt{\frac{1}{m!}} \tag{8.37}$$

In Problem 8.5, you found in Eq. 8.28 that $\hat{a}^\dagger$ increases the eigenvalue from E_m to $E_m + \hbar\omega$. Hence, if it operate m times, it will increase the eigenenergy to $E_m + m\hbar\omega$; furthermore the state $(\hat{a}^\dagger)^m\,|E_0\rangle$ is an eigenstate of $\widehat{H}$; i.e.,

$$\widehat{H}(\hat{a}^\dagger)^m\,|E_0\rangle = (E_0 + m\hbar\omega)\,(\hat{a}^\dagger)^m\,|E_0\rangle \tag{8.38}$$

Therefore we can conclude that

$$|E_m\rangle = C_m(\hat{a}^\dagger)^m\,|E_0\rangle \tag{8.39}$$

C_m is the normalization factor $C_m = \frac{1}{\sqrt{m!}}$ and

$$E_m = \left(m + \frac{1}{2}\right)\hbar\omega. \tag{8.40}$$

Because the operators $\hat{a}$ and $\hat{a}^\dagger$ change the eigenvalue by one unit of $\hbar\omega$ we refer to them as ***lowering and raising operators***, respectively. The operator $\hat{a}^\dagger\hat{a}$ in 8.34 also appears in the Hamiltonian and is the ***number operator***:

$$\hat{N} = \hat{a}^\dagger\hat{a}, \tag{8.41}$$

such that

$$\hat{a}^\dagger\hat{a}\ |E_m\rangle = m\ |E_m\rangle \tag{8.42}$$

m represents the energy level of the harmonic oscillator and is sometimes called the **occupation number** representing n quanta of energy in the oscillator and, of course, takes on integer values: $n = 0, 1, 2, \ldots$ The Hamiltonian for the radiation field is exactly like the harmonic oscillator and the raising and lowering operators become annihilation and creation operators for photons, as we will see in a later chapter.

As we did before, it is possible to expand any state vector in this space in terms of the eigenvectors of the problem. Hence, as before:

$$|\psi\rangle = \sum_n C_n |E_n\rangle \tag{8.43}$$

and

$$|\psi(t)\rangle = \sum_n C_n\, e^{-\frac{iE_n t}{\hbar}}\, |E_n\rangle \tag{8.44}$$

where as usual

$$C_n = \langle E_n|\psi\,(t=0)\rangle \tag{8.45}$$

We can now evaluate $\langle x(t)\rangle = \langle \hat{x}\rangle$. We know in Dirac notation:

$$\langle \hat{x}\rangle\,(t) = \langle\psi(t)|\hat{x}|\psi(t)\rangle = \sum_{nm} C_n^* C_m e^{\frac{i(E_n-E_m)t}{\hbar}}\,\langle E_n|\hat{x}|E_m\rangle = \sum_{nm} C_n^* C_m e^{i\omega_{nm}t}\,\langle E_n|\hat{x}|E_m\rangle \tag{8.46}$$

where

$$\hbar\omega_{nm} = E_n - E_m \tag{8.47}$$

For clarity in the operations we have replaced $|E_n\rangle$ with $|n\rangle$ but the eigenvalue remains $E_n = \left(n + \frac{1}{2}\right)\hbar\omega$.

In Chapter 2, we had eigenfunctions and operators and operators expressed in terms of the coordinate x. Here, we have only eigenvectors and operators. Then $\hat{x}$ and $\hat{p}$ remain as operators. The only information we have about them comes from their commutator given in the postulates: $[\hat{x}, \hat{p}] = i\hbar$.

Knowing the Schrödinger solution, we could move over to Schrödinger's description. However, consider Eqs 8.11 and 8.12

$$\hat{a} = \left(\sqrt{\frac{k}{2\hbar\omega}}\hat{x} + i\frac{\hat{p}}{\sqrt{2m\hbar\omega}}\right) \tag{8.48}$$

and

$$\hat{a}^\dagger = \left(\sqrt{\frac{k}{2\hbar\omega}}\hat{x} - i\frac{\hat{p}}{\sqrt{2m\hbar\omega}}\right) \tag{8.49}$$

We can solve these equations for either $\hat{x}$ or $\hat{p}$ in terms of $\hat{a}$ and $\hat{a}^\dagger$:

$$\hat{x} = \sqrt{\frac{\hbar\omega}{2k}}\left(\hat{a} + \hat{a}^\dagger\right) \tag{8.50}$$

and

$$\hat{p} = -i\sqrt{\frac{m\hbar\omega}{2}}\left(\hat{a} - \hat{a}^\dagger\right) \tag{8.51}$$

Returning now to Eq. 8.46 we can insert 8.50 and use 8.34 and 8.35:

$$\langle\hat{x}\rangle(t) = \sqrt{\frac{\hbar\omega}{2k}}\sum_{nm} C_n^* C_m e^{i\omega_{nm}t}\langle E_n|\left(\hat{a} + \hat{a}^\dagger\right)|E_m\rangle \tag{8.52}$$

$$= \sqrt{\frac{\hbar\omega}{2k}}\sum_{nm} C_n^* C_m e^{i\omega_{nm}t}\left(\langle E_n|\hat{a}|E_m\rangle + \langle E_n|\hat{a}^\dagger|E_m\rangle\right)$$

$$= \sqrt{\frac{\hbar\omega}{2k}}\sum_{nm} C_n^* C_m e^{i\omega_{nm}t}\left(\sqrt{m}\langle E_n|E_{m-1}\rangle + \sqrt{m+1}\langle E_n|E_{m+1}\rangle\right)$$

$$= \sqrt{\frac{\hbar\omega}{2k}}\sum_{n=0}\sqrt{n+1}\left(C_n^* C_{n+1} e^{-i\omega t} + C_{n+1}^* C_n e^{i\omega t}\right) \tag{8.53}$$

A remarkable result for its simplicity. It shows that only adjacent states are coupled by x and that all such couplings oscillate at the same frequency, the natural frequency of the classical harmonic oscillator.

Problem 8.7 *Starting with Eq. 8.51, use the same process leading to 8.53 to find* $\langle\hat{p}(t)\rangle$.

This approach is so powerful that even when working with eigenfunctions for the harmonic oscillator, such as when you have an arbitrary wave function in space and you expand in harmonic oscillator states, you would naturally switch to this simple picture, converting your eigenfunctions to eigenvector notation and working with operators rather doing spatial integrals.

Finally, it is now possible to show that the spatial information contained in the coordinate representation of the operators can be obtained more easily. The question is how can this approach be related to what was done in Chapter 2? This lies in returning to Eq. 8.21, repeated here:

$$\hat{a}\,|E_0\rangle = 0 \tag{8.21}$$

and then using the definition for the lowering operator in functional form, namely

$$\hat{a} = \sqrt{\frac{k}{2\hbar\omega}}\hat{x} + i\frac{\hat{p}}{\sqrt{2m\hbar\omega}} = \sqrt{\frac{k}{2\hbar\omega}}x + \sqrt{\frac{\hbar}{2m\omega}}\frac{d}{dx} \tag{8.54}$$

Eq. 8.21 becomes

$$\left(\sqrt{\frac{k}{2\hbar\omega}}x + \sqrt{\frac{\hbar}{2m\omega}}\frac{d}{dx}\right)u_0(x) = 0 \tag{8.55}$$

Using Eqs 2.29 and 2.30 repeated here,

$$\ell = \left(\frac{\hbar^2}{m\kappa}\right)^{1/4} \tag{8.56}$$

and

$$\xi = \frac{x}{\ell} \tag{8.57}$$

we can rewrite Eq. 8.55 in the dimensionless coordinate used in Chapter 2 to get

$$\left(\xi + \frac{d}{d\xi}\right)u_0(\xi) = 0 \tag{8.58}$$

Integrating and normalizing we get

$$u_0(\xi) = \frac{1}{\pi^{1/4}}e^{-\frac{1}{2}\xi^2} \tag{8.59}$$

This is the normalized ground state eigenfunction of the nano-vibrator that we found in Chapter 2. Furthermore, it follows from the above that since $|E_m\rangle = C_m(\hat{a}^\dagger)^m|E_0\rangle$ from Eq. 8.39, we have

$$u_m(\xi) = C_m(\hat{a}^\dagger)^m u_0(\xi) = \sqrt{\frac{1}{2^m m!\sqrt{\pi}}}\left(\xi - \frac{d}{d\xi}\right)^m e^{-\frac{1}{2}\xi^2} \tag{8.60}$$

Where we rewrote the expression for $\hat{a}^\dagger$ as

$$\hat{a}^\dagger = \left(\sqrt{\frac{k}{2\hbar\omega}}\hat{x} - i\frac{\hat{p}}{\sqrt{2m\hbar\omega}}\right) = \frac{1}{\sqrt{2}}\left(\xi - \frac{d}{d\xi}\right) \tag{8.61}$$

Hence it becomes possible to find all the corresponding spatial solutions without solving a second order differential equation.

8.3 Matrix Representation of Operators and Eigenvectors in Quantum Mechanics

There was discussion of the use of matrices to represent operators in Chapter 7. It was convenient to facilitate learning and understanding, but as we have seen here, it is not necessary. In that spirit, we consider the **matrix representation** of these operators in the Heisenberg approach.

Using Eq. 8.15, we have

$$\widehat{H}_0 |n\rangle = E_n |n\rangle \tag{8.62}$$

where $\widehat{H}_0 = \hbar\omega\left(\hat{a}^\dagger\hat{a} + \frac{1}{2}\right)$, and we have replaced $|E_n\rangle$ with $|n\rangle$.

The subscript 0 on $\widehat{H}$ is common when $\widehat{H}$ is diagonal. The matrix form of the operator is then given by

$$\widehat{H}_0 = \sum_{n,m} \left\langle m|\widehat{H}_0|m\right\rangle |m\rangle\langle m| \tag{8.63}$$

Again, the set of energy eigenvectors, $\{|n\rangle\}$ are **unit vectors** of the form

$$|1\rangle = \begin{bmatrix} 1 \\ 0 \\ 0 \\ \vdots \\ 0 \\ 0 \\ \vdots \end{bmatrix}, |2\rangle = \begin{bmatrix} 0 \\ 1 \\ 0 \\ \vdots \\ 0 \\ 0 \\ \vdots \end{bmatrix}, \ldots |n\rangle = \begin{bmatrix} 0 \\ 1 \\ 0 \\ \vdots \\ 1_n \\ 0 \\ \vdots \end{bmatrix} \tag{8.64}$$

The matrix for the Hamiltonian is then

$$\widehat{H}_0 = \begin{bmatrix} \square & |E_1\rangle & |E_2\rangle & |E_3\rangle & |E_4\rangle & \cdots \\ \langle E_1| & E_1 & 0 & 0 & 0 & \cdots \\ \langle E_2| & 0 & E_2 & 0 & 0 & \cdots \\ \langle E_3| & 0 & 0 & E_3 & 0 & \cdots \\ \langle E_4| & 0 & 0 & 0 & E_4 & \cdots \\ \vdots & \vdots & \vdots & \vdots & \vdots & \ddots \end{bmatrix} \equiv \begin{bmatrix} E_1 & 0 & 0 & 0 & \cdots \\ 0 & E_2 & 0 & 0 & \cdots \\ 0 & 0 & E_3 & 0 & \cdots \\ 0 & 0 & 0 & E_4 & \cdots \\ \vdots & \vdots & \vdots & \vdots & \ddots \end{bmatrix} \tag{8.65}$$

In the first matrix, we have included the first row and column to show the ordering of the eigenstates. Normally, this is not included, as shown in the second matrix. We have used the fact that the matrix elements of $\widehat{H}_0$ are given by

$$\left\langle E_i|\widehat{H}_0|E_j\right\rangle = E_j\delta_{ij} \tag{8.66}$$

The lowering operator is not diagonal but can still be represented as a matrix using the energy basis,

$$\hat{a} = \sum_{n,m} \langle n|\hat{a}|m\rangle |n\rangle\langle m| = \begin{bmatrix} 0 & \sqrt{1} & 0 & 0 & \cdots \\ 0 & 0 & \sqrt{2} & 0 & \cdots \\ 0 & 0 & 0 & \sqrt{3} & \cdots \\ 0 & 0 & 0 & 0 & \cdots \\ \vdots & \vdots & \vdots & \vdots & \ddots \end{bmatrix} \tag{8.67}$$

where now we have used the result from above that

$$\langle n|\hat{a}|m\rangle = \sqrt{m}\delta_{nm-1} \tag{8.68}$$

Problem 8.8 *We recognize the definition of the lowering operator as being linearly related to the condition for the minimum uncertainty state given in Eq. 7.95. An important quantum state is the eigenvector of this operator.*

$$\hat{a}\,|\alpha\rangle = \alpha\,|\alpha\rangle \tag{8.69}$$

$|\alpha\rangle$ *is often called the coherent state. Expand* $|\alpha\rangle$ *in terms of the number states above* $|m\rangle$ *where* $|m\rangle \equiv |E_m\rangle$ *and* $E_m = \hbar\omega\left(m + \frac{1}{2}\right)$: i.e., *start with*

$$|\alpha\rangle = \sum_m c_m\,|m\rangle$$

Show that, by substituting this into the eigenvalue equation above, projecting onto a specific state, say $|n\rangle$, *and using orthonormality, you will get a relationship between* c_m *and* c_{m+1} *that allows you to figure out the form of the expansion coefficients in terms of the first coefficient,* c_0. *Then require the expression to be normalized to figure out* c_0.

$$|\alpha\rangle = e^{-\frac{\alpha^2}{2}} \sum_n \frac{\alpha^n}{\sqrt{n!}}\,|n\rangle \tag{8.70}$$

Hint: Recall from Appendix B

$$e^x = 1 + x + \frac{x^2}{2!} + \ldots$$

8.4 The Quantum LC Circuit

The above approach completely eliminated the need of solving a differential equation in space, working only with the operators. We can use this approach to solve a completely different problem, not associated with real space: the quantum LC circuit. This formalism forms the basis for superconducting qubits and the field of **circuit quantum electrodynamics** (cQED)

We begin by first showing that Hamilton's approach can be used to solve the classical problem yielding the same result as Kirchhoff's rules. Kirchhoff's law for circuits says that the sum of the voltages around any loop must be zero. Since $\mathrm{v}(t) = L\frac{di(t)}{dt}$ for inductor (L is the inductance) and $\mathrm{v}(t) = \frac{1}{C}\int_0^t i\,dt$ for the capacitor (C is the capacitance), the circuit equation for Fig. 8.1 becomes

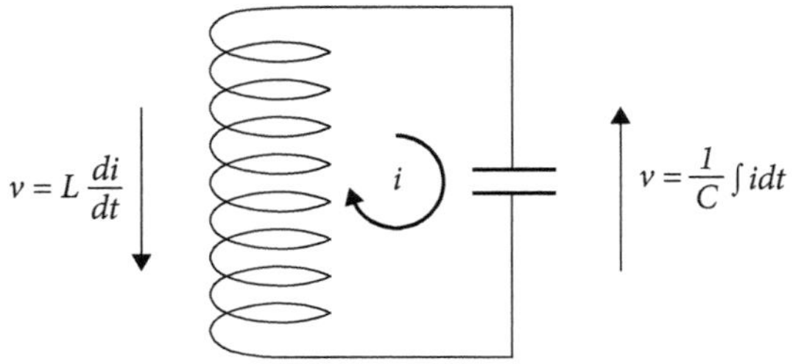

Fig. 8.1 LC circuit for the voltage and the current.

$$L\frac{di(t)}{dt} + \frac{1}{C}\int_0^t idt = 0 \tag{8.71}$$

or by taking the time derivative in Eq. 8.71

$$\frac{d^2i(t)}{dt^2} + \frac{1}{LC}i(t) = 0 \tag{8.72}$$

with the resonant frequency being given by

$$\omega_0^2 = \frac{1}{LC} \tag{8.73}$$

The solution is given by

$$i(t) = Acos\omega_0 t + Bsin\omega_0 t \tag{8.74}$$

where A and B are determined by the initial conditions. The charge on the capacitor is also an observable. The charge, q, is now time dependent, $q(t)$, and a dynamical variable (same as $i(t)$ and $v(t)$). Since $i = \frac{dq}{dt}$, the integral of $i(t)$ is $q(t)$, the charge on the capacitor given by

$$q(t) = A'sin\omega_0 t + B'cos\omega_0 t \tag{8.75}$$

As usual, the classical result gives the observable as a function of time.

It is possible to approach this whole problem using Hamilton's formulation. The remarkable result of Hamilton's approach is that it applies to problems other than mechanics of massive particles. To do this, we have to know the Hamiltonian. As usual, it is necessary to guess the Hamiltonian, which is the total energy expressed in terms of the canonical coordinates and their conjugate momenta.

To get the total energy needed to begin Hamilton's approach, you know from introductory physics texts that the energy in an inductor is

$$E_{ind} = \frac{1}{2}Li^2 = \frac{1}{2}L\dot{q}^2 \tag{8.76}$$

and in a capacitor, it is

$$E_{cap} = \frac{1}{2}Cv^2 \tag{8.77}$$

and since $C \equiv \frac{q}{v}$

$$E_{cap} = \frac{1}{2}Cv^2 = \frac{1}{2C}q^2 \tag{8.78}$$

The total energy E is now given by

$$E = \frac{1}{2}L\dot{q}^2 + \frac{1}{2C}q^2 \tag{8.79}$$

If q is taken to be a canonical coordinate, we note that Eq. 8.76 looks like a kinetic energy, similar to the nano-vibrator, while Eq. 8.78 behaves as a potential energy. Making this assumption, the Lagrangian from Chapter 7 would be $L\left(q, \dot{q}, t\right) = T - V = \frac{1}{2}L\dot{q}^2 - \frac{1}{2C}q^2$. Then the conjugate momentum would $p = \frac{\partial L}{\partial \dot{q}} = L\dot{q}$. This would allow us to modify Eq. 8.79 to get the Hamiltonian:

$$H = \frac{p^2}{2L} + \frac{1}{2C}q^2 \tag{8.80}$$

The proof that you have the right Hamiltonian for the description of the LC circuit is that, using Hamilton's equation, we get what we measure in the laboratory. Rather than reverting to the lab, Kirchhoff's law has already been validated, so it is simply necessary to show that Hamilton's equations give us the same result as Kirchhoff's laws.

Problem 8.9 *Use Hamilton's equations (Eqs 7.73 and 7.74) and show that you can recover the same equation of motion for q, as we did using Kirchhoff's law. We are working with q because it was easy to transform to a Hamiltonian form.*

The work here will be important first in the study of the quantum LC circuit and then in the discovery of the photon.

It is certainly possible and reasonable to substitute $-i\hbar\frac{d}{dq}$ for $\hat{p}$ and q for $\hat{q}$. You could solve the differential equation and recover the eigenfunctions that we developed before. However, the function in that picture has no physical meaning. Hence, it is more straightforward to use the Heisenberg approach. We will simply reproduce the results above but now in the context of the quantum LC circuit. Specifically, we again begin with the time independent Schrödinger equation and solve for the eigenvalues and eigenvectors,

$$\widehat{H}\left|\psi\right\rangle = E\left|\psi\right\rangle$$

and again rewrite the Hamiltonian operator:

$$\begin{aligned}\widehat{H} &= \left(\frac{\hat{p}^2}{2L} + \frac{1}{2C}\hat{q}^2\right) = \hbar\omega\left(\frac{\hat{p}^2}{2L\hbar\omega} + \frac{1}{2C\hbar\omega}\hat{q}^2\right) \\ &= \hbar\omega\left(\sqrt{\frac{1}{2C\hbar\omega}}\hat{q} + i\frac{\hat{p}}{\sqrt{2L\hbar\omega}}\right)\left(\sqrt{\frac{1}{2C\hbar\omega}}\hat{q} - i\frac{\hat{p}}{\sqrt{2L\hbar\omega}}\right) - \frac{\hbar\omega}{2} = \hbar\omega\left(\hat{a}\hat{a}^\dagger - \frac{1}{2}\right)\end{aligned} \tag{8.81}$$

where

$$\hat{a} = \left(\sqrt{\frac{1}{2C\hbar\omega}}\hat{q} + i\frac{\hat{p}}{\sqrt{2L\hbar\omega}}\right) \tag{8.82}$$

and

$$\hat{a}^{\dagger} = \left(\sqrt{\frac{1}{2C\hbar\omega}}\hat{q} - i\frac{\hat{p}}{\sqrt{2L\hbar\omega}}\right). \tag{8.83}$$

and

$$\omega = \frac{1}{\sqrt{LC}} \tag{8.84}$$

All of the operator algebra is repeated again and we have that the states are quantized and designated by $|E_n\rangle$ with

$$E_n = \hbar\omega\left(n + \frac{1}{2}\right); n = 0, 1, 2, ... \tag{8.85}$$

and

$$\widehat{H} = \hbar\omega\left(\hat{a}^{\dagger}\hat{a} + \frac{1}{2}\right) \tag{8.86}$$

Problem 8.10 *Show how the voltage across the capacitor varies with increasing energy for a classical and a quantum LC circuit. Make a plot of the maximum voltage on the capacitor as a function of the energy. Note the classical graph will be a continuous line and the quantum graph will be discrete points.*

Problem 8.11 *For the usual arbitrary state $|\psi\rangle$ expand in the basis states and use the results in Eqs 8.4–8.8 to find $|\psi\rangle(t)\rangle$ and then find $\langle i\rangle(t)$.*

Problem 8.12 *In the classical circuit in the absence of a sinewave generator or initially charged capacitor, the current and voltage are zero, nothing is happening. However, the quantum circuit is different. Take the system to be initially in the ground state $|E_0\rangle$. Experimentally, we will take the approach of measuring the current. To do this, we would prepare the system in $|E_0\rangle$ and measure i and i^2. Then reinitialize and repeat.*

a) Find the variance of the current, which is given by

$$(\Delta i)^2 \equiv \left\langle\psi|(\hat{\imath} - \langle\hat{\imath}\rangle)^2|\psi\right\rangle = \left\langle\psi|\hat{\imath}^2|\psi\right\rangle - \langle\psi|\hat{\imath}|\psi\rangle^2 \tag{8.87}$$

Find $\hat{\imath}$ in terms of the $\hat{a}$ and $\hat{a}^{\dagger}$; remarkably, this is not zero! This is related to the Heisenberg Uncertainty relation that states for a canonical coordinate and its conjugate momentum:

$$(\Delta q)^2(\Delta p)^2 \geq \frac{\hbar^2}{4} \tag{8.88}$$

b) What does this mean for your measurements even when not "in use" (i.e., the system is in the ground state)? Specifically, what do you get in your instruments if you are monitoring the variance in these parameters. What would you expect in an ideal classical system (perfect components). The answer to this is mostly with words, not math.

In superconducting quantum computing, a nonlinearity in the circuit causes a deviation in the energy separation between the states from harmonic (equal energy separation) to anharmonic (not equal energy separation). By exploiting the concept of resonance (Chapter 9), it is possible to work with just the first two levels as we discuss in more detail in Chapter 9.

8.5 Summary

This chapter explored how to solve an operator equation without using an explicit representation for the canonical coordinate and the conjugate momentum. This approach will get used again to analyze rotation (angular momentum) and the intrinsic spin of the electron where a coordinate representation for the angular momentum operators does not exist. The operator approach is central for development of the quantized electromagnetic field, as well as providing a powerful approach to studying dynamics in two-state systems.

Based on the Hamiltonian for the nano-vibrator

$$\hat{H} = \frac{\hat{p}^2}{2m} + \frac{1}{2}k\hat{x}^2$$

new operators were introduced

$$\hat{a} = \left(\sqrt{\frac{k}{2\hbar\omega}}\hat{x} + i\frac{\hat{p}}{\sqrt{2m\hbar\omega}}\right)$$

and

$$\hat{a}^\dagger = \left(\sqrt{\frac{k}{2\hbar\omega}}\hat{x} - i\frac{\hat{p}}{\sqrt{2m\hbar\omega}}\right).$$

where using $[\hat{x}, \hat{p}] = i\hbar$, we found

$$\hat{H} = \hbar\omega\left(\hat{a}^\dagger\hat{a} + \frac{1}{2}\right)$$

with eigenvectors $|E_n\rangle$ where

$$\hat{H}\,|E_n\rangle = \hbar\omega\left(n + \frac{1}{2}\right)|E_n\rangle$$

Furthermore, through the problems it was found that:

$$\hat{a}\,|E_m\rangle = \sqrt{m}\,|E_m - \hbar\omega\rangle$$
$$\hat{a}^\dagger|E_m\rangle = \sqrt{m+1}\,|E_m + \hbar\omega\rangle$$

and

$$|E_m\rangle = \sqrt{\frac{1}{m!}}\left(\hat{a}^{\dagger}\right)^m |E_0\rangle$$

Using this approach, the analysis recovers the eigenenergies of the problem and enables evaluation of all the quantities identified in the postulates, including the expectation values of $\hat{x}$and $\hat{p}$ without use of the eigenfunctions. This was done by solving for operators in terms of the raising and lowering operators:

$$\hat{x} = \sqrt{\frac{\hbar\omega}{2k}}\left(\hat{a} + \hat{a}^{\dagger}\right)$$

and

$$\hat{p} = -i\sqrt{\frac{m\hbar\omega}{2}}\left(\hat{a} - \hat{a}^{\dagger}\right)$$

Vocabulary (page) and Important Concepts

- Heisenberg approach 118
- operator approach 118
- Hermitian 120
- adjoint 121
- lowering operator 124
- raising operator 124
- number operator 124
- occupation number 125
- matrix representation 127
- unit vectors 128
- quantum LC circuit 129
- circuit quantum electrodynamics 129

9 Quantum Dynamics: Resonance and a Quantum Flip-Flop

9.1 Introduction

The most exciting aspect of quantum mechanics that is relevant to modern technology is the dynamic aspect, i.e., how the system evolves in time. Systems that can be in two states, such as on and off, or state 1 and state 2, are also highly nonlinear in the driving term. We will explore the case of time evolution of the isolated system, such as was introduced in Chapter 2 and then examine the behavior of a system coupled to a classical driving source.

9.2 Time Evolution Operator

We have discussed the time evolution as simply an additional detail in solving the equation of motion. When the Hamiltonian is time independent, it has eigenfunctions or eigenvectors that are called **stationary state solutions** because they are time independent. The wave function or state vector is still generally time dependent even when the Hamiltonian is time independent. The time dependence comes through the multiplicative phase factor, which becomes important when the initial state of the system is described by a superposition of two or more states. For example, when we add two states together and look at the spatial probability density of finding a particle in dx about x, the time dependence of the phase of each state is different, and leads to a time dependence in the location of the constructive and destructive interference of the spatial probability density that is observable in the laboratory. Such behavior was discussed in Chapter 2, Fig. 2.3.

We consider now the problem of time dependence more closely because Heisenberg made an additional major contribution to parallel Schrödinger's description (Schrödinger's approach with time independent operators is known as the **Schrödinger picture**).

We begin by returning to the equation of motion with Schrödinger's equation. We work in Dirac notation for convenience, but everything carries over if you are working with real-space functions. We start again with Schrödinger's equation:

$$\widehat{H}\,|\psi(t)\rangle = i\hbar\frac{d\,|\psi(t)\rangle}{dt} \tag{9.1}$$

Since Schrödinger's equation is first order in time, then if $|\psi(t_0)\rangle$ is known, $|\psi(t)\rangle$ is known for all time through a **time evolution operator**, $\widehat{U}(t, t_0)$, such that

$$|\psi(t)\rangle = \widehat{U}(t, t_0)\,|\psi(t_0)\rangle \tag{9.2}$$

Introduction to Quantum Nanotechnology: A Problem Focused Approach. Duncan Steel, Oxford University Press (2021).
 DOI: 10.1093/oso/9780192895073.003.0009

and

$$\langle \psi(t)| = \langle \psi(t_0)|\, \widehat{U}^{\dagger}(t, t_0) \tag{9.3}$$

Given how we approached dynamics in Chapter 2, it would seem that this is an unnecessary formality. However, this discussion will show that it enables a parallel formalism that has additional power in doing calculations. We show below that Eq. 9.2 is true by showing that $|\psi(t)\rangle = \widehat{U}(t, t_0)\, |\psi(t_0)\rangle$ is a solution to Schrödinger's equation if $\widehat{U}(t, t_0)$ is also a solution.

We have the requirement that $|\psi(t)\rangle$ is normalized for all time, hence

$$\langle \psi(t)|\psi(t)\rangle = 1 \tag{9.4}$$

and therefore we require

$$\langle \psi(t_0)|\, \widehat{U}^{\dagger}(t, t_0)\, \widehat{U}(t, t_0)\, |\psi(t_0)\rangle = 1 \tag{9.5}$$

This must be true for any $|\psi(\mathrm{t}_0)\rangle$. Therefore

$$\widehat{U}^{\dagger}(t, t_0)\, \widehat{U}(t, t_0) = 1 \tag{9.6}$$

We then see that a consequence is

$$\widehat{U}^{\dagger}(t, t_0) = \widehat{U}^{-1}(t, t_0) \tag{9.7}$$

This is the definition of a **unitary operator**. The meaning of unitary is seen in the expression

$$\langle \psi(t)|\psi(t)\rangle = \langle \psi(t_0)|\, \widehat{U}^{\dagger}(t, t_0)\, \widehat{U}(t, t_0)\, |\psi(t_0)\rangle = \langle \psi(t_0)\, |\psi(t_0)\rangle \tag{9.8}$$

This means that $\widehat{U}(t, t_0)\ \ |\psi(t_0)\rangle$ does not change the magnitude of the length of the original vector $|\psi(t_0)\rangle$, given in Eq. 9.4.

To find an equation of motion for $U(t, t_0)$, we substitute Eq. 9.2 into Eq. 9.1:

$$i\hbar\frac{\partial}{\partial t}\widehat{U}(t, t_0)\, |\psi(t_o)\rangle = \widehat{H}(t)\widehat{U}(t, t_0)\, |\psi(t_o)\rangle \tag{9.9}$$

Since this is true for any $|\psi(t_0)\rangle$, it follows that

$$i\hbar\frac{\partial}{\partial t}\widehat{U}(t, t_0) = \widehat{H}(t)\widehat{U}(t, t_0) \tag{9.10}$$

and also

$$-i\hbar\frac{\partial}{\partial t}\widehat{U}^{\dagger}(t, t_0) = \widehat{U}^{\dagger}(t, t_0)\, \widehat{H}(t) \tag{9.11}$$

If $\widehat{H}(t) = \widehat{H}$, i.e., the Hamiltonian is time independent, then Eq. 9.10 may be integrated to give

$$\widehat{U}(t, t_0) = e^{-i\frac{\widehat{H}(t-t_0)}{\hbar}} \tag{9.12}$$

where $\widehat{U}(t, t_0)$ must satisfy the initial condition

$$\widehat{U}(t = t_0, t_0) = 1 \quad (9.13)$$

In Eq. 9.12 we have $e^{-i\frac{\widehat{H}(t-t_0)}{\hbar}}$ which is a *function of an operator*. Functions of operators are defined by the power series (see Appendix B) for that function. So

$$e^{-i\frac{\widehat{H}(t-t_0)}{\hbar}} = 1 + \left(\frac{-i\widehat{H}(t-t_0)}{\hbar}\right) + \frac{1}{2!}\left(\frac{-i\widehat{H}(t-t_0)}{\hbar}\right)^2 + \frac{1}{3!}\left(\frac{-i\widehat{H}(t-t_0)}{\hbar}\right)^3 + \cdots \quad (9.14)$$

If $\widehat{H}\ |E_n\rangle = E_n\ |E_n\rangle$ then

$$\begin{aligned}\widehat{U}(t, t_0)\ |E_n\rangle &= e^{-i\frac{\widehat{H}(t-t_0)}{\hbar}}\ |E_n\rangle \\ &= \left[1 + \left(\frac{-i\widehat{H}(t-t_0)}{\hbar}\right) + \frac{1}{2!}\left(\frac{-i\widehat{H}(t-t_0)}{\hbar}\right)^2 + \frac{1}{3!}\left(\frac{-i\widehat{H}(t-t_0)}{\hbar}\right)^3 + \cdots\right] |E_n\rangle \\ &= \left[1 + \left(\frac{-iE_n(t-t_0)}{\hbar}\right) + \frac{1}{2!}\left(\frac{-iE_n(t-t_0)}{\hbar}\right)^2 + \frac{1}{3!}\left(\frac{-iE_n(t-t_0)}{\hbar}\right)^3 + \cdots\right] |E_n\rangle\end{aligned} \quad (9.15)$$

After collapsing the power series, we get for Eq. 9.15:

$$\widehat{U}(t, t_0)\ |E_n\rangle = e^{-i\frac{E_n(t-t_0)}{\hbar}}\ |E_n\rangle \quad (9.16)$$

Recalling now that $|\psi(t = 0)\rangle$ can be expanded in terms of the eigenfunctions of $\widehat{H}$:

$$|\psi(t = 0)\rangle = \sum_n C_n\ |E_n\rangle \quad (9.17)$$

we have then that (with $t_0 = 0$)

$$|\psi(t)\rangle = \widehat{U}(t)\sum_n C_n\ |E_n\rangle = e^{-i\frac{\widehat{H}t}{\hbar}}\sum_n C_n\ |E_n\rangle = \sum_n C_n e^{-i\frac{\widehat{H}t}{\hbar}}\ |E_n\rangle = \sum_n C_n e^{-i\frac{E_n t}{\hbar}}\ |E_n\rangle \quad (9.18)$$

Similarly, for wave functions

$$\psi(r, t) = \widehat{U}(t)\sum_n C_n\psi_n(r) = e^{-i\frac{\widehat{H}t}{\hbar}}\sum_n C_n\psi_n(r) = \sum_n C_n e^{-i\frac{\widehat{H}t}{\hbar}}\psi_n(r) = \sum_n C_n e^{-i\frac{E_n t}{\hbar}}\psi_n(r) \quad (9.19)$$

Returning to Dirac notation, we have for the expectation value of an observable

$$\langle\hat{A}\rangle(t) = \langle\psi(t)|\hat{A}|\psi(t)\rangle = \langle\psi(t = 0)|\widehat{U}^{\dagger}(t)\hat{A}\widehat{U}(t)|\psi(t = 0)\rangle \quad (9.20)$$

where again $\widehat{U}^{\dagger}(t)$ is the adjoint (complex transpose).

In Chapter 2, we looked at the dynamics of a **coherent superposition** state for a nanovibrator. Here, we repeat this problem for a coherent superposition state of the two lowest energy eigenstates of the infinite box potential, using the time evolution operator. The wave function at $t = 0$ is given by

$$\psi(x, t = 0) = \frac{1}{\sqrt{L}}(\cos k_0 x + \sin k_1 x) \tag{9.21}$$

where we have assumed the probability amplitudes of each state are equal. To find the time dependence, we use the Hamiltonian for the infinite box potential and apply the time evolution operator, giving

$$\psi(x, t) = \hat{U}(t)\psi(x, t = 0) = \frac{1}{\sqrt{L}}\left(e^{-\frac{iE_0 t}{\hbar}} \cos k_0 x + e^{-\frac{iE_1 t}{\hbar}} \sin k_1 x\right) \tag{9.22}$$

We can factor out the first time evolution factor, and then, *unless you are using quantum interference effects with another wave function, the overall phase of the wave function does not matter*, it can be simply removed, since when you take $|\psi(x, t)|^2$, it vanishes. Hence, we get

$$\psi(x, t) = \frac{1}{\sqrt{L}}\left(\cos k_0 x + e^{-\frac{i(E_1 - E_0)t}{\hbar}} \sin k_1 x\right) \tag{9.23}$$

$$\psi(x, t) = \frac{1}{\sqrt{L}}\left(\cos k_0 x + e^{-i\omega_0 t} \sin k_1 x\right) \tag{9.24}$$

where from Eqs 3.41 and 3.43 we have

$$k_0 = \frac{\pi}{L} \tag{9.25}$$

$$k_1 = \frac{2\pi}{L} \tag{9.26}$$

The dipole moment, P, is given classically by

$$P = qx \tag{9.27}$$

So, if the particle is an electron ($q = -e$), we see that quantum mechanically

$$P \to \hat{P} = -e\hat{x} = -ex \tag{9.28}$$

and

$$\langle \hat{P} \rangle (t) = -e \langle \hat{x} \rangle (t) = -e \langle x \rangle (t) = -e \langle \psi(t)|x|\psi(t) \rangle = -e \int_{-\frac{L}{2}}^{\frac{L}{2}} dx\, \psi^* (x, t)\, x \psi (x, t) \tag{9.29}$$

Again the time dependence is shown outside the symbol for the expectation value to avoid giving the impression that either $\hat{x}$ or x have become time dependent.

There is a time dependence in the oscillation of the polarization (polarization reflects a displacement of the charge) in space where the charge is periodically localized on the left and then on the right. Such an oscillation results in radiation when it is inserted into Maxwell's equations. The nano-vibrator has the same result in Chapter 2, except that the solutions to the time independent Schrödinger equation lead to different forms for the spatial part of the wave functions.

Problem 9.1 *Find $\langle \hat{P} \rangle (t)$ from Eq. 9.29 using Eq. 9.24 for the wave function. Recall that for this problem the cosine function is even and sine is odd. Clearly x is odd. The product of an even function and an even function is even, the product of an odd function and an odd function is an even function, but the product of an even function and an odd function is odd. Integration over the symmetric space of an odd function is zero. Also, note that $\langle \hat{P} \rangle (t)$ can be measured, and so the answer must be entirely a real function. For perspective, recall that in Maxwell's equations the polarization is a source term for electromagnetic fields. This system is a radiator. For a 10 nm box and an electron, determine the frequency in hertz for the above state, given in Eq. 9.24.*

Problem 9.2 *At t = 0, a nano-vibrator is described by the wave function:*

$$\psi(\xi, t=0) = \frac{2}{\sqrt{3\sqrt{\pi}}}\xi^2 e^{-\frac{\xi^2}{2}}$$

This is not an eigenstate of the oscillator. Use the time evolution operator to find $\psi(\xi, t)$. Hint: use the table of normalized nano-vibrator eigenfunctions in Chapter 2 *to rewrite $\psi(\xi, t=0)$ as a linear combination of eigenfunctions. You will not need to do projection integrals to get the explanation coefficients. They will emerge naturally in your calculation.*

The discussion up to this point has used the Schrödinger operators for observables as defined in the postulates, and the time evolution of the system develops using the time evolution operator. This is traditionally called the **Schrödinger picture**. Below, we show how Heisenberg came up with an equivalent approach that sometimes can be much more powerful and easy to use than the Schrödinger picture. This is the **Heisenberg picture**.

9.3 Heisenberg Picture of Dynamics

Heisenberg looked at the result of Eq. 9.20 and defined a new operator, $\hat{A}_H(t)$, where

$$\hat{A}_H(t) = \hat{U}^\dagger(t)\hat{A}_S\hat{U}(t) \tag{9.30}$$

Here, $\hat{A}_S$ has the subscript S to indicate that this operator is the usual time independent Schrödinger operator. We note that, unlike the time independent Schrödinger operator, the **Heisenberg operator** is *time dependent* and its form depends on the Hamiltonian. Eq. 9.20 becomes

$$\langle \hat{A} \rangle (t) = \langle \psi(t=0) | \hat{A}_H | \psi(t=0) \rangle \tag{9.31}$$

If we knew how to evaluate Eq. 9.30, this would improve our situation in terms of calculations because knowing the form of $\hat{A}_{\mathrm{H}}$ for a specific Hamiltonian we would be able to find $\langle \hat{A} \rangle (t)$ knowing only the initial value of the state vector. Furthermore, once $\hat{A}_{\mathrm{H}}$ is determined for a specific Hamiltonian, it does not have to be evaluated again whereas $\hat{A}_S$ does not depend on the form of the Hamiltonian. It is often the case that the critical physics is more easily seen by just evaluating $\hat{A}_H(t)$ rather than having to evaluate $\langle \hat{A}(t) \rangle$, thus saving another step in the analysis. Finally, we note that by the definition of $\hat{A}_H(t)$ in Eq. 9.30, the initial condition for $\hat{A}_H\,(t=0)$ is

$$\hat{A}_H\,(t=0) = \hat{A}_S \tag{9.32}$$

The importance of this operator will emerge with problems of increasing complexity. It is common in the teaching of quantum to expand 9.30 in a Taylor series and represent the result in terms of commutators involving $\widehat{H}$ and $\hat{A}_S$.[1] This is very effective, but there is a way that is often more useful. To see this, we develop a **Heisenberg equation of motion** for $\hat{A}_H(t)$. We start with the time derivative of Eq. 9.30:

$$\frac{d}{dt}\hat{A}_H(t) = \frac{d}{dt}\left(\widehat{U}^\dagger(t)\hat{A}_S\widehat{U}(t)\right) = \left(\frac{d}{dt}\widehat{U}^\dagger(t)\right)\hat{A}_S\widehat{U}(t) + \widehat{U}^\dagger(t)\hat{A}_S\left(\frac{d}{dt}\widehat{U}(t)\right) \tag{9.33}$$

In the unusual case of the Schrödinger operator having an explicit time dependence, then the corresponding term in the chain rule must be included. Using Eqs 9.10 and 9.11, we get

$$\begin{aligned}\frac{d}{dt}\hat{A}_H(t) &= \frac{i}{\hbar}\widehat{U}^\dagger(t)\widehat{H}(t)\hat{A}_S\widehat{U}(t) - \frac{i}{\hbar}\widehat{U}^\dagger(t)\hat{A}_S\widehat{H}(t)\widehat{U}(t) = \frac{i}{\hbar}\widehat{U}^\dagger(t)\widehat{H}(t)\widehat{U}(t)\widehat{U}^\dagger(t)\hat{A}_S\widehat{U}(t)\\ &\quad - \frac{i}{\hbar}\widehat{U}^\dagger(t)\hat{A}_S\widehat{U}(t)\widehat{U}^\dagger(t)\widehat{H}(t)\widehat{U}(t) = \frac{i}{\hbar}\left[\widehat{H}_H(t), \hat{A}_H(t)\right]\end{aligned} \tag{9.34}$$

or in the usual form,

$$-i\hbar\frac{d}{dt}\hat{A}_H(t) = \left[\widehat{H}_H(t), \hat{A}_H\right] \tag{9.35}$$

This is the Heisenberg equation of motion subject to the initial condition from Eq. 9.32: $\hat{A}_H\,(t=0) = \hat{A}_S$.

In the Schrödinger picture, the behavior of the operators is determined exclusively by the commutation relations for the canonical coordinates and conjugate momenta. However, in the Heisenberg picture the form of the operators also depends on the Hamiltonian for the problem. Hence while the Schrödinger operators are the same for every problem, the Heisenberg operators will depend on the Hamiltonian for the problem.

[1] Evaluation of Eq. 9.30 can be done for a time independent Hamiltonian by expanding in a power series:

$$\begin{aligned}\hat{A}_H(t) &= \widehat{U}^\dagger(t)\hat{A}_S\widehat{U}(t) = \hat{A}_H(t) = e^{+i\frac{\widehat{H}t}{\hbar}}\hat{A}_S e^{-i\frac{\widehat{H}t}{\hbar}} = \hat{A}_S + \left(\frac{it}{\hbar}\right)\left[\widehat{H}, \hat{A}_S\right] + \frac{1}{2!}\left(\frac{it}{\hbar}\right)^2\left[\widehat{H}, \left[\widehat{H}, \hat{A}_S\right]\right]\\ &\quad + \frac{1}{3!}\left(\frac{it}{\hbar}\right)^3\left[\widehat{H}, \left[\widehat{H}, \left[\widehat{H}, \hat{A}_S\right]\right]\right] + \ldots\end{aligned}$$

We use the nano-vibrator as an example. In general, n*either the form of the Hamiltonian nor the form of the commutator relations at the same time changes in the Heisenberg picture*. Hence:

$$\widehat{H} = \frac{\hat{p}_H^2}{2m} + \frac{1}{2}k\hat{x}_H^2 \tag{9.36}$$

$$[\hat{x}_H, \hat{p}_H] = i\hbar \tag{9.37}$$

The commutator relations will not hold if the operators are at different times. The two operators of interest then are for x and p. Using 9.35 with 9.36 and 9.37, we find

$$\begin{aligned} -i\hbar\frac{d}{dt}\hat{x}_H(t) &= \left[\widehat{H}, \hat{x}_H\right] = \frac{1}{2m}\left[\hat{p}_H^2, \hat{x}_H\right] = \frac{1}{2m}\left(\hat{p}_H^2\hat{x}_H - \hat{x}_H\hat{p}_H^2\right) \\ &= \frac{1}{2m}\left(\hat{p}_H\hat{p}_H\hat{x}_H - \hat{p}_H\hat{x}_H\hat{p}_H + \hat{p}_H\hat{x}_H\hat{p}_H - \hat{x}_H\hat{p}_H\hat{p}_H\right) \\ &= \frac{1}{2m}\hat{p}_H\left(\hat{p}_H\hat{x}_H - \hat{x}_H\hat{p}_H\right) + \frac{1}{2m}\left(\hat{p}_H\hat{x}_H - \hat{x}_H\hat{p}_H\right)\hat{p}_H \\ &= \frac{1}{2m}\hat{p}_H\left[\hat{p}_H, \hat{x}_H\right] + \frac{1}{2m}\left[\hat{p}_H\hat{x}_H\right]\hat{p}_H = -\frac{i\hbar}{m}\hat{p}_H \end{aligned} \tag{9.38}$$

that is

$$\frac{d}{dt}\hat{x}_H = \frac{1}{m}\,\hat{p}_H \tag{9.39}$$

So, the equation of motion for $\hat{x}_H(t)$ depends on $\hat{p}_H$. Hence, we need the equation of motion for $\hat{p}_H(t)$.

$$\begin{aligned} -i\hbar\frac{d}{dt}\hat{p}_H(t) &= \left[\widehat{H}, \hat{p}_H\right] = \frac{k}{2}\left[\hat{x}_H^2, p_H\right] = \frac{k}{2}\left(\hat{x}_H^2\hat{p}_H - \hat{p}_H\hat{x}_H^2\right) \\ &= \frac{k}{2}\,\hat{x}_H\left[\,x_H, p_H\right] + \frac{k}{2}\left[\,\hat{x}_H p_H\,\right]\,\hat{x}_H = i\hbar k\,\hat{x}_H \end{aligned} \tag{9.40}$$

that is

$$\frac{d}{dt}\hat{p}_H(t) = -k\hat{x}_H(t) \tag{9.41}$$

These equations are easy to solve, but already the result is remarkable. If you thought of the Heisenberg operators as classical observables, these would be the exact same equations you would get classically. Indeed, Eq. 9.39 resembles $v = \dot{x} = \frac{p}{m}$ and Eq. 9.41 resembles Hooke's law for the harmonic oscillator (recall classically $F \equiv \frac{dp}{dt} = -kx$). To solve Eq. 9.39, take the second derivative and substitute 9.41 on the right-hand side:

$$\frac{d^2}{dt^2}\hat{x}_H(t) = -\frac{k}{m}\hat{x}_H(t) = -\omega_0^2\hat{x}_H(t) \tag{9.42}$$

$$\hat{x}_H(t) = A\cos\omega_0 t + B\sin\omega_0 t \tag{9.43}$$

Using Eq. 9.39 to get $\hat{p}_H$ we find

$$\hat{p}_H = -Am\omega_0 \sin \omega_0 t + Bm\omega_0 \cos \omega_0 t \tag{9.44}$$

So, we have solved the coupled equations of motion, but now we have to evaluate A and B from initial conditions. The initial condition is given at t = 0 in Eq. 9.32 as $\hat{A}_H(t = 0) = \hat{A}_S$. Hence, we find A and B in Eqs 9.43 and 9.44 giving,

$$\hat{x}_H(t) = \hat{x}_S \cos \omega_0 t + \frac{\hat{p}_S}{m\omega_0} \sin \omega_0 t \tag{9.45}$$

$$\hat{p}_H(t) = -m\omega_0 \hat{x}_S \omega_0 t + \hat{p}_S \cos \omega_0 t \tag{9.46}$$

Unlike in the Schrödinger picture, the Heisenberg picture involves two coupled first order differential equations. The solution requires the result to satisfy the initial condition for each operator which is that the Heisenberg operator must reduce to the Schrödinger operator at time $t = 0$.

One of the important features of the Heisenberg formulation is that many problems which can be very difficult in the Schrödinger picture become more doable in the Heisenberg picture. An example is the driven harmonic oscillator with the Hamiltonian given by

$$\hat{H} = \frac{\hat{p}^2}{2m} + \frac{1}{2}k\hat{x}^2 + \alpha\hat{x} \sin \upsilon t \tag{9.47}$$

In the Schrödinger picture this takes many pages of algebra (see Merzbacher), but it takes just a few lines in the Heisenberg picture.

Problem 9.3 *Find the Heisenberg operators for x and p for the driven harmonic oscillator Hamiltonian described by Eq. 9.47. Show that the operators increase in magnitude linearly with the driving field. Because of the linear response, it is not possible to switch a nano-vibrator with a driving field, unlike say states in an atom or electronic states in a dot. This is why the quantum flip-flop must be built with another kind of quantum structure, such as a dot. Nevertheless, the nano-vibrator linearity can be an asset; it is a remarkable result and only occurs for a harmonic oscillator type Hamiltonian. The nano-vibrator, for example, could be a quantum sensor when the item to be sensed (e.g., toxic gas, proximity to a surface for lithography, etc.) causes it to be nonlinear since such deviations can be detected with modern techniques.*

Many problems of importance to the work of nano-technology are described by the same Hamiltonian as for the nano-vibrator. The preferred form is usually with the raising and lowering operators. Amazingly the Hamiltonian for the electromagnetic field is the nano-vibrator Hamiltonian (Chapter 15) and the raising and lowering operators are renamed the annihilation and creation operators.

In the Heisenberg form,

$$\hat{a}_H(t) = \hat{a}_S e^{-i\omega_0 t} \tag{9.48}$$

$$\hat{a}^\dagger_H(t) = \hat{a}^\dagger_S e^{i\omega_0 t} \tag{9.49}$$

Problem 9.4 *Find the Heisenberg form for the raising and lowering operators (Eqs 8.11 and 8.12) for the nano-oscillator (not independent, i.e., no driving field) using the Heisenberg equation of motion. Begin by finding the Heisenberg equations of motion for the raising and lowering operators and by then solving them.*

Problem 9.5 *Assume a particle is described at* $t = 0$ *by a localized wave packet of the form* $\psi(x, t = 0) = Ae^{-\frac{x^2}{2\ell^2}}$*. The particle has been localized inside a harmonic potential. Since the particle is localized symmetrically around the origin, it is obvious that the expectation value for* $\langle x(t) \rangle = 0$*. It is then relatively straightforward to calculate the variance,* $\Delta x^2(t)$*. Use the Heisenberg operators to do this.*

Problem 9.6 *Use the Hamiltonian for a free particle and find the Heisenberg representation for the operators* $\hat{x}$ *and* $\hat{p}$*. Using Hamilton's equations, show that the equations of motion for the quantum operators are identical to the equations of motion for the classical observables. Find the variance* $\Delta x^2(t)$ *for a free particle localized in a wave packet given by* $\psi(x, t = 0) = Ae^{-\frac{x^2}{2\ell^2}}$*. Just to be clear, the* ℓ *here is not the same as the as the* ℓ *in the nano-vibrator used in Problem 9.5.*

9.4 Interaction Picture

In the Heisenberg picture, the operators become time dependent, and the wave function is time independent. In some cases, there is a need to find the time dependence of the wave function, but only the part associated with the time dependent term in the Hamiltonian. This is because the time independent term just contributes to a time dependent phase which can be included later. Hence, the equations to be solved become simpler. This is done by introducing the **interaction picture**.

Consider a Hamiltonian of the form:

$$\hat{H} = \hat{H}_0 + \hat{V}(t) \tag{9.50}$$

where $\hat{H}_0$ has been diagonalized with eigenstates and eigenenergies: $\hat{H}_0 \,|n\rangle = E_n \,|n\rangle$. So $\hat{V}(t)$ may or may not be time dependent, but it is shown as time dependent for generality.

We are now interested in solving Schrödinger's equation:

$$i\hbar \frac{d}{dt} \mid \psi_S(t)\rangle = \hat{H} \mid \psi_S(t)\rangle \tag{9.51}$$

We introduce the subscript S again on ψ_S to indicate that it is the solution to Schrödinger's equation. We next define an operator $\hat{U}_I$:

$$\hat{U}_I = e^{i\frac{\hat{H}_0 t}{\hbar}} \tag{9.52}$$

the inverse of the time evolution operator if $\hat{V}(t) = 0$. Note also that

$$\left[\hat{U}_I, \hat{H}_o\right] = 0 \tag{9.53}$$

because a function of an operator commutes with that operator and therefore

$$\hat{U}_I \hat{H}_o \hat{U}_I^{-1} = \hat{U}_I \hat{U}_I^{-1} \hat{H}_o = \hat{H}_o \tag{9.54}$$

we then define

$$|\psi_I(t)\rangle = \hat{U}_I \, |\psi_S(t)\rangle \tag{9.55}$$

Then substituting

$$|\psi_S(t)\rangle = \hat{U}_I^{-1} \, |\psi_I(t)\rangle \tag{9.56}$$

into Schrödinger's equation

$$i\hbar \frac{d}{dt} \hat{U}_I^{-1} \, |\psi_I(t)\rangle = \hat{H} \hat{U}_I^{-1} \, |\psi_I(t)\rangle = \left(\hat{H}_0 + \hat{V}(t)\right) \hat{U}_I^{-1} \, |\psi_I(t)\rangle \tag{9.57}$$

We now multiply both sides by $\hat{U}_I$, take the time derivative on the left and use Eq. 9.54 on the right:

$$i\hbar \hat{U}_I \frac{d}{dt} \hat{U}_I^{-1} \, |\psi_I(t)\rangle = \hat{U}_I \left(\hat{H}_0 + \hat{V}(t)\right) \hat{U}_I^{-1} \, |\psi_I(t)\rangle \tag{9.58}$$

$$i\hbar \frac{d}{dt} \, |\psi_I(t)\rangle = \hat{U}_I \hat{V}(t) \hat{U}_I^{-1} \, |\psi_I(t)\rangle \tag{9.59}$$

From this, we define $\hat{V}_I(t)$

$$\hat{V}_I(t) = \hat{U}_I \hat{V}(t) \hat{U}_I^{-1} \tag{9.60}$$

Hence, in the interaction representation, Schrödinger's equation becomes

$$i\hbar \frac{d}{dt} \, |\psi_I(t)\rangle = \hat{V}_I(t) \, |\psi_I(t)\rangle \tag{9.61}$$

One can now find $\hat{V}_I(t)$ in the usual way by expanding the $\hat{U}_I$ and $\hat{U}_I^{-1}$ in a power series, as done for the Heisenberg picture (see footnote 1). It is also possible to do this in an easier way, by solving the equation of motion for $\hat{V}_I(t)$, again paralleling the Heisenberg discussion. To find the equation of motion, we begin by evaluating the time derivative of $\hat{V}_I(t)$:

$$\frac{d}{dt} \hat{V}_I(t) = i\hat{U}_I \frac{\hat{H}_o}{\hbar} \hat{V}(t) \hat{U}_I^{-1} - i\frac{1}{\hbar} \hat{U}_I \hat{V}(t) \hat{H}_o \hat{U}_I^{-1} + \hat{U}_I \left(\frac{d}{dt} \hat{V}(t)\right) \hat{U}_I^{-1} \tag{9.62}$$

or

$$\frac{d}{dt} \hat{V}_I(t) = i\hat{U}_I \frac{\hat{H}_o}{\hbar} \hat{U}_I^{-1} \hat{U}_I \hat{V}(t) \hat{U}_I^{-1} - i\frac{1}{\hbar} \hat{U}_I \hat{V}(t) \hat{U}_I^{-1} \hat{U}_I \hat{H}_o \hat{U}_I^{-1} + \hat{U}_I \left(\frac{d}{dt} \hat{V}(t)\right) \hat{U}_I^{-1} \tag{9.63}$$

Again using $\hat{U}_I \hat{H}_o \hat{U}_I^{-1} = \hat{H}_o$

$$-i\hbar \frac{d}{dt}\hat{V}_I(t) = \hat{H}_o \hat{V}_I(t) - \hat{V}_I(t)\hat{H}_o - i\hbar \hat{U}_I \left(\frac{d}{dt}\hat{V}(t)\right)\hat{U}_I^{-1} = \left[\hat{H}_o, \hat{V}_I(t)\right] - i\hbar \hat{U}_I \left(\frac{d}{dt}\hat{V}(t)\right)\hat{U}_I^{-1} \tag{9.64}$$

Therefore

$$-i\hbar \frac{d}{dt}\hat{V}_I(t) = \left[\hat{H}_o, \hat{V}_I(t)\right] - i\hbar \hat{U}_I \left(\frac{d}{dt}\hat{V}(t)\right)\hat{U}_I^{-1} \tag{9.65}$$

Subject to the initial condition that

$$\hat{V}_I(t=0) = \hat{V}(t=0) \tag{9.66}$$

In some problems where this is used, $\hat{V}(t) = \hat{V}$ is time independent, hence

$$-i\hbar \frac{d}{dt}\hat{V}_I(t) = \left[\hat{H}_o, \hat{V}_I(t)\right] \tag{9.67}$$

In the interaction representation, as in the Heisenberg representation, the form of $\hat{H}_o$ does not change in the new representation because $\hat{U}_I \hat{H}_o \hat{U}_I^{-1} = \hat{H}_o$.

To see this explicitly, consider a specific example of the harmonic oscillator (or from Ch. 15, the radiation field):

$$\hat{H}_o = \hbar\Omega\left(a^\dagger a + \frac{1}{2}\right) \tag{9.68}$$

Then

$$\hat{H}_{o_I} = \hbar\Omega\left(\hat{U}_I a^\dagger a \hat{U}_I^{-1} + \frac{1}{2}\right) = \hbar\Omega\left(\hat{U}_I a^\dagger \hat{U}_I^{-1}\hat{U}_I a \hat{U}_I^{-1} + \frac{1}{2}\right) = \hbar\Omega\left(\hat{a}_I^\dagger \hat{a}_I + \frac{1}{2}\right) \tag{9.69}$$

where

$$-i\hbar \frac{d}{dt}\hat{a}_I^\dagger = \left[\hat{H}_o, \hat{a}_I^\dagger\right] = \hbar\Omega\left[\hat{a}_I^\dagger \hat{a}_I, \hat{a}_I^\dagger\right] = \hbar\Omega\left(\hat{a}_I^\dagger \hat{a}_I \hat{a}_I^\dagger - \hat{a}_I^\dagger \hat{a}_I^\dagger \hat{a}_I\right) = \hbar\Omega a^\dagger\left[\hat{a}_I, \hat{a}_I^\dagger\right] = \hbar\Omega \hat{a}_I^\dagger \tag{9.70}$$

and

$$-i\hbar \frac{d}{dt}\hat{a}_I = \left[\hat{H}_o, \hat{a}_I\right] = -\hbar\Omega \hat{a}_I$$

Therefore:

$$\hat{a}_I^\dagger = a^\dagger e^{i\Omega t} \tag{9.71}$$

$$\hat{a}_I = \hat{a} e^{-i\Omega t} \tag{9.72}$$

Hence

$$\widehat{H}_{0_I} = \hbar\Omega\left(\hat{a}_I^\dagger \hat{a}_I + \frac{1}{2}\right) \equiv \hbar\Omega\left(a^\dagger \hat{a} + \frac{1}{2}\right) \tag{9.73}$$

One final note that will help in understanding the discussion of the quantized radiation field notes. Recall the usual Schrödinger form:

$$\widehat{H} = \widehat{H}_0 + \widehat{V}(t) \tag{9.74}$$

$$|\psi(t)\rangle = \sum_n c_n(t)\,|E_n\rangle = \sum_n C_n(t)e^{-i\frac{E_n t}{\hbar}}\,|E_n\rangle \tag{9.75}$$

The last form of the expansion in Eq. 9.75 is called an interaction picture because, by inserting the time evolving phase in the exponential (the same phase appears after using the time evolution operator for time independent Hamiltonian), the effect of the $\widehat{H}_0$ will be eliminated in the equations of motion for the expansion coefficients. This is the same result as happens in the development of the interaction picture where we have

$$|\psi_I(t)\rangle = \sum_n C_n(t)\,|E_n\rangle \tag{9.76}$$

9.5 Quantum Flip-Flop: Coherent Control of a Two-Level System and Rabi Oscillations

Thinking classically: the simplest switch can either be on or off. This is a **two-state system**. The on and off can represent a 1 and 0 if you are thinking about computers. If you think about the system being off, and you apply some kind of energy to switch it on, the response is highly nonlinear in the strength of the amount of driving energy, because if you increase the driving energy from 0 to a threshold where it switches, after you switch it on, you can apply more drive energy but nothing happens. Since the response is not linear with the energy, we say the system is nonlinear. Above, you saw that the nano-vibrator (including an LC circuit) is linear in the driving potential which you considered in the problem of the driven harmonic oscillator (in the Heisenberg picture, Problem 9.3). In information control, manipulation, transmission, and storage almost all the devices are highly nonlinear in this way. The system would not function correctly if they were linear. A quantum system becomes nonlinear in the driving potential if the energy level spacings between the subsequent levels is not equal (i.e., anharmonic, and not harmonic as in the harmonic oscillator) or if the coupling changes. So, two-state quantum systems are very important for technological applications, especially if you are working on digital type problems.

In the later discussion on spin (Chapter 10), we will discuss an intrinsic two-state system critical to modern applications. You could also engineer a potential well of the type we encountered in Chapter 3 to have just two discrete states, but it would still have an infinite number of scattering states. It could be one-, two- or three-dimensional, but even if the system has many levels, it is possible to consider it to be just a two-level system. The reason is that it takes energy to change the probability of state occupation from one state to another.

Since energy is conserved, if the energy coupled into the system matches the energy of the transition, that transition can happen but the other transitions are at different energies so, in most cases, they will not be affected. This is the basic idea behind resonance, which we will explore below. *(It is important to remember that energy conversation is not a postulate of classical physics; it results from the existing postulates. The same is true for quantum mechanics: energy conservation and other common conservation laws are not used in a calculation, they result from the calculation.)*

To see how this works, we begin by assuming that we have solved the time independent Schrödinger equation to get the eigenstates, $|E_n\rangle$ and eigenenergies, E_n, for a specific system. It could be a simple system describing a three-dimensional infinite potential, for example. The Hamiltonian is designated H_0 with $H_0\,|E_n\rangle = E_n\,|E_n\rangle$. We assume, also, that the system is not a nano-vibrator, thus ensuring that the energy level spacings will not be harmonic: i.e., $E_n - E_{n-1} \neq E_{n+1} - E_n$.

We then assume that the problem of interest is for a charged particle (e.g., an electron). Such a system will interact with an electric field (for example, the oscillating electric field in an electromagnetic wave) which can be used to drive the system. The coupling is through the **induced electric dipole moment**, given in three dimensions by the vector $\boldsymbol{\mu} = q\boldsymbol{r}$, where $\boldsymbol{r}$ measures the displacement of the charge from equilibrium due to the applied field. For a hydrogen atom, that would be the distance between the nucleus and the electron. Equilibrium in this case means that the system is in the lowest energy state. The applied oscillating electric field exerts a force on the electron, causing the displacement. The potential energy for this is given by[2]

$$V = -\hat{\boldsymbol{\mu}} \cdot \boldsymbol{E}(\boldsymbol{R}, t) \tag{9.77}$$

written with the quantum operator for the electron dipole momentum $\hat{\boldsymbol{\mu}} = -e\hat{\boldsymbol{r}}$ where $\hat{\boldsymbol{r}}$ is the displacement of the charge and $\boldsymbol{E}(\boldsymbol{R})$ is the classical electric field associated with the electromagnetic plane wave located at $\boldsymbol{R} = \boldsymbol{R}_C + \boldsymbol{r}$. Then $\boldsymbol{R}_C$ locates the quantum system where, in an atom, it would be the center of mass. With the classical field, we call Eq. 9.77 the **semiclassical description**. Later, we will consider the quantum mechanical description of the field. We assume that, for evaluation of the matrix element, the spatial extent of the associated eigenfunctions is small (~<nm) compared to the wavelength of the electric field (recall that $k = {}^{2\pi}\!/_{\lambda}$ and typically $\lambda > 100$ nm). This is called the **dipole approximation**. So, for an atom, the Bohr radius is the characteristic length. In this case, we can set $\boldsymbol{E}(\boldsymbol{R}, t) = \boldsymbol{E}(\boldsymbol{R}_C, t)$. The implication is that then $\boldsymbol{E}(\boldsymbol{R}_C, t)$ *is constant over the spatial integration in the matrix element.* Often in the notes that follow, when the location of the quantum system is not important, we set $\boldsymbol{E}(\boldsymbol{R}_C, t) = \boldsymbol{E}(t)$. In the discussion that follows, we will drop the subscript on $\boldsymbol{R}$ since $\boldsymbol{R}_C \sim \boldsymbol{R}$.

We also assume that in this discussion we can ignore the vector nature of the coupling since that basically determines when the matrix element is non-zero or 0, in which case we just

[2] Because of the addition of the electromagnetic field, a cautious reader might ask if, by asserting that the potential is given by Eq. 9.78, whether in fact the system is written in terms of canonical coordinate momentum. In fact, the form of the Hamiltonian $\hat{H} = \hat{H}_0 + \hat{V} = \hat{H}_0 - \hat{\boldsymbol{\mu}} \cdot \boldsymbol{E}(t)$ is the result of starting with a correct Hamiltonian followed by a unitary transformation. The details are presented in Appendix E.

assume it is non-zero. This is called the **scalar approximation.** The vector nature is important in two- and three-dimensional quantum systems like atoms or semiconductor dots (see Appendix G).

Since the systems considered in these notes are predominantly 1-D, we now reduce the potential to the simple form

$$V = -\hat{\mu}E(\mathbf{R}, t) = e\hat{x}E(t). \tag{9.78}$$

The **electromagnetic interaction Hamiltonian** in one dimension becomes:

$$\hat{H} = \hat{H}_0 + \hat{V} = \hat{H}_0 + e\hat{x}E(t) \tag{9.79}$$

whereas stated above, $\hat{H}_0 \, |E_n\rangle = E_n \, |E_n\rangle$ and defines the Hilbert space spanned by $\{|E_n\rangle\}$.

We are interested in matrix elements of the form $\langle E_{n'}|ex|E_n\rangle$. Since problems of interest here are typically associated with a symmetric Hamiltonians meaning the states have well-defined parity (i.e., even or odd), we know that because x is an odd function, that the matrix elements $\langle E_{\mathrm{n}}|ex|E_n\rangle = 0$. We then write the matrix elements for some state n and n' of interest as:

$$\begin{aligned} \hat{\mu}_{n'n} &\equiv \langle E_{n'}| - e\hat{x}|E_n\rangle = -e\int dx\, \psi^*_{n'}(x)x\psi_n(x) \neq 0 \text{ for } n' \neq n \text{ and} \\ &\langle E_n|ex|E_n\rangle = 0 \end{aligned} \tag{9.80}$$

If E is a static electric field, there are **stationary state solutions** (i.e., solutions to the *time independent* Schrödinger equation) that show that it shifts the energy levels and creates new eigenstates in terms of superpositions of old eigenstates. This is called the Stark effect which will be discussed later in Chapter 11. Since the Hamiltonian of interest is time dependent because of the oscillating field, there are no stationary state solutions. While $\hat{H}_0$ is diagonal in the energy representation, V_0 is not diagonal as we discussed above. As discussed in Chapter 7 (starting with Eq. 7.23 and following), the off-diagonal terms mix the states to which it is coupled. *Here we will see in the time dependent case the oscillating field can cause a transition of the electron from one state to another.*

The electric field, $E(t)$ where the position is now suppressed, has a sinusoidal form given by[3]

$$E(t) = \frac{1}{2}\left(\mathcal{E}_0 e^{-i\omega t} + \mathcal{E}_0^* e^{i\omega t}\right); \omega > 0 \tag{9.81}$$

It could be from a RF (Radio Frequency) or microwave generator or maybe from a laser. What will be important is the $\hbar\omega \sim E_2 - E_1$ where E_1 is a lower energy eigenstate and the state where the electron exists in the absence of the field and E_2 is a higher energy state where we wish to excite the electron. We then recall that since the basis is complete, we can expand the state vector along the lines of the interaction picture in Eq. 9.75:

[3] More completely for a traveling wave $E(\mathbf{R}, t) = \frac{1}{2}\left(\mathcal{E}_0 e^{i\mathbf{k}\cdot\mathbf{R} - i\omega t} + \mathcal{E}_0^* e^{-i\mathbf{k}\cdot\mathbf{R} + i\omega t}\right)$; and $\mathcal{E}_0$ is a complex constant assumed to be independent of time and space unless otherwise stated, such as for a pulse where it is time dependent. $\boldsymbol{k}$ is the **k-vector** that determines the direction of propagation and $k = |\boldsymbol{k}| = \frac{2\pi}{\lambda}$ and λ is the wavelength and in vacuum $\lambda = \frac{c}{\nu} = \frac{2\pi c}{\omega}$. Since we are not concerned with the spatial part, we set $A_0 e^{i\mathbf{k}\cdot\mathbf{R}} = \mathcal{E}_0$.

$$|\psi(t)\rangle = \sum_n C_n(t)e^{-\frac{iE_nt}{\hbar}}\,|E_n\rangle \tag{9.82}$$

The big change here is that the expansion coefficients are time dependent. Up until this point, they have always been time independent, but so was the Hamiltonian. Now the Hamiltonian is explicitly a function time. In some discussions, the exponential factor is suppressed. It reappears in the final expression for the expansion coefficients if you do it that way.

We now take Eq. 9.80 along with 9.81 and insert 9.82 into the full time dependent Schrödinger equation, $i\hbar\frac{d}{dt}\,|\psi(t)\rangle = \widehat{H}\,|\psi(t)\rangle$:

$$i\hbar\frac{d}{dt}\sum_n C_n(t)e^{-\frac{iE_nt}{\hbar}}\,|E_n\rangle = \left(\widehat{H}_0 - \hat{\mu}\frac{1}{2}\left(\mathcal{E}_0e^{-i\omega t} + \mathcal{E}_0^*e^{i\omega t}\right)\right)\sum_n C_n(t)e^{-i\frac{iE_nt}{\hbar}}\,|E_n\rangle \tag{9.83}$$

We proceed to find an equation of motion for the $C_n(t)$. The beauty of the solution that will emerge is that while this looks fairly complicated (at least very detailed), the orthogonality of the eigenvectors will eliminate the sums and result in a very simple form. This process of developing equations of motion for the expansion coefficients will be used frequently in the remaining text.

We begin by taking the time derivative on the left-hand side of the equation and operate with $\widehat{H}$ on the right-hand side. Then, we project the result onto an arbitrary state, $\langle E_m|$ and use the orthonormality of the states. We recall too that the diagonal elements of interaction potential are zero.

$$\langle E_m|\sum_n\left(i\hbar\frac{d}{dt}C_n(t) + E_nC_n(t)\right)e^{-\frac{iE_nt}{\hbar}}\,|E_n\rangle =$$

$$\langle E_m|\sum_n\left(E_n - \hat{\mu}\frac{1}{2}\left(\mathcal{E}_0e^{-i\omega t} + \mathcal{E}_0^*e^{i\omega t}\right)\right)C_n(t)e^{-\frac{iE_nt}{\hbar}}\,|E_n\rangle \tag{9.84}$$

Note that the E_n term on the left-hand side, resulting from application of the chain rule on the time derivative of $C_n(t)e^{-\frac{iE_nt}{\hbar}}$, cancels the E_n term resulting from $\widehat{H}_0\,|E_n\rangle = E_n\,|E_n\rangle$ on the right-hand side. This simplification results from including the $e^{-\frac{iE_nt}{\hbar}}$ in Eq. 9.82 and is the same result as if we had moved into the interaction pictures in the previous section. Hence, Eq. 9.82 is called the interaction picture, and Eq. 9.84 becomes:

$$\langle E_m|\sum_n i\hbar\left(\frac{d}{dt}C_n(t)\right)e^{-\frac{iE_nt}{\hbar}}\,|E_n\rangle = \langle E_m|\sum_n\left(-\hat{\mu}\frac{1}{2}\left(\mathcal{E}_0e^{-i\omega t} + \mathcal{E}_0^*e^{i\omega t}\right)\right)C_n(t)e^{-\frac{iE_nt}{\hbar}}\,|E_n\rangle \tag{9.85}$$

After using the orthonormality of the basis states, $\langle E_m|E_n\rangle = \delta_{mn}$, Eq. 9.85 becomes

$$\dot{C}_m(t) = -\frac{1}{2i\hbar}\sum_{n\neq m}C_n(t)\,\langle E_m|\hat{\mu}|E_n\rangle\left(\mathcal{E}_0e^{-i(\omega_{nm}+\omega)t} + \mathcal{E}_0^*e^{-i(\omega_{nm}-\omega)t}\right) \tag{9.86}$$

where $\hbar\omega_{nm} = E_n - E_m$, which can be positive or negative.

At this point, we have a large (typically infinite) set of coupled first order differential equations to solve. There are two time-varying phase factors that oscillate between -1 and 1. To get a sense of the mathematical behavior following integration of Eq. 9.86, we ignore the time dependence of $C_n(t)$. Then integrating the two exponential terms would give terms like $\frac{1}{(\omega_{nm}\pm\omega)}$. Assuming near **resonance** operation, meaning, $|\omega_{nm}| \sim \omega$ (recall that $\omega \geq 0$), then if $\omega_{nm} > 0$, the second term will dominate, and if $\omega_{nm} < 0$ the first term will dominate. Retaining only the dominant term is called the **rotating wave approximation.**[4] In addition, since at resonance $|\omega_{nm}| \sim \omega$, then usually ω is not near resonance for any other combinations. Hence, we retain only the two specific levels, n and m. This is called the **two-level approximation.**

We will designate the n and m levels as level 1 and 2 such that $E_2 - E_1 = \hbar\omega_{21} > 0$. We will also replace ω_{12} with $-\omega_{21}$ and $\langle E_m|\hat{\mu}|E_n\rangle$ with $\hat{\mu}_{mn}$ the matrix element of the dipole moment defined above in Eq. 9.79. This then reduces the above equations to two relatively simple coupled first order differential equations:

$$\dot{C}_1(t) = -\frac{1}{2i}\Omega_R^* e^{i(\omega-\omega_{21})t} C_2(t) \tag{9.87}$$

$$\dot{C}_2(t) = -\frac{1}{2i}\Omega_R e^{-i(\omega-\omega_{21})t} C_1(t) \tag{9.88}$$

where $\mu_{21} = \mu_{12}^*$ because the dipole moment operator is Hermitian, $\Omega_R = \frac{\mu_{21}\mathcal{E}_0}{\hbar}$ is a complex number and

$$|\Omega_R| = \frac{|\mu_{21}\mathcal{E}_0|}{\hbar} \tag{9.89}$$

is the **Rabi frequency**.

Figure 9.1 shows the energy level structure of the two-level system we are discussing, which is driven by a monochromatic field oscillating at frequency ω, close to the resonant frequency ω_{21}. Before working to solve these coupled equations, it is important to understand the important impact of resonance. If we assume that the applied oscillating electric field is sufficiently weak, then the effect of the field on the magnitude of the change in probability amplitude will be very small. Being more specific, suppose that for the ground state probability amplitude at $t = 0$, $C_1(0) = 1$, and therefore $C_2(0) = 0$. Then based on Eq. 9.87, the change in $C_1(t)$ will be small if Ω_R is sufficiently small. In other words, if Ω_R is small enough to be a perturbation, the effect

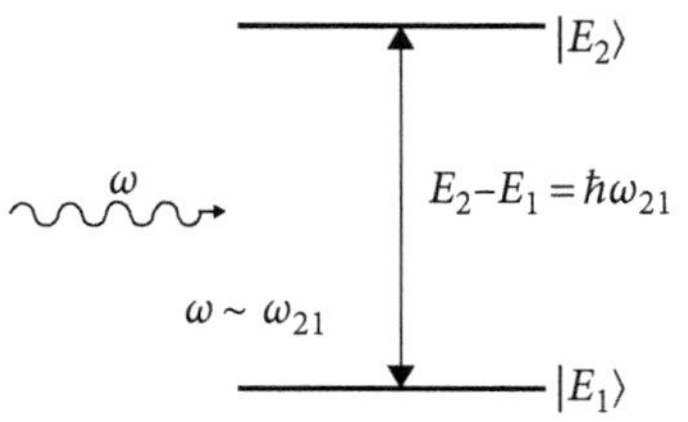

Fig. 9.1 Quantum system in the two-level approximation driven near resonance with a monochromatic field.

[4] The language arises because in the complex plane with axes x and iy with $z = x + iy = re^{i\theta}$, the phasor $e^{\pm i(|\omega_{nm}|-\omega)t}$ rotates more slowly than $e^{\pm i(|\omega_{nm}|+\omega)t}$.

of Ω_R is a perturbation (small change) of the probability amplitudes. Hence $C_1(t) \sim 1$ and $0 \le C_2(t) \ll 1$.

For the initial conditions we then use

$$\begin{aligned} C_1(t=0) &= C_1^{(0)} = 1 \\ C_2(t=0) &= C_2^{(0)} = 0 \end{aligned} \tag{9.90}$$

The superscript means zeroth order, as discussed in Chapter 11. Then in the limit of weak interactions, we assume $C_1(t) \sim C_1(t=0) = 1$. In which case, we can set $C_1(t=0) = 1$ in the integral in Eq. 9.88 and remove it from the integral to give

$$\begin{aligned} \int_0^t dt' \dot{C}_2(t') &= -\frac{1}{2i}\Omega_R \int_0^t dt' e^{-i(\omega-\omega_{21})t'} C_1(t') = -\frac{1}{2i}\Omega_R C_1(0) \int_0^t dt' e^{-i(\omega-\omega_{21})t'} \\ &= -\frac{1}{2}\Omega_R \frac{e^{-i(\omega-\omega_{21})t} - 1}{(\omega-\omega_{21})} \end{aligned} \tag{9.91}$$

giving

$$C_2(t) = C_2^{(1)} = -\frac{1}{2}\Omega_R \frac{e^{-i(\omega-\omega_{21})t} - 1}{(\omega-\omega_{21})} \tag{9.92}$$

The superscript 1 means to first order in perturbation (see Chapter 11). The probability as a function of time of being in state 2 is then

$$|C_2(t)|^2 = \left(\frac{|\Omega_R|}{2}\right)^2 2\frac{1-\cos(\omega-\omega_{21})t}{(\omega-\omega_{21})^2} = |\Omega_R|^2 \frac{\sin^2\frac{(\omega-\omega_{21})}{2}t}{(\omega-\omega_{21})^2} \tag{9.93}$$

where the second equality results from using the trigonometric identity $1-\cos(\omega-\omega_{21})t = 2\sin^2\frac{(\omega-\omega_{21})}{2}t$.

Furthermore, you will show in Problem 9.8 that if we assume $\omega \sim \omega_{21}$, then

$$|C_2(t)|^2 = \frac{|\Omega_R|^2}{4}t^2 \tag{9.94}$$

We see from Eq. 9.94 that the probability of being in state 2 quickly increases to greater than unity, which is impossible. Further, the result for $|C_2(t)|^2$ is shown in Fig. 9.2. A striking result is that as the frequency of the oscillating field approaches resonance, i.e., as $\omega - \omega_{21} \to 0$, the probability diverges for large time. This is unphysical and is a failure of the approach (i.e., using perturbation theory) because we assumed that the change in $C_2(t)$ is small, meaning that since $C_2(t=0) = 0$, then $C_2(t) \ll 1$. This is obviously not the case. However, it shows the effect of a resonance: *resonance strongly (by orders of magnitude) increases the strength of coupling to the system*. This coupling is so strong that many groups are looking to replace using voltage and wires to controlling an electronic circuit with electromagnetic radiation.

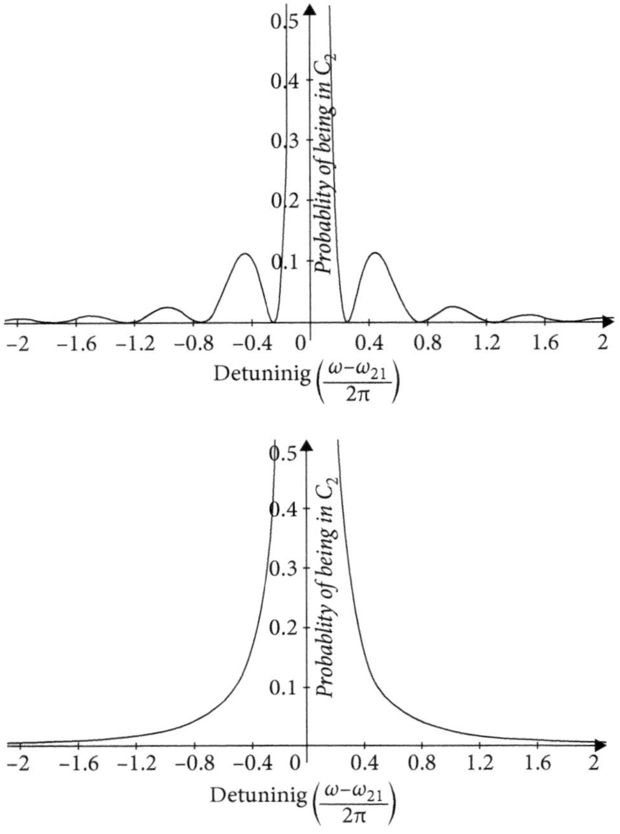

Fig. 9.2 Plot of the oscillating $|C_2(t)|^2$ at a fixed time as a function of detuning $\omega - \omega_{21}$.

Problem 9.7 *Consider the initial conditions in Eq. 9.90. It is clear that the ground state amplitude is initially unity since* $C_2(t=0) = 0$. *Using perturbation theory, we assume that any changes would be small. Hence, the change in* C_1 *to first order is zero.*

a) Write down the wave function for $|\psi(t)\rangle$ *for the two-state problem above, with the initial conditions in Eq. 9.90. You need to apply Eq. 9.92 and then the results of perturbation theory for* $C_2(t)$.

b) Find $\langle x\rangle(t)$ *assuming* $\langle E_i|x|E_i\rangle = 0$ *and* $\left\langle E_i|x|E_{j\neq i} = \xi_{ij}\right\rangle$.

Problem 9.8 *Working with the same equations, integrate 9.88, after setting the detuning equal to zero (i.e.,* $\omega - \omega_{21} = 0$*), again assuming a weak driving field and the same initial conditions in Eq. 9.90 and Eq. 9.91.*

a) Show that you get

$$|C_2(t)|^2 = \frac{|\Omega_R|^2}{4}t^2$$

b) Find the inequality now required to keep perturbation theory valid.

Problem 9.9 *Assume now that in* Eq. 9.81, *$\mathcal{E}_0$ becomes a pulse in time, i.e., $\mathcal{E}_0 \rightarrow \mathcal{E}_0\tau\delta(t)$ where τ is an effective pulse width. Assuming the same boundary conditions in* Eq. 9.89:
a) Find $C_2(t)$ immediately after the pulse.
b) Derive the inequality required to keep perturbation theory valid.

The approach above of assuming a small perturbation in the Hamiltonian will be rigorously developed and generalized in Chapter 11: the solution here is based on **first order time dependent perturbation** theory. Here, the interaction, given by the Rabi frequency, appears to the first power (i.e., linear) only.

The question now is "How small is small?". This means the interaction term in the Hamiltonian $\hat{\mu}E(t)$ must be small compared to something. The requirement is that there not be a significant change in the probability amplitudes. Looking at the expression above for $|C_2(t)|^2$, Eqs 9.93 or 9.94, we see that for the theory to be valid, we require

$$|\Omega_R|^2 \ll 2(\omega - \omega_{21})^2 \tag{9.95}$$

i.e., the Rabi frequency must be small compared to the detuning.

In starting this discussion, we made the rotating wave approximation. This is equivalent in Eqs 9.87 and 9.88 to ignoring the terms that vary in time as $e^{\pm i(\omega+\omega_{21})t}$ (note by definition that $\omega_{21} > 0$.) The reason is easier to see now. Considering even just the equation for $C_2(t)$ (Eq. 9.91), keeping the other term would have resulted in

$$\begin{aligned} C_2(t) &= -\frac{1}{2i}\Omega_R \int_0^t dt' e^{-i(\omega-\omega_{21})t'} - \frac{1}{2i}\Omega_R^* \int_0^t dt' e^{-i(\omega+\omega_{21})t'} \\ &= -\frac{1}{2}\Omega_R \frac{e^{-i(\omega-\omega_{21})t} - 1}{(\omega - \omega_{21})} - \frac{1}{2}\Omega_R^* \frac{e^{-i(\omega+\omega_{21})t} - 1}{(\omega + \omega_{21})} \end{aligned} \tag{9.96}$$

In most experiments of interest, $|\Omega_R| \ll \omega_{21}$. Furthermore, to justify the rotating wave approximation, we required $\omega - \omega_{21} \ll \omega + \omega_{21} \sim \omega_{21}$. Taken together, this also justifies the two-level approximation. Based on the inequality above, for perturbation theory, we see that the strong part of the resonance effect is in the limit that $\omega - \omega_{21} \rightarrow 0$.

The meaning of the language of rotating wave approximation can now be made a little more obvious. Thinking of the two phasors, $e^{-i(\omega\pm\omega_{21})t}$, as each being the product of two phasors $e^{-i\omega t} * e^{\mp i\omega_{21}t}$, then if you imagined looking at the second phasor while you are rotating (in complex space $z = x + iy = ra^{i\theta}$) with the first phasor, then in the case of the minus sign, $\omega - \omega_{21}$, it would appear that the second phasor was moving slowly relative to you on the first phasor, while in the case of the plus sign, $\omega + \omega_{21}$, the second phase would be moving extremely fast. Hence by suppressing the second term, the calculation appears in a rotating frame.

Eqs 9.87 and 9.88 are now easily solved exactly, but we will make a few more changes (no more approximations) just to make them appear even simpler. We can eliminate the explicit time dependence in the exponentials by introducing two new variables:

$$C_1(t) = \widetilde{C}_1(t)e^{\frac{i(\omega-\omega_{21})t}{2}} \tag{9.97}$$

$$C_2(t) = \widetilde{C}_2(t) e^{-\frac{i(\omega-\omega_{21})t}{2}} \tag{9.98}$$

This is often called the **field interaction picture**. This is will eliminate the explicit time dependence of the coefficients. Substituting we get

$$\frac{d}{dt}\widetilde{C}_1(t) = -\frac{i\Delta}{2}\widetilde{C}_1(t) + \frac{i\Omega_R^*}{2}\widetilde{C}_2(t) \tag{9.99}$$

$$\frac{d}{dt}\widetilde{C}_2(t) = \frac{i\Omega_R}{2}\widetilde{C}_1(t) + \frac{i\Delta}{2}\widetilde{C}_2(t) \tag{9.100}$$

where $\Delta = \omega - \omega_{21}$. To solve these equation, we write the above two equations in matrix form as

$$\frac{d}{dt}\begin{bmatrix} \widetilde{C}_1(t) \\ \widetilde{C}_2(t) \end{bmatrix} = \begin{bmatrix} -\frac{i\Delta}{2} & \frac{i\Omega_R^*}{2} \\ \frac{i\Omega_R}{2} & \frac{i\Delta}{2} \end{bmatrix}\begin{bmatrix} \widetilde{C}_1(t) \\ \widetilde{C}_2(t) \end{bmatrix} \tag{9.101}$$

The use of the matrix here has nothing to do with quantum behavior. We are simply exploiting a simple method for solving coupled first order differential equations. Taking $\widetilde{C}_1(t)$ and $\widetilde{C}_2(t)$ to vary as $e^{i\nu t}$ we find an eigenvector equation of the form:

$$\begin{bmatrix} -\frac{i\Delta}{2} & \frac{i\Omega_R^*}{2} \\ \frac{i\Omega_R}{2} & \frac{i\Delta}{2} \end{bmatrix}\begin{bmatrix} \widetilde{C}_1(t) \\ \widetilde{C}_2(t) \end{bmatrix} = i\nu\begin{bmatrix} \widetilde{C}_1(t) \\ \widetilde{C}_2(t) \end{bmatrix} \tag{9.102}$$

The result (see Appendix A, section A.3) is that

$$\nu = \pm\frac{1}{2}\left[\Delta^2 + |\Omega_R|^2\right]^{\frac{1}{2}} \equiv \pm\frac{\Omega_{GR}}{2} \tag{9.103}$$

where Ω_{GR} is the ***generalized Rabi frequency***. To complete the solution, we assume, that at $t = 0$, $\widetilde{C}_1(0) = 1$ and therefore $\widetilde{C}_2(0) = 0$. Therefore

$$\widetilde{C}_1(t) = Ae^{i\frac{\Omega_{GR}t}{2}} + Be^{-i\frac{\Omega_{GR}t}{2}} \tag{9.104}$$

which at $t = 0$ gives

$$A + B = 1 \tag{9.105}$$

From Eq. 9.99 we get

$$\widetilde{C}_2(t) = Ae^{i\frac{\Omega_{GR}t}{2}}\frac{1}{\Omega_R^*}(\Omega_{GR} + \Delta) - Be^{-i\frac{\Omega_{GR}t}{2}}\frac{1}{\Omega_R^*}(\Omega_{GR} - \Delta) \tag{9.106}$$

at $t = 0$, we get

$$-A\left(\frac{\Omega_{GR}}{2} + \frac{\Delta}{2}\right) + B\left(\frac{\Omega_{GR}}{2} - \frac{\Delta}{2}\right) = 0 \tag{9.107}$$

Therefore

$$A = \frac{\left(\frac{\Omega_{GR}}{2} - \frac{\Delta}{2}\right)}{\Omega_{GR}} \text{ and } B = \frac{\left(\frac{\Omega_{GR}}{2} + \frac{\Delta}{2}\right)}{\Omega_{GR}} \tag{9.108}$$

To make sense of what we have learned, we consider the case when the monochromatic driving field is exactly on resonance; i.e., $\Delta = 0$. Then

$$A = B = \frac{1}{2} \tag{9.109}$$

$$\Omega_{GR} = \mid \Omega_R \mid \tag{9.110}$$

and therefore:

$$\widetilde{C}_1(t) = \cos\frac{\mid \Omega_R \mid t}{2} \text{ or } \left|\widetilde{C}_1(t)\right|^2 = \frac{1}{2}(1 + \cos|\Omega_R|t) \tag{9.111}$$

$$\widetilde{C}_2(t) = i\sin\frac{\mid \Omega_R \mid t}{2} \text{ or } \left|\widetilde{C}_2(t)\right|^2 = \frac{1}{2}(1 - \cos|\Omega_R|t) \tag{9.112}$$

This shows that the probability oscillates at the Rabi frequency (called a **Rabi oscillation**). Figure 9.3 shows a plot of the probability of being in the ground state and excited state as a function of time for typical parameters in the optical region. The Rabi oscillation is clearly visible. Time scales are much longer if the system is, for example, a nuclear spin state. This is a rather remarkable result. First, at optical frequencies, it is not hard to arrange for $\mid \Omega_R \mid$ to approach a terahertz. That is a very fast switch. If the system is initially in the lower lying state, the oscillating field can drive the system entirely to the upper state when $\mid \Omega_R \mid t = \pi$. Not only that, but it has been shown recently that it can be done with just a few photons. One photon is about $10-^{19}$ joules. This calculation is done with classical fields, but it is easy to do with a quantized field when you study quantum optics. *The point is that you can create a switch to go between two states that is toggled by using a gated/modulated resonant drive frequency.*

You can do this with a semiconductor and a diode laser and $LiNbO_3$ modulators, standard equipment of the telecom world. The response time of the switch is set by the Rabi frequency. Finally, notice something that is entirely non-classical. If you set $\mid \Omega_R \mid t = \frac{\pi}{2}$, your switch is in a dynamically controllable **coherent superposition** of state 1 and state 2 (or a "0" and a "1"). **Dynamic control** in a quantum system is called **coherent control**. So, since a classical system with a zero and a 1 is called a "bit", a quantum system that can be in both states at the same time is quantum: a quantum bit or ***qubit***. The qubit is a remarkable concept because it can store an infinite number of combinations of 0 and 1. And the driving interaction above shows you how to go from an arbitrary initial state to any other state. Specifically returning to Eq. 9.102 and Eq. 9.104, for arbitrary initial conditions at time t_1 given by $C_1(t_1)$and $C_2(t_1)$, a matrix can be developed from these results that gives you $C_1(t_1+\tau)$ and $C_2(t_1+\tau)$, i.e.,

$$\begin{bmatrix} \widetilde{C}(t_1+\tau) \\ \widetilde{C}_2(t_1+\tau) \end{bmatrix} = \begin{bmatrix} A & B \\ C & D \end{bmatrix} \begin{bmatrix} \widetilde{C}_1(t_1) \\ \widetilde{C}_2(t_1) \end{bmatrix} \tag{9.113}$$

where we assume that some time $t = 0$, the entire system was in the ground state.

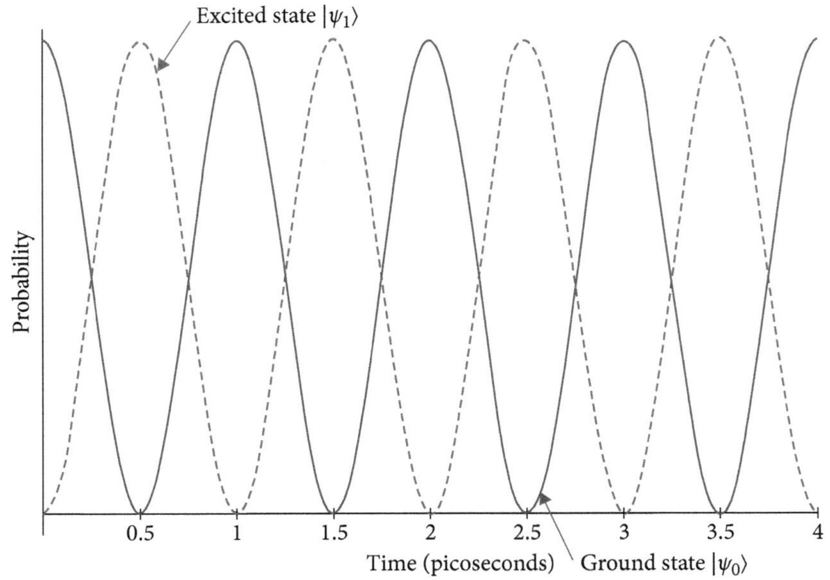

Fig. 9.3 Rabi oscillations when the system is excited on resonance. The system starts out in the ground state and oscillates as a function of time between the ground state and the excited state. The calculation is done for typical parameters in the visible region of the spectrum. Recall that in the visible region, the period of light (500 nm) is about 2 fsec.

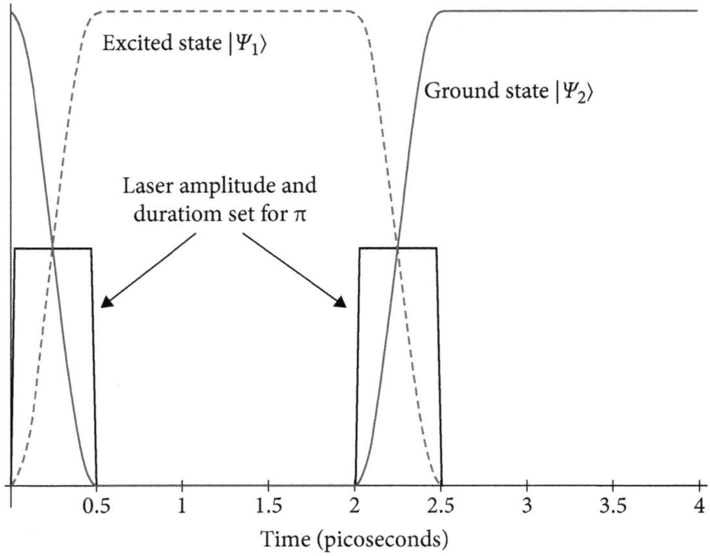

Fig. 9.4 Using a fast gate typical for telecom applications, a square pulse takes the system from ground to excited state. A second pulse returns the system to the initial condition. This functions like a traditional flip-flop.

Problem 9.10

a) With the information above, develop a transfer matrix to relate an initial qubit that is in an arbitrary superposition state to the state produced by a square pulse driving field with temporal width, τ, i.e., find the matrix in Eq. 9.102. For simplicity, set $\Delta = 0$ and take Ω_R to be real. By square pulse, this means that if the field turns on at time t' it turns off at time $t' + \tau$. Referring back to Eq. 9.81, this means that now $E = \frac{1}{2}\left(\mathcal{E}_0(t)e^{-i\omega t} + \mathcal{E}_0^*(t)e^{i\omega t}\right)$ and $\mathcal{E}_0(t) = \mathcal{E}_0$ for $t \leq t' \leq t + \tau$

b) For $\Omega_R(t - t_0) = \frac{\pi}{2}$ find $\begin{bmatrix} \widetilde{C}_1(t - t_0) \\ \widetilde{C}_2(t - t_0) \end{bmatrix}$ and write out the corresponding state vector $|\,\psi(t)$.

This is a coherent superposition state between states 1 and 2, if initially you had taken say $\begin{bmatrix} \widetilde{C}_1(0) \\ \widetilde{C}_2(0) \end{bmatrix} = \begin{bmatrix} 1 \\ 0 \end{bmatrix}$

c) Find the condition on $|\,\Omega_R\,|\,\tau$ so that pulse transforms $\begin{bmatrix} \widetilde{C}_1(t_1) \\ \widetilde{C}_2(t_1) \end{bmatrix} = \begin{bmatrix} 1 \\ 0 \end{bmatrix}$ *to* $\begin{bmatrix} \widetilde{C}_1(t_1 + \tau) \\ \widetilde{C}_2(t_1 + \tau) \end{bmatrix} = \begin{bmatrix} -1 \\ 0 \end{bmatrix}$. *This is called a **geometrical phase** change. Such a change is important for converting various operations for quantum information into the needed gate operation.*

It is now within your ability, with a little guidance, to show that if you had two of these systems side by side in a device, close enough so that their excited states could interact, you could produce a two-qubit device where you controlled each system separately but the state of one system affected where you would or would not control the state of the second system. This is called a controlled NOT gate. In addition, if you add this knowledge to the knowledge you will learn about spin, you will have the foundation of nuclear magnetic resonance used to get three-dimensional structures of large organic molecules and magnetic resonance imaging of biomedical importance.

9.6 **Summary**

The full Schrödinger equation:

$$i\hbar\frac{\partial\ |\psi(t)\rangle}{\partial t} = \widehat{H}\ |\psi(t)\rangle$$

is a first order time dependent equation of motion. Hence, if the state vector or wave function is known at some time $t = t_0$, then the entire time dependence of the state vector or wave function is given by

$$|\psi(t)\rangle = \widehat{U}(t, t_0)\ |\psi\ (t_0)\rangle$$

Where $\widehat{U}(t, t_0)$ is also a solution to Schrödinger's equation. When the Hamiltonian is time independent, the time evolution operator is given by:

$$e^{-i\frac{\widehat{H}(t-t_0)}{\hbar}}$$

where functions of operators are defined by their power series expansion. If the eigenvectors of $\hat{H}$ are known then an arbitrary state vector at t_o, $|\psi(t_o)\rangle$, can be expanded in terms of the eigenvectors, $|E_n\rangle$,

$$|\psi(t_o)\rangle = \sum_n C_n \, |E_n\rangle$$

and then $|\psi(t)\rangle$ is given as

$$|\psi(t)\rangle = \hat{U}(t, t_0) \, |\psi(t_0)\rangle = \sum_n C_n e^{-i\frac{E_n(t-t_0)}{\hbar}} \, |E_n\rangle$$

Using the time evolution operator, a different approach to quantum dynamics was examined by defining the Heisenberg form of the Schrödinger operator. The Schrödinger operator, $\hat{A}_S$, is usually time independent and results from application of the postulates. Using the time evolution operator, the Heisenberg operator was defined as:

$$\hat{A}_H(t) = \hat{U}^{\dagger}(t)\hat{A}_S\hat{U}(t)$$

and satisfies the equation of motion

$$-i\hbar\frac{d}{dt}\hat{A}_H(t) = \left[\hat{H}, \hat{A}_H\right]$$

which follows from Schrödinger's equation and satisfies the initial condition

$$\hat{A}_H(t = 0) = \hat{A}_S$$

In the discussion of dynamics, this section was concluded by examining one of the most well-known problems in quantum mechanics, the problem of a two-level system driven by a sinusoidal field. In this discussion, the idea of the two-level approximation was introduced with a field that was resonant, or nearly resonant, with the corresponding frequency associated with the transition energy, ω_0. The rotating wave approximation was introduced to keep terms that reflected near resonant operation, namely $\omega - \omega_0$ and terms that varied like $\omega + \omega_0$ are ignored. In this case, the system was seen to oscillate the Rabi frequency, $\Omega_R = \frac{|\mu_{21}\mathcal{E}_0|}{\hbar}$.

A very powerful system for technology is the basic two-level structure in any quantum system. Such a system is possible for most Hamiltonians of interest, except for the harmonic oscillator. The two-level system can be controlled with oscillating electromagnetic fields where the frequency of the field is very close (near or in resonance) to the transition frequency given by the energy difference between the two states over h-bar: $\frac{E_2-E_1}{\hbar}$. In fact, it is possible to completely switch the system with a short pulse of radiation. This is the basis of not only many devices under development now, but also the basis for magnetic resonance imaging.

Vocabulary (page) and Important Concepts

- time evolution operator 135
- stationary state solutions 135
- Schrödinger picture 15
- time evolution operator 135
- unitary operator 136
- coherent superposition 138
- Heisenberg picture 139
- Heisenberg operator 139
- Heisenberg equation of motion 140
- interaction picture 143
- semiclassical description 147
- electric dipole moment 147
- dipole approximation 147
- two-state system 146
- scalar approximation 148
- stationary state solutions 148
- electromagnetic interaction Hamiltonian 148
- resonance 150
- two-level approximation 150
- rotating wave approximation 150
- two-level approximation 150
- Rabi frequency 150
- first order time dependent perturbation 153
- field interaction picture 154
- generalized Rabi frequency 154
- Rabi oscillation 155
- coherent superposition 155
- dynamic control 155
- coherent control 155
- qubit 115
- geometrical phase 157

10 Angular Momentum and the Quantum Gyroscope: The Emergence of Spin

10.1 Introduction

Understanding that **angular momentum** is central to understanding quantum systems with spherical symmetry, such as atoms, as discussed in Chapter 6, or spherical quantum dots made from semiconductors. However, the physics of angular momentum is also central to the performance of inertial guidance systems such as found historically in submarines and airplanes. An inertial guidance system is composed of a gyroscope, accelerometers, and an accurate clock. There is a growing need for smaller and more accurate gyroscopes for such guidance systems on platforms such as drones and autonomous vehicles, in case the global positioning system fails or is compromised. Given the importance of the gyroscope in this technology, and the fact that the performance of these systems is being pushed to the quantum regime, we consider a more powerful approach to understanding angular momentum in this chapter, using the Heisenberg operator approach described in Chapter 8.

The basic gyroscope is shown in Fig. 10.1. In modern technology, the system shown is replaced with other kinds of more suitable rotating systems, but in all cases, the physics of systems related to guidance is associated with rotation, and hence angular momentum.

We are not going to focus heavily on the technology, but with the motivation that quantum gyros have a place in the engineer's quantum toolbox, this chapter is going to revisit rotation and angular momentum that was introduced very briefly in the discussions in Chapter 6 on systems with spherical symmetry such as atoms and quantum dots. We will use a Heisenberg type approach as we did in our second encounter with the nano-vibrator and LC circuit (Chapter 8) rather than solving differential equations as we did in Chapter 6 (by looking up the solution). A remarkable result will be that the math predicts the existence of a new kind of angular momentum called spin, which does not correspond to any kind of rotation in real space. This is a profound result and is the basis of quantum information, as we discuss below as well as accounting for the statistical description of electrons in common semiconductors. The algebra we develop to understand spin is surprisingly directly applicable to any two-level (on-off) quantum system, independent of whether spin is present or not.

10.2 The Heisenberg Approach

We begin by recalling the definition of angular momentum: ***Angular momentum*** *is a vector that points in a direction perpendicular to the* ***plane of rotation*** *of a massive object*. Specifically:

Introduction to Quantum Nanotechnology: A Problem Focused Approach. Duncan Steel, Oxford University Press (2021).
 DOI: 10.1093/oso/9780192895073.003.0010

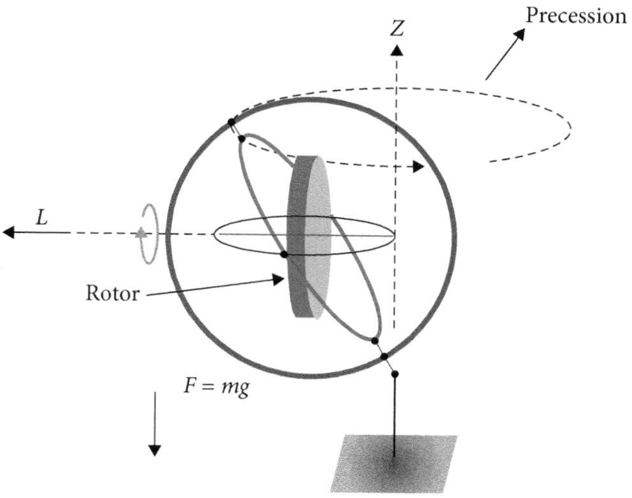

Fig. 10.1 The classical gyroscope. This or its variants are critical in inertial guidance systems. The rotor is a solid disk here, with mass m and a momentum of inertia $I = \frac{1}{2}mr^2$ in the direction perpendicular to rotation. If L is the angular momentum of the rotating disc, the kinetic energy of the disk is $E = \frac{L^2}{2I}$. Rotation, looking at the disk from the right side of the figure is clockwise. The torque is out of the page and precession is counter-clockwise, looking down on the gyroscope.

$$\boldsymbol{L} = \boldsymbol{r} \times \boldsymbol{p} \tag{10.1}$$

with components, L_x, L_y, and L_z evaluated in the usual way with the components of a cross product:

$$\begin{aligned} L_x &= yp_z - zp_y \\ L_y &= zp_x - xp_z \\ L_z &= xp_y - yp_x \end{aligned} \tag{10.2}$$

or

$$L_u = vp_w - wp_v \text{ under cyclic permutation so } x, y, z \to y, z, x \to z, x, y \tag{10.3}$$

In quantum, all of these dynamical observables are operators, and so

$$\begin{aligned} \hat{L}_x &= \hat{y}\hat{p}_z - \hat{z}\hat{p}_y \\ \hat{L}_y &= \hat{z}\hat{p}_x - \hat{x}\hat{p}_z \\ \hat{L}_z &= \hat{x}\hat{p}_y - \hat{y}\hat{p}_x \end{aligned} \tag{10.4}$$

As you might recall or can see intuitively in Fig. 10.1, a classical gyroscope precesses around the z-axis when subject to a force. In spite of the **precession**, there are two quantities that do not change. That is to say they are constants of the motion. One is the magnitude of $\boldsymbol{L}$ and the other is the projection of the angular momentum along the z-axis, L_z. The magnitude of L is a scalar and represented by $\boldsymbol{L} \cdot \boldsymbol{L} = L^2$. The corresponding operators define the entire Hilbert space associated with angular momentum and hence comprise a **complete set of commuting operators (CSCO)**. The choice of the z-axis is arbitrary but it is the preferred axis by convention and is called the ***axis of quantization***.

In terms of the operator form, we have now (from the postulates) three commutators in three dimensions associated with the three canonical coordinates, x, y, z, and three corresponding conjugate momentum p_x, p_y, p_z. Namely, $[\hat{u}, \hat{p}_u] = i\hbar$ for $u = x, y, z$:

$$\begin{aligned} [\hat{x}, \hat{p}_x] &= i\hbar \\ [\hat{y}, \hat{p}_y] &= i\hbar \\ [\hat{z}, \hat{p}_z] &= i\hbar \end{aligned} \tag{10.5}$$

From these commutators, it is now straightforward to show that

$$\left[\hat{L}_x, \hat{L}_y\right] = i\hbar \hat{L}_z \text{ under cyclic permutation.} \tag{10.6}$$

And it is easy to show that

$$\left[\hat{L}_u, \hat{L}^2\right] = 0 \text{ where } \hat{L}^2 = \hat{L}_x^2 + \hat{L}_y^2 + \hat{L}_z^2 \text{ and } u = x, y, z \tag{10.7}$$

Problem 10.1 *Show*

$$\left[\hat{L}_x, \hat{L}_y\right] = i\hbar \hat{L}_z \text{ under cyclic permutation (meaning } x \to y, y \to z, z \to x)$$

Problem 10.2 *Show*

$$\left[\hat{L}_u, \hat{L}^2\right] = 0 \text{ where } \hat{L}^2 = \hat{L}_x^2 + \hat{L}_y^2 + \hat{L}_z^2.$$

The problem is to find the eigenvalues and eigenvectors of this problem. Since classically, once L^2 and L_z are fixed, they are constants of the motion (do not change in time in the absence of loss) and since they form a complete set of commuting observables, they represent all that is knowable about the system. We initially identify this state with Dirac notation as, $|L, L_z\rangle$, where the eigenvector equations are

$$\hat{L}^2 \mid L, L_z\rangle = L^2 \mid L, L_z\rangle \tag{10.8}$$

$$\hat{L}_z \mid L, L_z\rangle = L_z \mid L, L_z\rangle \tag{10.9}$$

L^2 and L_z are just symbols representing scalars. Only after the analysis do we know their relationship to the eigenvalues and what values can be measured.

From here, it appears challenging to make progress without returning to the differential equation form for these operators that we examined in Chapter 6. However, recall the approach used with a quadratic Hamiltonian in Chapter 8. We take a similar path here.

By definition,

$$\hat{L}^2 = \hat{L}_x^2 + \hat{L}_y^2 + \hat{L}_z^2 \tag{10.10}$$

This is sum of quadratics where $\hat{L}_z$ is special in the sense that we chose it to be a part of the complete set of commuting operators to define the Hilbert space. Recalling that $\hat{L}_x$ and $\hat{L}_y$ are Hermitian, we define an operator $\hat{L}_+$ and its adjoint, $\hat{L}_-$, as we did in Chapter 8:

$$\hat{L}_\pm \equiv \hat{L}_x \pm i\hat{L}_y \tag{10.11}$$

The next thing to do now is to see if this definition leads to new information. We start by evaluating the various new commutators. Noting that

$$\hat{L}_\pm \hat{L}_\mp = \hat{L}^2 - \hat{L}_z^2 \pm \hbar \hat{L}_z \tag{10.12}$$

then the relevant commutators are

$$[\hat{L}_+, \hat{L}_-] = \hat{L}^2 - \hat{L}_z^2 + \hbar\hat{L}_z - (\hat{L}^2 - \hat{L}_z^2 - \hbar\hat{L}_z) = 2\hbar\hat{L}_z \tag{10.13}$$

$$[\hat{L}^2, \hat{L}_\pm] = 0 \tag{10.14}$$

$$[\hat{L}_z, \hat{L}_\pm] = \pm\hbar\hat{L}_\pm \tag{10.15}$$

Again, the use of $\pm$ signs means the equation holds for the upper sign and for the lower sign.

Since $L^2 = L_x^2 + L_y^2 + L_z^2$, then $L^2 \geq L_z^2$. This is not saying anything other than the z-projection of the angular momentum vector cannot be bigger than the total vector itself. Then, given Eqs 10.8 and 10.9 and using the upper sign of 10.15, we have that

$$\begin{aligned}\hat{L}_z\hat{L}_+ \mid L, L_z\rangle &= (\hat{L}_+\hat{L}_z + \hbar\hat{L}_+) \mid L, L_z\rangle = \hat{L}_+ (\hat{L}_z + \hbar) \mid L, L_z\rangle = \hat{L}_+ (L_z + \hbar) \mid L, L_z\rangle \\ &= (L_z + \hbar) \hat{L}_+ \mid L, L_z\rangle = (L_z + \hbar) C_+ \mid L, L_z + \hbar\rangle \end{aligned} \tag{10.16}$$

This is a new eigenvalue equation and shows that $\hat{L}_+ \mid L, L_z\rangle$ corresponds to a state with a projection of the angular momentum along the z-axis that is 1 unit of $\hbar$ larger than the original state. In addition, because of Eq. 10.14,

$$\hat{L}^2\hat{L}_+ \mid L, L_z\rangle = L^2\hat{L}_+ \mid L, L_z\rangle \tag{10.17}$$

meaning that $\hat{L}_+ \mid L, L_z\rangle$ remains an eigenstate of $\hat{L}^2$ with the same eigenvalue as for $\mid L, L_z\rangle$. We could continue to apply $\hat{L}_+$ now on $\hat{L}_+ \mid L, L_z\rangle$ and then on $\hat{L}_+\hat{L}_+ \mid L, L_z\rangle$ and so on, each time increasing the value of the projection of the angular momentum on the z-axis. At some point, we violate the condition above that $L^2 \geq L_z^2$. Let $L_z = a$ be that maximum value. Then, as in the case of the nano-vibrator, we require

$$\hat{L}_+ \mid L, a\rangle = 0 \tag{10.18}$$

We now operate with $\hat{L}_-$ on $\hat{L}_+ \mid L, a\rangle = 0$, and use the lower sign in Eq. 10.12 to get

$$\hat{L}_-\hat{L}_+ \mid L, a\rangle = (\hat{L}^2 - \hat{L}_z^2 - \hbar\hat{L}_z) \mid L, a\rangle = (L^2 - a^2 - \hbar a) \mid L, a\rangle = 0 \tag{10.19}$$

Therefore, $L^2 - a^2 - \hbar a = 0$ or

$$L^2 = a(a + \hbar) \tag{10.20}$$

Similarly for L_- and using the lower sign in Eq. 10.15:

$$\hat{L}_z \hat{L}_- \mid L, L_z\rangle = (L_z - \hbar)\,\hat{L}_- \mid L, L_z\rangle = (L_z - \hbar)\,C_- \mid L, L_z - \hbar\rangle \tag{10.21}$$

As with $\hat{L}_+$, we can operate repeatedly, lowering the projection of angular momentum by 1 unit of $\hbar$ each time. We consider the result of operating n times on the state $\mid L, a\rangle$:

$$\hat{L}_z \hat{L}_-^n \mid L, a\rangle = (a - n\hbar)\,\hat{L}_-^n \mid L, a\rangle \tag{10.22}$$

As some point, n will be large enough that $(L_z - n\hbar)^2$ will also violate the condition that $L^2 \geq L_z^2$. Let n correspond to the largest value still satisfying the requirement that $L^2 \geq L_z^2$. Then

$$\hat{L}_-\left(\hat{L}_-^n \mid L, a\rangle\right) = 0 \tag{10.23}$$

and

$$\hat{L}_+\hat{L}_-\left(\hat{L}_-^n \mid L, a\rangle\right) = \left(\hat{L}^2 - \hat{L}_z^2 + \hbar\hat{L}_z\right)\left(\hat{L}_-^n \mid L, a\rangle\right) = 0 \tag{10.24}$$

or

$$\left[L^2 - (a - n\hbar)^2 + \hbar(a - n\hbar)\right]\left(\hat{L}_-^n \mid L, a\rangle\right) = 0 \tag{10.25}$$

Implying $\left[L^2 - (a - n\hbar)^2 + \hbar(a - n\hbar)\right] = 0$ leading to

$$L^2 = (a - n\hbar)^2 - \hbar(a - n\hbar) \tag{10.26}$$

Using Eqs 10.20 and 10.26, we solve for a, and we get

$$a = n\frac{\hbar}{2}, n = 1, 2, 3, \tag{10.27}$$

Using Eq. 10.20, we get that the eigenvalues of L^2 are

$$L^2 = \frac{n}{2}\left(\frac{n}{2} + 1\right)\hbar^2 \tag{10.28}$$

and the eigenvalues of L_z are

$$L_z = m\hbar \tag{10.29}$$

where m is either integer or half integer, but the change in m is always an integer.

We have found that eigenvectors of $\hat{L}^2$ have eigenvalues of $\frac{n}{2}\left(\frac{n}{2} + 1\right)\hbar^2$ and for a given eigenvector of $\hat{L}^2$, the projection along the z-axis corresponding to the eigenvalues of $\hat{L}_z$ can range from $\frac{n}{2}$ to $-\frac{n}{2}$.

However, in the previous discussion in Chapter 6, *the value of m was restricted to integer to ensure that a rotation of the system of 2π gave the same analytical result*. The implication here, then, is that $\frac{n}{2}$ must be integer m. With the restriction on m, this is similar to our results in Chapter 6 when we were discussing the coordinate space solution to the eigenvalue problem where the eigenfunctions corresponded to spherical harmonics, Y_{lm}, and where the quantum numbers, l and m, are integers and

$$\hat{L}^2 Y_{lm} = l(l+1)\hbar^2 Y_{lm} \tag{10.30}$$

and

$$\hat{L}_z Y_{lm} = m\hbar Y_{lm}. \tag{10.31}$$

We then replace the notation $|L, L_z\rangle$ with the notation $|l, m\rangle$, where m is the sometimes called the quantum number for the **magnetic substate**. The magnetic substates are degenerate, but you see in Problem 10.4, that the magnetic field couples these states, and in Chapter 11 you will see that applying a magnetic field lifts the degeneracy.

One very important result that will be useful below is that

$$\hat{L}_\pm\, |l, m\rangle = \hbar\sqrt{(l \mp m)(l \pm m + 1)}\; |l, m \pm 1\rangle \tag{10.32}$$

$\hat{L}_\pm$ are called ladder operators, or raising and lowering operators, for obvious reasons. A summary of the important results so far is given in Table 10.1 where the symbol $\hat{L}$ is replaced by the symbol $\hat{J}$. This is because $\hat{L}$ is usually associated with orbital angular momentum and the symbol $\hat{J}$ stands for any kind of angular momentum, even if it is not orbital. However, care is

Table 10.1 Summary of important relationships for angular momentum operators. The symbol J is used to represent any kind of angular momentum, as discussed below.

$\hat{J}^2 \mid j, m\rangle = j(j+1)\hbar^2 \mid j, m\rangle$
$\hat{J}_z \mid j, m\rangle = m\hbar \mid j, m\rangle$
$[\hat{J}_x, \hat{J}_y] = i\hbar\hat{J}_z$ under cyclic permutation
$\mathrm{J}_\pm \equiv \hat{J}_x \pm i\mathrm{J}_y$
$\hat{J}_x = \frac{1}{2}(\mathrm{J}_+ + \mathrm{J}_-)$
$\hat{J}_y = \frac{1}{2i}(\mathrm{J}_+ - \mathrm{J}_-)$
$\hat{J}_\pm\hat{J}_\mp = \hat{J}^2 - \hat{J}_z^2 \pm \hbar\hat{J}_z$
$[\hat{J}_+, \hat{J}_-] = 2\hbar\hat{J}_z$
$[\hat{J}_z, \hat{J}_\pm] = \pm\hbar\hat{J}_\pm$
$\hat{J}_\pm \mid j, m\rangle = \hbar\sqrt{(j \mp m)(j \pm m + 1)} \mid j, m \pm 1\rangle$

needed because $\hat{J}$ can also represent the total angular momentum, when including the angular momentum due to spin. Usually, authors will include a comment as to the meaning.

Problem 10.3 *In Eqs 10.16 and 10.21, you found that operating with $\hat{L}_\pm$ changed the eigenvalue by one unit of $\hbar$. So it must be the case that $\hat{L}_\pm |l, m\rangle = C_\pm |l, m \pm 1\rangle$. Similarly $\langle l, m | \hat{L}_\mp = C_\pm^* \langle l, m \pm 1|$. Use these two results to verify Eq. 10.32.*

Problem 10.4 *A quantum gyroscope. A small solid disk or even a simple diatomic molecule can be a rotating object, sometimes also called a **rigid rotor**. Such an object rotating about its center of mass is shown in Fig. 10.2. If we assume that it is somehow anchored in space so that it does not behave as a free particle, the Hamiltonian is*

$$\hat{H} = \frac{\hat{L}^2}{2I}$$

*$I = \sum_{j=1}^{N} m_j r_j^2$ is the usual **moment of inertia** where m_j is the mass at j. Of course, the eigenvectors are just $|l, m\rangle$ (m is the magnetic quantum number, not the mass, m_j). So, for each value of l, there are $2l + 1$ values of m, but the energy does not depend on m. This makes sense since the energy of a rotating disc cannot depend on how the axis of coordinates is chosen. We now assume that the disc is magnetized, so that it has with it a magnetic moment (a north and south pole) pointing along the axis of rotation. The **magnetic moment** for an orbiting electron is given by $\boldsymbol{\mu}_M = \mu_M \mathbf{L}$, $\mu_M = -\frac{e}{2m}$ and e is the magnitude of the electron charge. In this case, we will just take the disk to have a magnetic moment of μ_D. In a magnetic field, the potential energy of a magnetic moment is given by*

$$\hat{V} = -\boldsymbol{\mu}_M \cdot \mathbf{B} = -\mu_M \mathbf{L} \cdot \mathbf{B}$$

so that the total Hamiltonian is now

$$\hat{H} = \frac{\hat{L}^2}{2I} - \mu_M \mathbf{L} \cdot \mathbf{B} \tag{10.33}$$

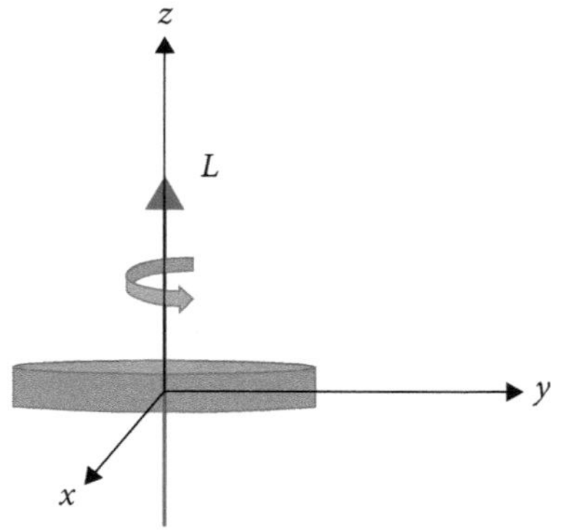

Fig. 10.2 A disc of radius *R* with mass *m* has a **moment of inertia** given by $I = \frac{1}{2} mR^2$.

The axis of quantization is along the axis of rotation. Note that, while the axis of quantization is completely arbitrary, some choices are easier to use for calculations than others. We take that axis to be z. The corresponding other axes are x and y. Assume $l = 1$ to make this simple. Therefore, there are three basis states (m has three values, 0, ±1). Write out the matrix form of the Hamiltonian where the columns are labeled by $|1,-1\rangle$, $|1,0\rangle$, and $|1,1\rangle$ and the rows by the corresponding bra. Show that

$$\widehat{H} = \frac{\hat{L}^2}{2I} - \mu_M \mathbf{L} \cdot \mathbf{B} = \begin{bmatrix} \frac{\hbar^2}{I} + \mu_M \hbar B_z & -\mu_M \frac{\hbar}{\sqrt{2}} (B_x + iB_y) & 0 \\ -\mu_M \frac{\hbar}{\sqrt{2}} (B_x - iB_y) & \frac{\hbar^2}{I} & -\mu_M \hbar \frac{1}{\sqrt{2}} (B_x + iB_y) \\ 0 & -\mu_M \hbar \frac{1}{\sqrt{2}} (B_x - iB_y) & \frac{\hbar^2}{I} - \mu_M \hbar B_z \end{bmatrix}$$

You will need Eq. 10.30.

10.3 Intrinsic Angular Momentum: Spin

In Eq. 10.28, we found that $L^2 = \frac{n}{2}\left(\frac{n}{2}+1\right)\hbar^2$. We then said that $\frac{n}{2}$ is an integer because in the discussion on orbital angular momentum in Chapter 6, we said that m had to be an integer so that $e^{im\phi} = e^{im(\phi+2\pi)}$. If m could be half-integer, then with m being a half-integer, $e^{im\phi}$ would be periodic only in 4π. An object would have to turn around twice to get back to the same value of the eigenstate. This is clearly unphysical.

But *what if the angular momentum was not associated with a rotating object? What if, somehow, angular moment was intrinsic, something a part of a seemingly point particle? Then it would not be associated with rotation in real space.* What you see here is that using a purely operator approach without recourse to a real space representation for the operators and using only the commutator relations, the formalism has predicted, or shown that in principle, we could have something with units of half angular momentum if it was *not* associated with a real space rotation. This is quite remarkable.

Experimental data came next. We will not go over the details of the Stern–Gerlach experiment and advanced atomic and molecular spectroscopy measurements. However, it is important to know that many charged particles have an **intrinsic angular momentum** that gives rise to a magnetic field that, like the needle on a compass, interacts with an applied magnetic field. The interaction with an applied magnetic field, B, with the orbiting electrons is given by

$$V = -\boldsymbol{\mu}_L \cdot \mathbf{B} \tag{10.34}$$

where, as discussed chapter 6 (see problem 6.4),

$$\boldsymbol{\mu}_L = -\frac{e}{2m_e}\mathbf{L} = -\mu_B \frac{\mathbf{L}}{\hbar} \tag{10.35}$$

wherem_e is the mass of the electron and

$$\mu_B = \frac{|e|\hbar}{2m_e}$$

is the **Bohr magneton.**

But here is where the problem lies. If the magnetic field, $\boldsymbol{B}$, is taken in the z-direction then $V = -\frac{eB_z}{2m_e} L_z$. Since $L_z = m$ and m is an integer, then if the total angular moment is $l = 1$, then we see that there would be three different possible energies for the state, $m = 0, \pm 1$. In fact, for any integer l, there are an *odd number of energy eigenstates*. The details of this calculation are presented in Chapter 11. However, when they did measurements on an electron system with $l = 0$, they discovered two different states. It meant that there was something in the system that had a half integer value for the total angular momentum quantum number and had a magnetic moment but was *not* associated with rotation in real space, i.e., not associated with orbital motion. Furthermore, the spacing of the two energies was wrong.

This was a major problem and took years—with many people involved—to resolve this problem. But ultimately, it was addressed by postulating that the electron had associated with it an **intrinsic angular momentum**, with value $\boldsymbol{S}$ (same units as L, units of $\hbar$). This feature is called the electron **spin.** The existence of spin represents an additional **postulate** in quantum mechanics. The postulate is that particles can have intrinsic angular momentum. It turns out that other particles such as the proton, neutron, and nuclei also have spin. Dirac eventually showed that the spin of the electron naturally emerges in relativistic quantum mechanics. But since this is a non-relativistic approach to quantum in these notes, we simply take it as a **postulate of spin** as done originally in the development of quantum theory.

The operators associated with $\boldsymbol{S}$ follow the same algebra as the operators associated with orbital angular momentum (Table 10.1), but the eigenvector is designated as $|s, m_s\rangle$ compared to $|l, m_l\rangle$. The subscript is added to the magnetic quantum number differentiate the two kinds of angular momentum. The spin, s, is 1/2 for the electron, proton, and neutron and $m_s = \pm\frac{1}{2}$. Nuclei can have integer or half-integer spin, depending on the isotope. Therefore, following the rules above for the behavior of the eigenstates,

$$\hat{S}^2 \, |s, m_s\rangle = s(s+1)\hbar^2 \, |s, m_s\rangle \tag{10.36}$$

and

$$\hat{S}_z \, |s, m_s\rangle = m_s\hbar \, |s, m_s\rangle \tag{10.37}$$

The other relationships involving operators such as $\hat{S}_\pm$ also hold for the spin. Part of the postulate included a correction to the value of the magnetic moment, in order to get the correct energy spacing. This number is called the electron spin **g-factor**. So, for the spin, the electron magnetic moment is given by

$$\boldsymbol{\mu}_s = -g_s\mu_B \frac{\boldsymbol{S}}{\hbar} \tag{10.38}$$

where $g_s = 2$ for the electron (to a very good approximation). Ultimately, this physics came out of a unified development of relativistic quantum mechanics. However, it is common to just accept this as another postulate. Protons, neutrons, nuclei, and other systems can also have intrinsic angular momentum. Nuclear spin is the basis for magnetic resonance imaging.

There are more issues for this new concept. *There is no eigenfunction representation of the system here because there is no spatial function associated with the spin.* The spin is intrinsic.

Dirac notation is required for this. It is not necessary to identify explicit vectors for the eigenvectors or matrices for the operators, but it is sometimes helpful, especially if the reader is comfortable with matrix algebra. Since there are two eigenvectors, it is easy to identify them for $|s, m_s\rangle$ as

$$\left|\frac{1}{2},+\frac{1}{2}\right\rangle \equiv \left|+\frac{1}{2}\right\rangle |\uparrow\rangle\, |+\rangle \equiv |\uparrow\rangle = \begin{bmatrix} 1 \\ 0 \end{bmatrix} \tag{10.39}$$

and

$$\left|\frac{1}{2},-\frac{1}{2}\right\rangle \equiv \left|-\frac{1}{2}\right\rangle |\downarrow\rangle\, |-\rangle \equiv |\downarrow\rangle = \begin{bmatrix} 0 \\ 1 \end{bmatrix} \tag{10.40}$$

The eigenvectors are sometimes called **eigenspinors**. We see that, unlike problems such as the nano-vibrator of Chapter 2 which has an infinite number of states to define the Hilbert space, spin has only *two* states and forms a two-dimensional Hilbert space.

To see the operators in matrix form, we create the **spin matrices** using the eigenspinors as the basis; namely, a matrix in the eigenspinors basis is given by $\widehat{M} = \sum_{m_s m_s'} \langle m_s|\widehat{M}|m_s'\rangle |m_s\rangle\langle m_s'|$, where $|m_s\rangle$ is shorthand for $|s, m_s\rangle$. The order of the column labels from left to right is $m_s = +\frac{1}{2}$ and $m_s = -\frac{1}{2}$, respectively. The rows are labeled the same, from top to bottom. The matrix elements are evaluated using the results shown in Table 10.1.

$$\hat{S}_z = \sum_{m_s m_s'} \langle m_s|\hat{S}_z|m_s'\rangle\, |m_s\rangle\langle m_s'| = \frac{1}{2}\hbar \begin{bmatrix} 1 & 0 \\ 0 & -1 \end{bmatrix} \tag{10.41}$$

$$\hat{S}_x = \sum_{m_s m_s'} \langle m_s|\hat{S}_x|m_s'\rangle\, |m_s\rangle\langle m_s'| = \frac{1}{2}\hbar \begin{bmatrix} 0 & 1 \\ 1 & 0 \end{bmatrix} \tag{10.42}$$

$$\hat{S}_y = \sum_{m_s m_s'} \langle m_s|\hat{S}_y|m_s'\rangle\, |m_s\rangle\langle m_s'| = \frac{1}{2}\hbar \begin{bmatrix} 0 & -i \\ i & 0 \end{bmatrix} \tag{10.43}$$

$$\hat{S}_+ = \sum_{m_s m_s'} \langle m_s|\hat{S}_+|m_s'\rangle\, |m_s\rangle\langle m_s'| = \hbar \begin{bmatrix} 0 & 1 \\ 0 & 0 \end{bmatrix} \tag{10.44}$$

$$\hat{S}_- = \sum_{m_s m_s'} \langle m_s|\hat{S}_-|m_s'\rangle\, |m_s\rangle\langle m_s'| = \hbar \begin{bmatrix} 0 & 0 \\ 1 & 0 \end{bmatrix} \tag{10.45}$$

There is a corresponding set of matrices called the Pauli spin matrices, defined corresponding to Eqs 10.41–10.43. **Pauli matrices** are dimensionless and do not include the factor of $\frac{\hbar}{2}$:

$$\begin{aligned} \hat{\sigma}_z &= \begin{bmatrix} 1 & 0 \\ 0 & -1 \end{bmatrix} \\ \hat{\sigma}_x &= \begin{bmatrix} 0 & 1 \\ 1 & 0 \end{bmatrix} \\ \hat{\sigma}_y &= \begin{bmatrix} 0 & -i \\ i & 0 \end{bmatrix} \end{aligned} \tag{10.46}$$

These operators are often useful for a number of calculations and generalized to an even more useful set of operators called atomic operators (Chapter 16) for historical reasons. They are central to many problems including quantum information.

An electron in an energy eigenstate of the hydrogen atom would be written as $u_{nlm_l s m_s} = NR_{nl}(r)Y_{lm_l}(\theta,\varphi)\,|s,m_s\rangle$. The equivalent eigenvector could be written as $|n,l,m_l,s,m_s\rangle = |n,l,m_l\rangle\,|s,m_s\rangle$. The Hilbert space formed by the product of the Hilbert space for the Coulomb component, $|n,l,m_l\rangle$, and the Hilbert space for the spin, $|s,m_s\rangle$, is a **product space**.

In terms of spin dynamics, the Dirac notation and the Heisenberg approach provide all the information that can be had in the quantum formalism. It is a remarkable achievement. The formalism for which this is the basis is even more powerful. Some of that power will be used in later discussions.

Problem 10.5 *Develop equations 10.41–10.45. The easiest approach is to use Eq. 10.11 to write out $\hat{S}_x$ and $\hat{S}_y$ and then use the results of $\hat{S}_\pm$ and the results of 10.32 adapted for spin as*

$$\hat{S}_\pm \mid s,m_s\rangle = \hbar\sqrt{(s\mp m_s)(s\pm m_s+1)}\mid s,m_s\pm 1\rangle$$

Recall from Appendix A that the outer product of, say, two two-dimensional vectors is given by:

$$|A\rangle\langle B| = \begin{bmatrix} a_1 \\ a_2 \end{bmatrix}\begin{bmatrix} b_1^* & b_2^* \end{bmatrix} = \begin{bmatrix} a_1 b_1^* & a_1 b_2^* \\ a_2 b_1^* & a_2 b_2^* \end{bmatrix}$$

Problem 10.6 *Nuclei have spin, also. The operator for nuclear spin is $\hat{I}$, but we will just continue to use $\hat{S}$. For a proton and neutron the spin is ½, but the total angular momentum for the different isotopes has to be measured. For simplicity, we work with just hydrogen with spin ½. The point of this problem is to get the basic equations of motion and their behavior that are used to develop images in magnetic resonance imaging (MRI). The math is identical to that in the two-level Rabi problem in Chapter 10. The Hamiltonian of interest consists of a DC magnetic field that leads to the spin states $|\uparrow\rangle$ and $|\downarrow\rangle$ having different energies and an oscillating magnetic field oscillating on resonance in the x-direction:*

$$\hat{H} = -\boldsymbol{\mu}_S \cdot B_0\hat{\mathbf{z}} - \boldsymbol{\mu}_S \cdot B\hat{\mathbf{x}}\cos\omega t$$

Note that the charge on the proton is positive so $\boldsymbol{\mu}_S = +g\frac{|e|}{2m_e}\hat{\mathbf{S}}$ where $|e|$ remains the magnitude of the charge.

Show that this reduces to

$$\hat{H} = -\boldsymbol{\mu}_S \cdot B_0\mathbf{z} - \boldsymbol{\mu}_S \cdot B\mathbf{x}\cos\omega t = -\hbar\frac{g|e|B_0}{4m_p}\begin{bmatrix} 1 & 0 \\ 0 & -1 \end{bmatrix} - \hbar\frac{g|e|B}{4m_p}\begin{bmatrix} 0 & 1 \\ 1 & 0 \end{bmatrix}\cos\omega t$$

$$= -\hbar\frac{\omega_0}{2}\begin{bmatrix} 1 & 0 \\ 0 & -1 \end{bmatrix} - \hbar\Omega\begin{bmatrix} 0 & 1 \\ 1 & 0 \end{bmatrix}\cos\omega t$$

where $\omega_0 = g\frac{|e|B_0}{2m_p}$ and $\Omega = g\frac{|e|B}{4m_p}$

Assume now that at $t=0$, $|\psi(t=0)\rangle = |\downarrow\rangle = \begin{bmatrix} 0 \\ 1 \end{bmatrix}$. Find the equations of motion using the full time dependent Schrödinger equation for the probability amplitudes in the expansion for $|\psi(t)\rangle$, namely starting with

$$|\psi(t)\rangle = C_1(t)e^{-i\frac{\omega_0}{2}t} \mid \downarrow\rangle + C_2(t)e^{i\frac{\omega_0}{2}t} \mid \uparrow\rangle$$

find the equations $\frac{dC_1(t)}{dt}$ *and* $\frac{dC_2(t)}{dt}$ *and solve for* $C_1(t)$ *and* $C_2(t)$*. Note that your work parallels the discussion in* Chapter 9. *Assume that you are on resonance to make the algebra simpler.*

A frequent problem in devices for measurement of quantum systems, especially spin based systems, is that the system is prepared in some state or superposition state, oriented say along the z-axis. Then measurements are made along the x-axis. So in the simplest problem, say $|\psi\rangle = |\downarrow\rangle_z$. Without a doubt, a measurement of $\hat{S}_z$ will give $\downarrow$. But a measurement is going to be made along x which will yield eigenvalues of $\hat{S}_x$. Below, you will calculate the new eigenstates, but this discussion is more than about the math. In preparing the system in state $|\downarrow_z\rangle$, we identified the z-axis as the axis of quantization. But in making a measurement, we are measuring eigenvalues of $\hat{S}_x$. This is a common task. You will consider this in Problem 10.7 below, and determine the probability of finding the particle in the $|\uparrow_x\rangle$ or $|\downarrow_x\rangle$ state corresponding to the magnetic momentum pointing in the $+x$ or $-x$ direction. There, you will need to express the $|\downarrow_z\rangle$ in terms of eigenstates of $\hat{S}_x$. That is to say, you will need to show that

$$|\uparrow\rangle_x = \frac{1}{\sqrt{2}}(|\uparrow\rangle_z + |\downarrow\rangle_z) \tag{10.47}$$

$$|\downarrow\rangle_x = \frac{1}{\sqrt{2}}(|\uparrow\rangle_z - |\downarrow\rangle_z) \tag{10.48}$$

This is the usual expansion of the unknown vector of interest, in terms of the complete basis vectors. Only in this case, the determination of the expansion coefficients is not done by integration. We will have to use the properties of the operators given in Table 10.1, which can then be inverted to expand $|\uparrow\rangle_z$ in terms of the eigenstates for the measurement of $\hat{S}_x$.

Problem 10.7 *Show that* $|\uparrow\rangle_x$ *and* $|\downarrow\rangle_x$ *are given by Eqs 10.47* and *10.48.*

Problem 10.8 *An electron with spin is in a static magnetic field applied along z. This lifts the degeneracy of the two spin states. They are still stationary eigenstates of the Hamiltonian, but there is a time evolution of the phase between them. So the Hamiltonian from above is used without the time varying component:*

$$\hat{H}_0 = -\boldsymbol{\mu}_S \cdot B_0\hat{\mathbf{z}} = \hbar\omega_0\hat{S}_z$$

where

$$\omega_0 = \frac{g\,|e|\,B_0}{2m_e}$$

Recall now that

$$\hat{H}_0\,|\uparrow\rangle_z = \hat{H}\begin{bmatrix} 1 \\ 0 \end{bmatrix} = \hbar\frac{\omega_0}{2}\,|\uparrow\rangle_z$$

and

$$\hat{H}_0 \left|\downarrow\right\rangle_z = \hat{H}\begin{bmatrix} 0 \\ 1 \end{bmatrix} = -\hbar\frac{\omega_0}{2} \left|\downarrow\right\rangle_z$$

At $t = 0$, *the system is prepared in an eigenstate of* $\hat{S}_x$ *given by*

$$\left|\psi\,(t=0)\right\rangle = \left|\uparrow\right\rangle_x = \frac{1}{\sqrt{2}}\left(\left|\uparrow\right\rangle_z + \left|\downarrow\right\rangle_z\right)$$

a) Use the time evolution operator to find $|\psi(t)\rangle$.
b) Find the times when $|\psi(t)\rangle$ *is an eigenstate of* $\hat{S}_y$,
where for $m_{s_y} = -\frac{1}{2}$

$$|\downarrow_y\rangle = \frac{1}{\sqrt{2}}\left(|\downarrow_z\rangle + i|\uparrow_z\rangle\right)$$

and for $m_{s_y} = +\frac{1}{2}$

$$|\uparrow_y\rangle = \frac{1}{\sqrt{2}}\left(|\downarrow_z\rangle - i|\uparrow_z\rangle\right)$$

10.4 The Bloch Sphere and Spin

The electron spin is an example of a two-dimensional Hilbert space describing the state of the system with an arbitrary orientation in three-dimensional real space. If measurements are made along the z-axis, only two values of the angular momentum can be measured corresponding to a projection along the z-axis of $\pm\frac{1}{2}\hbar$. To get a better understanding of the relationship between the quantum description and the vector, the **Bloch sphere** was introduced, as shown in Fig. 10.3.

The Bloch sphere shows the three components of the spin vector. Since each component is an observable, the vector component corresponds to an eigenspinor or superposition of eigenspinors. The sphere has a radius of $\frac{1}{2}\hbar$. The arrow then represents the direction of the spin. If a measurement is made along the z-axis, it will only give a value of $\pm\frac{1}{2}\hbar$ along z (Fig. 10.2, upper left panel) corresponding to spin up $\left(+\frac{1}{2}\right)$ and spin down $\left(-\frac{1}{2}\right)$. If it is measured along the x-axis, the only allowed measurements are $\pm\frac{1}{2}\hbar$ along x (Fig. 10.2, upper right panel) because what is measured must be in an eigenstate of $\hat{S}_x$. The two eigenvectors of $\hat{S}_z$ form a complete set and hence can be used to write the eigenvectors $\hat{S}_x$ or $\hat{S}_y$ as we discussed above. Problem 10.7 will allow you to derive the result for $\hat{S}_x$. In some applications such as a quantum sensor for a magnetic field, the magnetic field lifts the energy degeneracy between the spin states. In this case when the field is along z, the eigenstates of $\hat{S}_x$ are not stationary, because the basis states along the z-axis have different energies. When this happens, the eigenvector

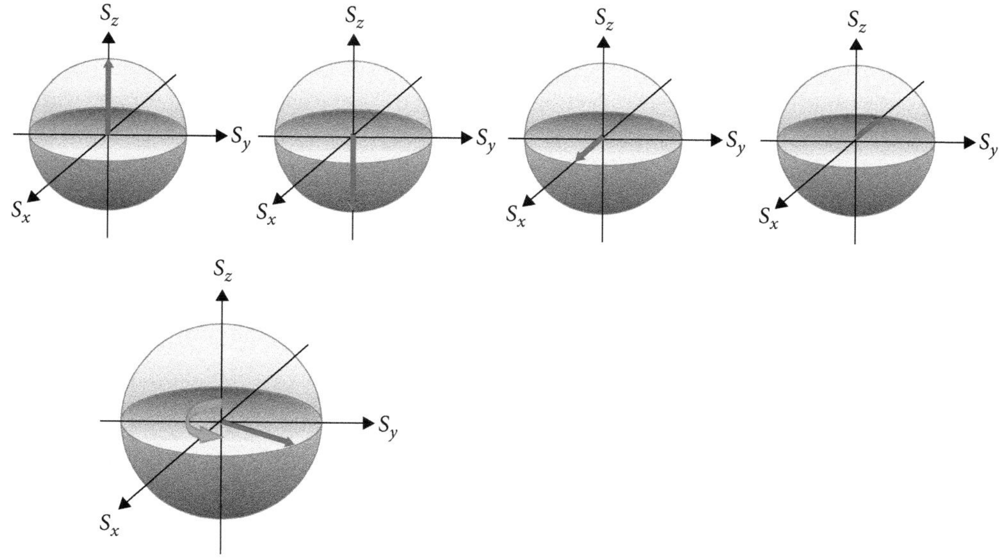

Fig. 10.3 The Bloch sphere with radius $\frac{1}{2}\hbar$. In the upper left frame, measurements are made along z and must give eigenvalues of $\hat{S}_z$. In the upper right frame, measurements are made along x. Now the measurements can only give eigenvalues of $\hat{S}_x$ which, of course, have the same magnitude as for $\hat{S}_z$. If a magnetic field lies along z, the degeneracy between the spin states in the z-basis is lifted. A measurement of $\hat{S}_x$ would find that the spin is precessing around the z-axis, shown in the lower left image.

precesses around the z-axis as shown in Fig. 10.3 (lower left) at a frequency given by the energy difference between the two basis states. Problem 10.8 will allow you to show this.

10.5 Addition of Angular Momentum

It is easy to imagine cases where there are two separate electrons, each with its own angular momentum that might interact with each other. In fact, because they each have a magnetic moment, they create a magnetic field so that the magnetic moment of particle 1, $\boldsymbol{\mu}_{M_1} = -g\frac{|e|}{2m_e}\boldsymbol{L}_1$ creates magnetic field $\boldsymbol{B}_1 = f(\boldsymbol{r})\,\boldsymbol{L}_1$ that interacts with the magnetic moment of particle 2, giving rise to a term in the Hamiltonian that is $\alpha f(\boldsymbol{r})\,\boldsymbol{L}_1 \cdot \boldsymbol{L}_2 + \textit{other terms}$, or maybe a term like $\boldsymbol{L} \cdot \boldsymbol{S}$, if the spin interacts with the orbital angular moment (Chapter 11). The problems can be a little more complicated than just these two kinds of terms, but these terms show the importance of being able to add two angular momenta.

To evaluate this we note that what applies to the vectors applies to the operators. To keep this most general, we let the vector symbol $\boldsymbol{J}$ represent an arbitrary angular momentum (such as $\boldsymbol{L}$ or $\boldsymbol{S}$ or some other angular momentum.) Then we can write the total angular momentum as

$$\boldsymbol{J} = \boldsymbol{J}_1 + \boldsymbol{J}_2 \tag{10.49}$$

where in the case of above, $\boldsymbol{J}_1$and $\boldsymbol{J}_2$ could correspond to $\boldsymbol{L}_1$ and $\boldsymbol{L}_2$. The operator associated with the square of the total angular momentum is then

$$\hat{\boldsymbol{J}}^2 = \hat{\boldsymbol{J}}_1^2 + \hat{\boldsymbol{J}}_2^2 + 2\hat{\boldsymbol{J}}_1 \cdot \hat{\boldsymbol{J}}_2 \tag{10.50}$$

and

$$\hat{\boldsymbol{J}}_z = \hat{\boldsymbol{J}}_{z_1} + \hat{\boldsymbol{J}}_{z_2} \tag{10.51}$$

Therefore,

$$m = m_1 + m_2 \tag{10.52}$$

Furthermore, for obvious physical reasons, $|\, \boldsymbol{J}_1 + \boldsymbol{J}_2 \,|$ is constrained to the inequality:

$$|\, J_1 - J_2 \,| \leq |\, \boldsymbol{J}_1 + \boldsymbol{J}_2 \,| \leq J_1 + J_2 \tag{10.53}$$

So, in addition to Eq. 15.52, we conclude that for the total angular moment, the corresponding complete set of commuting operators for the total angular momentum are, $\hat{\boldsymbol{J}}^2$, and the projection, $\hat{\boldsymbol{J}}_z$, subject to the constraints for Eqs 10.28 and 10.29 and described by eigenvectors, $|\, jm\rangle$, where

$$\hat{\boldsymbol{J}}^2 \,|\, jm\rangle = j\,(j+1)\,\hbar^2 \,|\, jm\rangle$$

and

$$\hat{\boldsymbol{J}}_z \,|\, jm\rangle = m\hbar \,|\, jm\rangle$$

The Hilbert space spanned by $\{|jm\rangle\}$ must be the result of the **product space** defined by $|J_1 m_1\rangle\, |J_2 m_2\rangle$, where with the set of all allowed values of m_1 and m_2, $\{|J_1 m_1\rangle\, |J_2 m_2\rangle\}$ form a complete basis. We can then expand the $|jm\rangle$ state in terms of the $|J_1 m_1\rangle\, |J_2 m_2\rangle$ states which form a complete set. From Eq. 10.49, j_1 and l_2 are fixed, j is fixed at an allowable value constrained by 10.53. We then form a sum over all possible states subject to Eqs 10.51 and 10.52 since j and m are fixed:

$$|\, jm\rangle = \sum_{\substack{m_1, m_2 \\ m_1 + m_2 = m}} C\left(j_1, j_2, m_1, m_2 \middle| jm\right) \,|\, j_1 m_1\rangle \,|\, j_2 m_2\rangle \tag{10.54}$$

The sum notation means a double sum over m_1 and m_2 subject to the condition $m_1 + m_2 = m$

The expansion coefficient, $C\ (j_1, j_2, m_1, m_2 | jm)$, is called a **Clebsch–Gordan coefficient**. These coefficients are calculated using the results of a much more complete analysis. It is typically most convenient to use the standard tables to look up the specific coefficients needed. Many software packages make these coefficients available for calculations. For many systems of interest, the problem is a spin ½ particle interacting with another particle with some arbitrary spin or with its own orbital angular momentum. Both cases involve the coupling of a spin ½ to another component of angular momentum. In this case, the Clebsch–Gordan coefficients can be described by two simple formulas,[1] in Eqs 10.55 and 10.56.

$$C\left(j_1, \frac{1}{2}, m_j - \frac{1}{2}, \frac{1}{2}\middle| j_1 \pm \frac{1}{2}, m_j\right) = \pm\left(\frac{j_1 \pm m_j + \frac{1}{2}}{2j_1 + 1}\right)^{\frac{1}{2}} \tag{10.55}$$

$$C\left(j_1, \frac{1}{2}, m_j + \frac{1}{2}, -\frac{1}{2}\middle| j_1 \pm \frac{1}{2}, m_j\right) = \left(\frac{j_1 \mp m_j + \frac{1}{2}}{2j_1 + 1}\right)^{\frac{1}{2}} \tag{10.56}$$

Table 10.2 A few of the most common Clebsch-Gordan coefficients.

j_1	j_2	$\lvert jm\rangle = \lvert jm\rangle$	$C(j_1, j_2, m_1, m_2 \lvert j, m)$	
$j_1 = \frac{1}{2}$;	$j_2 = \frac{1}{2}$;	$\lvert jm\rangle = \lvert 0,0\rangle$	$C\left(\frac{1}{2}, \frac{1}{2}, -\frac{1}{2}\frac{1}{2}\lvert 0,0\right)$	$-1/\sqrt{2}$
$j_1 = \frac{1}{2}$;	$j_2 = \frac{1}{2}$;	$\lvert jm\rangle = \lvert 0,0\rangle$	$C\left(\frac{1}{2}, \frac{1}{2}, \frac{1}{2} - \frac{1}{2}\lvert 0,0\right)$	$1/\sqrt{2}$
$j_1 = \frac{1}{2}$;	$j_2 = \frac{1}{2}$;	$\lvert jm\rangle = \lvert 1,-1\rangle$	$C\left(\frac{1}{2}, \frac{1}{2}, -\frac{1}{2}, -\frac{1}{2}\lvert 1,-1\right)$	1
$j_1 = \frac{1}{2}$;	$j_2 = \frac{1}{2}$;	$\lvert jm\rangle = \lvert 1,0\rangle$	$C\left(\frac{1}{2}, \frac{1}{2}, -\frac{1}{2}, \frac{1}{2}\lvert 1,0\right)$	$\frac{1}{\sqrt{2}}$
$j_1 = \frac{1}{2}$;	$j_2 = \frac{1}{2}$;	$\lvert jm\rangle = \lvert 1,0\rangle$	$C\left(\frac{1}{2}, \frac{1}{2}, \frac{1}{2}, -\frac{1}{2}\lvert 1,0\right)$	$\frac{1}{\sqrt{2}}$
$j_1 = \frac{1}{2}$;	$j_2 = \frac{1}{2}$;	$\lvert jm\rangle = \lvert 1,1\rangle$	$C\left(\frac{1}{2}, \frac{1}{2}, \frac{1}{2}, \frac{1}{2}\lvert 1,1\right)$	1
$j_1 = 1$;	$j_2 = \frac{1}{2}$;	$\lvert jm\rangle = \left\lvert \frac{1}{2}, -\frac{1}{2}\right\rangle$	$C\left(1, \frac{1}{2}, -1, \frac{1}{2}\lvert \frac{1}{2}, -\frac{1}{2}\right)$	$-\sqrt{2/3}$
$j_1 = 1$;	$j_2 = \frac{1}{2}$;	$\lvert jm\rangle = \left\lvert \frac{1}{2}, -\frac{1}{2}\right\rangle$	$C\left(1, \frac{1}{2}, 0, -\frac{1}{2}\lvert \frac{1}{2}, -\frac{1}{2}\right)$	$\sqrt{1/3}$
$j_1 = 1$;	$j_2 = \frac{1}{2}$;	$\lvert jm\rangle = \left\lvert \frac{1}{2}, \frac{1}{2}\right\rangle$	$C\left(1, \frac{1}{2}, 0, \frac{1}{2}\lvert \frac{1}{2}, \frac{1}{2}\right)$	$-\sqrt{1/3}$
$j_1 = 1$;	$j_2 = \frac{1}{2}$;	$\lvert jm\rangle = \left\lvert \frac{1}{2}, \frac{1}{2}\right\rangle$	$C\left(1, \frac{1}{2}, 1, -\frac{1}{2}\lvert \frac{1}{2}, \frac{1}{2}\right)$	$\sqrt{2/3}$
$j_1 = 1$;	$j_2 = \frac{1}{2}$;	$\lvert jm\rangle = \left\lvert \frac{3}{2}, -\frac{3}{2}\right\rangle$	$C\left(1, \frac{1}{2}, -1, -\frac{1}{2}\lvert \frac{3}{2}, -\frac{3}{2}\right)$	1
$j_1 = 1$;	$j_2 = \frac{1}{2}$;	$\lvert jm\rangle = \left\lvert \frac{3}{2}, -\frac{1}{2}\right\rangle$	$C\left(1, \frac{1}{2}, -1, \frac{1}{2}\lvert \frac{3}{2}, -\frac{1}{2}\right)$	$\sqrt{1/3}$
$j_1 = 1$;	$j_2 = \frac{1}{2}$;	$\lvert jm\rangle = \left\lvert \frac{3}{2}, -\frac{1}{2}\right\rangle$	$C\left(1, \frac{1}{2}, 0, -\frac{1}{2}\lvert \frac{3}{2}, -\frac{1}{2}\right)$	$\sqrt{2/3}$
$j_1 = 1$;	$j_2 = \frac{1}{2}$;	$\lvert jm\rangle = \left\lvert \frac{3}{2}, \frac{1}{2}\right\rangle$	$C\left(1, \frac{1}{2}, 0, \frac{1}{2}\lvert \frac{3}{2}, \frac{1}{2}\right)$	$\sqrt{2/3}$
$j_1 = 1$;	$j_2 = \frac{1}{2}$;	$\lvert jm\rangle = \left\lvert \frac{3}{2}, \frac{1}{2}\right\rangle$	$C\left(1, \frac{1}{2}, 1, -\frac{1}{2}\lvert \frac{3}{2}, \frac{1}{2}\right)$	$\sqrt{1/3}$
$j_1 = 1$;	$j_2 = \frac{1}{2}$;	$\lvert jm\rangle = \left\lvert \frac{3}{2}, -\frac{3}{2}\right\rangle$	$C\left(1, \frac{1}{2}, -1 - \frac{1}{2}\lvert \frac{3}{2}, -\frac{3}{2}\right)$	1

A few of the most commonly used coefficients are tabulated in Table 10.2. The tables and formulas are very useful for quick determination of the corresponding numerical value. They were derived in a more unified mathematical picture that gives a closed form solution for an arbitrary Clebsch–Gordan coefficient.

The Clebsch–Gordan coefficients can also be derived for each specific case. As an example consider the case of two spin ½ particles, where the total spin is given by

[1] E. Merzbacher, Quantum Mechanics 3rd ed. Wiley, New York (1997).

$$\boldsymbol{S} = \boldsymbol{S}_1 + \boldsymbol{S}_2 \tag{10.57}$$

The total angular momentum state is then given by

$$|s, m\rangle = \sum_{\substack{m_1, m_2 \\ m_1 + m_2 = m}} C(s_1, s_2, m_1, m_2|sm)\ |s_1 m_1\rangle\ |s_2 m_2\rangle \tag{10.58}$$

s can then be 0 or 1. The four states can be written then as:

$$|0,0\rangle = C\left(\frac{1}{2},\frac{1}{2},\frac{1}{2},-\frac{1}{2}|00\right)\left|\frac{1}{2}\frac{1}{2}\right\rangle\left|\frac{1}{2}-\frac{1}{2}\right\rangle + C\left(\frac{1}{2},\frac{1}{2},-\frac{1}{2},\frac{1}{2}|0,0\right)\left|\frac{1}{2}-\frac{1}{2}\right\rangle\left|\frac{1}{2}\frac{1}{2}\right\rangle \tag{10.59}$$

$$|1,-1\rangle = C\left(\frac{1}{2},\frac{1}{2},-\frac{1}{2},-\frac{1}{2}|1,-1\right)\left|\frac{1}{2}-\frac{1}{2}\right\rangle\left|\frac{1}{2}-\frac{1}{2}\right\rangle = \left|\frac{1}{2}-\frac{1}{2}\right\rangle\left|\frac{1}{2}-\frac{1}{2}\right\rangle \tag{10.60}$$

$$|1,0\rangle = C\left(\frac{1}{2},\frac{1}{2},\frac{1}{2},-\frac{1}{2}|1,0\right)\left|\frac{1}{2}\frac{1}{2}\right\rangle\left|\frac{1}{2}-\frac{1}{2}\right\rangle + C\left(\frac{1}{2},\frac{1}{2},-\frac{1}{2},\frac{1}{2}|1,0\right)\left|\frac{1}{2}-\frac{1}{2}\right\rangle\left|\frac{1}{2}\frac{1}{2}\right\rangle \tag{10.61}$$

$$|1,1\rangle = C\left(\frac{1}{2},\frac{1}{2},\frac{1}{2},\frac{1}{2}|1,1\right)\left|\frac{1}{2}\frac{1}{2}\right\rangle\left|\frac{1}{2}\frac{1}{2}\right\rangle = \left|\frac{1}{2}\frac{1}{2}\right\rangle\left|\frac{1}{2}\frac{1}{2}\right\rangle \tag{10.62}$$

The second and forth equations are obvious because of normalization, leaving Eqs 10.60 and 10.61. In the case of Eq. 10.59, we operate with either $\hat{S}_+$ or $\hat{S}_-$, defined by

$$\hat{S}_\pm = \hat{S}_{1_\pm} + \hat{S}_{2_\pm} \tag{10.63}$$

Choosing the upper sign:

$$\begin{aligned}\hat{S}_+\ |0,0\rangle &= (\hat{S}_{1_+} + \hat{S}_{2_+})\, C\left(\frac{1}{2},\frac{1}{2},\frac{1}{2},-\frac{1}{2}|00\right)\left|\frac{1}{2}\frac{1}{2}\right\rangle\left|\frac{1}{2}-\frac{1}{2}\right\rangle \\ &+ (\hat{S}_{1_+} + \hat{S}_{2_+})\, C\left(\frac{1}{2},\frac{1}{2},-\frac{1}{2},\frac{1}{2}|0,0\right)\left|\frac{1}{2}-\frac{1}{2}\right\rangle\left|\frac{1}{2}\frac{1}{2}\right\rangle\end{aligned} \tag{10.64}$$

Using the identities in Table 10.1 and the fact that $\hat{S}_+\ |0,0\rangle = \hat{S}_+\left|\frac{1}{2}\frac{1}{2}\right\rangle = 0$, we get for Eq. 10.64

$$0 = C\left(\frac{1}{2},\frac{1}{2},\frac{1}{2},-\frac{1}{2}|00\right)\left|\frac{1}{2}\frac{1}{2}\right\rangle\left|\frac{1}{2}\frac{1}{2}\right\rangle + C\left(\frac{1}{2},\frac{1}{2},-\frac{1}{2},\frac{1}{2}|0,0\right)\left|\frac{1}{2}\frac{1}{2}\right\rangle\left|\frac{1}{2}\frac{1}{2}\right\rangle \tag{10.65}$$

or, after projecting with $\left\langle\frac{1}{2}\frac{1}{2}\right|\left\langle\frac{1}{2}\frac{1}{2}\right|$ we get

$$C\left(\frac{1}{2},\frac{1}{2},\frac{1}{2},-\frac{1}{2}|00\right) = -C\left(\frac{1}{2},\frac{1}{2},-\frac{1}{2},\frac{1}{2}|0,0\right) \tag{10.66}$$

Using the requirement of normalization, we have

$$C\left(\frac{1}{2},\frac{1}{2},\frac{1}{2},-\frac{1}{2}|00\right) = -C\left(\frac{1}{2},\frac{1}{2},-\frac{1}{2},\frac{1}{2}|1,0\right) = \frac{1}{\sqrt{2}} \tag{10.67}$$

To get the coefficients in Eq. 10.61 we start with Eq. 10.62 and operate with $\hat{S}_-$

$$\hat{S}_-\,|1,1\rangle = \hbar\sqrt{2}\,|1,0\rangle = \left(\hat{S}_{1\_} + \hat{S}_{2\_}\right)\left|\frac{1}{2}\frac{1}{2}\right\rangle\left|\frac{1}{2}\frac{1}{2}\right\rangle = \hbar\left|\frac{1}{2} - \frac{1}{2}\right\rangle\left|\frac{1}{2}\frac{1}{2}\right\rangle + \hbar\left|\frac{1}{2}\frac{1}{2}\right\rangle\left|\frac{1}{2} - \frac{1}{2}\right\rangle \tag{10.68}$$

or

$$|1,0\rangle = \frac{1}{\sqrt{2}}\left|\frac{1}{2} - \frac{1}{2}\right\rangle\left|\frac{1}{2}\frac{1}{2}\right\rangle + \frac{1}{\sqrt{2}}\left|\frac{1}{2}\frac{1}{2}\right\rangle\left|\frac{1}{2} - \frac{1}{2}\right\rangle \tag{10.69}$$

where we used the result developed earlier and tabulated in Table 10.1 that

$$\hat{J}_\pm\,|j,m\rangle = \hbar\sqrt{(j \mp m)(j \pm m + 1)}\,|j,m \pm 1\rangle \tag{10.70}$$

Hence:

$$C\left(\frac{1}{2},\frac{1}{2},\frac{1}{2},-\frac{1}{2}\middle|1,0\right) = C\left(\frac{1}{2},\frac{1}{2},-\frac{1}{2},\frac{1}{2}\middle|1,0\right) = \frac{1}{\sqrt{2}} \tag{10.71}$$

Notice that the state where $S = 1$ then has three m-states. The $S = 1$ state is called a **triplet**. The $S = 0$ state has only one m-state and is called a **singlet**. Two-electron systems are critical in both semiconductors and organic materials.

Problem 10.9 *Using Eq. 10.54, find the state,* $\left|l,s,j,m_j\right\rangle = \left|1,\frac{1}{2},\frac{1}{2},-\frac{1}{2}\right\rangle$ *in terms of the basis states* $|l,m_l\rangle$ *and* $|s,m_s\rangle$*. The* $\left|l,s,j,m_j\right\rangle$ *is the spin and orbital angular momentum part of an electron in an atom.*

a) Use Eqs 10.55 and 10.56, or the table, to find the necessary Clebsch–Gordan coefficients to express the state $\left|1,\frac{1}{2},\frac{1}{2},-\frac{1}{2}\right\rangle$ *of product states of spin and orbital angular momentum.*

b) Now do this problem using just the operators as was illustrated for two spin ½ particles. For part (b), use the upper sign and operate with $\hat{J}_\pm = \hat{L}_\pm + \hat{S}_\pm$ *twice. Notice then that the left-hand side is 0. Recall that the requirement of normalization is part of the solution.*

10.6 Angular Momentum and the Rotation Operator

In Chapter 4, we found that the linear momentum operator is a generator of infinitesimal translations. From this, we were able to construct a translation operator. Here we will see that the angular momentum operator is the **generator of infinitesimal rotations** about an arbitrary axis, pointing in the direction of its unit vector, $\check{\boldsymbol{u}}$ by an amount $\Delta\alpha$. Using the same approach as we did in Chapter 4 when we developed the translation operator that depended on the momentum operator, $\hat{P}$, we will construct a **rotation operator** for finite rotations of the coordinate system. As we discussed in Chapter 4, when the problem is characterized by a symmetry, it is often helpful in revealing new physics and for also doing calculations to find an operator that represents that symmetry that commutes with the Hamiltonian. The rotation operator obviously is then important in systems with spherical symmetry since a rotation of the system leaves the system unchanged. But it is also useful in dealing with rotations of systems with intrinsic spin that have no real space function.

To illustrate the approach, we first consider a more limited kind of rotation, one around the z-axis. The direction of rotation is described by $\hat{\boldsymbol{z}} \times \boldsymbol{r}$, which is a vector perpendicular to both $\hat{\boldsymbol{z}}$ and $\boldsymbol{r}$. Recall that $\hat{z}$ is a unit vector in the z-direction. In the unrotated system, we have $\boldsymbol{r}$, and if the system is rotated the same vector is now described by $\boldsymbol{r}'$:

$$\boldsymbol{r} \overset{rot.}{\rightarrow} \boldsymbol{r}' = \boldsymbol{r} - \Delta\alpha\hat{\boldsymbol{z}} \times \boldsymbol{r} = \boldsymbol{r} - \Delta\boldsymbol{r} \tag{10.72}$$

Hence for the function, $f(\boldsymbol{r})$ in the unrotated system, the function in the rotated system is evaluated at

$$\boldsymbol{r}' = \boldsymbol{r} - \Delta\boldsymbol{r} \tag{10.73}$$

Therefore

$$\begin{aligned} f(\boldsymbol{r} - \Delta\boldsymbol{r}) &= f(\boldsymbol{r}) - \Delta\boldsymbol{r} \cdot \nabla f(\boldsymbol{r}) = f(\boldsymbol{r}) - (\Delta\alpha\hat{\boldsymbol{z}} \times \boldsymbol{r}) \cdot \nabla f(\boldsymbol{r}) = f(\boldsymbol{r}) - \Delta\alpha\hat{\boldsymbol{z}} \cdot (\boldsymbol{r} \times \nabla) f(\boldsymbol{r}) \\ &= f(\boldsymbol{r}) - \frac{i}{\hbar}\Delta\alpha\hat{\boldsymbol{z}} \cdot (\boldsymbol{r} \times \hat{\boldsymbol{p}}) f(\boldsymbol{r}) = f(\boldsymbol{r}) - \frac{i}{\hbar}\Delta\alpha\hat{\boldsymbol{z}} \cdot \hat{\boldsymbol{L}} f(\boldsymbol{r}) = \left(1 - \frac{i}{\hbar}\Delta\alpha\hat{\boldsymbol{z}} \cdot \hat{\boldsymbol{L}}\right) f(\boldsymbol{r}) \end{aligned} \tag{10.74}$$

since $\hat{\boldsymbol{L}} = \boldsymbol{r} \times \hat{\boldsymbol{p}}$. This is a rotation around $\check{\boldsymbol{z}}$, but we can easily generalize this by replacing $\check{\boldsymbol{z}}$ with $\check{\boldsymbol{u}}$, where now $\check{\boldsymbol{u}}$ is a unit vector in an arbitrary direction:

$$f(\boldsymbol{r} - \Delta\boldsymbol{r}) = \left(1 - \frac{i}{\hbar}\Delta\alpha\check{\boldsymbol{u}} \cdot \hat{\boldsymbol{L}}\right) f(\boldsymbol{r}) \tag{10.75}$$

We define the operator to the left of $f(\boldsymbol{r})$ as the rotation operator for an infinitesimal rotation around an arbitrary axis $\check{u}$:

$$\hat{R}_{\hat{u}}(\Delta\alpha) = \left(1 - \frac{i}{\hbar}\Delta\alpha\check{\boldsymbol{u}} \cdot \hat{\boldsymbol{L}}\right) \tag{10.76}$$

$\hat{\boldsymbol{L}}$ is a **generator of infinitesimal rotations.** To make a finite rotation, α,we apply this operator n times, in the limit as $n \to \infty$, and define the infinitesimal rotation angle $\Delta\alpha$ as

$$\Delta\alpha = \frac{\alpha}{n} \tag{10.77}$$

Therefore, the **rotation operator** for finite rotations is

$$\hat{R}_{\hat{u}}(\alpha) = \lim_{n\to\infty}\left(1 - \frac{i}{\hbar}\frac{\alpha}{n}\hat{\boldsymbol{u}} \cdot \hat{\boldsymbol{L}}\right)^n = e^{-i\frac{\alpha\hat{\boldsymbol{u}}\cdot\hat{\boldsymbol{L}}}{\hbar}} \tag{10.78}$$

where the equality follows from the same math as in Chapter 4 (see Eq. 4.8).

An important example is the **spin ½ rotation operator** where $\hat{S} = \frac{\hbar}{2}\hat{\boldsymbol{\sigma}}$

$$\hat{R}_{\hat{u}}^{1/2}(\alpha) = \hat{I}\cos\frac{\alpha}{2} - i\hat{\boldsymbol{\sigma}}\cdot\hat{\boldsymbol{u}}\sin\frac{\alpha}{2} \tag{10.79}$$

where $\boldsymbol{\sigma} = \hat{\sigma}_x\hat{x}+\hat{\sigma}_y\hat{y}+\hat{\sigma}_z\hat{z}$ and $\hat{\sigma}_j$ is the Pauli operator, a 2×2 matrix. Since a bit of information can be described by 2×1 column vector, the rotation operator can be used to manipulate classical or quantum information. It is also an important operator in the study and use of quantum entangled states.

Problem 10.10

a. *Show for the Pauli spin matrices in Eq. 10.46 that*

$$\hat{\sigma}_x^2 = \hat{\sigma}_y^2 = \hat{\sigma}_z^2 = \hat{I}$$

b. *Show for the Pauli spin matrices that*

$$\hat{\sigma}_x\,\hat{\sigma}_y = i\hat{\sigma}_z;\text{ and } \left\{\hat{\sigma}_x, \hat{\sigma}_y\right\} = 0$$

$$\hat{\sigma}_y\hat{\sigma}_z = i\hat{\sigma}_x;\text{ and } \left\{\hat{\sigma}_y, \hat{\sigma}_z\right\} = 0$$

$$\hat{\sigma}_z\hat{\sigma}_x = i\hat{\sigma}_y;\text{ and } \{\hat{\sigma}_z, \hat{\sigma}_x\} = 0$$

where $\left\{\hat{\sigma}_i, \hat{\sigma}_j\right\} \equiv \left[\hat{\sigma}_i, \hat{\sigma}_j\right]_+ \equiv \hat{\sigma}_i\hat{\sigma}_j + \hat{\sigma}_j\hat{\sigma}_i$ *is the* ***anticommutator.***

c. *Now show that Eq. 10.79 follows from the definition in Eq. 10.78 by expanding in a power series (Appendix B), using identities for the Pauli spin matrices, and then collecting terms that represent the power series expansion for the corresponding trigonometric functions.*

10.7 **Summary**

In this chapter, we recovered the eigenvalues and eigenvectors for the orbital angular momentum presented in Chapter 6 by using the same kind of operator algebra that we used in Chapter 8 to study the nano-vibrator. However, we found that using just the postulates, the angular momentum quantum numbers could take on half-integer values. This contradicted the results in Chapter 6, which eliminated half-integer values because it was required that if the coordinate system was rotated by 2π, it had to revert to the same state as before the rotation. However, it was in experiments that such half-integer angular momentum states could exist in what seemed like point particles such as the electron and nucleons. This angular momentum is called intrinsic angular momentum and is not associated with rotation of mass in real space. Finally, we developed the rules that show how to add two different angular momentum together. This is very important for many problems, but especially for the electron in atoms, since the electron has both intrinsic and orbital angular momentum. The coupling between the two controls many properties in electronic materials.

Vocabulary (page) and Important Concepts

- angular momentum 160
- plane of rotation 160
- precession 161
- complete set of commuting operators (CSCO) 162
- axis of quantization 162
- magnetic substate 165
- rigid rotor 166
- moment of inertia 166
- magnetic moment 166
- Bohr magneton 167
- spin 168
- intrinsic angular momentum 168
- postulate of spin 168
- electron spin g-factor 168
- eigenspinor 169
- spin matrices 169
- Pauli matrices 169
- product space 170
- Bloch sphere 172
- addition of angular momentum 173
- Clebsch–Gordan coefficient 174
- triplet 177
- singlet 177
- generator of infinitesimal rotations 177
- rotation operator 177
- spin ½ rotation operator 179
- anticommutator 179

11 Approximation Techniques for Stationary and Dynamic Problems

11.1 Introduction

In many applications of quantum systems the basic approach, especially in technology, begins with understanding the simplest representation of that system. The earlier chapters have focused on that. Those ideas form the foundation of a quantum toolbox. Some of these discussions can be made more relevant to a specific problem by possibly expanding to higher dimensions, but this only leads to a different analytical form with more degrees of freedom. The analytical form rather than a numerical model provides valuable power for establishing an initial concept or design before moving on to more detailed and often numerical calculations. The value of the intuitive physical insight obtained by careful study of these simple systems for someone working to create a new idea cannot be overstated. Hence, it is not surprising that simple methods have been established that enable us to extend the analytical results to more complex problems, thus delaying the need for advanced computation and providing a means to extend physical understanding using simple analytical tools.

The methods in this present section typically provide all that is needed to tackle most problems that are encountered. In many cases, additional information from an exact solution would have no impact. For sensors and detectors, for example, the tools in this section are often enough. For switches, as in the discussion of quantum flip-flop in Chapter 9, the full nonlinear response is required, but even there we made numerous approximations (two-level, rotating wave, and dipole approximation). A more exact solution would again have no impact. Hence, the ability to make approximations in quantum is an essential component in the quantum toolbox.

In this approach, we start with the Hamiltonian for the basic system, e.g., a nano-vibrator or an electron in a trap. We assume that we have previously diagonalized the Hamiltonian, and recall that we have previously solved the time independent Schrödinger equation:

$$\hat{H}_0|E_n\rangle = E_n|E_n\rangle \tag{11.1}$$

where $\hat{H}_0$ is the Hamiltonian for the basic system. This gives us the energy eigenvalues but also a complete basis. The change in the basic system is then represented by an additional potential, $\hat{V}$, in the same Hilbert space, which gives a new Hamiltonian:

$$\hat{H} = \hat{H}_0 + \hat{V} \tag{11.2}$$

Introduction to Quantum Nanotechnology: A Problem Focused Approach. Duncan Steel, Oxford University Press (2021).
© Duncan Steel. DOI: 10.1093/oso/9780192895073.003.0011

If $\widehat{V}$ is time independent, then we need to find the solutions (i.e., **stationary states**) to the new time independent Schrödinger equation:

$$\left(\widehat{H}_0 + \widehat{V}\right) | E_n' \rangle = E_n' \, | E_n' \rangle \tag{11.3}$$

If $\widehat{V}$ is time dependent, $\widehat{V}(t)$, then we need to find a solution to the time dependent Schrödinger equation:

$$\left(\widehat{H}_0 + \widehat{V}(t)\right) |\psi\rangle = i\hbar \frac{d}{dt} |\psi\rangle \tag{11.4}$$

Our approach to both problems will be similar. We require only that the Hilbert space described by the new Hamiltonian is the same Hilbert space defined by $\widehat{H}_0$. This means, for example, that if we have an $\widehat{H}_0$ that describes a one-dimensional system along the x-axis, the new potential is not a function of y. Or if $\widehat{H}_0$ is in real space, the new potential does not include the effects of spin. We will use the idea of expanding the new state of the system in terms of the original basis states. The results of the additional potential will be contained in the calculations of the associated expansion coefficients, where in the time dependent case, the expansion coefficients will be time dependent. In both cases, we assume that the change in the state of the system is small compared to its original system. Such an approach is called a perturbation. The idea of "perturbation" may suggest "not useful". However, as you will see, what constitutes small in perturbation theory does not translate into "small impact" or an effect that can be ignored even if "small". Many sensors and detectors are described quite adequately by this approach as well as most quantum systems.

11.2 Time Independent Perturbation Theory

Non-degenerate Perturbation Theory

When V is independent of time, we need to find the new stationary state solutions, $|E_n'\rangle$, to the time independent Schrödinger equation. This amounts to rediagonalizing the new Hamiltonian. In general, finding the exact analytical solution is often difficult or impossible. It may be necessary to resort to a numerical approach, but in many cases of practical interest, the interaction potential is relatively weak compared to the energies associated with the original Hamiltonian. In this case, standard approximation methods can enable much easier solutions with more physical insight. Here we consider the approach based on perturbation theory.

We begin by writing the Hamiltonian in Eq. 11.3 as

$$\widehat{H} = \left(\widehat{H}_0 + \lambda \widehat{V}\right) \tag{11.5}$$

where λ is an arbitrary parameter between 0 and 1. The time independent Schrödinger equation is now

$$\left(\widehat{H}_0 + \lambda \widehat{V}\right) | E_n' \rangle = E_n' \, | E_n' \rangle \tag{11.6}$$

When $\lambda = 0$, we have $|E_n'\rangle = |E_n\rangle$. To find the result when $\lambda \neq 0$, we expand $|\psi_n'\rangle$ and E_n' in powers of λ:

$$|E_n'\rangle = \lambda^0 \left|E_n^{(0)}\right\rangle + \lambda^1 \left|E_n^{(1)}\right\rangle + \lambda^2 \left|E_n^{(2)}\right\rangle + \lambda^3 \left|E_n^{(3)}\right\rangle + \cdots \tag{11.7}$$

where $|E_n^{(0)}\rangle = |E_n\rangle$ and $|E_n^{(j)}\rangle$ is the jth order correction to the eigenvector. E_n' is also expanded in a power series of λ

$$E_n' = \lambda^0 E_n^{(0)} + \lambda^1 E_n^{(1)} + \lambda^2 E_n^{(2)} + \lambda^3 E_n^{(3)} + \cdots \tag{11.8}$$

where again $E_n^{(0)} = E_n$, and $E_n^{(j)}$ is the jth order correction to the energy.

Inserting Eqs 11.7 and 11.8 into Eq. 11.6 we have

$$\begin{aligned}&\left(\hat{H}_0 + \lambda \hat{V}\right)\left(\lambda^0 \left|E_n^{(0)}\right\rangle + \lambda^1 \left|E_n^{(1)}\right\rangle + \lambda^2 \left|E_n^{(2)}\right\rangle + \lambda^3 \left|E_n^{(3)}\right\rangle + \cdots\right)\\ &= \left(\lambda^0 E_n^{(0)} + \lambda^1 E_n^{(1)} + \lambda^2 E_n^{(2)} + \lambda^3 E_n^{(3)} + \cdots\right)\left(\lambda^0 \left|E_n^{(0)}\right\rangle + \lambda^1 \left|E_n^{(1)}\right\rangle + \lambda^2 \left|E_n^{(2)}\right\rangle + \lambda^3 \left|E_n^{(3)}\right\rangle + \cdots\right)\end{aligned} \tag{11.9}$$

Expanding this out, we group the left side and the right side into terms having the same power of λ. $\hat{H}_0$ has the factor of λ^0 in front also, but typically this is suppressed in writing since $\lambda^0 = 1$. Since λ is arbitrary, the terms on the left associated with a given power λ must be equal to terms on the right associated with the same power of λ. This results in a series of new equations, one for each power of λ:

$$\lambda^0 \quad \hat{H}_0 \left|E_n^{(0)}\right\rangle = E_n^{(0)} \left|E_n^{(0)}\right\rangle \tag{11.10}$$

$$\lambda^1 \quad \hat{H}_0 \left|E_n^{(1)}\right\rangle + \hat{V}\left|E_n^{(0)}\right\rangle = E_n^{(1)} \left|E_n^{(0)}\right\rangle + E_n^{(0)} \left|E_n^{(1)}\right\rangle \tag{11.11}$$

$$\lambda^2 \quad \hat{H}_0 \left|E_n^{(2)}\right\rangle + \hat{V}\left|E_n^{(1)}\right\rangle = E_n^{(0)} \left|E_n^{(2)}\right\rangle + E_n^{(1)} \left|E_n^{(1)}\right\rangle + E_n^{(2)} \left|E_n^{(0)}\right\rangle \tag{11.12}$$

etc.

First Order Time Independent Perturbation Theory

Equation 11.10 is the same as Eq. 11.1. We then work to use Eq. 11.11 to find $|E_n^{(1)}\rangle E_n^{(1)}$ to **first order** (i.e., in λ^1 or $\hat{V}$). The language of first order arises from the fact that Eq. 11.11 comes from the first power of λ. The result will be corrections that are linear in matrix elements of $\hat{V}$. The approach is to expand $|E_n^{(1)}\rangle$ in terms of the complete set of eigenvectors for $\hat{H}_0$, since both $\hat{H}_0$ and $\hat{V}$ are necessarily in the same Hilbert space and the basis states of $\hat{H}_0$ are complete:

$$\left|E_n^{(1)}\right\rangle = \sum_k a_k^{(1)} \left|E_k^{(0)}\right\rangle \tag{11.13}$$

which we substitute into Eq. 11.11 and use $\hat{H}_0|E_k^{(0)}\rangle = E_k^{(0)}|E_k^{(0)}\rangle$, which will then lead to the first order correction. We first get

$$\sum_k a_k^{(1)} E_k^{(0)} \left|E_k^{(0)}\right\rangle + \widehat{V}\left|E_n^{(0)}\right\rangle = E_n^{(0)} \sum_k a_k^{(1)} \left|E_k^{(0)}\right\rangle + E_n^{(1)} \left|E_n^{(0)}\right\rangle \tag{11.14}$$

We now multiply by $\langle E_j|$ and use orthonormality:

$$a_j^{(1)} E_j^{(0)} + \left\langle E_j^{(0)}\right| \widehat{V} \left|E_n^{(0)}\right\rangle = E_n^{(0)} a_j^{(1)} + E_n^{(1)} \delta_{jn} \tag{11.15}$$

When $j \neq n$ and assuming the eigenvalues are non-degenerate, we can solve for $a_j^{(1)}$:

$$a_j^{(1)} = \frac{\left\langle E_j^{(0)}\right| \widehat{V} \left|E_n^{(0)}\right\rangle}{\left(E_n^{(0)} - E_j^{(0)}\right)} \quad j \neq n \tag{11.16}$$

When $j = n$, Eq. 11.15 gives

$$E_n^{(1)} = \left\langle E_n^{(0)}\right| V \left|E_n^{(0)}\right\rangle \tag{11.17}$$

We can summarize the eigenvector and eigenenergy correction to first order in the potential as

$$|E_n'\rangle \approx \left|E_n^{(0)}\right\rangle + \sum_{j\neq n} \frac{\left\langle E_j^{(0)}\right| \widehat{V} \left|E_n^{(0)}\right\rangle}{\left(E_n^{(0)} - E_j^{(0)}\right)} \left|E_j^{(0)}\right\rangle \tag{11.18}$$

$$E_n' \approx E_n^{(0)} + \left\langle E_n^{(0)}\right| \widehat{V} \left|E_n^{(0)}\right\rangle \tag{11.19}$$

It is important to note again that there is a problem in the solution if for some n and j, $E_n^{(0)} - E_j^{(0)} = 0$, meaning that states $|E_j^{(0)}\rangle$ and $|E_n^{(0)}\rangle$ have degenerate eigenstates, and also when $\langle E_j^{(0)}|\widehat{V}|E_n^{(0)}\rangle \neq 0$. Hence, we must require that for the current calculation, the states must be non-degenerate or $\langle E_j^{(0)}|\widehat{V}|E_n^{(0)}\rangle = 0$. The process above is then sometimes called **non-degenerate perturbation theory**. A discussion on how to proceed with states that are degenerate is given at the end of this section. If the matrix element $\langle E_j^{(0)}|\widehat{V}|E_n^{(0)}\rangle = 0$, then this theory still works for degenerate states, as in the case of **Zeeman splitting** below.

The above results in Eqs 11.18 and 11.19 are important for design. Physically, Eq. 11.18 shows the $|E_n'\rangle$ is no longer a pure $|E_n\rangle$ state. Instead, $|E_n\rangle$ includes a contribution from the other states. The strength of the **state mixing** is given by $\frac{\langle E_j^{(0)}|\widehat{V}|E_n^{(0)}\rangle}{\left(E_n^{(0)}-E_j\right)} |E_j^{(0)}\rangle$. The matrix element $\langle E_j^{(0)}|\widehat{V}|E_n^{(0)}\rangle$ shows that $\widehat{V}$ couples $|E_n^{(0)}\rangle$ to $|E_j^{(0)}\rangle$; i.e., $\langle E_j^{(0)}|\widehat{V}|E_n^{(0)}\rangle$ leads to **state coupling** or state mixing of the two original eigenstates $E_n^{(0)}$ and $E_j^{(0)}$. If the state E_n is populated in the absence of the potential, the probability amplitude for being in any other state in the presence of the potential is $\frac{\langle E_j^{(0)}|\widehat{V}|E_n^{(0)}\rangle}{\left(E_n^{(0)}-E_j\right)}$.

State mixing can lead to important new properties. For example, suppose that a problem is symmetric about the origin. The solutions are even or odd, meaning parity is a good quantum number. Suppose now that some kind of potential, such as strain or an electric field, is added to the system. These interactions do not have the same symmetry, so the interaction is responsible for **symmetry breaking**. The importance of this is that for some additional interaction that you might need for, say a sensor, if that interaction is described by V, it is maybe that in the original basis $\langle E_j^{(0)}|V|E_n^{(0)}\rangle = 0$. V might represent detection say of some airborne toxic molecule or maybe polarized electromagnetic radiation. So the system will not work. But, if you break symmetry, you can design the system such that in the new energy basis, $\langle E_j{}'|V|E_n{}'\rangle \neq 0$. The objective in design becomes one of engineering the system to achieve the objectives of the problem while minimizing collateral side effects that may have unintended consequences.

Problem 11.1 *Use first order perturbation theory to find the effect of a square bump in the infinite potential well shown in Fig. 11.1, from Chapter 3, used to describe a point defect. Examine the effect on both the ground and first excited states. Recall that for the problem in Chapter 3, for even states:*

$$\psi_{even}(x) = \sqrt{\frac{2}{L}}\cos\left(\left(n+\frac{1}{2}\right)\frac{2\pi x}{L}\right) \qquad n = 0, 1, 2, \cdots$$

And

$$E_{n_{even}} = E_{box}\left(n+\frac{1}{2}\right)^2 \text{ and } n = 0, 1, 2, \cdots \quad E_{box} = \frac{\hbar^2}{2m}\left(\frac{2\pi}{L}\right)^2$$

For odd states:

$$\psi_{odd}(x) = \sqrt{\frac{2}{L}}\sin\left(n\frac{2\pi x}{L}\right) \qquad n = 1, 2, \cdots$$

$$E_{n_{odd}} = E_{box}\, n^2 \qquad n = 1, 2, \cdots$$

In the two problems that follow, the question pertains to the hydrogen atom, where there is considerable degeneracy. However, for reasons that will become more obvious in the following section on degenerate perturbation theory, non-degenerate perturbation theory

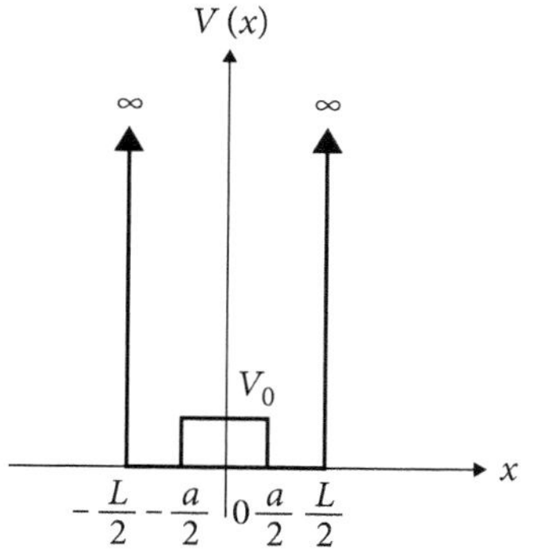

Fig. 11.1 An infinite potential well with a small barrier in the middle.

is appropriate for these problems since the applied field does not couple the degenerate states.

Hydrogen Atom in a Magnetic field: Zeeman Splitting

To set up the first problem below (Problem 11.2), we consider the case of a spherically symmetric quantum system in a homogeneous magnetic field pointing along the axis of quantization, the z-axis. For convenience, we assume a simple hydrogenic atom. This calculation works extremely well in even more complex systems such as excitons in semiconductors. The Hamiltonian is the hydrogen atom Hamiltonian, $\hat{H}_0$, from Chapter 6. The eigenfunctions are $R_{nl}(r)Y_{lm}(\theta, \varphi)$. Repeating some of the discussion from Chapter 6 and Chapter 10, the interaction with the magnetic field is through the orbital motion of the electron (Eq. 10.34):

$$V = -\boldsymbol{\mu}_L \cdot \boldsymbol{B} \tag{11.20}$$

For an orbiting electron in spherically a symmetric central potential, the orbital **magnetic moment** (Eq. 10.35) is

$$\boldsymbol{\mu}_L = -\frac{|e|}{2m_e}\boldsymbol{L} = -\frac{\mu_B}{\hbar}\boldsymbol{L} \tag{11.21}$$

The symbol "e" is the magnitude of the charge of the electron, and the minus sign is because the electron is negative. μ_B is the **Bohr magneton** given by

$$\mu_B = \frac{|e|\hbar}{2m_e} \tag{11.22}$$

The potential energy for such a magnetic moment is then

$$V = \frac{\mu_B}{\hbar}\boldsymbol{L} \cdot \boldsymbol{B} \tag{11.23}$$

Understanding the effect of the magnetic field on such a system to first order in the applied field leads to **Zeeman splitting**, as you will see in Problem 11.2

Problem 11.2 *A hydrogen atom is a magnetic field in the z-direction ignoring the electron spin. Hence, the eigenfunctions are* $u(r, \theta, \varphi) = NR_{nl}(r)Y_{lm}(\theta, \varphi)$ *and the Hamiltonian is given by*

$$H = H_0 + \frac{\mu_B}{\hbar}\hat{\boldsymbol{L}} \cdot \boldsymbol{B} = H_0 + \frac{\mu_B}{\hbar}\hat{L}_z B_z \tag{11.24}$$

The states for different m are called ***magnetic substates*** *and are degenerate for different m and l within a given n-manifold. Show first that the states are not coupled by the magnetic field. This means that the matrix elements for the perturbation between different states are zero. Then find the splitting between adjacent m-states as a function of the field strength using non-degenerate perturbate theory. The energy difference between adjacent m-states that results from a finite magnetic field is called* ***Zeeman splitting****. This effect in quantum devices enables improved control* *over individual magnetic substates that can hold information. In general, it is important to include the effect of spin.*[1]

This is a remarkably simple result. It is a perturbation in the sense that it is small compared to energy level spacings arising from Coulomb coupling. But it has the effect of taking a state with a given value of l which is $2l + 1$ degenerate and completely removing the degeneracy. To be useful in engineering design, the change in energy needs to be larger than the linewidth of the transition, which is at least of order the inverse of the state lifetime (to be discussed in Chapters 14 and 15). However, most linewidths of interest for applications are very narrow, and hence the first order correction is quite adequate.

Spin-Orbit Coupling

Another important problem that can now be solved is the problem of the spin of an electron interacting with the magnetic field that it experiences by its motion through the Coulomb electric field. This leads to a new term in the Hamiltonian due to **spin-orbit coupling**. This term can dramatically alter the simpler properties of many materials important for quantum devices, and must be considered in any successful design. Specifically, from introductory electricity and magnetism, the magnetic field experienced by a particle moving through an electric field is given by:

$$\boldsymbol{B} = -\frac{1}{c^2}\boldsymbol{v} \times \boldsymbol{E} \tag{11.25}$$

The electric field arises from the gradient of the Coulomb potential (Eq. 6.10) between the electron and the nucleus, $V_C = -\frac{1}{4\pi\epsilon_0}\frac{e^2}{\mathrm{r}}$ and points in the radial direction, hence

$$\boldsymbol{E} = \boldsymbol{r}\left|\frac{E}{r}\right| = \frac{1}{|\,e\,|}\frac{\boldsymbol{r}}{r}\left|\frac{\partial V_C(r)}{\partial r}\right| \tag{11.26}$$

[1] To include spin, Eq. 11.23 is modified to become $\widehat{V} = \frac{\mu_B}{\hbar}\left(\boldsymbol{L} + g_S\boldsymbol{S}\right) \cdot \boldsymbol{B}$. Based on the discussion so far, evaluating $\left\langle nlsm_lm_s|\widehat{V}|nlsm_lm_s\right\rangle$ is rather detailed. However, using the Wigner–Eckart theorem (Appendix G), it is straightforward to show that

$$\left\langle nlsm_lm_s|\widehat{V}|nlsm_lm_s\right\rangle = \langle nlsjm_j|\widehat{V}|nlsjm_j\rangle$$

and then $\left\langle nlsjm_j\,|V|\,nlsjm_j\right\rangle = \frac{\frac{\mu_B}{\hbar}\left\langle nlsjm_j|(\hat{\boldsymbol{L}}\cdot\hat{\boldsymbol{J}}+g_S\hat{\boldsymbol{S}}\cdot\hat{\boldsymbol{J}})|nlsjm_j\right\rangle\left\langle nlsjm_j|\hat{\boldsymbol{J}}\cdot\boldsymbol{B}|nlsjm_j\right\rangle}{j(J+1)}$.

Starting with $\hat{\boldsymbol{J}} = \hat{\boldsymbol{L}} + \hat{\boldsymbol{S}}$, then $\langle nlsjm_j|\hat{\boldsymbol{L}}\cdot\hat{\boldsymbol{J}}|nlsjm_j\rangle = \langle nlsjm_j|\frac{1}{2}(\hat{\boldsymbol{J}}^2 + \hat{\boldsymbol{L}}^2 - \hat{\boldsymbol{S}}^2)|nlsjm_j\rangle = \frac{1}{2}(j(j+1) + l(l+1) - \frac{3}{4})$ and $\langle nlsjm_j|\hat{\boldsymbol{S}}\cdot\hat{\boldsymbol{J}}|nlsjm_j\rangle = \langle nlsjm_j|\frac{1}{2}(\hat{\boldsymbol{J}}^2 + \hat{\boldsymbol{S}}^2 - \hat{\boldsymbol{L}}^2)|nlsjm_j\rangle = \frac{1}{2}(j(j+1) - l(l+1) + \frac{3}{4})$. For $\boldsymbol{B}$ in the z-direction, this gives $\langle nlsjm_j|\widehat{V}|nlsjm_j\rangle = \mu_B\left[\frac{\left(\left(j(j+1)+l(l+1)-\frac{3}{4}\right)+g_S\left(j(j+1)-l(l+1)+\frac{3}{4}\right)\right)}{2j(j+1)}\right]B_zm_j$, where the term in the brackets is the **Landé g-factor**, g_J. We have used the rules in Chapter10 for the addition of angular momentum. (See Appendix G or the discussion of the Wigner–Eckart theorem in I.I. Sobelman, Introduction to Atomic Spectra, Oxford 1972.)

The magnetic field seen by the electron is then

$$\boldsymbol{B} = -\frac{1}{2c^2}\boldsymbol{v}\times\boldsymbol{E} = \frac{1}{2m_e\,|\,e\,|\,c^2}\frac{1}{r}\left|\frac{\partial V(r)}{\partial r}\right|\boldsymbol{r}\times\boldsymbol{p} = \frac{1}{2m_e|e|c^2}\frac{1}{r}\left|\frac{\partial V(r)}{\partial r}\right|\boldsymbol{L} \tag{11.27}$$

where the factor of ½ comes from a relativistic correction for evaluating the field in the rest frame of the electron.[2] Then the fine structure potential energy of a magnetic moment associated with the spin of the electron in a magnetic field is (see Eq. 10.34)

$$V = -\boldsymbol{\mu}_{M_s}\cdot\boldsymbol{B} \tag{11.28}$$

The magnetic moment arising from the electron spin is (Eq. 10.38)

$$\boldsymbol{\mu}_{M_s} = -g_s\frac{|\,e\,|}{2m_e}\boldsymbol{S} \tag{11.29}$$

where $g_s = 2$ in Eq. 10.38. The form of the potential for spin orbit coupling, becomes

$$V_{s-o} = \frac{1}{2}\frac{1}{m_e^2c^2}\left|\frac{1}{r}\frac{\partial V(r)}{\partial r}\right|\boldsymbol{L}\cdot\boldsymbol{S} = \frac{1}{8\pi\epsilon_0}\frac{e^2}{m_e^2c^2}\frac{1}{r^3}\boldsymbol{L}\cdot\boldsymbol{S} \tag{11.30}$$

In line with Eq. 11.19,

$$E_n' = E_n + \langle E_n|\,\hat{V}\,|E_n\rangle \tag{11.31}$$

It becomes necessary to evaluate $\langle E_n|\hat{V}|E_n\rangle$. We again replace $|E_n\rangle$ with $|n, l, s, j, m_j\rangle$. It is helpful if you have studied the rules in Chapter 10 for adding two angular momentum together, but not essential. Physically, both $\boldsymbol{L}$ and $\boldsymbol{S}$ are vectors and hence the total angular momentum, designated $\boldsymbol{J}$, defined as $\boldsymbol{J} = \boldsymbol{L} + \boldsymbol{S}$. $\hat{\boldsymbol{J}}$ and $\hat{\boldsymbol{J}}_z$ have eigenstates given by $|\,l, s, j, m_j\rangle$ where

$$\hat{\boldsymbol{J}}^2\,|\,l, s, j, m_j\rangle = j\,(j+1)\,\hbar^2\,|\,l, s, j, m_j\rangle \tag{11.32}$$

and

$$\hat{\boldsymbol{J}}_z\,|\,l, s, j, m_j\rangle = m_j\hbar\,|\,l, s, j, m_j\rangle \tag{11.33}$$

and

$$\langle l', s, j', m_j'|l, s, j, m_j\rangle = \delta_{ll'}\delta_{jj'}\delta_{m_jm_j'}. \tag{11.34}$$

The spin s is taken to be the same on both sides since the spin for the electron is ½ and cannot change. As discussed in Chapter 10, and also on physical grounds, the magnitude of $\boldsymbol{L}$ and $\boldsymbol{S}$ set the magnitude of $\boldsymbol{J}$, the maximum being when they are co-aligned (i.e., $j = l + s$) and the minimum is when they are oppositely aligned, giving $j = |l-s|$. Of course, the algebra of angular momentum showed that the differences in the projection of $\boldsymbol{J}_z$ along the z-axis are separated by 1 unit of $\hbar$. For a spin ½ system, it is then clear that j can only take on two values, $j = l \pm \frac{1}{2}$.

[2] Robert B. Leighton, *Principles Modern Physics*, McGraw Hill (1959).

Returning now to Eq. 11.31, we have for $\langle E_n | \hat{V} | E_n \rangle$

$$\langle E_n | \hat{V} | E_n \rangle = \frac{1}{2}\frac{1}{m_e^2 c^2}\left\langle n, l, s, j, m_j \right| \left| \frac{1}{r}\frac{\partial V(r)}{\partial r} \right| \hat{\boldsymbol{L}} \cdot \hat{\boldsymbol{S}} \left| n, l, s, j, m_j \right\rangle$$

$$= \frac{1}{2}\frac{e^2}{4\pi\epsilon_0 m_e^2 c^2}\left\langle n, l, s, j, m_j \right| \frac{1}{r^3}\hat{\boldsymbol{L}} \cdot \hat{\boldsymbol{S}} \left| n, l, s, j, m_j \right\rangle \tag{11.35}$$

Because of the presence of spin, the eigenfunction of the hydrogen atom cannot be rewritten entirely in the coordinate representation anymore. However, we have seen that everything associated with angular momentum that is important with this problem can be handled without the coordinate representation. Hence, we separate the eigenstate $|n, l, s, j, m_j\rangle$ into the radial part, $R_{nl}(r)$, and the part containing the coupled angular momentum, hence $|n, l, s, j, m_j\rangle \rightarrow R_{nl}(r)|l, s, j, m_j\rangle$. Therefore the matrix element for the spin orbit coupling is

$$\left\langle n, l, s, j, m_j \left| \frac{1}{r^3}\boldsymbol{L} \cdot \boldsymbol{S} \right| n, l, s, j, m_j \right\rangle = \left\langle l, s, j, m_j \left| \hat{\boldsymbol{L}} \cdot \hat{\boldsymbol{S}} \right| l, s, j, m_j \right\rangle \int_0^\infty r^2 dr\, R_{nl}^*(r)\frac{1}{r^3}R_{nl}(r) \tag{11.36}$$

The second term on the right-hand side is an integral[3]

$$\int_0^\infty r^2 dr\, R_{nl}^*(r)\frac{1}{r^3}R_{nl}(r) = \frac{1}{a_0^3 n^3 l\left(l + \frac{1}{2}\right)(l+1)} \tag{11.37}$$

Of course the first term on the right-hand side could now be written in terms of the products of individual orbital and spin eigenvectors summed with the appropriate Clebsch–Gordan coefficients, discussed in Chapter 10. However, it is much easier to recognize that

$$\hat{\boldsymbol{J}}^2 = \hat{\boldsymbol{L}}^2 + \hat{\boldsymbol{S}}^2 + 2\hat{\boldsymbol{L}} \cdot \hat{\boldsymbol{S}} \tag{11.38}$$

and hence

$$\hat{\boldsymbol{L}} \cdot \hat{\boldsymbol{S}} = \frac{1}{2}\left(\hat{\boldsymbol{J}}^2 - \hat{\boldsymbol{L}}^2 - \hat{\boldsymbol{S}}^2\right) \tag{11.39}$$

The term is now easily evaluated without such complexity.

Problem 11.3

a) Show that

$$\left\langle l, s, j, m_j \left| \hat{\boldsymbol{L}} \cdot \hat{\boldsymbol{S}} \right| l, s, j, m_j \right\rangle = \left[\begin{array}{c} l \text{ for } j = l + \frac{1}{2} \\ l - 1 \text{ for } j = l - \frac{1}{2} \end{array} \right. \tag{11.40}$$

b) Combine the results above now to find the change, ΔE, in the eigenenergy for the nth state of hydrogen with orbital angular momentum, l. The ΔE is then known to first order in the spin orbit interaction.

[3] Hans A. Bethe and Edwin E. Salpeter, *Quantum Mechanics of One- and Two-Electron Atoms*, Plenum 1977; I.I. Sobelman, *Introduction to Atomic Spectra*, Pergamon Press, Oxford 1972.

Second Order Time Independent Perturbation Theory: The Stark Effect

There are important problems where sometimes the first order correction is zero. Consider a spherically symmetric system such as the hydrogen atom. Since the Coulomb potential is symmetric, the solutions are even or odd for $\boldsymbol{r} \to -\boldsymbol{r}$. The potential energy for an electric dipole $\boldsymbol{\mu}_E = e\boldsymbol{r}$ sitting in a uniform electric field, $\boldsymbol{E}$ is given by

$$V = -\boldsymbol{\mu}_E \cdot \boldsymbol{E} \tag{11.41}$$

From Eq. 11.19, first order perturbation theory gives the change in energy as

$$\langle E_{nlm}| \widehat{V} |E_{nlm}\rangle = -e\,\langle E_{nlm}| \boldsymbol{r} |E_{nlm}\rangle \cdot \boldsymbol{E} = 0 \tag{11.42}$$

because $\boldsymbol{r}$ is an odd function. Hence, to get the effect of the electric field on the atom or other quantum system characterized by reflection symmetry, called the **Stark effect**, we have to go to **second order time independent perturbation theory**.

We begin as before by expanding $|E_n^{(2)}\rangle$ in Eq. 11.12:

$$\left|E_n^{(2)}\right\rangle = \sum_k a_k^{(2)} |E_k\rangle \tag{11.43}$$

and insert this in our results for the second order correction along with our first order results into Eq. 11.12:

$$\begin{aligned}&\sum_k a_k^{(2)} E_k |E_k\rangle + \sum_{k\neq n} \frac{\langle E_k|\widehat{V}|E_n\rangle}{(E_n - E_k)} \widehat{V}|E_k\rangle \\ &\quad = E_n \sum_k a_k^{(2)} |E_k\rangle + \langle E_n| \widehat{V} |E_n\rangle \sum_{k\neq n} \frac{\langle E_k| \widehat{V} |E_n\rangle}{(E_n - E_k)} |E_k\rangle + E_n^{(2)} |E_n\rangle\end{aligned} \tag{11.44}$$

We again project onto $\langle E_j|$ to give

$$a_j^{(2)} E_j + \sum_{k\neq n} \frac{\langle E_j|\widehat{V}|E_k\rangle\langle E_k|\widehat{V}|E_n\rangle}{(E_n - E_k)} = E_n a_j^{(2)} + \left\langle E_n|\widehat{V}|E_n\right\rangle \frac{\langle E_j|\widehat{V}|E_n\rangle}{\left(E_n - E_j\right)} \left(1 - \delta_{nj}\right) + E_n^{(2)} \delta_{nj} \tag{11.45}$$

When $j \neq n$, we get an expression for the second order correction $a_j^{(2)}$:

$$a_j^{(2)} = \sum_{k\neq n} \frac{\langle E_j|\widehat{V}|E_k\rangle\langle E_k|\widehat{V}|E_n\rangle}{\left(E_n - E_j\right)(E_n - E_k)} - \left\langle E_n|\widehat{V}|E_n\right\rangle \frac{\langle E_j|\widehat{V}|E_n\rangle}{\left(E_n - E_j\right)^2} \tag{11.46}$$

When $j = n$, we get the expression for the second order correction to the energy:

$$E_n^{(2)} = \sum_{k\neq n} \frac{\langle E_n|\widehat{V}|E_k\rangle\langle E_k|\widehat{V}|E_n\rangle}{(E_n - E_k)} \tag{11.47}$$

We add the second order correction to the first order solution in Eqs 11.18 and 11.19 to get the solution to second order in the potential:

$$|E_n'\rangle = |E_n\rangle + \sum_{j\neq n}\left\{\frac{\langle E_j|\hat{V}|E_n\rangle}{(E_n - E_j)} - \langle E_n|\hat{V}|E_n\rangle \frac{\langle E_j|\hat{V}|E_n\rangle}{(E_n - E_j)^2} + \sum_{k\neq n}\frac{\langle E_j|\hat{V}|E_k\rangle\langle E_k|\hat{V}|E_n\rangle}{(E_n - E_j)(E_n - E_k)}\right\}|E_j\rangle \quad (11.48)$$

$$E_n' = E_n + \langle E_n|V|E_n\rangle + \sum_{k\neq n}\frac{\langle E_n|V|E_k\rangle\langle E_k|V|E_n\rangle}{(E_n - E_k)} \quad (11.49)$$

In the coordinate picture, these results would be rederived to give a parallel expression for the corrected eigenfunctions:

$$U'_{E_n}(x) = U_{E_n}(x) + \sum_{j\neq n}\left\{\frac{\int dx U^*_{E_j}(x)\hat{V}(x)U_{E_n}(x)}{(E_n - E_j)} - \int dx U^*_{E_j}\hat{V}(x)U_{E_n}(x)\frac{\int dx U^*_{E_j}(x)\hat{V}(x)U_{E_n}(x)}{(E_n - E_j)^2}\right.$$
$$\left. + \sum_{k\neq n}\frac{\int dx U^*_{E_n}\hat{V}(x)U_{E_k}(x)\int dx U^*_{E_k}(x)\hat{V}(x)U_{E_n}(x)}{(E_n - E_j)(E_n - E_k)}\right\}U_{E_j}(x) \quad (11.50)$$

$$E_n' = E_n + \int dx U^*_{E_n}(x)\hat{V}(x)U_{E_n}(x) + \sum_{k\neq n}\frac{\int dx U^*_{E_n}(x)\hat{V}(x)U_{E_k}(x)\int dx U^*_{E_k}(x)\hat{V}(x)U_{E_n}(x)}{(E_n - E_k)} \quad (11.51)$$

Problem 11.4 *Examining the Stark effect in a spherically symmetric system like an atom is beyond the scope of this text. However, the Stark effect is easily seen in simple one-dimensional problem. Specifically, find the change in energy of the ground state due to coupling to the first excited state due to a dc electric field in an infinite quantum well containing a single electron. Calculate the result to second order. Using parity, show that the first order term is zero. The result is the* ***quadratic Stark effect****, the magnitude of the shift growing with the square of the applied electric field. Recall the Hamiltonian is given in one dimension by:*

$$H = H_0 - qxE \quad (11.52)$$

Using the form of the eigenfunctions for odd and even developed in Chapter 3 but repeated in problem 11.1 above, show:

$$\Delta E_{ground} = -\frac{4(qE)^2}{3LE_{box}}\left(\int_{-\frac{L}{2}}^{\frac{L}{2}} dx\, x\left(\sin\frac{3\pi x}{L} + \sin\left(\frac{\pi x}{L}\right)\right)\right)^2$$

The following trigonometric identities might be useful

$$\sin u \cos v = \frac{1}{2}(\sin(u+v) + \sin(u-v))$$

$$\cos u \sin v = \frac{1}{2}(\sin(u+v) - \sin(u-v))$$

Degenerate Perturbation Theory

Above we assumed that the states were non-degenerate. Recalling Eq. 11.11 we had developed the analysis to the point where we had found to first order:

$$a_j^{(1)} E_j + \langle E_j | \widehat{V} | E_n \rangle = E_n^{(1)} \delta_{jn} + E_n a_j^{(1)} \tag{11.53}$$

If for two (or more) states j and n, the eigenvalues are degenerate but the matrix element between the states is non-zero we have a singularity in Eq. 11.16:

$$a_j^{(1)} = \frac{\left\langle E_j^{(0)} \middle| \widehat{V} \middle| E_n^{(0)} \right\rangle}{\left(E_n^{(0)} - E_j^{(0)}\right)} \tag{11.54}$$

Note that if the matrix element is zero, i.e., $\langle E_j | V | E_n \rangle = 0$, *there is no problem with the application of non-degenerate perturbation theory, as we discussed above.*

To see the problem better and the solution more evident, we set the problem up in the matrix representation discussed in Chapter 7.[4]

The original Hamiltonian is

$$\widehat{H}_0{}' = \begin{bmatrix} E_{1_0} & 0 & 0 \\ 0 & E_{2_0} & 0 \\ 0 & 0 & E_{3_0} \end{bmatrix} \tag{11.55}$$

With eigenkets are

$$|E_{1_0}\rangle = \begin{bmatrix} 1 \\ 0 \\ 0 \end{bmatrix}, |E_{2_0}\rangle = \begin{bmatrix} 0 \\ 1 \\ 0 \end{bmatrix}, |E_{3_0}\rangle = \begin{bmatrix} 0 \\ 0 \\ 1 \end{bmatrix} \tag{11.56}$$

We take E_{1_0} and E_{2_0} to be degenerate: $E_{1_0} = E_{2_0} = a$ and include an interaction potential to couple various states:

$$\widehat{H}' = \begin{bmatrix} a & \widehat{V}_{1_o 2_o} & \widehat{V}_{1_o 3_o} \\ \widehat{V}_{2_o 1_o} & a & \widehat{V}_{2_o 3_o} \\ \widehat{V}_{3_o 1_o} & \widehat{V}_{3_o 2_o} & E_{3_0} \end{bmatrix} \tag{11.57}$$

For simplicity, we assume the matrix elements are real (therefore $\widehat{V}_{ij} = \widehat{V}_{ji}$. We work with a 3×3 rather than just a 2×2 to see the impact of the modifications on the remaining eigenstates *even if they are not degenerate with the first two.* We now rewrite $\hat{H}'$ as the sum of two matrices:

[4] This discussion is adapted from Paul R. Berman, "Introductory Quantum Mechanics" Springer (2018).

$$\hat{H}_0' = \begin{bmatrix} a & \hat{V}_{1_o2_o} & 0 \\ \hat{V}_{2_o1_o} & a & 0 \\ 0 & 0 & E_{3_0} \end{bmatrix} + \begin{bmatrix} 0 & 0 & \hat{V}_{1_o3_o} \\ 0 & 0 & \hat{V}_{2_o3_o} \\ \hat{V}_{3_o1_o} & \hat{V}_{3_o2_o} & 0 \end{bmatrix} \tag{11.58}$$

We use the results in Chapter 7 following the diagonalization process in starting with Eq. 7.23 and find new eigenvalues $E_1 = a - |\hat{V}_{1_o2_o}|$ and $E_2 = a + |\hat{V}_{1_o2_o}|$ with eigenkets:

$$|E_1\rangle = \frac{1}{\sqrt{2}}\begin{bmatrix} 1 \\ -1 \\ 0 \end{bmatrix}; |E_2\rangle = \frac{1}{\sqrt{2}}\begin{bmatrix} 1 \\ 1 \\ 0 \end{bmatrix}; |E_3\rangle = \begin{bmatrix} 0 \\ 0 \\ 1 \end{bmatrix} \tag{11.59}$$

In line with Chapter 7 (Eq. 7.33 and the following text), we form $\hat{\mathbf{S}}^{\dagger}$ and $\hat{\mathbf{S}}$ using the kets in Eq. 11.59 to make the columns in $\hat{\mathbf{S}}^{\dagger}$ and proceed to transform Eq. 11.58 such that the first matrix is diagonal.

$$\hat{\mathbf{S}}^{\dagger} = \begin{bmatrix} \frac{1}{\sqrt{2}} & \frac{1}{\sqrt{2}} & 0 \\ -\frac{1}{\sqrt{2}} & \frac{1}{\sqrt{2}} & 0 \\ 0 & 0 & 1 \end{bmatrix} \tag{11.60}$$

We now apply the diagonalization procedure to $\hat{H}_0'$:

$$\hat{H} = \hat{\mathbf{S}}\hat{H}_0'\hat{\mathbf{S}}^{\dagger} = \begin{bmatrix} \frac{1}{\sqrt{2}} & -\frac{1}{\sqrt{2}} & 0 \\ \frac{1}{\sqrt{2}} & \frac{1}{\sqrt{2}} & 0 \\ 0 & 0 & 1 \end{bmatrix}\begin{bmatrix} a & \hat{V}_{1_o2_o} & 0 \\ \hat{V}_{2_o1_o} & a & 0 \\ 0 & 0 & E_{3_0} \end{bmatrix}\begin{bmatrix} \frac{1}{\sqrt{2}} & \frac{1}{\sqrt{2}} & 0 \\ -\frac{1}{\sqrt{2}} & \frac{1}{\sqrt{2}} & 0 \\ 0 & 0 & 1 \end{bmatrix}$$

$$+ \begin{bmatrix} \frac{1}{\sqrt{2}} & -\frac{1}{\sqrt{2}} & 0 \\ \frac{1}{\sqrt{2}} & \frac{1}{\sqrt{2}} & 0 \\ 0 & 0 & 1 \end{bmatrix}\begin{bmatrix} 0 & 0 & \hat{V}_{1_o3_o} \\ 0 & 0 & \hat{V}_{2_o3_o} \\ \hat{V}_{3_o1_o} & \hat{V}_{3_o2_o} & 0 \end{bmatrix}\begin{bmatrix} \frac{1}{\sqrt{2}} & \frac{1}{\sqrt{2}} & 0 \\ -\frac{1}{\sqrt{2}} & \frac{1}{\sqrt{2}} & 0 \\ 0 & 0 & 1 \end{bmatrix}$$

$$= \begin{bmatrix} a- |\hat{V}_{1_o2_o}| & 0 & 0 \\ 0 & a+ |\hat{V}_{1_o2_o}| & 0 \\ 0 & 0 & E_{3_0} \end{bmatrix}$$

$$+ \frac{1}{\sqrt{2}}\begin{bmatrix} 0 & 0 & \hat{V}_{1_o3_o} - \hat{V}_{3_o4_o} \\ 0 & 0 & \hat{V}_{1_o3_o} + \hat{V}_{2_o3_o} \\ \hat{V}_{3_o1_o} - \hat{V}_{3_o2_o} & \hat{V}_{3_o1_o} + \hat{V}_{3_o2_o} & 0 \end{bmatrix} \tag{11.61}$$

It is now possible to proceed with using non-degenerate perturbation theory to pursue the correction to second order or higher.

Few-level approximation

For some problems, perturbation theory is not adequate. It may then be necessary to solve the problem exactly. However,it is often possible to get a solution that is much better than that in perturbation theory but not quite exact. The approach is based on recognizing from the results of perturbation theory, the significance of states far away in energy from the states of interest may not contribution significantly because of the increasing size of the denominators in Eq. 11.18 and Eq. 11.50. In addition, the matrix elements are usually getting smaller because the spatial overlap of the eigenfunctions is decreasing as the distance in energy between states grows. In this case, we can select the closest states and proceed to find an exact solution by the usual expansion methods used above. If this involves n-states, this means diagonalizing an $n\times n$ matrix. For anything greater than $n = 2$, it is easier to use a computer assisted solution such as Mathematica. Nevertheless, here we show important behavior by just considering coupling between two adjacent states. As an example, we consider the Stark effect on two levels and solve the problem exactly.

To see this, we consider a quantum system (e.g., an atom, a quantum dot, well, etc.) with a number of eigenstates described by a Hamiltonian that has been diagonalized. However, of interest is the behavior of two states that are coupled by the addition of a potential to the Hamiltonian. We consider the case of the Stark effect when an electric field (or some other interaction that couples two or more levels) is added. Then, assuming coupling to other states can be ignored because the effects (matrix elements) are small, it is possible to understand and predict the behavior, exactly. The Hamiltonian is given by

$$\widehat{H} = \widehat{H}_0 - \mu\mathcal{E} = \widehat{H}_0 - q\hat{x}\mathcal{E} \tag{11.62}$$

We assume the time independent Schrödinger equation has been solved for the quantum system in the absence of the field. That system is described by $\widehat{H}_0$. Since we know the eigenenergies and are only concerned about two specific states we identify as state 1, $|E_1\rangle$ *and state* 2 $|E_2\rangle$, we can write $\widehat{H}_0$ as

$$\widehat{H}_0 = \begin{bmatrix} E_1 & 0 \\ 0 & E_2 \end{bmatrix} \tag{11.63}$$

Then, we can write down the matrix for the potential energy as:

$$q\hat{x}\mathcal{E} = e\mathcal{E}\begin{bmatrix} 0 & x_{12} \\ x_{21} & 0 \end{bmatrix} \tag{11.64}$$

Problem 11.5 *Show that new eigenenergies are given by*

$$\begin{aligned} E^2_{H\pm} &= \pm\sqrt{\left(\frac{\hbar\omega_0}{2}\right)^2 + \mu_{21}\mu_{12}\,\mathcal{E}^2} \\ &= \pm\frac{\hbar\omega_0}{2}\sqrt{1+\left(\frac{2\,|\mu_{12}\,|\mathcal{E}}{\hbar\omega_0}\right)^2} \underset{lim\ \mathcal{E}\to 0}{\longrightarrow} \pm\frac{\hbar\omega_0}{2}\left(1+\frac{1}{2}\left(\frac{2\,|\mu_{12}\,|\mathcal{E}}{\hbar\omega_0}\right)^2\right) \end{aligned} \tag{11.65}$$

And the new eigenstates are given by

$$|E_{H\pm}\rangle = \frac{1}{\sqrt{1+\left(\frac{\hbar\omega_0}{2} - E_{H\pm}\right)}}\left(|1_{H_0}\rangle + \frac{\mu_{21}\mathcal{E}}{\left(\frac{\hbar\omega_0}{2} - E_{H\pm}\right)}|2_{H_0}\rangle\right) \tag{11.66}$$

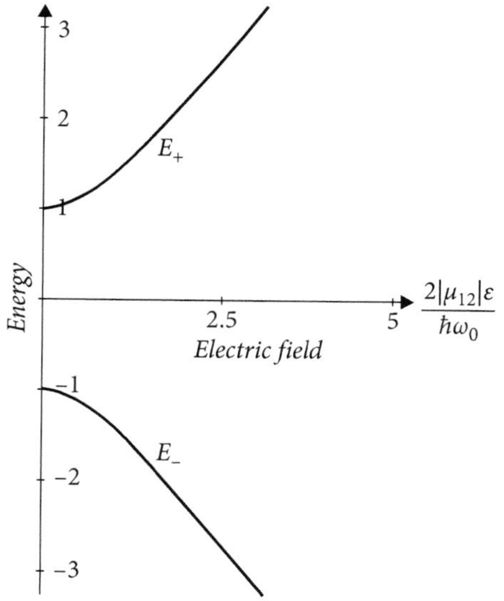

Fig. 11.2 The exact solution for the energy (units of $\frac{\hbar\omega_0}{2}$) as a function of applied field is shown for the DC Stark effect. The dependence on the applied field is quadratic for small bias, as predicted by perturbation theory, and then becomes linear for large bias.

You can see that new energy eigenstates in terms of the eigenstates without the field and the new energy eigenvalues. The DC Stark effect is quadratic for small fields as we saw in second order perturbation theory and linear with large fields (linear in this case does not imply first order perturbation theory). Figure 11.2 shows how the eigenenergies change as a function of the applied electric field. The electric field increases the energy splitting. This is a general result. For large fields, the levels separate linearly with the applied field. Imagine the case where the fields are so large that the zero field splitting appears small. The states then would appear to intersect as the field decreases. This can happen only if there is no interaction that lifts the degeneracy. In this case, a confining potential lifts the degeneracy. As a result of the interaction, the levels do not cross. When viewed as an energy map, the interaction between the two levels gives rise to an **avoidance crossing**. The field represents just an applied voltage bias and is routinely used to tune the energy level spacing (**Stark tuning**) in electronic and optical devices.

11.3 Time Dependent Perturbation Theory: Fermi's Golden Rule

Time dependent Hamiltonians dominate most devices of interest at the nano-scale as various electric, magnetic, strain, and RF/optical fields or other terms in the Hamiltonian change with time for various operations. Earlier we examined the physics of Rabi oscillations which required the exact solution of a two-level system interacting with a two-level electronic system. Most problems do not require such solutions. Perturbation theory is often quite adequate for calculating important features such as carrier mobility, sensor sensitivities, etc. Here we will develop solutions to the time dependent Schrödinger equation to second order, following the same approach above. The results from first order perturbation theory work in problems where the system performance is optimum if the strength of the perturbation is small enough that the system remains in the linear response region of operation such as a quantum system applied to metrology. However, some processes for different devices require a stronger coupling and are necessarily then nonlinear in the perturbation. The second order theory includes such important behavior as two-photon absorption that can lead to quantum entangled states and or stimulated Raman coherence between two electron spins. In practice, calculations may often be carried out to 3 rd order and higher. While it might seem more useful to just solve the problem numerically, it is extremely difficult if not impossible to understand the role of specific physics that can be tuned or optimized for a given application. However, once that understanding is had, a numerical solution may be very appropriate.

The difference between this problem and the above discussion is that $V = V(t)$, hence the Hamiltonian is time dependent:

$$\hat{H}(t) = \left(\hat{H}_0 + V(t)\right) \tag{11.67}$$

As in the case of time independent perturbation theory, it was the coupling between the basis states of the original Hamiltonian by the addition of a potential that caused the changes in the stationary states. The changes had no impact on the probability amplitude of being in a given state since that would have been a time dependent change. In **time dependent perturbation theory**, it is again the coupling between the states of the original Hamiltonian by the addition of the perturbation that is the origin of new behavior. Now, because the coupling is time dependent, it becomes possible to change the probability of being in different states, causing transitions from state to another. This is what happens when you turn the device "on".

As above, we assume the basic system is described by a Hamiltonian $\hat{H}_0$ that has already been diagonalized, meaning we have solved the time independent Schrödinger equation and have identified the energy eigenvalues E_n and the energy eigenstates $|E_n\rangle$. In the time independent case, we found solutions that gave us a revised set of eigenstates $|E'_n\rangle$ that amounted to an approximation of the basis states that would have resulted if the new Hamiltonian had been rediagonalized in the presence of the added potential. The new states are stationary (time independent), as required for eigenstates. In this section, we find that under the right conditions time dependent interactions can lead to **transitions** from one state to another. Time dependent interactions that lead to no change in the probability amplitudes because it is too slowly varying are said to be due to **adiabatic coupling** implying no change in energy. A slowly varying electric field of the kind discussed in Eq. 11.52 or Eq. 11.60 leads to a change in the energy levels but no change in the probability amplitudes.

Hence, the challenge now is to find a general means to solve the time dependent Schrödinger equation:

$$i\hbar\frac{d}{dt}|\psi(t)\rangle = \hat{H}(t)|\psi(t)\rangle \tag{11.68}$$

We know that for $\widehat{H}_0$ we have:

$$\widehat{H}_0 \left|E_n\right\rangle = E_n \left|E_n\right\rangle \tag{11.69}$$

Then the time dependent solution, in the absence of a $V(t)$, is given by:

$$|\psi(t)\rangle = e^{-i\frac{\widehat{H}_0 t}{\hbar}} \left|\psi\left(t=0\right)\right\rangle = \sum_k a_n e^{-i\frac{E_n t}{\hbar}} \left|E_n\right\rangle \tag{11.70}$$

where

$$a_n = \langle E_n|\psi\left(t=0\right)\rangle \tag{11.71}$$

Returning to Eq. 11.67, we use the solution given in Eq. 11.70 but allow the expansion coefficients in Eq. 11.70 to become time dependent:

$$|\psi(t)\rangle = \sum_n a_n(t) e^{-i\frac{E_n t}{\hbar}} \left|E_n\right\rangle \tag{11.72}$$

subject to the initial condition that

$$a_n\left(t=0\right) = \langle E_n|\psi\left(t=0\right)\rangle \tag{11.73}$$

Recall that Eq. 11.72 is equivalent to the interaction picture discussed in Chapter 9. Now substitute the expansion from Eq. 11.72 into the time dependent Schrödinger equation:

$$i\hbar\frac{d}{dt}\sum_n a_n(t)e^{-i\frac{E_n t}{\hbar}}\left|\psi_n\right\rangle = \left(\widehat{H}_0 + \widehat{V}(t)\right)\sum_n a_n(t)e^{-i\frac{E_n t}{\hbar}}\left|E_n\right\rangle \tag{11.74}$$

Giving:

$$i\hbar\sum_n\left(\frac{d}{dt}a_n(t) - i\frac{E_n}{\hbar}a_n(t)\right)e^{-i\frac{E_n t}{\hbar}}\left|E_n\right\rangle = \sum_n a_n(t)\left(E_n + \widehat{V}(t)\right)e^{-i\frac{E_n t}{\hbar}}\left|E_n\right\rangle \tag{11.75}$$

Projecting onto $\langle E_j|$ and collecting terms we get the equation of motion for the probability amplitude:

$$\frac{d}{dt}a_j(t) = \frac{1}{i\hbar}\sum_n a_n(t)\langle E_j|\widehat{V}(t)|E_n\rangle e^{i\frac{(E_j-E_n)t}{\hbar}} \tag{11.76}$$

This is an infinite set of coupled first order differential equations. This derivation and result is a slightly more general discussion to that resulting in Eq. 9.86.

To develop a perturbation theory approach, we insert an expansion parameter, λ, in the Hamiltonian as we did in time independent perturbation theory. This approach formalizes the informal but accurate discussion in Chapter 9 in the introduction to Rabi oscillations. We start with the Hamiltonian including the expansion parameter λ:

$$\widehat{H}(t) = \left(\widehat{H}_0 + \lambda V(t)\right) \tag{11.77}$$

where again λ is ab arbitrary number between 0 and 1.

$$0 \leq \lambda \leq 1 \tag{11.78}$$

We then expand the $a_n(t)$ in a power series of λ as we did in Eq. 11.7. Again setting terms on the left with a given power of λ equal to the terms right side of Eq. 11.75 associated with the same power of λ, we get:

$$\lambda^0 : \quad \frac{d}{dt}a_j^{(0)}(t) = 0 \tag{11.79}$$

$$\lambda^1 : \quad \frac{d}{dt}a_j^{(1)}(t) = \frac{1}{i\hbar}\sum_n a_n^{(0)}(t)\langle E_j|\widehat{V}(t)|E_n\rangle e^{i\frac{(E_j-E_n)t}{\hbar}} \tag{11.80}$$

$$\lambda^2 : \quad \frac{d}{dt}a_j^{(2)}(t) = \frac{1}{i\hbar}\sum_n a_n^{(1)}(t)\langle E_j|\widehat{V}(t)|E_n\rangle e^{i\frac{(E_j-E_n)t}{\hbar}} \tag{11.81}$$

Or most generally:

$$\frac{d}{dt}a_j^{(m)}(t) = \frac{1}{i\hbar}\sum_n a_j^{(m-1)}(t)\langle E_j|\widehat{V}(t)|E_n\rangle e^{i\frac{(E_j-E_n)t}{\hbar}} \tag{11.82}$$

From Eq. 11.79, we see that $a_j^{(0)}(t)$ is a constant for all time, in the limit of small perturbations, and since $a_j\,(t=0)$ must satisfy the initial condition, we have that

$$a_j^{(0)}(t) = a_j\,(t=0). \tag{11.83}$$

Since we are doing perturbation theory, we require the coupling of the perturbation of the system is sufficiently weak so that changes to all $a_j(t)$ are small compared to 1. So if for a given j, $a_j\,(t=0) = 1$ then by conservation of probability,

$$a_k^{(0)}(t) = a_k\,(t=0) = 0; k \neq j \tag{11.84}$$

First Order Time Dependent Perturbation Theory

In order to find the first order correction, we next integrate Eq. 11.80, where $a_j^{(0)}$ is removed from the integral since it is a constant according to Eq. 11.79:

$$a_k^{(1)}(t) = \frac{1}{i\hbar}\sum_{j\neq k} a_j^{(0)} \int_0^t dt'\langle E_k|\widehat{V}(t')\,|E_j\rangle e^{-i\frac{(E_j-E_k)t'}{\hbar}} \tag{11.85}$$

This is a general result from **first order time dependent perturbation theory** (taking $\lambda = 1$) and the first order correction to the time dependent wave function. This also now provides the solution to $|\psi(t)\rangle$ for all time to first order in time. By recalling Eq. 11.72, we have then

$$|\psi^{(1)}(t)\rangle = \sum_j a_j^{(1)}(t)e^{-i\frac{E_jt}{\hbar}}\,|E_j\rangle \tag{11.86}$$

To get the entire $|\psi(t)\rangle$ we must sum $|\psi^{(0)}(t)\rangle$ and $|\psi^{(1)}(t)\rangle$

$$|\psi(t)\rangle = |\psi^{(0)}(t)\rangle + |\psi^{(1)}(t)\rangle \tag{11.87}$$

where

$$|\psi^{(0)}(t)\rangle = \sum_k a_k^{(0)} e^{-i\frac{E_k t}{\hbar}} |E_k\rangle \tag{11.88}$$

In calculations, we might be interested in calculating the expectation value of some observable. For example, $\langle x(t)\rangle$ to first order in time. The would then be written as:

$$\begin{aligned}\langle x(t)\rangle &= \langle\psi(t)|\hat{x}|\psi(t)\rangle = \left(\langle\psi^{(0)}| + \langle\psi^{(1)}|\right)\hat{x}\left(|\psi^{(0)}(t)\rangle + |\psi^{(1)}(t)\rangle\right)\\ &= \langle\psi^{(0)}|\hat{x}|\psi^{(0)}\rangle + \langle\psi^{(0)}|\hat{x}|\psi^{(1)}\rangle + \langle\psi^{(1)}|\hat{x}|\psi^{(0)}\rangle\end{aligned} \tag{11.89}$$

The last two terms represent the first order contribution due to the time dependent perturbation. The term $\langle\psi^{(1)}|\hat{x}|\psi^{(1)}\rangle$ is not included because that would contribute at second order and if that was included, we would need to calculate the last two terms to second order to be consistent.

However, much more information can be extracted by this analysis if we use a common form for $\widehat{V}(t)$. We write $\widehat{V}(t) = \widehat{V}f(t)$ and take for $f(t)$

$$f(t) = \begin{cases} e^{i\epsilon\omega t} & t \geq 0, \epsilon = 0, \pm 1, \omega \geq 0 \\ 0 & t < 0 \end{cases} \tag{11.90}$$

where ω is real and $\omega \geq 0$. When $\omega = 0$, this corresponds to a step function representing the sudden turn on of a constant potential, such as the bias in a diode. Since the potential must be real, both phasors must be included. Often, only one will contribute to a calculation, the other phasor contribution is usually a negligible contribution. We further assume that $\langle E_k|\widehat{V}|E_k\rangle = 0$ to simplify issues, if it is not, see the procedure described in Davydov.[5] Integrating Eq. 11.80, we get:

$$a_k^{(1)}(t) = \frac{1}{i\hbar}\sum_j a_j^{(0)}\widehat{V}_{kj}\int_0^t dt' e^{i(\epsilon\omega+\omega_{kj})t'} = \frac{1}{i\hbar}\sum_j a_j^{(0)}\widehat{V}_{kj}\frac{e^{i(\epsilon\omega+\omega_{kj})t} - 1}{i\left(\epsilon\omega + \omega_{kj}\right)} \tag{11.91}$$

where the matrix element $\hat{V}_{kj} = \langle E_k|\hat{V}(t')|E_j\rangle$ and $\omega_{kj} = \frac{(E_k - E_j)}{\hbar}$.

Eq. 11.91 is complete to first order and a useful description of the time evolution of the probability amplitude. While $a_k^{(1)}(t = 0) = 0$, $|a_k^{(1)}(t)| > 0$, implying a **transition** has occurred from state j to state k, provided the matrix element connecting state j and k, $\widehat{V}_{kj} \neq 0$. $\left|a_k^{(1)}(t)\right|^2$ then represents the probability of undergoing a transition to state k when starting in state j. We designate this transition probability, $P_{k\leftarrow j}$:

[5] A.S. Davydov, Quantum Mechanics, 2 nd ed., Pergamon Press, Oxford (1965).

$$P_{k \leftarrow j} :\equiv \left|a_k^{(1)}(t)\right|^2 = \frac{2}{\hbar^2}\left|\hat{V}_{jk}\right|^2 \left(\frac{1 - cos\left(\epsilon\omega + \omega_{kj}\right) t}{\left(\epsilon\omega + \omega_{jk}\right)^2}\right) \tag{11.92}$$

with $\left|a_k^{(0)}\right|^2 = 1$.

For large t, we see in Appendix C that (after changing the variable to match the current discussion)

$$\frac{1}{\pi}\lim_{t\to\infty} \frac{1 - cos\left[\left(\epsilon\omega + \omega_{jk}\right) t\right]}{t\left(\epsilon\omega + \omega_{jk}\right)^2} = \delta\left(\epsilon\omega + \omega_{kj}\right) \tag{11.93}$$

Substituting this into Eq. 11.92 we get

$$P_{kj}(t) = \frac{2\pi t}{\hbar^2}\left|\hat{V}_{jk}\right|^2 \delta\left(\epsilon\omega + \omega_{kj}\right) \tag{11.94}$$

Finally, the **transition rate**, given by the **transition probability per unit time**, T_{kj}, is had by taking the time derivative of $P_{kj}(t)$ giving

$$T_{kj}(t) = \frac{2\pi}{\hbar}\left|\hat{V}_{kj}\right|^2 \delta\left(\epsilon\hbar\omega + \left(E_k - E_j\right)\right) \tag{11.95}$$

where to emphasize the physical significance of the argument in the Dirac δ-function, it has been changed from frequency to energy, recalling that $\delta(ax) = \frac{1}{|a|}\delta(x)$. Recalling from earlier studies that $\hbar\omega$ is an energy associated with a vibration, the δ-function ensures conservation of energy, where, as is the case in classical physics, conservation of energy is not a postulate of physics but a result of an appropriate set of initial postulates. This is the most general form of **Fermi's golden rule** (some texts designate a later version (below) as Fermi's Golden Rule). Usually, the language of Fermi's Golden Rule implies first order perturbation theory.

Density of States

From the standpoint of measurement or impact on experiments or technology, the analysis leading to 11.95 cannot be complete because the answer includes a Dirac delta-function. As stated before, the Dirac delta-function only has physical meaning under an integral and cannot be associated with the measurement of an observable. The problem is that the variables ω and ω_{kj} have a well-defined value. Trying to adjust ω so that it would be identical ω_{kj} is impossible since, even if we could generate a perfectly monochromatic field at ω, infinite precision in adjustment is impossible. So, it is necessary to have some variable in the argument of the delta-function that is continuous.

For example, if we were thinking about the **photoionization** or the **photoelectric effect**, purely monochromatic radiation excites an electronic system from a bound state (bound states are "infinitely" sharp) to a state in the **continuum** (i.e., a **scattering state**, such as we used in Chapter 5) above the confining potential barrier. The continuum state is, by definition, a continuum of energies. The reverse process could also happen where an electron in the continuum recombines with an electronic system by emitting energy and returning to a bound state. Alternatively, an excited electronic system may undergo a transition from an excited bound state to a lower lying bound state, emitting radiation into the vacuum. The radiation that is emitted is emitted into a continuum of allowed states (different frequencies, directions of propagation, polarization). It is also possible to consider using a broad spectrum of incoherent radiation (e.g., a thermal source) described by a distribution $f(\omega)$, to excite (up or down) an electron from bound state to another bound state. The result in Eq. 11.95 describes a probability of going from an initial state to a final state. From the standpoint of probabilities, it is required here that the *total probability is the sum over all final states and an average over all possible initial states*. When the final or initial states are a continuum, the sum becomes an integral allowing an evaluation of Eq. 11.95.

Hence it is clear that while the delta-function in Eq. 11.95 puts a pause in the calculation, it is only because the question of interest, the transition probability per unit time, has not yet been fully described in terms of the details of a measurement. In general, care is needed to correctly apply Fermi's Gold Rule. As an example we will consider photoionization from a bound state to a continuum with purely monochromatic radiation. We are focused on an electron being excited to the continuum and so we must consider the number of states in the continuum that are available for excitation. This means we will need to sum over all possible final states of the electron which corresponds to state $|k\rangle$ above. To do this, we must now find the number of states between $E_k$and $E_k + dE_k$.

From this point on, we will suppress the state j designation and assume that E corresponds to that state, designated $|E\rangle$. The question then is what is the Hamiltonian for the continuum. Under the assumption that we are exciting to an energy considerably higher than the confinement potential in the problem (for hydrogen, that means much above 0, for a trap with a barrier height that would mean V_0, etc.) In this case, we assume the effect of the binding potential is negligible and the Hamiltonian is that of a free particle which we studied at the beginning of Chapter 5.

$$\hat{H}_0 = \frac{\hat{p}^2}{2m} \tag{11.96}$$

Any corrections to this assumption just means the mathematics becomes a little more complicated with usually small quantitative changes in the result. The qualitative behavior remains the same.

The usual way to approach this problem is take the electron to be confined to a three-dimensional box that is large compared to the localized wave function of the initial state. We assume periodic boundary conditions for a cavity length L. The idea is that at the end of the problem, we will let L go to infinity. It is also easier to maintain physical insight when the system is in a finite volume. Periodic boundary conditions ensure that there is no reflection or cavity buildup of field strength.

In one dimension, the normalized eigenfunction for the particle is given by:

$$u(k, x) = \sqrt{\frac{1}{L}} e^{ikx} \tag{11.97}$$

where

$$E = \frac{\hbar^2 k^2}{2m}; \hbar k = p \tag{11.98}$$

With periodic boundary conditions:

$$e^{ikx} = e^{ik(x+L)} \tag{11.99}$$

Requiring

$$k = \frac{2n\pi}{L}, n = \pm 1, \pm 2, \cdots \tag{11.100}$$

Both signs are included for k since the particle can move in either direction. This is easily generalized to 2 or 3 dimensions. For 3 dimensions,

$$k \to k_x, k_z, k_z \text{ and } n \to n_x, n_z, n_z \tag{11.101}$$

$$u(\mathbf{k}, \mathbf{r}) = \sqrt{\frac{1}{V}} e^{i\mathbf{k}\cdot\mathbf{r}} \tag{11.102}$$

The volume $V = L_x L_y L_z$.

Returning now to Eq. 11.95 and to account for that fact that there are multiple final states, we must sum over those states. We do that by summing over all states n_x, n_z, and n_z. However, that sum becomes an integral as V becomes large. In three dimensions, we get

$$\sum_{n_x=-\infty}^{\infty} \sum_{n_y=-\infty}^{\infty} \sum_{n_Z=-\infty}^{\infty} \to \lim_{L_x,L_y,L_z\to\infty} \int_{-\infty}^{\infty} dn_x \int_{-\infty}^{\infty} dn_y \int_{-\infty}^{\infty} dn_z$$
$$= \lim_{V\to\infty} \frac{V}{8\pi^3} \int_{-\infty}^{\infty} dk_x \int_{-\infty}^{\infty} dk_x \int_{-\infty}^{\infty} dk_x = \lim_{V\to\infty} \frac{V}{8\pi^3} \int_0^{2\pi} d\varphi \int_0^{\pi} \sin\theta \, d\theta \int_0^{\infty} k^2 dk \tag{11.103}$$

where the last expression results from conversion to a spherical form since the shape of the cavity does not matter for $V \to \infty$. Since the argument of the delta-function is energy, we need to convert the integral from one over k to one over E. Since $E = \frac{\hbar^2 k^2}{2m}$, we have $k = \left(\frac{2mE}{\hbar^2}\right)^{\frac{1}{2}}$ giving for dk

$$dk = \left(\frac{m}{2\hbar^2 E}\right)^{\frac{1}{2}} dE \tag{11.104}$$

And hence

$$\frac{V}{8\pi^3} k^2 dk = \frac{V}{8\pi^3} \frac{2mE}{\hbar^2} \left(\frac{m}{2\hbar^2 E}\right)^{\frac{1}{2}} dE = \frac{1}{8}\sqrt{2} \frac{V}{\pi^3 \hbar^3} (m)^{\frac{3}{2}} \sqrt{E} dE \tag{11.105}$$

Hence,

$$\sum_{n_x}\sum_{n_y}\sum_{n_z} \rightarrow \lim_{V\rightarrow\infty}\frac{1}{8}\sqrt{2}\frac{V}{\pi^3\hbar^3}(m)^{\frac{3}{2}}\iiint \sqrt{E}dEd\Omega \tag{11.106}$$

where $d\Omega \equiv sin\,\theta d\vartheta d\varphi$ is the differential solid angle The term

$$n_\Omega(E) = \frac{1}{8}\sqrt{2}\frac{V}{\pi^3\hbar^3}(m)^{\frac{3}{2}}\sqrt{E} \tag{11.107}$$

is referred to as the **differential density of states** for free particles in three dimensions. The word differential is there because we have not yet integrated over the solid angle $d\Omega \equiv sin\,\theta d\theta d\varphi$. Many texts write down

$$n(E) = \frac{1}{\sqrt{2}}\frac{V}{\pi^2\hbar^3}(m)^{\frac{3}{2}}\sqrt{E} \tag{11.108}$$

by carrying out the integration over 4π (i.e., multiply by 4π.) This is the **density of states.** However, *if there is a vector dependence in the matrix element, e.g.,* $\boldsymbol{\mu}\cdot\mathbf{E}$, *then that angle must be included in the integral.* Hence, for the most general discussion, it is best to use $n_\Omega(E)$.

If the calculation involves an electron but not explicitly involving spin in the Hamiltonian, then a factor of 2 would multiply Eq. 11.105 and Eq. 11.106.

We return now to the first order result above, Eq. 11.95, and we consider photoionization, then the final state is a continuum and we must sum over those states:[6]

$$\begin{aligned} T_{kj}(t) &= \frac{2\pi}{\hbar}\int_{-\infty}^{\infty} dE_k \int_{4\pi} d\Omega \left|a_j^{(0)}\right|^2 \left|\hat{V}_{kj}\right|^2 \frac{1}{8}\sqrt{2}\frac{V}{\pi^3\hbar^3}(m)^{\frac{3}{2}}\sqrt{E_k}\delta\left(\epsilon\hbar\omega + \left(E_k - E_j\right)\right) \\ &= \frac{2\pi}{\hbar}\left|a_k^{(0)}\right| \int_{4\pi} d\Omega^2 \left|\hat{V}_{kj}\right|^2 \frac{1}{8}\sqrt{2}\frac{V}{\pi^3\hbar^3}(m)^{\frac{3}{2}}\sqrt{E_j + \hbar\omega} \end{aligned} \tag{11.109}$$

Since the final state is a higher energy state than the initial state, we have that $E_k - E_j > 0$. Therefore, since $\omega > 0$, we must set $\epsilon = -1$, such that

$$E_k = E_j + \hbar\omega, \omega > 0 \tag{11.110}$$

and represents the energy in the continuum state.

We note at this point, there is a volume in the numerator from the density states. This is problematic as it tends to infinity. However, the resolution of this is evident in the evaluation of the matrix element. The potential goes like $V = -q\mathbf{E}\cdot\mathbf{r}$.[7] Assuming the electron is initially in the ground state of the hydrogen atom, $\psi_{nlm}(\mathbf{r}) = \psi_{100}(\mathbf{r}) = \frac{1}{\sqrt{\pi a_0^3}}e^{\frac{r}{a_0}}$ and the final state is a free particle in the three-dimensional box discussed above, $\psi(\mathbf{k}, \mathbf{r}) = \frac{1}{\sqrt{V}}e^{i\mathbf{k}\cdot\mathbf{r}}$ where $\boldsymbol{k}$ is the momentum and direction of the emitted electron, the matrix element becomes

6 $\int_{4\pi} d\Omega$ means by definition $\int_0^{2\pi} d\varphi \int_0^{\pi} sin\,\theta d\theta$.

7 This form is only valid in the dipole approximation and the dipole approximation may not be appropriate for some situations. In this case, a more complete form of the Hamiltonian must be used and is usually written in terms of the vector potential in the Coulomb gauge.

$$\left|\hat{V}_{kj}\right|^2 = \frac{e^2}{V}\left|\boldsymbol{E}\cdot\int_{all\ space} e^{i\boldsymbol{k}\cdot\boldsymbol{r}}\boldsymbol{r}\frac{1}{\sqrt{\pi a_0^3}}e^{-\frac{r}{a_0}}r^2 \sin\theta drd\theta d\varphi dr\right|^2 \tag{11.111}$$

The details of completing the integral are not relevant for this discussion. However, what is important is to note that because the calculation was done in a box, the free particle wave function is normalized with a factor of $\frac{1}{\sqrt{V}}$. That becomes $\frac{1}{V}$ after the matrix element is squared and that cancels the V in the numerator of the density of states in Eq. 11.119. So, the answer becomes independent of V, as expected. The transition probability per unit time is a rate and can be used in rate equations which are often phenomenological, though in Chapter 18 we see that in quantum systems these equations can be developed from quantum mechanics. This is one more important reason for developing these solutions.

Problem 11.6 *Find the density of states in three dimensions for the radiation field. The solution to the wave equation for a classical radiation field in a cavity with periodic boundary conditions is the same as Eq. 11.79.*

Problem 11.7 *Consider the problem now of photoionization from the ground state in a one-dimensional system. The energy to ionize the system is given by $E_i = V_0 - E_1$. Photoionization occurs when an oscillating electric field with a sufficiently high frequency that $\hbar\omega > E_i$. Using Fig. 11.3, we see that the energy of the electron E,* and the electron kinetic energy after photoionization is $E - E_i$. *Assume that a particle is in a one-dimensional square trap of the type shown in Fig. 11.3, similar to that in Chapter 3. Note that $\hat{V} = -ex\frac{E_0}{2}$. The final state is a free particle state which we assume has the form given by $\psi_E(x) = \frac{1}{\sqrt{\mathcal{L}}}e^{ikx}$ where as usual $k = \sqrt{\frac{2m(E-E_i)}{\hbar^2}}$. Periodic boundary conditions give $k = \frac{2n\pi}{\mathcal{L}}$, $n = \pm 1, \pm 2, \cdots$ with the normalization factor as shown. $\mathcal{L}$ is chosen to be distinct from the size of the square potential, L. To keep the math as simple but realistic as possible, we assume that the square trap is very deep, and that very little of the wave function for the ground state is in the forbidden region. Hence, we assume that the state $\psi_{E_1}(x)$ is given approximate as the ground state of the infinite potential well of Chapter 3:*

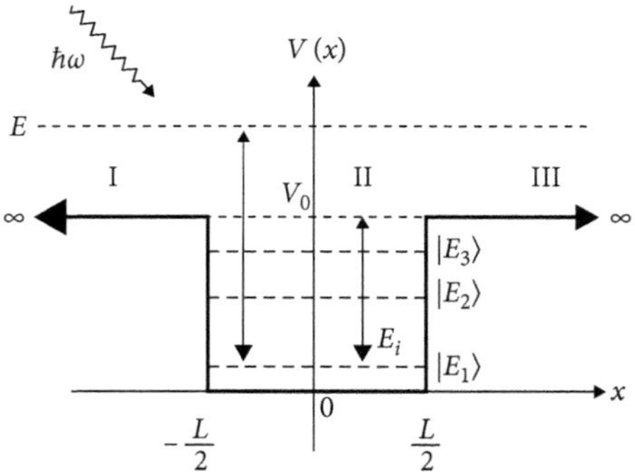

Fig. 11.3 A potential well is an excellent model for a defect: a quantum well or even a quantum dot, if generalized to three dimensions.

$$\psi_{E_1}(x) = \sqrt{\frac{2}{L}}\cos\left(\frac{\pi x}{L}\right)$$

a) Evaluate the matrix element.

b) Determine the differential density of states for a particle in one dimension in this problem. Show, not including spin, that

$$\rho_E = \frac{\mathcal{L}}{4\pi}\sqrt{\frac{2m}{\hbar^2 (E - V_0)}}$$

c) Now use the expression from Eq. 11.95 to find the transition probability per unit time and in the process show that the size of the box used in calculating the density of states is cancelled.

The first order time dependent perturbation calculation above is extremely powerful for many problems in devices. However, second order perturbation has historically been important to understand things like harmonic generation, solid state laser behavior and more. More recently, second order type processes have been exploited in new schemes to control quantum systems.

Second Order Time Dependent Perturbation Theory

To evaluate the time evolution of the system at second order, we integrate Eq. 11.81 and substitute the result from first order perturbation theory from Eq. 11.85:

$$\begin{aligned} a_l^{(2)}(t) &= \frac{1}{i\hbar}\int_0^t dt' \sum_k \langle E_l|\hat{V}(t')|E_k\rangle e^{i\frac{(E_l-E_k)t'}{\hbar}} a_k^{(1)}(t') \\ &= \left(\frac{1}{i\hbar}\right)^2 \int_0^t dt' \sum_k \hat{V}_{lk} e^{i(\epsilon\omega+\omega_{ik})t'} \int_0^{t'} dt'' \sum_n \hat{V}_{kn} e^{i(\epsilon'\omega'+\omega_{kn})t''} a_n^{(0)} \\ &= \left(\frac{1}{i\hbar}\right)^2 \int_0^t dt' \sum_{k,n} \hat{V}_{lk}\hat{V}_{kn} \frac{e^{i(\epsilon\omega+\omega_{lk}+\epsilon'\omega'+\omega_{kn})t'} - e^{i(\epsilon\omega+\omega_{lk})t'}}{i(\epsilon'\omega'+\omega_{kn})} a_n^{(0)} \end{aligned} \tag{11.112}$$

The second term, $e^{i(\epsilon\omega+\omega_{lk})t'}$, is usually ignored at this point because it arises from the $t = 0$ limit of the first integral, which is a considered a transient arising from the first order part of the calculation and will not contribute for long times (not evident from this calculation).[8] This leaves

$$a_i^{(2)}(t) = \left(\frac{1}{\hbar}\right)^2 \sum_j \hat{V}_{lk}\hat{V}_{kn} \frac{e^{i(\epsilon'\omega'+\epsilon\omega+\omega_{ln})t} - 1}{(\epsilon'\omega' + \epsilon\omega + \omega_{ln})(\epsilon'\omega' + \omega_{kn})} a_k^{(0)} \tag{11.113}$$

This is now the **second order time dependent perturbation theory** result and it gives us the second order correction to probability amplitude. The strongest contributions in Eq. 11.112 are usually when the denominators are the smallest, corresponding to the rotating wave approximation introduced in Chapter 9. In the problems below, the sign of ϵ' and ϵ will depend

[8] See J.J. Sakurai, *Advanced Quantum Mechanics*, Benjamin/Cummings Publishing Co., Menlo Park, CA (1967).

on whether a given $\hbar\omega_{nm} = E_n - E_m$ is positive or negative since the electromagnetic driving frequencies, ω and ω' are always positive.

Finding the probability of undergoing a transition from state l to n, we have

$$P_{l\leftarrow n}(t) = \frac{2t}{\hbar^2}\left|\sum_j \frac{\hat{V}_{lk}\hat{V}_{kn}}{(\epsilon'\omega' + \omega_{kn})}\right|^2 \frac{1 - cos\,(\epsilon'\omega' + \epsilon\omega + \omega_{ln})\,t}{t(\epsilon'\omega' + \epsilon\omega + E_{ln})^2}$$
$$= \frac{2\pi t}{\hbar}\left|\sum_j \frac{\hat{V}_{ij}\hat{V}_{jk}}{\left(\epsilon\omega + \omega_{jk}\right)}\right|^2 \delta\left(\hbar\epsilon'\omega' + \hbar\epsilon\omega + E_{ln}\right) \tag{11.114}$$

where the last equality is the limit for large t (Eq. 11.93). Finally, transition probability per unit time, T_{jk}, is had by again dividing $P_{jk}(t)$ by time, giving

$$T_{l\leftarrow n}(t) = \frac{2\pi}{\hbar}\left|\sum_j \frac{\hat{V}_{lk}\hat{V}_{kn}}{(\epsilon'\omega' + \omega_{kn})}\right|^2 \delta\left(\hbar\epsilon'\omega' + \hbar\epsilon\omega + E_{ln}\right) \tag{11.115}$$

This is the second order contribution to Fermi's golden rule. Note that if both the first and second order contributions contribute to $a_i(t)$, it is essential to add the first (Eq. 11.91) and second order (Eq. 11.113) amplitudes together first, before squaring to calculate the probability and transition probability per unit time.

Problem 11.8 *Two-photon processes have been an important part of many technologies. For example, two-photon absorption can limit performance of semiconductor lasers. It has also been used to enable real-time holography for correction of optical aberrations in imaging systems. The inverse process can result in production of quantum entangled states of the photon. Find the probability amplitude for near resonant excitation of a* ***cascade-up three-level system*** *shown in Fig. 11.4. Assume that frequency ω_1 is only near resonant with the transition from state $|1\rangle \rightarrow |2\rangle$ transition and ω_2 is only near resonant with the transition from state $|2\rangle \rightarrow |3\rangle$ transition. Define ω_{21} and ω_{32} such that $\hbar\omega_{21} = E_2 - E_1$ and $\hbar\omega_{32} = E_3 - E_2$.*

Problem 11.9 *The energy level structure on the right is called a* ***three-level Λ-system,*** *where the Greek symbol, Λ, is a capital gamma. Again assume that frequency ω_1 is only near resonant with*

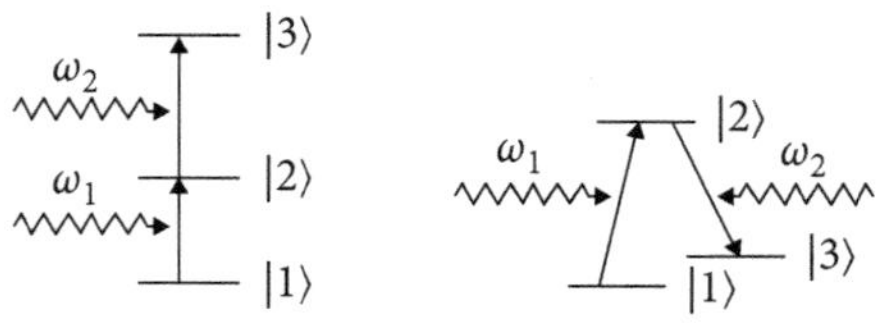

Fig. 11.4 On the left side, a cascade-up three-level system is used to calculate near resonant excitation. The arrows show on-resonance but the calculation is limited to near resonance–a cascade-up three-level system for examining near resonant two-photon absorption. The diagram on the right is a three-level Λ-system, where the Greek symbol, Λ, is a capital gamma. The wavy arrows represent classical electromagnetic radiation. The arrows show on-resonance but the calculation is limited to near resonance. For resonant excitation, coupling to the quantum vacuum must be considered and is discussed in the chapters on the operator approach (Chapter 16) and the density matrix approach (Chapter 18).

the transition from state $|1\rangle \rightarrow |2\rangle$ *and* ω_2 *is only near resonant with the transition from state* $|2\rangle \rightarrow |3\rangle$*. Define* ω_{21} *and* ω_{32} *such that* $\hbar\omega_{21} = E_2 - E_1$ *and* $\hbar\omega_{32} = E_3 - E_2$*. Find the state vector that results from excitation of this system assuming that at* $t = 0$*, the system is in the ground state. This is a highly relevant system for quantum devices. In the appropriate system accompanied by spin-orbit coupling discussed above, such as an atom or quantum dot, state* $|1\rangle$ *could correspond to a spin down state and state* $|3\rangle$ *could correspond to a spin up state. The resulting state vector is a coherent superposition of spin states. If state* $|2\rangle$ *can spontaneously emit a photon to either states* $|1\rangle$ *or* $|3\rangle$*, the result will be a quantum entangled state between the polarization or frequency of the emitted photon and the spin states. These ideas will be explored more in later chapters.*

11.4 Summary

It is the dynamical processes that make technology function; this means a time dependent Hamiltonian. Solving such equations is generally difficult and, fortunately, not typically needed. In that case, perturbation theory is usually the analytical tool of choice. Of course, there are many software platforms that can provide full and (usually) accurate answers. However, it is difficult to build up physical intuition without seeing the physics in the solutions. In the case of the above analysis, we first encountered Eq. 11.36 in the study of the Rabi oscillations. Here we have the possibility of many more states and more complex driving (time dependent) terms in the Hamiltonian. Only on long time scales with constant or oscillating potentials that turn on at $t = 0$ is it possible to extract the rather simple result embodied in Fermi's golden rule. The matrix elements we encounter in this work are those of the potential in the energy basis. Because the potential is time dependent, there is no stationary (steady state) solution to Schrödinger's equation, so it is not possible to think about diagonalizing such a Hamiltonian. So the matrix elements, e.g., $\langle E_j|\hat{V}(t')|E_k\rangle$, show a coupling (any matrix element between separate states) represents a coupling between the states caused by the physics represented by the operator, like an electric field on the charge in an atom.

Vocabulary (page) and Important Concepts

- stationary states 182
- time independent perturbation theory 182
- non-degenerate perturbation theory 182
- first order time independent perturbation theory 183
- state coupling 184
- state mixing 184
- symmetry breaking 185
- magnetic moment 186
- magnetic substates 186

12 Bosons and Fermions: Indistinguishable Particles with Intrinsic Spin

12.1 Introduction

Information in technology is traditionally transmitted and manipulated using elementary particles such as electrons and photons. The discussion so far has focused on the quantum behavior of a single particle described by some Hamiltonian. However, the particles of a specific kind (e.g., an electron) are indistinguishable. Having indistinguishable particles means that if two identical particles are interchanged, the system is unchanged. This is an exchange symmetry and, as seen in earlier chapters, symmetry can provide insight into physical behavior.

12.2 Eigenfunctions and Eigenvalues of the Exchange Operator

Then **exchange operator** for two particles, $\hat{P}_{12}$, is defined such that if $\psi(1,2)$ is a wave function describing a system with two identical particles identified as 1 and 2 then

$$\hat{P}_{12}\psi(1,2) = \psi(2,1) \tag{12.1}$$

Even if there were many other identical particles, the operator would only exchange particles 1 and 2 and the rest would be left unchanged. The result would be same; hence, we focus on a simple two particle system.

As before, now that an operator that represents the symmetry has been defined, we proceed to find the eigenfunctions of the operator. That is, we find λ such that

$$\hat{P}_{12}\psi(1,2) = \lambda\psi(1,2) \tag{12.2}$$

Given Eq. 12.2, we then know that

$$\hat{P}_{12}^{2}\psi(1,2) = \lambda^2\psi(1,2) \tag{12.3}$$

But we also know from Eq. 12.1 that

$$\hat{P}_{12}^{2}\psi(1,2) = \psi(1,2) \tag{12.4}$$

Combining Eqs 12.3 and 12.4, we see that

$$\lambda^2 = 1 \tag{12.5}$$

Introduction to Quantum Nanotechnology: A Problem Focused Approach. Duncan Steel, Oxford University Press (2021).
 DOI: 10.1093/oso/9780192895073.003.0012

or

$$\lambda = \pm 1 \tag{12.6}$$

and therefore

$$\hat{P}_{12}\psi(1,2) = \pm\psi(2,1) \tag{12.7}$$

Hence, since the Hamiltonian is unaffected by the interchange of the two particles, i.e., $\left[\hat{H}, \hat{P}_{12}\right] = 0$, the wave function for two identical particles must have either **even (+1) or odd (−1) exchange symmetry**. **Even and odd exchange symmetry** are also referred to as **symmetric** and **antisymmetric** functions or vectors, respectively.

The time independent Schrödinger equation for each particle is the same form since the particles are identical. However, the operators for the Hamiltonian for each particle are specific to that particle. We write the corresponding time independent Schrödinger equation for particle j then as

$$\hat{H}_j \psi_\alpha\left(r_j\right) = E_{\alpha_j}\psi_\alpha\left(r_j\right) \tag{12.8}$$

where E_{α_j} is an eigenenergy for particle j in state ψ_α. The Hamiltonian for the two particles is then

$$\hat{H} = \hat{H}_1 + \hat{H}_2 \tag{12.9}$$

and the eigenstates of $\hat{H}$ are given by:

$$\hat{H}\psi_{\alpha\beta}(r_1, r_2) = \left(\hat{H}_1 + \hat{H}_2\right)\psi_\alpha(r_1)\psi_\beta(r_2) = \left(E_\alpha + E_\beta\right)\psi_\alpha(r_1)\psi_\beta(r_2) = E_{\alpha\beta}\,\psi_{\alpha\beta}(r_1, r_2) \tag{12.10}$$

where $\psi_{\alpha\beta}(r_1, r_2)$ is a **product space** of $\psi_\alpha(r_1)$ and $\psi_\beta(r_2)$.

However, the requirement is that the overall wave function be symmetric or antisymmetric under exchange of the particles. Eq. 12.10 satisfies the time independent Schrödinger equation, but the system does not have the symmetry property required to be an eigenstate of the symmetry operator. To see this we write out the effect of the exchange operator on the wave function of two particles if

$$\psi_{\alpha\beta}(r_1, r_2) = \psi_\alpha(r_1)\psi_\beta(r_2) \tag{12.11}$$

then

$$\hat{P}_{12}\psi_{\alpha\beta}(r_1, r_2) = \psi_{\alpha\beta}(r_2, r_1) = \psi_\alpha(r_2)\psi_\beta(r_1) \neq \pm\psi_\alpha(r_1)\psi_\beta(r_2) \tag{12.12}$$

This is solved by writing $\psi_{\alpha\beta}(r_1, r_2)$ as a linear combination of the $\psi_\alpha(r_1)\psi_\beta(r_2)$ and $\psi_\alpha(r_2)\psi_\beta(r_1)$, both of which have the same energy eigenvalue $\left(E_\alpha + E_\beta\right)$:

$$\psi_{\pm_{\alpha\beta}}(r_1, r_2) = \left(\psi_\alpha(r_1)\psi_\beta(r_2) \pm \psi_\alpha(r_2)\psi_\beta(r_1)\right) \tag{12.13}$$

which now satisfies the requirement

$$\hat{P}_{12}\psi_{\pm\alpha\beta}(r_1, r_2) = (\psi_\alpha(r_2)\psi_\beta(r_1) \pm \psi_\alpha(r_1)\psi_\beta(r_2)) = \pm\psi_{\pm\alpha\beta}(r_1, r_2) \quad (12.14)$$

For normalization, using the shorthand notation

$$(f(r_1)g(r_2)|h(r_1)\ell(r_2)) \equiv \int_{-\infty}^{\infty}\int_{-\infty}^{\infty} dr_1 dr_2 \, (f^*(r_1)g^*(r_2)\ell(r_2)h(r_1)) \quad (12.15)$$

and that

$$(\psi_\alpha(r)|\psi_\beta(r)) = \delta_{\alpha\beta}$$

then

$$\begin{aligned}(\psi_\alpha(r_2)\psi_\beta(r_1) &\pm \psi_\alpha(r_1)\psi_\beta(r_2)|\psi_\alpha(r_2)\psi_\beta(r_1) \pm \psi_\alpha(r_1)\psi_\beta(r_2)) \\ &= (\psi_\alpha(r_2)\psi_\beta(r_1)|\psi_\alpha(r_2)\psi_\beta(r_1)) \pm (\psi_\alpha(r_2)\psi_\beta(r_1)|\psi_\alpha(r_1)\psi_\beta(r_2)) \\ &\pm (\psi_\alpha(r_1)\psi_\beta(r_2)|\psi_\alpha(r_2)\psi_\beta(r_1)) + (\psi_\alpha(r_1)\psi_\beta(r_2)|\psi_\alpha(r_1)\psi_\beta(r_2)) = 1 + 1 = 2\end{aligned} \quad (12.16)$$

Therefore, the final version of Eq. 12.12 is given by

$$\psi_{\pm\alpha\beta}(r_1, r_2) = \frac{1}{\sqrt{2}}(\psi_\alpha(r_1)\psi_\beta(r_2) \pm \psi_\alpha(r_2)\psi_\beta(r_1)) \quad (12.17)$$

Notice that if the state is antisymmetric (minus sign) and two particles are in the same state, then

$$\psi_\alpha(r_2)\psi_\alpha(r_1) - \psi_\alpha(r_1)\psi_\alpha(r_2) = 0$$

This is a mathematical statement of the **Pauli exclusion principle**, which was first found appropriate for atoms, as we discuss below, and was a major factor in developing an understanding of the periodic chart.

12.3 The Exchange Symmetry Postulate: Bosons and Fermions

Having explored the consequences of exchange, it was up to experiments to show that the exchange symmetry was relevant. The **exchange symmetrization postulate** says that, based on experiments, *if the intrinsic angular momentum (i.e., the spin) of a particle is half-integer, the wave function must be antisymmetric*. The particles, like the electron, are called **fermions**. *If the spin is integer, the wave function must be symmetric*, and the particles are called **bosons,** like the nucleon $He^{(4)}$. The **Pauli exclusion principle** then implies that no two electrons can be in the same state, the state being the product of the energy eigenstate and the spin state.

Extending the above approach for generating a symmetric or antisymmetric wave function for a more complex problem of two particles in two states is straightforward using the **Slater determinant** for fermions and the **Slater permanent** for bosons. A permanent is the determinant but with the minus signs converted to plus signs. For n fermions in n states, we form the matrix,

$$A = \begin{bmatrix} \psi_1(1) & \psi_2(1) & \cdots & \psi_\alpha(1) & \cdots & \psi_n(1) \\ \psi_1(2) & \psi_2(2) & \cdots & \psi_\alpha(2) & \cdots & \psi_n(2) \\ \vdots & \vdots & \vdots & \vdots & \vdots & \vdots \\ \psi_1(\alpha) & \psi_2(\alpha) & \vdots & \psi_\alpha(\alpha) & \cdots & \psi_n(\alpha) \\ \vdots & \vdots & \vdots & \vdots & \vdots & \vdots \\ \psi_1(n) & \psi_2(n) & \cdots & \psi_\alpha(n) & \cdots & \psi_n(n) \end{bmatrix} \tag{12.18}$$

and then evaluate the determinant. So for three identical fermions in three states:

$$A = \begin{bmatrix} \psi_1(1) & \psi_2(1) & \psi_3(1) \\ \psi_1(2) & \psi_2(2) & \psi_3(2) \\ \psi_1(3) & \psi_2(3) & \psi_3(3) \end{bmatrix} \tag{12.19}$$

For three identical fermions, we get

$$\begin{aligned} \psi(1,2,3) &= N\,det\,A \\ &= N\psi_1(1)det\begin{bmatrix} \psi_2(2) & \psi_3(2) \\ \psi_2(3) & \psi_3(3) \end{bmatrix} - N\psi_1(2)det\begin{bmatrix} \psi_2(1) & \psi_3(1) \\ \psi_2(3) & \psi_3(3) \end{bmatrix} + N\psi_1(3)det\begin{bmatrix} \psi_2(1) & \psi_3(1) \\ \psi_2(2) & \psi_3(2) \end{bmatrix} \\ &= N\left[\psi_1(1)\left(\psi_2(2)\psi_3(3) - \psi_3(2)\psi_2(3)\right) - \psi_1(2)\left(\psi_2(1)\psi_3(3)\right.\right. \\ &\quad \left.\left. -\psi_3(1)\psi_2(3)\right) + \psi_1(3)\left(\psi_2(1)\psi_3(2) - \psi_3(1)\psi_2(2)\right)\right] \end{aligned} \tag{12.20}$$

where is $N = \sqrt{\frac{1}{n!}}$ for n particles. For three identical bosons in three states, we find

$$\begin{aligned} \psi(1,2,3) &= N\,perm\,A \\ &= N\psi_1(1)perm\begin{bmatrix} \psi_2(2) & \psi_3(2) \\ \psi_2(3) & \psi_3(3) \end{bmatrix} + N\psi_1(2)perm\begin{bmatrix} \psi_2(1) & \psi_3(1) \\ \psi_2(3) & \psi_3(3) \end{bmatrix} + N\psi_1(3)perm\begin{bmatrix} \psi_2(1) & \psi_3(1) \\ \psi_2(2) & \psi_3(2) \end{bmatrix} \\ &= N\psi_1(1)\left(\psi_2(2)\psi_3(3) + \psi_3(2)\psi_2(3)\right) + N\psi_1(2)\left(\psi_2(1)\psi_3(3)\right. \\ &\quad \left. + \psi_3(1)\psi_2(3)\right) + N\psi_1(3)\left(\psi_2(1)\psi_3(2) + \psi_3(1)\psi_2(2)\right) \end{aligned} \tag{12.21}$$

Problem 12.1 *show in Eq. 12.20 that if two of the three particles are in the same state, then* $\psi(1,2,3) = 0$.

Problem 12.2 *For the symmetric and antisymmetric states in Eq. 12.17, evaluate* $\langle r_1 - r_2 \rangle$. *To do this, it is essential to recognize that*

$$\langle r_j \rangle = \left(\psi_\alpha\left(r_j\right)\right| r_j \left|\psi_\alpha\left(r_j\right)\right) = (\psi_\alpha(r)|\, r\, |\psi_\alpha(r)) = \langle r \rangle_\alpha \tag{12.22}$$

since the expectation is just a number, and not specific to the particle. The coordinate r_j becomes the integration variable once it is given by $\left(\psi_\alpha\left(r_j\right)\middle| r_j \middle| \psi_\alpha\left(r_j\right)\right)$.
Show that

$$\langle r_2 - r_1\rangle_B < \langle r_2 - r_1\rangle_D < \langle r_2 - r_1\rangle_F \tag{12.23}$$

for bosons, distinguishable (non-identical), and fermion, respectively.

The result of Problem 12.2 in Eq. 12.23 is an amazing result showing without any specifics on the Hamiltonian (e.g., for free particles), the average distance between them depends on whether or not they are distinguishable and, for identical particles, whether they are bosons or fermions. This **exchange interaction** becomes important when their wave functions spatially overlap.

Problem 12.3 *For two spin ½ electrons in an infinite square potential, use the results (Eqs 3.41–3.45) to write the eigenfunction of the two-electron system in the first four excited states of the two-electron system. An example for a state above the fourth state is shown in Fig 12.1. Hint: it is not necessary to use the Slater determinant for this.*

The fermionic nature of the electron is important in determining its performance in materials including conductors and semiconductors. Thinking of something as simple as an infinite well for a defect, it is clear that as more electrons are added to the system, the energy levels start to fill up, since only two electrons can be placed in each eigenstate, and that requires that their spins are in opposite directions. The problem even exists if the electrons are free particles, where the energy goes like $E = \frac{\hbar^2 k^2}{2m}$, meaning that it is a continuous variable, not discrete. The question that we ask is if we have N_F fermions, and they are in the lowest possible total energy state (at zero temperature), what is the energy of the highest state that is filled. Recall, from Chapter 11, that we introduced the differential density of states in Eq. 11.107:

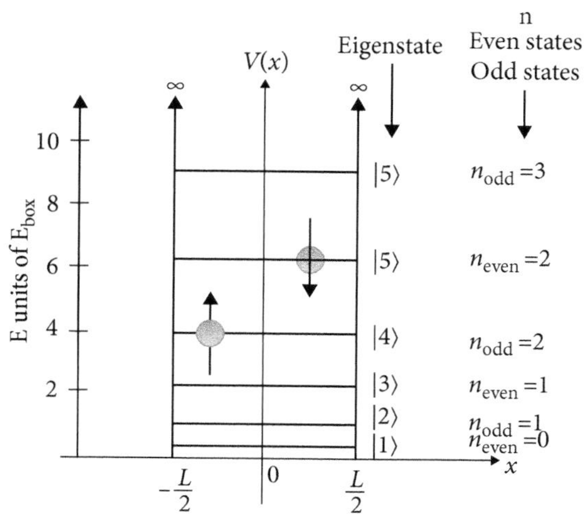

Fig. 12.1 Energy levels similar to that shown in Ch. 3. but with two noninteracting spin ½ electrons shown in specific states.

$$n_{\Omega}(E) = \frac{1}{8}\sqrt{2}\frac{L^3}{\pi^3\hbar^3}(m)^{\frac{3}{2}}\sqrt{E} \tag{12.24}$$

In this form, this is the number of states between E and $E + dE$ in solid angle $d\Omega$. In the problem, there is no angle dependence, so we integrate for over 4π. We must also multiply by two because each energy state can hold two oppositely aligned spins. This gives, for the total density of states in three dimensions,

$$n(E) = \sqrt{2}\frac{L^3}{\pi^2\hbar^3}(m)^{\frac{3}{2}}\sqrt{E} \tag{12.25}$$

The total number of states is then given by

$$\int_0^{E_F} dE\, n(E) = N \tag{12.26}$$

We set $N = N_F$ for the number for fermions and label the highest energy state as E_F, corresponding to what is called the **Fermi energy**. Then

$$N_F = \sqrt{2}\frac{L^3}{\pi^2\hbar^3}(m)^{\frac{3}{2}}\int_0^{E_F} dE\sqrt{E} = 2^{\frac{3}{2}}\frac{L^3}{\pi^2\hbar^3}(m)^{\frac{3}{2}}\frac{1}{3}E_F^{\frac{3}{2}} \tag{12.27}$$

Noting that $n_F = \frac{N_F}{L^3}$ is the density of fermions, we solve for the

$$E_F = \frac{\hbar^2}{2m}(3\pi^2 n_F)^{\frac{2}{3}} \tag{12.28}$$

12.4 The Heitler-London Model and the Heisenberg Exchange Hamiltonian

Early on in the development of quantum theory, the **Heitler-London model** emerged to explain the ground state structure of **molecular hydrogen**, H_2. This model has become important in the understanding of many crystalline materials, but it is also important in the area of nano-physics because various studies now involve interacting quantum systems that are so close to each other that the fermionic nature of the electrons or holes needs to be included. The model deals with two pairs of charged spin ½ particles, and each pair has a positive and negative particle, like H_2 with two electrons and two nuclei. Here, we are going to use a much simpler system to focus on the basic physics and the importance of the exchange interaction in determining electronic properties.

Like the original model, we start with two separate quantum systems, two one-dimensional quantum wells, each containing a single electron with spin ½. The system is shown in

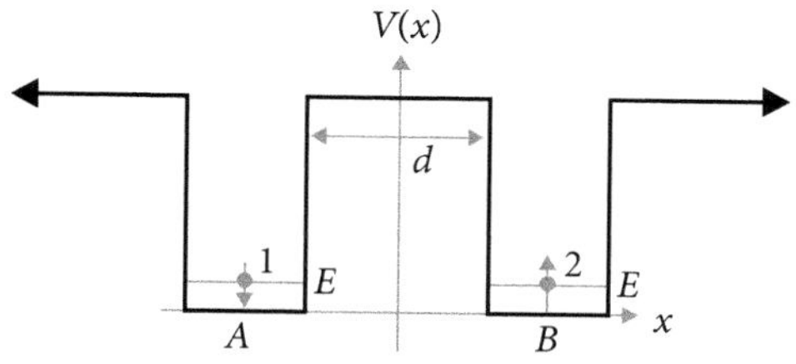

Fig. 12.2 Two identical finite quantum wells, each containing a single electron and separated by a variable distance, d. The particles are numbered 1 and 2 and the corresponding coordinate is labeled, x_1 and x_2. The spin orientation shown is just one of four possibilities. All four must be included in this discussion.

Fig. 12.2. Rather than having two positively charged nuclei, confinement of the electron is by the quantum wells.

When $d \to \infty$, the eigenfunctions of the two particles are given by $\phi_A(x_1)$ and $\phi_B(x_2)$, which we assume to real functions for convenience. The Hamiltonian for each is

$$\hat{H}_A = \frac{p_1^2}{2m} + V_A(x_1) \tag{12.29}$$

$$\hat{H}_B = \frac{p_2^2}{2m} + V_B(x_2) \tag{12.30}$$

and

$$\hat{H}_A\phi_A(x_1) = E\phi_A(x_1) \tag{12.31}$$

$$\hat{H}_B\phi_B(x_2) = E\phi_B(x_2) \tag{12.32}$$

Since there is no spatial overlap of the wave functions (as $d \to \infty$, the eigenfunction for each well penetrates into the barrier and goes asymptotically to 0), there is no need to consider their fermionic nature and no need to antisymmetrize the functions.

However, as d is decreased, we assume that the eigenfunctions are not significantly impacted. Now, a part of the function that extends into the barrier extends into the adjacent potential well. The eigenfunctions for the two particles spatially overlap in space where they can now have a significant value (i.e., not 0). In this case, the functions must be antisymmetrized. To analyze this, we assume that the Hamiltonian, for sufficiently small d, is given by

$$\hat{H} = \hat{H}_A + \hat{H}_B + V(x_1, x_2) \tag{12.33}$$

where $V(x_1, x_2)$ now accounts for the second well for each particle, as well as the repulsive Coulomb potential between the particles if they are charged.

Even though there is no spin dependence in the Hamiltonian, we must include the spinors with the notation s_1 and s_2. The total wave function is then given by $\psi(x_1s_1; x_2s_2) = \phi(x_1, x_2)$ $\chi(s_1, s_2)$. Because these are spin ½ particles, we require $\psi(x_1s_1; x_2s_2) = -\psi(x_2s_2; x_1s_1)$. Then

$\phi(x_1, x_2)$ can be symmetric or antisymmetric, as long as the corresponding $\chi(s_1, s_2)$ is antisymmetric or symmetric. The normalized $\phi(x_1, x_2)$ is then written as

$$\phi_\pm(x_1, x_2) = \frac{1}{\sqrt{2\left(1 \pm |\alpha|^2\right)}} \left(\phi_A(x_1)\,\phi_B(x_2) \pm \phi_A(x_2)\,\phi_B(x_1)\right) \tag{12.34}$$

where

$$\alpha = \int\!\!\int dx\,\phi_A(x)\phi_B(x) \tag{12.35}$$

ϕ_+ is symmetric and ϕ_- is antisymmetric.

Problem 12.4 *Derive the normalization constant for symmetric and antisymmetric states in 12.34 and show that* α is given by *Eq. 12.35.*

Problem 12.5 *Show for* $\phi_\pm(x_1, x_2)$ *given in Eq. 12.34 that,* in the absence of interaction,

$$\left\langle \hat{H}_A + \hat{H}_B \right\rangle = 2E \tag{12.36}$$

The result in Eq. 12.36 is that exchange effects do not impact the system when the particles do not interact.

To understand the impact of exchange symmetry, we calculate $\langle \hat{H} \rangle$ where $\hat{H}$ is given in Eq. 12.33. Having establish in Eq. 12.36 that $\left\langle \hat{H}_A + \hat{H}_B \right\rangle = 2E$, we ignore this term since it is the same as when the distance separating the two particles goes to infinity and exchange effects are unimportant. Hence, we need to evaluate $\langle V(x_1, x_2) \rangle$ in the presence of exchange effects. Using Eq. 12.34, we get

$$\langle V(x_1, x_2) \rangle = \int dx_2 \int dx_1 \phi_\pm(x_1, x_2)\, V(x_1, x_2)\, \phi_\pm(x_1, x_2) = E_0 \pm E_X \tag{12.37}$$

where

$$E_0 = \frac{1}{(1+\alpha^2)} \int dx_2 \int dx_1 V(x_1, x_2)\, \phi_A(x_1)^2 \phi_B(x_2)^2 \tag{12.38}$$

$$E_X = \frac{1}{(1+\alpha^2)} \int dx_2 \int dx_1 V(x_1, x_2)\, \phi_A(x_1)\, \phi_B(x_1)\, \phi_B(x_2)\, \phi_A(x_2) \tag{12.39}$$

The first term, E_0, corresponds to the usual first order correction for the effect of the well A (B) on well B (A) including Coulomb coupling. This is the same term that would result if the particles had no spin and without satisfying the requirements of exchange symmetry. The integral giving rise to E_X contributes because the eigenfunctions satisfy the requirements of exchange symmetry. It is only important when the spatial extent of the eigenfunction localized at say site A overlaps the function associated with site B, namely the product of $\phi_A(x_1)\,\phi_B(x_1)$ and $\phi_B(x_2)\,\phi_A(x_2)$ must each be non-zero over some range. We see then

$$E_{\uparrow\downarrow} = E_0 + E_X \tag{12.40}$$

where the spins are in the antisymmetric singlet (meaning the total spin $S = 0$) state because $\phi(x_1, x_2)$ is in a symmetric state, and

$$E_{\uparrow\uparrow} = E_0 - E_X \tag{12.41}$$

where the spins are in the symmetric triplet state (meaning the total spin $S = 1$) because $\phi(x_1, x_2)$ is in an antisymmetric state. Note the subscript notation $\uparrow\downarrow$ means the spins are in the antisymmetric spin state while the notation $\uparrow\uparrow$ means the spins are in a symmetric state.[1]

This four-state manifold (one singlet state and three triplet states) forms an important physical system that is involved in many quantum features involving two electron-based quantum systems such as two-quantum dots, two-atoms, molecules and solids. Though there are no interactions involving the spins of the system, the presence of spin has a profound effect on the energy level structure. For designing and understanding new systems that have this fundamental feature, the Hamiltonian in Eq. 12.37 is rewritten to reflect the impact of the exchange symmetry.

To develop the Hamiltonian, we start by considering the total spin of the system using the results in Section 10.5. $\hat{S}^2 = \left(\hat{S}_1 + \hat{S}_2\right)^2 = \hat{S}_1^2 + \hat{S}_2^2 + 2\hat{S}_1 \cdot \hat{S}_2 = \frac{3}{2} + 2\hat{S}_1 \cdot \hat{S}_2$ or

$$\hat{S}_1 \cdot \hat{S}_2 = \frac{1}{2}\left(\hat{S}^2 - \frac{3}{2}\right) \tag{12.42}$$

where $\hat{S}_1^2$ and $\hat{S}_2^2$ are replaced with their eigenvalues, namely the eigenvalue of $\hat{S}_i^2$ is $S_i(S_i + 1) = \frac{3}{4}$ since $S_i = \frac{1}{2}$. For the singlet $S = 0$ state, $\hat{S}_1 \cdot \hat{S}_2 = \frac{1}{2}\left(\hat{S}^2 - \hat{S}_1^2 - \hat{S}_2^2\right)$ takes on the value $-\frac{3}{4}$. For the triplet $S = 1$ state, $\hat{S}_1 \cdot \hat{S}_2$ takes on the value of $\frac{1}{4}$. We then see that Eqs 12.40 and 12.41 can be written as a single equation for the Hamiltonian as

$$\hat{H} = \frac{1}{4}\left(E_{\uparrow\downarrow} + 3E_{\uparrow\uparrow}\right)\hat{I} - \left(E_{\uparrow\downarrow} - E_{\uparrow\uparrow}\right)\hat{S}_1 \cdot \hat{S}_2 \tag{12.42}$$

The first term is common to both states and can be suppressed and included later. In general, the value of $\left(E_{\uparrow\downarrow} - E_{\uparrow\uparrow}\right)$ depends on the details of the quantum system. In the system above, it depends on the separation, d. In a more general system, like the hydrogen molecule, it depends on the distance (in three dimensions) between the two nuclei (corresponding to quantum well A and B), the scalar r. Hence, we set $\left(E_{\uparrow\downarrow} - E_{\uparrow\uparrow}\right) \equiv J(r)$, allowing us to write the **Heisenberg exchange Hamiltonian** as

$$\hat{H}_{ex} = -J(r)\hat{S}_1 \cdot \hat{S}_2 \tag{12.43}$$

Problem 12.6 *Write out the $\chi(s_1, s_2) = |s_1 s_2 s m_s\rangle$ in terms of $|s_1 m_{s_1}\rangle |s_2 m_{s_2}\rangle$ for $s_1 = s_2 = \frac{1}{2}$ for all four states. For this problem, you may need to review details for the addition of two angular momenta.*

[1] Recall from the discussion in Chapter 10 about the addition of two ½ spin system, that there is one $|sm_s\rangle$ state for $S = 0$ given by $|00\rangle = \frac{1}{\sqrt{2}}\left(|\uparrow\rangle_1 |\downarrow\rangle_2 - |\downarrow\rangle_1 |\uparrow\rangle_2\right)$. For the $S = 1$, there are three different states corresponding to $m_s = 0, \pm 1$ given by $|11\rangle = |\uparrow\rangle_1 |\uparrow\rangle_2, |10\rangle = \frac{1}{\sqrt{2}}\left(|\uparrow\rangle_1 |\downarrow\rangle_2 + |\downarrow\rangle_1 |\uparrow\rangle_2\right)$, and $|1-1\rangle = \frac{1}{\sqrt{2}} |\downarrow\rangle_1 |\downarrow\rangle_2$.

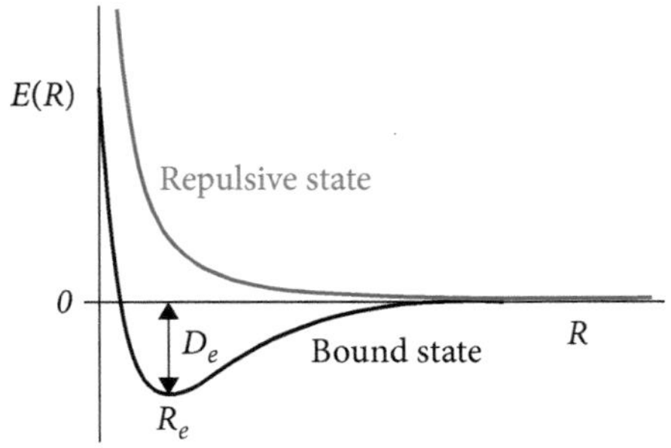

Fig. 12.3 The potential energy of two hydrogen atoms as a function of the distance between the nuclei. The singlet state is shown in the black curve while the behavior of the triplet state is seen in the red curve. Potential diverges as the distance between the two positively charge nuclei goes to 0. The potential minimum in the black curve is source of binding for the molecule. D_e is called the dissociation energy, the energy needed to separate the two atoms and is 4.746 for H_2. R_e is the location of the minimum energy and is called the equilibrium internuclear distance, 0.74 nm for H_2. Used with permission from Prof. S.R. Blinder http://www.umich.edu/~chem461/QMChap10.pdf

The original success of this theory was in explaining the existence of the hydrogen molecule. Figure 12.3 shows the potential energy of the system as a function of the distance (called the reaction coordinate in physical chemistry) between the nuclei. The potential diverges as the distance between the two positively charge nuclei goes to 0 due to repulsion. The potential minimum in the black curve is the source of binding for the molecule and occurs for the singlet state.

12.5 **Summary**

Exchange symmetry plays a central role in understanding the properties of semiconductors, metals, and the periodic chart. The Pauli exclusion principle is the origin of the early view for the periodic chart, that only two electrons could be in any orbital, corresponding to an antisymmetric spin state (a singlet). In quantum information, when the spin of a carrier (electron or hole) serves as a qubit, control gates often require the interaction between these qubits and that interaction must observe the appropriate symmetry. To be clear, the rule of symmetry only becomes important when the spatial extent of the wave function of each particle overlaps. If there is no overlap, the symmetry has little impact.

Vocabulary (page) and Important Concepts

- exchange operator 209
- even and odd exchange symmetry 210
- symmetric and antisymmetric states 210
- product space 210
- Pauli exclusion principle 211

- exchange symmetrization postulate 211
- fermions 211
- bosons 211
- Slater determinant 212
- Slater permanent 212
- exchange interaction 213
- Fermi energy 214
- Heitler–London model 214
- Heisenberg exchange Hamiltonian 217
- molecular hydrogen 214

13 Quantum Measurement and Entanglement: Wave Function Collapse

13.1 Introduction

The rules for quantum mechanics as given in Chapter 7 and supplemented in Ch. 10 and 12 have been very successful in describing many physical phenomena. However, there is no discussion in the rules of what happens in a **measurement**. Specifically, what happens to the state vector or wave function when we make a measurement of an observable. Classically, if a measurement is made of the momentum, then position, and then the momentum, the same value of the momentum is found in both measurements. The same with, say, the charge and the current in an LC circuit. In quantum, the only thing the rules say about this is that in the course of a measurement of an observable, the only values that you can observe are eigenvalues of the corresponding Hermitian operator. There is no mention of what happens to the system. In fact, as we see in the discussion of how the principle of superposition leads to quantum entanglement, the implications are seemingly even more mysterious from the standpoint of physical understanding and it remains a point of continuing discussion in research. In this chapter, we focus on what has so far been a successful extension of the rules of quantum mechanics.

13.2 Quantum Measurement

What does it mean to make a measurement on a quantum system? Referring to the postulates, we will only be able to measure observables and probabilities of being in a given eigenstate. Because of completeness, an arbitrary state $|\psi\rangle$ in some Hilbert space spanned by states, $|a_i\rangle$ can be described by a linear combination of basis states corresponding to eigenstates of an operator associated with some observable, namely,

$$|\psi\rangle = \sum_i c_i \, |a_i\rangle \tag{13.1}$$

This is the mathematical statement of the **principle of superposition,** which says that the most general state is a superposition of the basis states. This is just not possible in a classical system involving massive particles. However, as we have discussed throughout the text, it is similar in systems described by waves such as electromagnetic fields.

Let us consider the simplest superposition state which involves just two levels. We then write the state vector at $t = 0$ as:

$$|\psi\rangle = cos\ \theta \, |a_1\rangle + sin\ \theta \, |a_2\rangle \tag{13.2}$$

Introduction to Quantum Nanotechnology: A Problem Focused Approach. Duncan Steel, Oxford University Press (2021).
 DOI: 10.1093/oso/9780192895073.003.0013

We make measurements in the $|a_i\rangle$ basis and find eigenvalue of the corresponding operator, $\hat{A}$, where

$$\hat{A}\,|\,a_i\,\rangle = a_i\,|\,a_i\rangle \tag{13.3}$$

In this system, if we find the value a_2 for our measurement of the observable, it is then postulated and verified by experiment that the wave function in 13.2 then "collapses" into the corresponding eigenstate associated with the measured value a_2, namely it is now in the state $|\psi\rangle = |a_2\rangle$. If another measurement is immediately made, the same value will be found. The "measurement problem" is that based on the postulates of Chapter 7, *there is no prescription for how this happens.* Von Neumann, Wigner, and earlier Heisenberg developed a concept that led to the idea of **state vector collapse** or **wave function collapse**. More modern discussions may use the language of reinitialized rather than collapse. Regardless of the language, the math describes the idea that the wave function or state vector for the system becomes the eigenstate associated with the eigenvalue found in the measurement, and there is no discussion in this context of how this happens. This is sometimes called the **Von Neumann projection postulate.** If a subsequent measurement is made on the system, instantaneously, the same eigenvalue will be observed. On the other hand, if the system is reinitialized and prepared identically in state given by Eq. 13.2 and the measurement repeated, the measurement may now be either a_1 or a_2 according to Eq. 13.2 which is the result of reinitializing. The idea of wave function collapse is considered a hypothesis or a postulate.

This behavior is mathematically described with a **projection operator**, $|a_i\rangle\,\langle a_i|$, where in the above discussion $|a_i\rangle \to |a_2\rangle$:

$$|\,\psi\rangle_{before} = \cos\theta\;|\,a_1\rangle + \sin\theta\;|\,a_2\rangle \xrightarrow[\text{gives } a_2]{\text{measurement}} |\,\psi\rangle_{after} = \frac{|\,a_2\rangle\langle a_2\,|}{\sqrt{\langle\psi'|\psi'\rangle_{after}}}\;|\,\psi\rangle_{before} = \;|\,a_2\rangle \tag{13.4}$$

$|\,\psi'\rangle_{after}$ is $|\,\psi\rangle_{after}$ before re-normalizing with $\sqrt{\langle\psi'|\psi'\rangle_{after}}$. This is because $|\,a_2\rangle\langle a_2|\psi\rangle_{before} = \sin\theta\;\;|\;a_2\rangle$ and therefore the state $|\;a_2\rangle\langle a_2|\psi\rangle_{before}$ is not normalized. The probability of making this measurement $\sin^2\theta$, which we know from our understanding of the meaning of $|\psi\rangle_{\,before} = \cos\theta\,|a_1\rangle + \sin\theta\,|a_2\rangle$. However, mathematically, it comes from a **projection** (Born's rule in Ch. 7) onto the state $|\,a_2\rangle$, namely,

$$\left|\langle a_2|\psi\rangle_{before}\right|^2 = |\langle a_2|\,(\cos\theta\;|a_1\rangle + \sin\theta|a_2\rangle)|^2 = \sin^2\theta \tag{13.5}$$

Suppose now we make another measurement of the system after the measurement giving $|\,a_2\rangle$ but of an observable with operator $\hat{B}$, where $\hat{B}$ also defines the same Hilbert space as $\hat{A}$ but with eigenvectors $|\,b_i\rangle$ where

$$\hat{B}\,|\,b_i\,\rangle = b_i\,|\,b_i\rangle \tag{13.6}$$

The basis set is complete, and so

$$|a_i\rangle = \sum_j \langle b_j|a_i\rangle \, |b_j\rangle \tag{13.7}$$

After making a measurement in the $|a_i\rangle$ basis, which gives us a_2, we now decide to make a measurement in the $|b_j\rangle$ basis. In the new basis we find

$$|a_2\rangle = \cos\theta' \, |b_1\rangle + \sin\theta' \, |b_2\rangle \tag{13.8}$$

where $\cos\theta' = \langle b_1|a_2\rangle$ and $\sin\theta' = \langle b_2|a_2\rangle$. Assume this measurement gives b_2. Using 13.4 we get

$$|a_2\rangle = \cos\theta' \, |b_1\rangle + \sin\theta' \, |b_2\rangle \xrightarrow[\text{gives } b_2]{\text{measurement}} |\psi\rangle_{after} = \frac{|b_2\rangle\langle b_2|}{\sqrt{\langle\psi'|\psi'\rangle_{after}}} |\psi\rangle_{before} = |b_2\rangle \tag{13.9}$$

But now something non-classical happens. We now go back to the $|a_i\rangle$ basis. We expand $|b_2\rangle$ and get

$$|b_2\rangle = \cos\theta'' \, |a_1\rangle + \sin\theta'' \, |a_2\rangle \tag{13.10}$$

where now $\cos\theta'' = \langle a_1|b_2\rangle$ and $\sin\theta'' = \langle a_2|b_2\rangle$. So we found the system in $|a_2\rangle$ and then in $|b_2\rangle$ but when we go back to check the measurement of $\hat{A}$, we find that it is no longer guaranteed to be in $|a_2\rangle$. The probabilities are not even as given in Eq. 13.2. *The information contained in the first measurement is erased after we make the second measurement.* We also note that in the $\hat{B}$ basis, a measurement of $\hat{A}$ leads to a quantum coherence (superposition) in the $\hat{B}$ basis and a measurement of $\hat{B}$ leads to a coherence in the $\hat{A}$ basis. The loss or erasing of information in one basis following a measurement in another basis is certainly non-classical. *From the standpoint of information, if information is encoded in Eq. 13.2, it is lost in any subsequent measurement.* The implication is clear: to preserve information, measurements must be done at the end of the process, not in the middle.

13.3 Quantum Entanglement and the Impact of Measurement

We now consider something more complicated, based on the principle of superposition. To illustrate the physics, we will make the observables something we understand: linear momentum and spin. Consider a simple experiment. At some point in time, a system with zero angular momentum decays into two spin ½ particles moving in opposite directions.

Since the total angular momentum must remain zero, the only option for the spin part of the state vector is that it correspond to the singlet with $s = m_s = 0$ and $|sm_s\rangle = |00\rangle = \frac{1}{\sqrt{2}}\left(|\uparrow_z\rangle_1|\downarrow_z\rangle_2 - |\downarrow_z\rangle_1|\uparrow_z\rangle_2\right)$ for the two electrons where the arrows represent the projection of

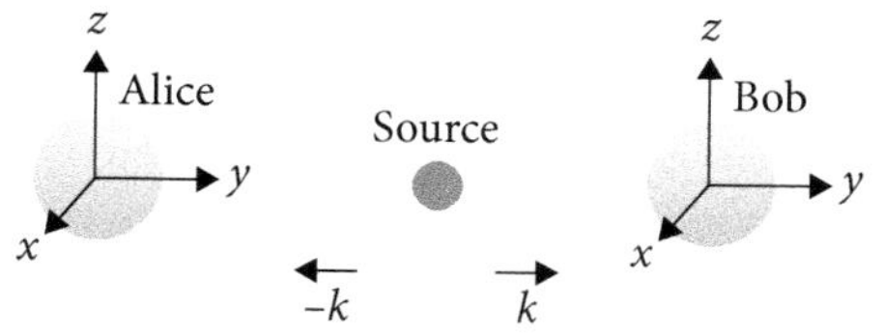

Fig. 13.1 A source emits two spin ½ particles with total angular momentum of zero, traveling in opposite direction to two different detectors operated by Alice and Bob.

the spin-½ on the z-axis. However, the state is also described by the linear momentum for each electron. Again, because the initial linear momentum is zero, the two electrons must move in opposite directions with the value of the momentum. We assume everything is in one dimension for simplicity, and the momentum state for each electron is $|-k\rangle_1$ and $|+k\rangle_2$, respectively. We change the language now, for reasons that will become obvious below, such that $|-k,\rangle_1 \to |-k,\rangle_A$ and $|k,\rangle_2 \to |k,\rangle_B$ where A and B stand for receivers Alice and Bob. The state vector for this system is then given by

$$|\psi\rangle = \frac{1}{\sqrt{2}}\left(|-k,\uparrow_z\rangle_A|k,\downarrow_z\rangle_B - |-k,\downarrow_z\rangle_A|k,\uparrow_z\rangle_B\right) \tag{13.11}$$

The subscripts are redundant with the sign of k, but remind us that that particle is heading to Alice or Bob since it contains information. Think of the z-projection of the spin corresponding to a 0 or 1, for example, corresponding to a **quantum bit (qubit)** of information.

Alice and Bob each have a measurement apparatus that determines if the particle they detect is spin up or spin down. If Alice measures $\uparrow_z$ then using the above measurement prescription, the state vector becomes

$$\begin{aligned}|-k,\uparrow_z\rangle_A\langle -k,\uparrow_{z_A}|\psi\rangle &= \frac{1}{\sqrt{2}\sqrt{\langle\psi'|\psi'\rangle_{after}}}\Big(|-k,\uparrow_z\rangle_A\langle -k,\uparrow_{z_A}|-k,\uparrow_z\rangle_A|k,\downarrow_z\rangle_B \\ &\quad -|-k,\uparrow_z\rangle_A\langle -k,\uparrow_z|-k,\downarrow_z\rangle_A|k,\uparrow\rangle_B\Big) \\ &= |-k,\uparrow_z\rangle_A\,|k,\downarrow_z\rangle_B\end{aligned} \tag{13.12}$$

Since the initial state vector is known, we see now that when Alice makes a measurement on the spin in her apparatus and finds the system in $|-k,\uparrow_z\rangle_A$ then she knows with certainty that Bob's measurement will find the particle he detects to be in state $|k,\downarrow_z\rangle_B$. The reverse is also true, if Alice finds the system in the spin $\downarrow_z$ state then she knows that Bob will find the spin in the $\uparrow_z$ state.

Even at this level it is interesting that the particle going toward Bob, which is in a linear combination of both spins (or a superposition of 0 and 1), loses information if Alice makes a measurement. It seems to lose this information even though the particle was on its way with its information when Alice decided to make a measurement. Furthermore, if Alice had not made a measurement, Bob had a 50% chance of finding the particle in either state. This kind of measurement is called a **quantum non-demolition measurement**: by making a measurement on her particle she knows with certainty the spin state of the particle with momentum k.

This kind of superposition state is an example of a **quantum entangled state**. Part of the definition of entanglement is that it cannot be written as the product of two Hilbert sub-spaces. In this case, it cannot be written as a product of space A and space B like:

$$\left(|-k,\uparrow_z\rangle_A + |-k,\downarrow_z\rangle_A\right)\left(|k,\downarrow\rangle_B - |k,\uparrow_z\rangle_B\right) \tag{13.13}$$

This state is not entangled: if Alice finds say spin up, Bob can still find his spin up OR spin down. So, we see that the system in 13.10 is highly correlated, and in this specific example, the effect is **non-local** because the spin information for Alice is identified with a negative momentum $(-\hbar k)$ and the information going to Bob is identified with positive momentum $(+\hbar k)$.

We note that, for the example above, there are a total of four possible quantum entangled states, called **Bell states**. These are summarized in Table 13.1.

However, the problem here is even more complex, but is analyzed using the procedure set up at the beginning of this chapter. This kind of manipulation is not "abstract" in that it is a part of the laboratory protocol in studies and applications of entanglement and is used in various encryption schemes for secure information transmission. The system is prepared as described in Eq. 13.11. (One approach to creating such a state is given in Chapter 17.) After the particles with spin are on their way, Alice and Bob decide to change their measurement base from measuring the spin projection along the z-axis to, say, the x-axis, where both x and z are orthogonal to the direction of propagation. To emphasize the paradox, the particles are prepared in the state given by Eq. 13.11. The question is what will Alice and Bob find? To predict this, we do exactly as was done in Problem 10.7. Since Alice and Bob will be measuring along the x-axis, they will be measuring eigenvalues of the $\hat{S}_x$ operator. Specifically, they will measure

$$\hat{S}_x\,|m_{s_x}\rangle = m_{s_x}\,|m_{s_x}\rangle \tag{13.14}$$

Prior to problem 10.7, we did not know, from a formal viewpoint, what values m_{s_x} could take. So we had to expand $|m_{s_x}\rangle$ in the z-basis and find the eigenvalues and eigenvectors, giving (Eqs 10.47 and 10.48)

Table 13.1 The four different quantum entangled Bell states.

$\lvert\psi^-\rangle = \frac{1}{\sqrt{2}}\left(\lvert -k,\uparrow_z\rangle_A\lvert k,\downarrow_z\rangle_B - \lvert -k,\downarrow_z\rangle_A\lvert k,\uparrow_z\rangle_B\right)$
$\lvert\psi^+\rangle = \frac{1}{\sqrt{2}}\left(\lvert -k,\uparrow_z\rangle_A\lvert k,\downarrow_z\rangle_B + \lvert -k,\downarrow_z\rangle_A\lvert k,\uparrow_z\rangle_B\right)$
$\lvert\Phi^-\rangle = \frac{1}{\sqrt{2}}\left(\lvert -k,\uparrow_z\rangle_A\lvert k,\uparrow_z\rangle_B - \lvert -k,\downarrow_z\rangle_A\lvert k,\downarrow_z\rangle_B\right)$
$\lvert\Phi^+\rangle = \frac{1}{\sqrt{2}}\left(\lvert -k,\uparrow_z\rangle_A\lvert k,\uparrow_z\rangle_B + \lvert -k,\downarrow_z\rangle_A\lvert k,\downarrow_z\rangle_B\right)$

$$|\uparrow_x\rangle = \frac{1}{\sqrt{2}}\left(|\downarrow_z\rangle + |\uparrow_z\rangle\right) \tag{13.15}$$

$$|\downarrow_x\rangle = \frac{1}{\sqrt{2}}\left(|\downarrow_z\rangle - |\uparrow_z\rangle\right) \tag{13.16}$$

with corresponding eigenvalues $\pm\frac{1}{2}\hbar$.

The question now is what does the state vector in Eq. 13.11 look like written in the x-basis? To see that, we have to now expand the z-basis states in the x-basis states, using Eqs 13.15 and 13.16 and find

$$|\downarrow_z\rangle = \frac{1}{\sqrt{2}}\left(|\uparrow_x\rangle + |\downarrow_x\rangle\right) \tag{13.17}$$

$$|\uparrow_z\rangle = \frac{1}{\sqrt{2}}\left(|\uparrow_x\rangle - |\downarrow_x\rangle\right) \tag{13.18}$$

Problem 13.1 *Find the projection of* $|\uparrow_z\rangle$ *on* $|\downarrow_x\rangle$. *What is the probability of finding this in the lab.*

So, we insert this into Eq. 13.11:

$$
\begin{aligned}
|\psi\rangle &= \frac{1}{\sqrt{2}}\left(|-k,\uparrow_z\rangle_A|k,\downarrow_z\rangle_B - |-k,\downarrow_z\rangle_A|k,\uparrow_z\rangle_B\right) = \\
&= \frac{1}{2\sqrt{2}}\Big(\left(|-k,\uparrow_x\rangle_A - |-k,\downarrow_x\rangle_A\right)\left(|k,\uparrow_x\rangle_B + |k,\downarrow_x\rangle_B\right) \\
&\quad - \left(|-k,\uparrow_x\rangle_A + |-k,\downarrow_x\rangle_A\right)\left(|k,\uparrow_x\rangle_B - |k,\downarrow_x\rangle_B\right)\Big) \\
&= \frac{1}{2\sqrt{2}}\left(|-k,\uparrow_x\rangle_A|k,\uparrow_x\rangle_B + |-k,\uparrow_x\rangle_A|k,\downarrow_x\rangle_B - |-k,\downarrow_x\rangle_A|k,\uparrow_x\rangle_B - |-k,\downarrow_x\rangle_A|k,\downarrow_x\rangle_B\right) \\
&\quad - \frac{1}{2\sqrt{2}}\left(|-k,\uparrow_x\rangle_A|k,\uparrow_x\rangle_B - |-k,\uparrow_x\rangle_A|k,\downarrow_x\rangle_B + |-k,\downarrow_x\rangle_A|k,\uparrow_x\rangle_B - |-k,\downarrow_x\rangle_A\,|k,\downarrow_x\rangle_B\right) \\
&= \frac{1}{\sqrt{2}}\left(|-k,\uparrow_x\rangle_A|k,\downarrow_x\rangle_B - |-k,\downarrow_x\rangle_A|k,\uparrow_x\rangle_B\right)
\end{aligned} \tag{13.19}
$$

This is a remarkable/incredible/astounding result. The correlation that was built into the z-basis remains when the measurement is flipped to the x-axis! This is what Einstein and many others thought had to be wrong. It is remarkable that particle B follows the correlation set by the measurement of A, even though the decision to make the measurement in the x-bases was done *after* the particles were correlated in the z-basis and on their way. This is called the **Einstein, Podolsky, and Rosen (EPR) paradox**. This quantum property is incorporated in one or more quantum encryption protocols for secure transfer of information. A key step is demonstrating quantum entangled states, a critical feature for quantum computing. Quantum entangled states enable a quantum computer to outperform a classical computer for Shore's factoring algorithm, because the size of the quantum computer scales more slowly with the size of the number to be factored than the size of a classical computer. The size of a classical computer scales exponentially with the size of the number. Proof that states are entangled requires us to demonstrate that the correlations are preserved in two separate bases and satisfy **Bell's inequality**, which is discussed elsewhere.[1]

Problem 13.2 *Show that if Alice makes a measurement in the x-basis that the information originally then sent to Bob is lost in the z-basis. In other words, show that while Alice would have known with certainty the answer Bob would get, if she measures in the x-basis and forgets to tell Bob to rotate, she no longer knows what answer Bob will get and that Bob can measure either spin up or spin down.*

Problem 13.3 *The fundamental fluctuations in their measurement is set by the Heisenberg uncertainty relation in Chapter 7. Namely, given two observables represented by operators $\hat{A}$ and $\hat{B}$ and*

$$[\hat{A}, \hat{B}] = i\hat{C}$$

then given the definition of the variances of an operator, e.g., for $\hat{A}$:

$$(\Delta\,\hat{A})^2 \equiv \langle\psi|\hat{A}^2 - \langle\hat{A}\rangle^2|\psi\rangle = \langle\hat{A}^2\rangle - \langle\hat{A}\rangle^2$$

we find

$$(\Delta\hat{A})^2(\Delta\hat{B})^2 \geq \frac{\langle\hat{C}\rangle^2}{4}$$

Alice and Bob are working in the x- and z-basis. The corresponding uncertainty relation for them following from the commutator $\left[\hat{S}_x, \hat{S}_y\right] = i\hbar\hat{S}_z$ under the usual cyclic permutation, hence for them

$$[\hat{S}_z, \hat{S}_x] = i\hbar\hat{S}_y$$

and the corresponding uncertainty relation is

$$(\Delta\hat{S}_z)^2(\Delta\hat{S}_x)^2 \geq \frac{\langle\hat{S}_y\rangle^2}{4}$$

Evaluate both sides of the uncertainty relation for the state in Eq. 13.11. Stay in the z-basis and express $\hat{S}_x$ and $\hat{S}_y$ in terms of $\hat{S}_+$ and $\hat{S}_-$. What does this mean for their intrinsic noise level.

13.4 **Quantum Teleportation**

One person, Alice (A), wants to send a piece of information (contained in a qubit, called C) securely to another person, Bob (B). We assume Alice and Bob share a maximally entangled Bell state. The situation is illustrated in Fig. 13.2.

[1] See discussion in J.J. Sakurai and Jim Napolitano, *Modern Quantum Mechanics 2nd edition*, Boston, Addison Wesley (1994) or Paul R. Berman, *Introductory Quantum Mechanics*, Springer (2018).

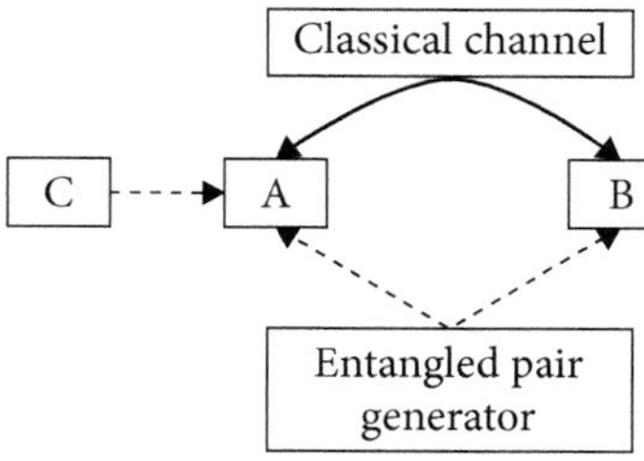

Fig. 13.2 A schematic of teleportation of information owned by Alice to information now

We take the information from C to be

$$|\psi\rangle_C = \alpha\,|0\rangle_C + \beta\,|1\rangle_C \tag{13.20}$$

Following Table 15.1, the Bell state for Alice and Bob can be any one of the four bipartite quantum entangled states:

$$|\Phi^+\rangle_{AB} = \frac{1}{\sqrt{2}}(|0\rangle_A|0\rangle_B + |1\rangle_A|1\rangle_B) \tag{13.21}$$

$$|\Phi^-\rangle_{AB} = \frac{1}{\sqrt{2}}(|0\rangle_A|0\rangle_B - |1\rangle_A|1\rangle_B) \tag{13.22}$$

$$|\Psi^+\rangle_{AB} = \frac{1}{\sqrt{2}}(|0\rangle_A|1\rangle_B + |1\rangle_A|0\rangle_B) \tag{13.23}$$

$$|\Psi^-\rangle_{AB} = \frac{1}{\sqrt{2}}(|0\rangle_A|1\rangle_B - |1\rangle_A|0\rangle_B) \tag{13.24}$$

We assume Alice and Bob share the first state: $|\Phi^+\rangle_{AB}$. The overall state vector for the combined three qubits is:

$$|\Phi^+\rangle_{AB}\,|\psi\rangle_C = \frac{1}{\sqrt{2}}(|0\rangle_A|0\rangle_B + |1\rangle_A|1\rangle_B)(\alpha|0\rangle_C + \beta|1\rangle_C) \tag{13.25}$$

Notice that we are keeping track of who owns which piece of information.

In general, for a bipartite entangled state, we can write each pair of qubits in a linear combination of the Bell states by choosing the correct combination of two states from Eqs 13.21–13.24. Hence:

$$|0\rangle_i\,|0\rangle_j = \frac{1}{\sqrt{2}}\left(|\Phi^+\rangle_{ij} + |\Phi^-\rangle_{ij}\right) \tag{13.26}$$

$$|1\rangle_i\,|1\rangle_j = \frac{1}{\sqrt{2}}\left(|\Phi^+\rangle_{ij} - |\Phi^-\rangle_{ij}\right) \tag{13.27}$$

$$|0\rangle_i\,|1\rangle_j = \frac{1}{\sqrt{2}}\left(|\Psi^+\rangle_{ij} + |\Psi^-\rangle_{ij}\right) \tag{13.28}$$

$$|1\rangle_i\,|0\rangle_j = \frac{1}{\sqrt{2}}\left(|\Psi^+\rangle_{ij} - |\Psi^-\rangle_{ij}\right) \tag{13.29}$$

Using this, we can rewrite the state vector above for the entire system in terms of an entangled state involving Alice, A, and the qubit, C. Starting with Eq. 13.25:

$$\begin{aligned} |\Phi^+\rangle_{AB}\,|\psi\rangle_C &= \frac{1}{\sqrt{2}}\left(|0\rangle_A|0\rangle_B + |1\rangle_A|1\rangle_B\right)\left(\alpha|0\rangle_C + \beta|1\rangle_C\right) \\ &= \frac{1}{\sqrt{2}}\left(\alpha|0\rangle_A|0\rangle_B|0\rangle_C + \alpha|1\rangle_A|1\rangle_B|0\rangle_C + \beta|0\rangle_A|0\rangle_B|1\rangle_C + \beta|1\rangle_A|1\rangle_B|1\rangle_C\right) \end{aligned} \tag{13.30}$$

and using the forms for the different bipartite states above, we can rewrite Eq. 13.30 in terms of Bell states involving A and C.

$$\begin{aligned} |\Phi^+\rangle_{AB}\,|\psi\rangle_C =& \frac{1}{2}\left(|\Phi^+\rangle_{AC}\left[\alpha|0\rangle_B + \beta|1\rangle_B\right] + |\Phi^-\rangle_{AC}\left[\alpha|0\rangle_B - \beta|1\rangle_B\right]\right. \\ &\left. + |\Psi^+\rangle_{AC}\left[\beta|0\rangle_B + \alpha|1\rangle_B\right] + |\Psi^-\rangle_{AC}\left[\beta|0\rangle_B - \alpha|1\rangle_B\right]\right) \end{aligned} \tag{13.31}$$

So far, no measurements have been made. Alice now determines the state vector describing her two qubits. The result is a collapse into one of the four states:

$$|\Phi^+\rangle_{AC}\left(\alpha|0\rangle_B + \beta|1\rangle_B\right) \tag{13.32}$$

$$|\Phi^-\rangle_{AC}\left(\alpha|0\rangle_B - \beta|1\rangle_B\right) \tag{13.33}$$

$$|\Psi^+\rangle_{AC}\left(\beta|0\rangle_B + \alpha|1\rangle_B\right) \tag{13.34}$$

$$|\Psi^-\rangle_{AC}\left(\beta|0\rangle_B - \alpha|1\rangle_B\right) \tag{13.35}$$

Alice is now allowed to simply report to Bob, through an open classical channel, which state she is in. Bob then knows the state of his qubit. For the first above, the information has been transferred. For the other three states, Bob has to perform a unitary rotation to recover the correct state. In the case of the second state above, the change of sign is a change in the **geometrical phase**. This is achieved by a rotation of that state via a cycle of, say, going to a third state and back, coherently.

We have now taken the information in state C in the possession of Alice and, without transporting it, given that information to Bob. This is the basic concept in **quantum teleportation**.

13.5 Summary

In this section, the discussion has focused on some of the complexity regarding quantum measurement. In the context of the original postulates in Chapter 7, the Von Neumann hypothesis is that the wave function collapses or is reinitialized. The effect seen in quantum entanglement caused considerable concern among physicists, and many still find it troubling, though no one doubts the reality. Most seem content to except the simple predictions that are in agreement with experiment, the ultimate proof. We did not discuss another serious problem which is still of concern. The quantum to classical transition remains a subject of current interest. Related to this, and perhaps of more immediate concern for quantum device engineering, is to understand the quantum/classical interface. This is what is created when a classical system, like a voltmeter, makes a measurement on a quantum system. The problem remains an open question, where research will certainly create improved understanding.

Vocabulary (page) and Important Concepts

- measurement 220
- principle of superposition 20
- measurement problem 221
- Von Neumann postulate 221
- state vector collapse 221
- wave function collapse 221
- projection operator 221
- projection 221
- quantum bit 223
- qubit 223
- quantum non-demolition measurement 224
- quantum entangled state 224
- non-local 224
- Bell states 24
- Einstein, Podolsky, and Rosen (EPR) paradox 225
- Bells inequality 226
- quantum teleportation 228
- geometrical phase 228

14 Loss and Dissipation: The RLC Circuit

14.1 Introduction

The *concept of classical decay represents the non-reversible loss of energy* into a "bath." Friction is an example, as is ohmic loss in a resistor. It is ohmic loss that is responsible for heating in electronic devices. In quantum systems this is only part of the story. If we have two energy eigenstates at different energies, the higher energy state, if excited, could lose its energy to a bath, and the particle in that state would then decay to the lower energy state. But unlike a classical system where the energy in the original model decays continuously, quantum systems with discrete levels can only decay by discrete transitions. Furthermore, consider a coherent superposition state of two states. These two states have a well-defined, even if time dependent, phase difference. Interactions with the bath can destroy the coherence without changing the probabilities of being in either state.

In this section, we will look at a model for decay that reveals a remarkable fundamental result of a quantum system that is quite similar to a classical system.[1] We will then use the mathematical tools and the physical features of that problem to understand a resistor and proceed then to consider the RLC circuit. This same mathematics will then be used to understand spontaneous emission, after we learn about the quantized radiation field in Chapter 15.

14.2 Coupling to a Continuum of States: The Weisskopf–Wigner Approximation

The problem begins with a very simple statement about what seems like a hypothetical problem.[2] We imagine a Hilbert space defined by a Hamiltonian describing a single discrete state and a continuum of states. The continuum of states is often referred to as a **reservoir** or **bath**. The details of a physical system that gives rise to this are not important for this discussion. It is sufficient to specify the features and basis states of the Hamiltonian. The bandwidth of

[1] The basic approach here can be found in William H. Louisell, *Radiation and noise in quantum electronics*, (1964) McGraw-Hill and Claude Cohen-Tannoudji, Bernard Diu, and Frank Laloë. *Quantum mechanics* (1977), John Wiley and Sons, New York.

[2] This part of the discussion relies heavily on a similar discussion in Claude Cohen-Tannoudji, Bernard Diu, Frank Laloë, *Quantum Mechanics*, Wiley (1977).

Introduction to Quantum Nanotechnology: A Problem Focused Approach. Duncan Steel, Oxford University Press (2021).
 DOI: 10.1093/oso/9780192895073.003.0014

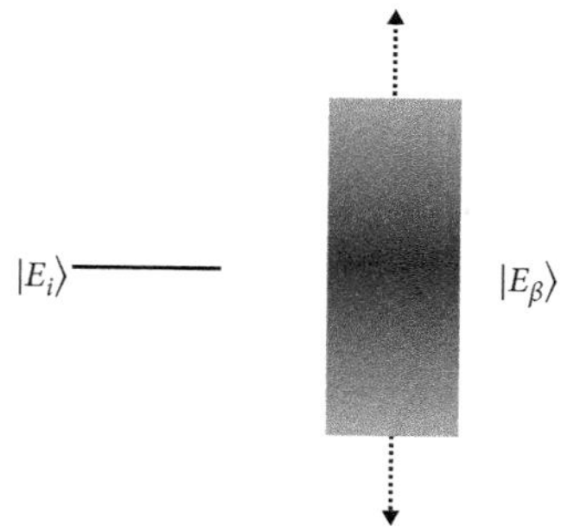

Fig. 14.1 Energy level diagram for a continuum of energy states that overlap the energy of a discrete state.

the continuum states overlap the single discrete state. The single state is given by $|E_i\rangle$. The continuum states are labeled as $|E_\beta\rangle$. The energy level diagram is illustrated in Fig. 14.1.

In the absence of coupling between $|E_i\rangle$ and $|E_\beta\rangle$, this entire system is described by a Hamiltonian operator, $\widehat{H}_0$, such that as usual

$$\widehat{H}_0|E_i\rangle = E_i|E_i\rangle \tag{14.1}$$

and

$$\widehat{H}_0|E_\beta\rangle = E_\beta|E_\beta\rangle \tag{14.2}$$

We assume that the eigenstates must also satisfy orthonormality

$$\langle E_i|E_i\rangle = 1 \tag{14.3}$$

$$\langle E_\beta|E_{\beta'}\rangle = \delta(\beta - \beta') \tag{14.4}$$

$$\langle E_i|E_\beta\rangle = 0 \tag{14.5}$$

and completeness

$$|E_i\rangle\langle E_i| + \int d\beta |E_\beta\rangle\langle E_\beta| = \hat{I} \tag{14.6}$$

We now include coupling by a potential, V, between the discrete state and the continuum states where in the simplest case we assume

$$\langle E_i|\widehat{V}|E_i\rangle = \langle E_\beta|\widehat{V}|E_\beta\rangle = \langle E_\beta|\widehat{V}|E_{\beta'}\rangle = 0 \tag{14.7}$$

which has very little impact on the result. However, importantly, we also assume

$$\langle E_i|\widehat{V}|E_\beta\rangle \neq 0 \tag{14.8}$$

If this is zero, the systems are decoupled and the states remain stationary. Physically, Eq. 14.7 means that $\widehat{V}$ does not change the energy of state $|E_i\rangle$ or state $|E_\beta\rangle$ to first order and there is no coupling by $\widehat{V}$ between different continuum states, i.e., between $|E_\beta\rangle$ and $|E_{\beta'}\rangle$. But, Eq. 14.8 means that there *is* coupling between $|E_i\rangle$ and $|E_\beta\rangle$.

To understand this problem, we think of this as an initial value problem where at $t = 0$, the system is initially state $|E_i\rangle$, and then determine how the system evolves over time. Up to this point in the discussion, if the Hamiltonian is not a function of time, and a quantum system is in a superposition of eigenstates with $\langle E_i|\widehat{V}|E_\beta\rangle = 0$, there is no evolution in time of the probability amplitude other than the usual phase associated with the time evolution operator. The probability, in that case, of being in any one state is time independent. However, in the case of coupling of an eigenstate to a continuum of eigenstates $\left(\langle E_i|\widehat{V}|E_\beta\rangle \neq 0\right)$, we will see that this is no longer true.

We begin by generating the equations of motion using the methods we developed earlier such as in Chapters 9 and 11. We start first with the general state vector of the problem. We expand in the usual way using the basis sets and letting the expansion coefficients be time dependent. We then solve the time dependent Schrödinger equation for the expansion coefficients, as we did, say, in Chapter 9 when we looked at the Rabi solution. In the interaction representation discussed in Chapter 9, meaning that we explicitly include the time dependent phase factors, the general state vector is given by

$$|\psi(t)\rangle = C_i(t)\, e^{-\frac{iE_it}{\hbar}}|E_i\rangle + \int d\beta\; C(\beta,t)\, e^{-\frac{iE_\beta t}{\hbar}}|E_\beta\rangle \tag{14.9}$$

Since we must sum over all the states, we have a discrete state, and then an integral over the continuum states. Now we insert this into Schrödinger's equation:

$$i\hbar\frac{d}{dt}|\psi(t)\rangle = \left(\widehat{H}_0 + \widehat{V}\right)|\psi(t)\rangle \tag{14.10}$$

giving

$$\begin{aligned} &i\hbar\frac{d}{dt}\left(C_i(t)e^{-\frac{iE_it}{\hbar}}|E_i\rangle + \int d\beta\; C(\beta,t)\, e^{-\frac{iE_\beta t}{\hbar}}|E_\beta\rangle\right) \\ &= \left(\widehat{H}_0 + \widehat{V}\right)\left(C_i(t)\,|E_i\rangle + \int d\beta\; C(\beta,t)\, e^{-\frac{iE_\beta t}{\hbar}}|E_\beta\rangle\right) \end{aligned} \tag{14.11}$$

After taking the time derivative on the left and operating with H_0 on the right we get

$$\begin{aligned} &E_iC_i(t)e^{-\frac{iE_it}{\hbar}}|E_i\rangle + i\hbar e^{-\frac{iE_it}{\hbar}}\frac{d}{dt}C_i(t)|E_i\rangle \\ &+ \int d\beta\left(C(\beta,t)\, e^{-\frac{iE_\beta t}{\hbar}}E_\beta|E_\beta\rangle + i\hbar e^{-\frac{iE_\beta t}{\hbar}}\frac{d}{dt}C(\beta,t)\,|E_\beta\rangle\right) \\ &= \left(C_i(t)\,\left(E_i + \widehat{V}\right)e^{-\frac{iE_it}{\hbar}}|E_i\rangle + \int d\beta\; C(\beta,t)\left(E_\beta + \widehat{V}\right)e^{-\frac{iE_\beta t}{\hbar}}|E_\beta\rangle\right) \end{aligned} \tag{14.12}$$

which can be simplified in the usual way to give

$$
\begin{aligned}
&i\hbar e^{-\frac{iE_i t}{\hbar}} \frac{d}{dt} C_i(t)|E_i\rangle + i\hbar \int d\beta\, e^{-\frac{iE_\beta t}{\hbar}} \frac{d}{dt} C(\beta,t)\,|E_\beta\rangle \\
&= C_i(t)\widehat{V} e^{-\frac{iE_i t}{\hbar}} |E_i\rangle + \int d\beta\, C(\beta,t)\, \widehat{V} e^{-\frac{iE_\beta t}{\hbar}} |E_\beta\rangle
\end{aligned}
\tag{14.13}
$$

Mathematically, we are right where we have been before, for example Eq. 9.85, but it looks a little different because first we had a sinusoidal driving term in Eq. 9.85, whereas here we just have a time independent interaction and also because we had only a discrete set of basis states in Eq. 9.85 but here we have the case of a discrete state and continuous state where the sum is replaced by an integral.

A subtle point of interest is that in the previous problems we had a time dependent Hamiltonian resulting in a time dependence of the probability amplitudes. Now we have a seemingly time independent Hamiltonian and yet we still have time dependent probability amplitudes. The reason is that, in this case, because $\langle E_i|\widehat{V}|E_\beta\rangle \neq 0$, the initial state is no longer a pure eigenstate of the system. We have somehow put the system in state $|E_i\rangle$ initially. This becomes an initial problem, which we know even classically can lead to dynamical behavior without a time dependent driving term.

We project Eq. 14.13 onto $|E_i\rangle$ and $|E_\beta\rangle$. We first project onto the discrete state:

$$
\begin{aligned}
&i\hbar e^{-\frac{iE_i t}{\hbar}} \frac{d}{dt} C_i(t)\,\langle E_i|E_i\rangle + i\hbar \int d\beta\, e^{-\frac{iE_\beta t}{\hbar}} \frac{d}{dt} C(\beta,t)\,\langle E_i|E_\beta\rangle \\
&= C_i(t) e^{-\frac{iE_i t}{\hbar}} \langle E_i|\widehat{V}|E_i\rangle + \int d\beta\, C(\beta,t)\, e^{-\frac{iE_\beta t}{\hbar}} \langle E_i|\widehat{V}|E_\beta\rangle
\end{aligned}
\tag{14.14}
$$

Using orthonormality ($\langle E_i|E_\beta\rangle = 0$ and $\langle E_i|E_i\rangle = 1$) and recalling our assumption that $\langle E_i|\widehat{V}|E_i\rangle = 0$ we get

$$
\frac{d}{dt} C_i(t) = \frac{1}{i\hbar} \int d\beta\, C(\beta,t)\, e^{-\frac{i(E_\beta\rangle - E_i\rangle)t}{\hbar}} \langle E_i|\widehat{V}|E_\beta\rangle
\tag{14.15}
$$

And for the continuum, using the usual orthonormality relation when the spectrum of eigenvalues is continuous, $\langle E_{\beta'}|E_\beta\rangle = \delta\left(E_{\beta'} - E_\beta\right)$, we get

$$
\begin{aligned}
&i\hbar e^{-\frac{iE_i t}{\hbar}} \frac{d}{dt} C_i(t)\,\langle E_\beta|E_i\rangle + i\hbar \int d\beta' e^{-\frac{iE_{\beta'} t}{\hbar}} \frac{d}{dt} C(\beta',t)\,\langle E_\beta|E_{\beta'}\rangle \\
&= C_i(t) e^{-\frac{iE_i t}{\hbar}} \langle E_\beta|\widehat{V}|E_i\rangle + \int d\beta' C(\beta',t)\, e^{-\frac{iE_{\beta'} t}{\hbar}} \langle E_\beta|\widehat{V}|E_\beta{}'\rangle
\end{aligned}
\tag{14.16}
$$

Since $\langle E_\beta|E_i\rangle = \langle E_\beta|\widehat{V}|E_\beta{}'\rangle = 0$, we get

$$
i\hbar \int d\beta' e^{-\frac{iE_{\beta'} t}{\hbar}} \frac{d}{dt} C(\beta',t)\,\delta(\beta - \beta') = C_i(t) e^{-\frac{iE_i t}{\hbar}} \langle E_\beta|\widehat{V}|E_i\rangle
\tag{14.17}
$$

or

$$\frac{d}{dt}C(\beta,t) = \frac{1}{i\hbar}C_i(t)e^{\frac{i(E_\beta-E_i)t}{\hbar}}\left\langle E_\beta|\widehat{V}|E_i\right\rangle \tag{14.18}$$

Eqs 14.15 and 14.18 are the two coupled differential equations. We are interested in the evolution of the system assuming that, at $t = 0$, the initial state of the system is $|E_i\rangle$:

$$C_i(t=0) = 1$$
$$C(\beta, t=0) = 0$$

We want to find the time evolution of C_i. We begin by integrating 14.18 over time:

$$\int_0^t dt' \frac{d}{dt'}C(\beta',t') = \frac{1}{i\hbar}\int_0^t dt' \left\langle E_\beta|\widehat{V}|E_i\right\rangle C_i(t')\, e^{\frac{i(E_\beta-E_i)t'}{\hbar}} \tag{14.19}$$

$$C(\beta,t) = \frac{1}{i\hbar}\int_0^t dt' \left\langle E_\beta|\widehat{V}|E_i\right\rangle C_i(t')\, e^{\frac{i(E_\beta-E_i)t'}{\hbar}} \tag{14.20}$$

Inserting into Eq. 14.15, we get

$$\frac{d}{dt}C_i(t) = -\frac{1}{\hbar^2}\int_0^t dt' \int d\beta |\langle E_\beta|\widehat{V}|E_i\rangle|^2 C_i(t')\, e^{-\frac{i(E_\beta-E_i)(t-t')}{\hbar}} \tag{14.21}$$

β is an index used to identify specific energy states in the continuum. It is the same kind of number as "n" in Eq. 11.103, except here we are considering only one dimension. Hence, we replace $d\beta$ with the corresponding density of states per unit energy interval (see Eq. 11.103 and the following discussion):

$$d\beta = \rho(E_\beta)\, dE_\beta$$

The details of the density of states, $\rho(E_\beta)$ discussed in Chapter 11 are not important for the current discussion. Eq. 14.21 is then slightly reorganized to give

$$\frac{d}{dt}C_i(t) = -\frac{1}{\hbar^2}\int_0^t dt' \int dE_\beta\; \rho(E_\beta)\; |\langle E_\beta|\widehat{V}|E_i\rangle|^2 e^{-\frac{i(E_\beta-E_i)(t-t')}{\hbar}}\, C_i(t') \tag{14.22}$$

This is an integro-differential equation (more specifically, a Fredholm equation of the second kind). In general, this is hard to solve because it says that the behavior of $C_i(t)$ (on the left) depends on the entire history of $C_i(t')$ (on the right). However, Weisskopf and Wigner saw their way through the complexity. Remarkably their trail is not hard to follow. The idea is to understand the mathematical behavior of the integrand, which then allows us to see how to solve this equation. There is more than one approach to developing the solution. We follow the approach taken by Cohen-Tannoudji et al.[3]

[3] Claude Cohen-Tannoudji, Bernard Diu, Frank Laloë, *Quantum Mechanics*, Wiley (1977).

We focus on the function $\rho(E)|\langle E_\beta|\widehat{V}|E_i\rangle|^2 e^{-\frac{i(E_\beta-E_i)(t-t')}{\hbar}}$ contained in the integrand of Eq. 14.22 where we now include explicitly the implicit limits of integration over the density of states.

$$f(E_i\, t-t') = -\frac{1}{\hbar^2}\int_0^\infty dE_\beta\, \rho\left(E_\beta\right)|\langle E_\beta|\widehat{V}|E_i\rangle|^2 e^{-\frac{i(E_\beta-E_i)(t-t')}{\hbar}} \tag{14.23}$$

This is the $\int dE_\beta$ part of Eq. 14.22.

We assume the very reasonable expectation that $\rho(E)|\langle E_\beta|\widehat{V}|E_i\rangle|^2$is slowly varying over the region of interest. Assuming it to be relatively constant is not unreasonable, but to make it more physical, we will just assume a peaked function of the form shown in Fig. 14.2. That this is physically reasonable, recall that we are interested in understanding what happens when a discrete state couples to a continuum of energy states. This is simply saying that there is no structure in the distribution of states near the energy of state E_i. Note: innovative engineering of this function has allowed for new capabilities. An example will be given in later chapters.

We now consider for convenience the real part of the exponential (the imaginary part is similar) in Eq. 14.23 as a function of $\frac{E_\beta}{E_i}-1$ for a fixed value of $\frac{E_i}{\hbar}(t-t')$. Unless t is close to t', the exponential oscillates quickly between -1 and 1 and averages the function $\rho(E)|\langle E_\beta|\widehat{V}|E_i\rangle|^2$ in the integral to zero. This is seen in Fig. 14.2 for $\frac{E_i}{\hbar}(t-t')=100$. Hence, the only value where there is a significant contribution is for $t' \sim t$. This implies that the only value of $C_i(t')$ that

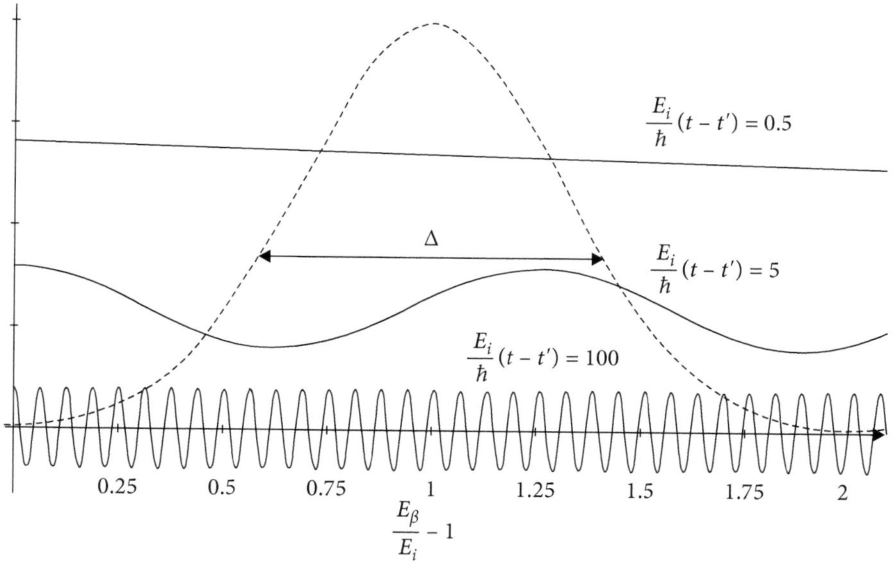

Fig. 14.2 A graphical display of the different terms in Eq. 14.23. The Gaussian is a plot of $\rho\left(E_\beta\right)|\langle E_\beta|\widehat{V}|E_i\rangle|^2$. The cosine function is a plot of $Re\, e^{-\frac{i(E_\beta-E_i)(t-t')}{\hbar}}$ for different values of $\frac{E_i}{\hbar}(t-t')$. The offset in the cosine function is added to make the plot clear. $\frac{E_\beta}{E_i}-1$. The y-axis units are arbitrary.

is important is when $t' \sim t$. This is the **Weisskopf–Wigner approximation:** it allows us to evaluate $C_i(t')$ at $t' = t$ and place it outside the integral in Eq. 14.22. This then gives

$$\frac{d}{dt}C_i(t) = -\frac{\Gamma}{2}C_i(t) \tag{14.24}$$

where:

$$\Gamma = \frac{2}{\hbar^2}\int_0^\infty dE_\beta \int_0^t dt' \rho\left(E_\beta\right) \left|\langle E_\beta|\hat{V}|E_i\rangle\right|^2 e^{-\frac{i\left(E_\beta - E_i\right)(t-t')}{\hbar}} \tag{14.25}$$

This means that the probability amplitude **decays** exponentially as a function of time. The integral over time is tabulated and is

$$\lim_{t\to\infty} Re \int_0^t dt' \, e^{-\frac{i\left(E_\beta - E_i\right)(t-t')}{\hbar}} = \pi\hbar\delta\left(E_\beta - E_i\right) \tag{14.26}$$

There is also an imaginary part, which gives rise to a small shift in the resonance frequency,[4] which we ignore. Inserting Eq. 14.26 this result into Eq. 14.25 gives

$$\Gamma = \frac{2\pi}{\hbar}\int dE_\beta \rho(E_\beta)\left|\langle E_\beta|\hat{V}|E_i\rangle\right|^2 \delta\left(E_\beta - E_i\right) \tag{14.27}$$

$$\Gamma = \frac{2\pi}{\hbar}\rho(E_i)\left|\langle E_\beta = E_i|\hat{V}|E_i\rangle\right|^2 \tag{14.28}$$

Equation 14.28 is the result that we could have anticipated from first order time dependent perturbation theory in terms of a transition rate per unit time. However, what has been gained here is that we have a rigorous derivation of the time evolution of the probability amplitude associated with this decay described by the equation of motion in Eq. 14.24 for all time. Note that the requirements for getting the exponential decay is a coupling to a continuum and the validity of the Weisskopf–Wigner approximation.

In 14.26, we took $t \to \infty$, but clearly we would like to know the behavior when $t \sim 0$. Note, however, that t^{-1} is the corresponding bandwidth of the measurement (just by Fourier transforming the corresponding exponential). However, we assumed in Eq. 14.23 that $\rho(E)\left|\langle E_\beta|\hat{V}|E_i\rangle\right|^2$ must be slowly varying over the range of $(t - t')^{-1}$ which we took to be most important when $t - t'$ is close to zero. Again, with $t - t' \sim 0$ this means that the corresponding bandwidth, $(t - t')^{-1}$, must be small compared the range of $\rho(E)\left|\langle E_\beta|\hat{V}|E_i\rangle\right|^2$. This translates into

$$(t - t')^{-1} \ll \Delta \tag{14.29}$$

[4] For details of this integral and the Lamb shift see Walter Heitler *The Quantum Theory of Radiation*, Oxford University Press, London (1935), 3rd ed. Dover; Paul R. Berman and Vladimir S. Malinovsky, *Principles of Spectroscopy and Quantum Optics*, Princeton (2011).

Hence, we cannot expect this answer to hold on time scales short compared to the bandwidth of the continuum states. However, this is typically extremely short, often femtoseconds or shorter.

Hence, we take 14.24 to describe the decay of the state when coupled to a continuum. Specifically, if at $t = 0$ we have some initial condition given by $C_i(t = 0)$, then

$$C_i(t) = C_i(t = 0)\, e^{-\frac{\Gamma}{2}t} \tag{14.30}$$

or

$$|C_i(t)|^2 = |C_i(t = 0)|^2 e^{-\Gamma t} \tag{14.31}$$

This is a remarkable and robust result, namely that *states coupled to a continuum decay irreversibly*. You see that there is no return of the probability amplitude over long times. The process is **irreversible**. This is very different from anything we have discussed. It is the source of dissipation in most quantum systems. Physically, the bath could be vibrations of a crystal lattice or the quantum radiation field (Chapter 15) or any other continuous system.

Moreover, suppose now you are designing a system characterized by a Hamiltonian $\hat{H}_0$ that is performing some function in the presence of a time dependent interaction of the forms, $\hat{V}_0(t)$, but the system connects to some bath of continuum states as above, then if the basic system is described by a complete set of basis states, $\{|E_n\rangle\}$, where

$$\hat{H}_0|E_n\rangle = E_n|E_n\rangle \tag{14.32}$$

and therefore

$$|\psi(t)\rangle = \sum_n C_n(t)e^{-\frac{iE_nt}{\hbar}}|E_n\rangle \tag{14.33}$$

Following the procedure in Chapter 9, the general equation of motion can be developed giving

$$\dot{C}_m(t) = \frac{1}{i\hbar}\sum_{n\neq m} C_n(t)\langle E_m|\hat{V}_0(t)|E_n\rangle \tag{14.34}$$

But now if we are to include the case that the states are coupled to a continuum, the equation of motion is modified to include the effect of reservoir coupling by

$$\dot{C}_m(t) = -\frac{\Gamma_m}{2}C_m(t) + \frac{1}{i\hbar}\sum_{n\neq m} C_n(t)\left\langle E_m|\hat{V}_0(t)|E_n\right\rangle e^{-i\frac{i(E_n-E_m)t}{\hbar}} \tag{14.35}$$

Problem 14.1 *Returning to Section 9.2 on the quantum flip-flop, you examined the effect of an oscillating field near resonance on the two-level system shown in Fig. 9.1 The equations of motion were developed then in Eqs 9.58 and 9.59. Use the results of Eq. 14.35 above to generalize* Eqs *9.58 and 9.59 and use Rabi's method of solution to find the new generalized Rabi frequency.*

Assuming that at $t = 0$, $C_1(t = 0) = 1$ and $C_2(t = 0) = 0$, find the solutions for all time for the probability amplitudes.

Plot the probability of being in the state $|E_n\rangle$ as a function of time, assuming $\Gamma_1 = \Gamma_2$, $\frac{\Gamma_1}{\omega_{21}} = 0.1$, $\omega = \omega_{21}$ and $\Omega_R = \frac{\mu E}{\hbar}$ and $\frac{\Omega_R}{\Gamma} = 10$, showing that it decays asymptotically to zero.

Importantly, coupling to the environment causing decay of the state also leads to loss of coherence. For a two-level system as above, the driving field at frequency ω gives rise to a state vector written in the usual form of

$$|\psi(t)\rangle = C_1|E_1\rangle + C_2 e^{-i\omega_{21}t}|E_2\rangle$$

where we have factored out the global phase $e^{-\frac{iE_1 t}{\hbar}}$ and $\hbar\omega_{21} = E_2 - E_1$. This is a coherent superposition state between the two states, and coherent superposition again means that there is a well-defined phase between the two states, even if it is time varying as $e^{-i\omega_{21}t}$. Experimentally, the quantum coherence is easily evident in the measurement of an observable that couples the two states. In the case of optical dipole as we are discussing here, such an observable is the induced polarization given by

$$P = e\langle r\rangle = C_1^* C_2 e^{-i\omega_{21}t}\langle E_1|r|E_2\rangle + C_1 C_2^* e^{i\omega_{21}t}\langle E_2|r|E_1\rangle$$

The polarization inserted into Maxwell's equations shows that an electromagnetic field is radiated by the two-state system. However, because of the decay, the polarization goes to zero, showing that decay destroys the coherent superposition.

Problem 14.2

a) Using the parameters in 14.1, find the polarization as a function of time, and plot the results.

b) Assuming $\Gamma_1 \neq \Gamma_2$, what is the decay rate of the coherence, defined as $C_1^* C_2$ or $C_1 C_2^*$ (they are complex conjugates of each other)?

While it is not evident in this analysis, the coherence can be destroyed by anything that destroys the phase, even if the interaction with the environment does not result in a loss of probability amplitude of either state. Something similar to a relative phase between the two states described by $e^{i\theta_{21}(t)}$, where $\theta_{21}(t)$ is a random fluctuation, will also lead to decay. However, these problems are best analyzed using a formalism based on the density matrix and the full Hamiltonian including the coupling to the source of decoherence.

As we see in Chapter 15, the above discussion exactly parallels the discussion for spontaneous emission.

14.3 Decay in the Nano-Vibrator Problem

The above analysis was done in the Schrödinger picture where we found the equation of motion for the probability amplitude a discrete state coupled to a continuum. However, we have found that examining the dynamics of some problems such as the nano-vibrator and LC circuit are more easily analyzed in the Heisenberg picture where the dynamical behavior is contained in the operators (see Chapter 9). We can now apply this to the problem of damped oscillations. This will be useful for either a damped mechanical vibrator as we discussed in Chapter 2 or a damped LC circuit of the type we considered in Chapter 8. For the nano-vibrator, the Hamiltonian in terms of a generalized coordinate, $q = x$, and conjugatemomentum, p, is given by

$$\hat{H} = \frac{\hat{p}^2}{2m} + \frac{1}{2}m\omega^2\hat{x}^2 \tag{14.36}$$

with $\omega = \sqrt{\frac{k}{m}}$ for a mechanical oscillator, and for the LC circuit is

$$\hat{H} = \frac{\hat{p}^2}{2L} + \frac{1}{2}L\omega^2\hat{q}^2 \tag{14.37}$$

where $p = L\dot{q}$, q is the charge and $\omega = \sqrt{\frac{1}{LC}}$. Written in terms of the raising and lowering operators, the form of the Hamiltonian is the same:

$$\hat{H} = \hbar\omega\left(\hat{a}^\dagger\hat{a} + \frac{1}{2}\right) \tag{14.38}$$

To include damping now, it is helpful to use an explicit form for the continuum states forming the reservoir and for the coupling potential. To imagine how this might be described mathematically based on simple physics, consider a classical system of a harmonic oscillator coupled to another harmonic oscillator. See Fig. 14.3. For the case of just two oscillators, if you now displace the mass of the first oscillator it will start to vibrate but will eventually diminish in amplitude as the second oscillator oscillates with increasing amplitude. Then the process goes backward and the amplitude of the first oscillator grows as the second oscillator decreases in amplitude. If you add now another harmonic oscillator, even at a different frequency, the motion will become more complicated, but the same process happens of a cycle of the energy in the first oscillator decreasing as the amplitudes in the other oscillators increase and then eventually returning to the first oscillator. However, the process takes longer. This is important. You can imagine now if you can do this with an infinite number of oscillators you will form a bath of continuum states like that illustrated in Fig. 14.1, and the energy in the first oscillator will be distributed over an infinite number of oscillators, the time scale for loss being set by the coupling to the other oscillators.[5]

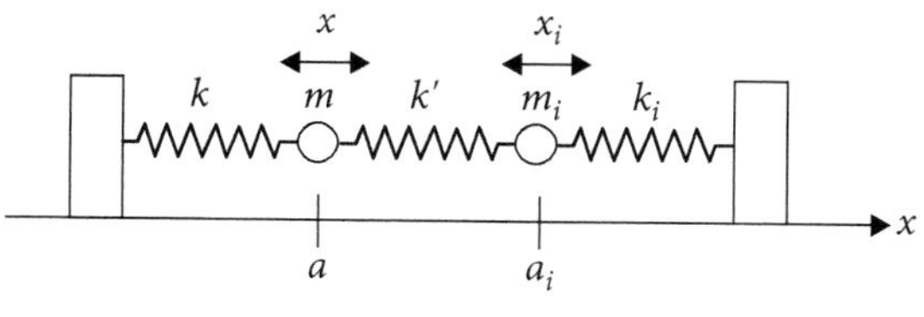

Fig. 14.3 The oscillator on the left is the same as in Chapter 2 and described by Eq. 14.36. The oscillator on the right represents mode *i* of an infinite number as indicated by the Hamiltonian in 14.39.

[5] The discussion here follows the lines presented by William Louisell, *Radiation and Noise in Quantum Electronics*, McGraw Hill 1964.

To make this quantitative, we develop a Hamiltonian based on the harmonic oscillator above, coupled to a bath of harmonic oscillators. So, the Hamiltonian for the bath will be

$$\hat{H}_B = \sum_i \hbar\omega_i \left(\hat{b}_i^\dagger \hat{b}_i + \frac{1}{2}\right) \tag{14.39}$$

The subscript i represents a different harmonic oscillator frequency that we refer to as a **mode**. The sum in Eq. 14.39 indicates that the modes are discrete. Eventually, this sum will be turned into an appropriate integral over the continuum of modes. By having an explicit model for the bath, unlike in the approach of the first section, we are able to move forward and solve the Heisenberg equations of motion, since we understand the various commutation relations that are needed. However, it is also an excellent model for a bath, since we nearly always observe that dissipation results in thermal heating which in crystals like semiconductors can result in vibrations of the crystal lattice.

We now need a coupling term. To arrive at that, we consider how a classical nano-vibrator is coupled to another vibrator through the equivalent of a spring. One way to see the coupling is to consider the simple coupled mechanical oscillator system shown in Fig. 14.3. The mass m, coordinate x, spring constant k, and at rest position a, correspond to the system of interest. The mass m_i, coordinate x_i, spring constant k_i, and at rest position a_i, correspond to a single mode i of the bath. Then x and x_i are the corresponding displacements from the at-rest locations, a and a_i. The potential energy of the coupled system is given by

$$V = \frac{1}{2}k'[(x-a)-(x_i-a_i)]^2 \tag{14.40}$$

We consider the entire Hamiltonian prior to writing it in terms of raising and lowering operators:

$$H = \frac{p^2}{2m} + \frac{1}{2}k(x-a)^2 + \frac{p_i^2}{2m_i} + \frac{1}{2}k_i(x_i-a_i)^2 + \frac{1}{2}k'[(x-a)-(x_i-a_i)]^2. \tag{14.41}$$

To make it more suitable for conversion to the raising and lowering operators, we change variables, defining

$$\tilde{x} = x - a; \quad \tilde{x}_i = x_i - a_i \tag{14.42}$$

$$H = \frac{p^2}{2m} + \frac{1}{2}k\tilde{x}^2 + \frac{p_i^2}{2m_i} + \frac{1}{2}k_i\tilde{x}_i^2 + \frac{1}{2}k'(\tilde{x}-\tilde{x}_i)^2 = \left(\frac{p^2}{2m} + \frac{1}{2}\tilde{k}\tilde{x}^2\right) + \left(\frac{p_i^2}{2m_i} + \frac{1}{2}\tilde{k}_i\tilde{x}_i^2\right) - k'\tilde{x}\tilde{x}_i \tag{14.43}$$

where

$$\tilde{k} = (k + k'), \quad \tilde{k}_i = (k_i + k') \tag{14.44}$$

We can now easily generalize this to an infinite number of oscillators coupled to the mass located at a by summing over i:

$$H = \left(\frac{p^2}{2m} + \frac{1}{2}\tilde{k}\tilde{x}^2\right) + \sum_i \left(\frac{p_i^2}{2m_i} + \frac{1}{2}\tilde{k}_i\tilde{x}_i^2\right) - \sum_i k'\tilde{x}\tilde{x}_i \tag{14.45}$$

We can convert this classical Hamiltonian to the quantum Hamiltonian by converting $p, p_i, \tilde{x}$, and $\tilde{x}_i$ to quantum operators, and recalling now from Eqs 8.11 and 8.12, that the definitions of the raising and lowering operator for the mechanical vibrator are given by

$$\hat{a} = \left(\sqrt{\frac{k}{2\hbar\omega}}\hat{x} + i\frac{\hat{p}}{\sqrt{2m\hbar\omega}}\right) \tag{14.46}$$

and

$$\hat{a}^\dagger = \left(\sqrt{\frac{k}{2\hbar\omega}}\hat{x} - i\frac{\hat{p}}{\sqrt{2m\hbar\omega}}\right). \tag{14.47}$$

We can solve these equations for $\hat{x}$ and $\hat{p}$ in terms of $\hat{a}$ and $\hat{a}^\dagger$:

$$\hat{x} = \sqrt{\frac{\hbar\omega}{2k}}\left(\hat{a} + \hat{a}^\dagger\right) \tag{14.48}$$

and

$$\hat{p} = -i\sqrt{\frac{m\hbar\omega}{2}}\left(\hat{a} - \hat{a}^\dagger\right) \tag{14.49}$$

We do the same for the bath operators. The total Hamiltonian becomes

$$\hat{H} = \hbar\omega\left(\hat{a}^\dagger\hat{a} + \frac{1}{2}\right) + \sum_i \hbar\omega_i\left(\hat{b}_i^\dagger\hat{b}_i + \frac{1}{2}\right) - \sum_i \frac{\hbar}{2}k'\sqrt{\frac{\omega\omega_i}{kk_i}}\left(\hat{a} + \hat{a}^\dagger\right)\left(\hat{b}_i + \hat{b}_i^\dagger\right) \tag{14.50}$$

where the first term corresponds to the Hamiltonian located at a, and the second term represents a sum over the Hamiltonians for mode i. The last term represents the coupling between the primary oscillator and the bath. Because k' is usually very small compared to the k and k sub i, we igore the small correction in Eq. 14.44 for simplicity. Of interest now is to solve the coupled Heisenberg equations of motion.

To make the process a little more transparent, it is convenient to think carefully now about the nature of Eq. 14.50, to make a simplifying assumption early, rather than doing more algebra and then making the same approximation at the end. Specifically, in the absence of coupling, we have four general equations of motion for each of the operators. We only need to consider two of the equations, for example the lowering operators, since the corresponding raising operators are just the Hermitian adjoint of the solution for the lowering operators. Specifically, we consider

$$-i\hbar\frac{d\hat{a}\left((t)\right)}{dt} = \left[\widehat{H}, \hat{a}(t)\right] = -\hbar\omega\hat{a}(t) \tag{14.51}$$

$$-i\hbar\frac{d\hat{b}_i(t)}{dt} = \left[\widehat{H}, \hat{b}_i(t)\right] = -\hbar\omega_i\hat{b}_i(t) \tag{14.52}$$

We have suppressed the usual H subscript used to designate Heisenberg operators and the explicit notation for time dependent operators. All operators in this discussion are time dependent Heisenberg operators unless otherwise noted such as $\hat{a}_S$ for the usual time independent Schrödinger lowing operator.

The solutions are simply $\hat{a}(t) = \hat{a}_S e^{-i\omega t}$ and $\hat{b}_i(t) = \hat{b}_{iS} e^{-i\omega_i t}$. It follows immediately that $\hat{a}^\dagger(t) = \hat{a}_S^\dagger e^{i\omega t}$ and $\hat{b}_i^\dagger(t) = \hat{b}_{i_S}^\dagger e^{i\omega_i t}$. We can now imagine roughly the time evolution of the third term in the Hamiltonian in Eq. 14.50:

$$\left(\hat{a} + \hat{a}^\dagger\right)\left(\hat{b}_i + \hat{b}_i^\dagger\right) = \hat{a}\hat{b}_i + \hat{a}^\dagger\hat{b}_i + \hat{a}\hat{b}_i^\dagger + \hat{a}^\dagger\hat{b}_i^\dagger = \hat{a}_S\hat{b}_{iS}e^{-i(\omega+\omega_i)t} + \hat{a}_S^\dagger\hat{b}_{iS}e^{i(\omega-\omega_i)t}$$
$$+ \hat{a}_S\hat{b}_{i_S}^\dagger e^{-i(\omega-\omega_i)t} + \hat{a}_S^\dagger\hat{b}_{i_S}^\dagger e^{i(\omega+\omega_i)t} \tag{14.53}$$

As we did in Chapter 9 in the discussion of resonant excitation of electromagnetic radiation, we ignore the $\hat{a}_S\hat{b}_{iS}e^{-i(\omega+\omega_i)t}$ term and the $\hat{a}_S^\dagger\hat{b}_{i_S}^\dagger e^{i(\omega+\omega_i)t}$ term since their contribution will be small after integration over time since the exponential oscillates at a high frequency $(\omega + \omega_i)$ rather than the much lower oscillation frequency at $(\omega - \omega_i)$. This is like the rotating wave approximation. Hence, the final form of the Hamiltonian in Eq. 14.50 is taken to be

$$\widehat{H} = \hbar\omega\left(\hat{a}^\dagger\hat{a} + \frac{1}{2}\right) + \sum_i \hbar\omega_i\left(\hat{b}_i^\dagger\hat{b}_i + \frac{1}{2}\right) - \sum_i \hbar\kappa_i\left(\hat{a}\hat{b}_i^\dagger + \hat{a}^\dagger\hat{b}_i\right) \tag{14.54}$$

where

$$\kappa_i = \frac{1}{2}k'\sqrt{\frac{\omega\omega_i}{kk_i}} \tag{14.55}$$

We now include the coupling to the bath (the third term in the Hamiltonian) and re-evaluate the equations of motion in Eqs 14.51 and 14.52. We use $\left[\hat{a}, \hat{a}^\dagger\right] = 1$ and find

$$\frac{d\hat{a}(t)}{dt} = -i\omega\hat{a}(t) + i\sum_i \kappa_i\hat{b}_i(t) \tag{14.56}$$

$$\frac{d\hat{b}_i(t)}{dt} = -i\omega_i\hat{b}_i(t) + i\kappa_i\hat{a}(t) \tag{14.57}$$

The objective is to understand the effect of the coupling to the bath on the equation of motion for $\hat{a}$. We first integrate Eq. 14.57 by introducing the usual integrating factor and using the initial condition that $\hat{b}_i\,(t = 0) = \hat{b}_{i_S}$:

$$\hat{b}_i(t) = e^{-i\omega_i t}\hat{b}_{i_S} + i\kappa_i\int_0^t dt'\,\hat{a}\left(t'\right)e^{-i\omega_i(t-t')} \tag{14.58}$$

Eq. 14.56 then becomes

$$\frac{d\hat{a}}{dt} + i\omega\hat{a} = +i\sum_i \kappa_i e^{-i\omega_i t}\hat{b}_{i_S} - \sum_i (\kappa_i)^2 \int_0^t dt' \hat{a}(t') e^{-i\omega_i(t-t')} \tag{14.59}$$

Let

$$\hat{\alpha}(t) = a(t)e^{i\omega t} \tag{14.60}$$

then

$$\begin{aligned}\frac{d\hat{\alpha}(t)}{dt} &= +i\sum_i \kappa_i e^{i(\omega-\omega_i)t}\hat{b}_{i_S} - \sum_i (\kappa_i)^2 \int_0^t dt' \hat{\alpha}(t') e^{i\omega(t-t')} e^{-i\omega_i(t-t')} \\ &= +i\sum_i \kappa_i e^{i(\omega-\omega_i)t}\hat{b}_{i_S} - \sum_i (\kappa_i)^2 \int_0^t dt' \hat{\alpha}(t') e^{i(\omega-\omega_i)(t-t')}\end{aligned} \tag{14.61}$$

This is the same form that we had in Eq. 14.21 where here the differential equation for $\hat{\alpha}(t)$ appears to depend on the entire history of $\hat{\alpha}(t)$ in the integral over time on the right-hand side. Following the early approach, because of the rapidly oscillating exponential as a function of t' for $\omega \neq \omega_i$, we make the Weisskopf–Wigner approximation by evaluating $\hat{\alpha}(t' = t)$ and remove it from the integral giving

$$\frac{d\hat{\alpha}}{dt} = +i\sum_i \kappa_i e^{i(\omega-\omega_i)t}\hat{b}_{i_S} - \frac{\Gamma}{2}\hat{\alpha}(t) \tag{14.62}$$

where

$$\frac{\Gamma}{2} = \sum_i (\kappa_i)^2 \int_0^t dt' e^{i(\omega-\omega_i)(t-t')} \tag{14.63}$$

To evaluate this, we follow the exact procedure above. Under the assumption of Eq. 14.29, we can effectively let $t \to \infty$ giving (Eq. 14.26):

$$\frac{\Gamma}{2} = \sum_i (\kappa_i)^2 \int_0^\infty dt' e^{i(\omega-\omega_i)(t-t')} = \pi\hbar\sum_i (\kappa_i)^2 \delta(E - E_i) \tag{14.64}$$

Recalling that the bath is a continuum, the $\sum_i$ becomes an integral over $E_i \rightarrow E_{bath}$; i.e., $\sum_i \rightarrow \int_0^\infty dE_{bath}\, \rho(E_{bath})$ giving

$$\frac{\Gamma}{2} = \pi\hbar\sum_i (\kappa_i)^2 \delta(E - E_i) = \pi\hbar \int_0^\infty dE_{bath}\, \rho(E_{bath}) \left[\kappa(E_{bath})\right]^2 \delta(E - E_{bath}) \tag{14.65}$$

giving, after the final integral,

$$\frac{\Gamma}{2} = \pi\hbar\, \rho(E)[\kappa(E)]^2$$

The equation of motion for $\hat{\alpha}$ becomes

$$\frac{d\hat{\alpha}(t)}{dt} = -\frac{\Gamma}{2}\hat{\alpha}(t) + i\sum_i \kappa_i e^{-i(\omega_i-\omega)t}\hat{b}_{i_S} = -\Gamma\hat{\alpha}(t) + i\hat{b}_{i_S}\int_0^\infty d\omega_i\, \rho(\omega_i)\,\kappa(\omega_i)\, e^{-i(\omega_i-\omega)t} \tag{14.66}$$

Since this is valid for $t > 0$, the last term will average to zero in this discussion if $\rho(\omega_i)\,\kappa(\omega_i)$ is sufficiently slowly varying, as required earlier. However, this term can be important for studying fluctuations. We ignore it here. Therefore

$$\frac{d\hat{\alpha}(t)}{dt} = -\frac{\Gamma}{2}\hat{\alpha}(t) \tag{14.67}$$

$$\hat{\alpha}(t) = \hat{\alpha}(0)e^{-\frac{\Gamma}{2}t} \tag{14.68}$$

and

$$\hat{a}(t) = \hat{\alpha}(0)e^{-\left(i\omega+\frac{\Gamma}{2}\right)t} = \hat{a}_S e^{-\left(i\omega+\frac{\Gamma}{2}\right)t} \tag{14.69}$$

Similar, the raising operator is given by

$$\hat{a}^\dagger(t) = \hat{\alpha}^\dagger(0)e^{\left(i\omega-\frac{\Gamma}{2}\right)t} = \hat{a}_S^\dagger e^{\left(i\omega-\frac{\Gamma}{2}\right)t} \tag{14.70}$$

From Eqs 14.48 and 14.49, we see in the Heisenberg picture,

$$\hat{x}_H(t) = \sqrt{\frac{\hbar\omega}{2k}}\left(\hat{a}_H(t) + \hat{a}_H^\dagger(t)\right) = \sqrt{\frac{\hbar\omega}{2k}}\left(\hat{a}_S e^{-i\omega t} + \hat{a}_S^\dagger e^{i\omega t}\right)e^{-\frac{\Gamma}{2}t} \tag{14.71}$$

and

$$\hat{p}_H(t) = -i\sqrt{\frac{m\hbar\omega}{2}}\left(\hat{a}_S e^{-i\omega t} - \hat{a}_S^\dagger e^{i\omega t}\right)e^{-\frac{\Gamma}{2}t} \tag{14.72}$$

Hence, it is clear that calculation of corresponding expectation values will show that the expectation values will go exponentially to zero with time.

Problem 14.3 *Working with the Heisenberg equations of motion for $\hat{a}^\dagger(t)$ and $\hat{b}^\dagger(t)$, show that by using 14.65, you get 14.70.*

Problem 14.4 *Using these results, find $\langle x(t)\rangle$ for a harmonic oscillator where*

$$|\psi(t=0)\rangle = \frac{1}{\sqrt{2}}\left(|E_0\rangle - |E_1\rangle\right)$$

Plot $\langle x(t)\rangle$.

14.4 RLC Circuit

Considering the case now for the RLC circuit in Fig. 14.4, compared to the discussion in Ch. 8, there is now a resistor in circuit. Classically, the resistor serves to dissipate energy in an electric circuit, in the same way that the coupled oscillator bath above dissipates the energy from the nano-vibrator. Hence, while we do not write down the Hamiltonian for the resistor (a complicated problem!), we can easily understand the behavior of the quantum RLC circuit by using the same formulation above. We do not need to redo the formulation.

The mapping from the nano-vibrator with loss to the RLC circuit follows directly from the application of the raising and lowering operators but defined in terms of the parameters of the LC circuit. From Chapter 8, we had from Eqs 8.82 and 8.83

$$\hat{a} = \left(\sqrt{\frac{1}{2C\hbar\omega}}\hat{q} + i\frac{\hat{p}}{\sqrt{2L\hbar\omega}}\right) \tag{14.73}$$

and

$$\hat{a}^{\dagger} = \left(\sqrt{\frac{1}{2C\hbar\omega}}\hat{q} - i\frac{\hat{p}}{\sqrt{2L\hbar\omega}}\right) \tag{14.74}$$

Recall that, classically, the coordinate operator $\hat{q}$ corresponds to the observable of the charge on the capacitor, q, and that the conjugate momentum operator, $\hat{p}$, corresponds to the inductance, and the current passing through it, $L\dot{q}$.

As the bath of oscillators above could be used to represent the dissipation of energy due to friction, and hence heat, the resistor behaves in a similar way, transferring energy from the moving electrons to the continuum of modes of vibrations (phonons) in the resistor, which leads to the formation of heat. The details of the transformation of the electron (charge carrier) motion through the resistor are beyond the scope of this book. However, we can be guided by experiments that show that, like the nano-vibrator, the electron loses energy, which is transformed into heat, which means vibrations of the atoms, and that once the system is energized (the initial condition), it oscillates with exponentially decaying amplitude.[6]

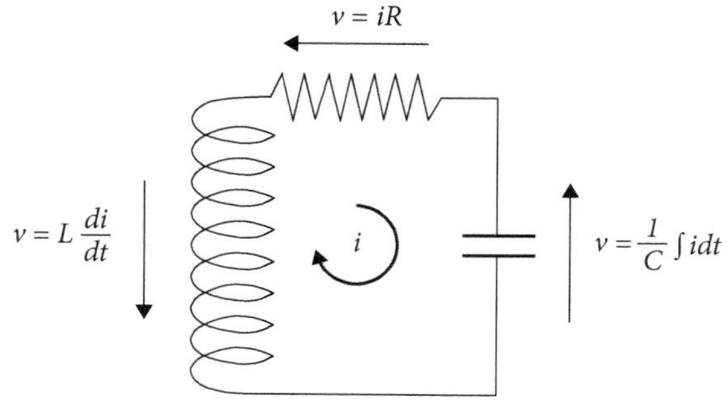

Fig. 14.4 Similar to the LC circuit in Chapter 8, we now consider how to understand dissipation in the RLC circuit.

[6] Summing the voltages around the loop

$$L\frac{di}{dt} + iR + \frac{1}{C}\int idt = 0$$

Hence, the form of the Heisenberg operators is given by

$$\hat{a}_H(t) = \hat{a}_S e^{-\left(i\omega+\frac{\Gamma}{2}\right)t} \tag{14.75}$$

$$\hat{a}_H^\dagger(t) = \hat{a}_S^\dagger e^{\left(i\omega-\frac{\Gamma}{2}\right)t} \tag{14.76}$$

Problem 14.5 *Using Eqs 14.73 and 14.74 show that the expectation value for the current has a form in time similar to the classical result. You will need to solve both the classical and the quantum case. To compare, evaluate the expectation value for the corresponding initial condition. Since the average current is zero since it is oscillating, you will need to evaluate the time dependence of* $\langle \hat{\imath} \rangle (t)$ *to compare to the classical case. The appropriate quantum state for this comparison is the coherent state where* $\hat{a}|\alpha\rangle = \alpha|\alpha\rangle$ *where* α *is a complex number* $\alpha = |\alpha|e^{i\theta}$. *Show that* $\langle \hat{\imath} \rangle (t) = -\frac{\hbar\omega}{L}|\alpha|e^{-\frac{\Gamma}{2}t}\sin(\omega t - \theta)\, t$. *Recall classically that for the series RLC circuit, Kirkoff's equations give:*

$$L\frac{di}{dt} + iR + \frac{1}{C}\int i dt = 0$$

14.5 **Summary**

This chapter showed that a fundamental cause of decay of a system is the coupling to a continuum of modes. The coupling to a continuum leads to an irreversible transfer of energy. Importantly, it was done by starting with a Hermitian Hamiltonian, but under the Weisskopf–Wigner approximation, the process became irreversible and now probability is not conserved. The Weisskopf–Wigner formalism is key to understanding spontaneous emission and ultimate many different decay channels. Mathematically, it requires making the transition from $\sum \to \int$ as we did in Chapter 9. Importantly, the inclusion of decay in the models can be maintained in other problems that now include other physical behaviors. Later, it will be seen that this approach leads to other important understandings, as some of the assumptions are removed.

Vocabulary (page) and Important Concepts

- reservoir 230
- bath 230
- decay 236
- Weisskopf–Wigner approximation 236
- irreversible 237
- mode 240

15 Quantum Radiation Field: Spontaneous Emission and Entangled Photons

15.1 Introduction

In nano-technology, things are becoming so small that it may be possible to switch a simple device with just a few quanta of light.[1] In other areas, this smallness means that pressure from the quantum vacuum may interfere with performance. In order to understand the basis of these ideas, it is essential to quantize the radiation field. This subject is a part of quantum electrodynamics. Normally, this field is presented through a rigorous foundation of Maxwell's equations. It is assumed here, however, that not all students in this field have yet developed a strong background in electromagnetics. The approach here will be to set aside as much of the foundation of electricity and magnetism as possible and build on what is typically covered in the first and second year physics courses.

Electromagnetic radiation includes the light you see with your eyes, but also radio waves and microwaves (lower energy) and X-rays and gamma rays (at a higher energy.) In all cases, the radiation contains a magnetic field and an electric field that are related to each other and that oscillate in time. Both the electric field and the magnetic field are observables, so the postulates provide a means to convert these fields to operators. However, before we can implement the postulates of quantum mechanics to convert the classical field to a quantum field, we have to identify the canonical coordinate and the conjugate momentum for the field and then, to get the Hamiltonian, we have to find the total energy in the field expressed in terms of these variables.

In learning this material, this chapter has been written so that it is possible to skip the development of the Hamiltonian (Section 15.2) and begin with Section 15.3 where the field is quantized.

15.2 Finding the Hamiltonian for the Transverse Electromagnetic Field

For this problem, we assume that the radiation is a plane wave propagating in the +z-direction in the vacuum and with the electric field, $\boldsymbol{E}(z,t)$ along the x- or y-axis (with vertical or horizontal polarization). The magnetic field, $\boldsymbol{B}(z,t)$, is then along the y- or x-axis. Figure 15.1 shows a typical field as a function of z for a specific time, t.

[1] Much of this discussion follows the ideas of Stig Stenholm in *Foundations of Laser Spectroscopy*, John Wiley and Sons, New York (1984) and J. J. Sakurai, *Advanced Quantum Mechanics*, Addison Wesley (1968).

Introduction to Quantum Nanotechnology: A Problem Focused Approach. Duncan Steel, Oxford University Press (2021).
 DOI: 10.1093/oso/9780192895073.003.0015

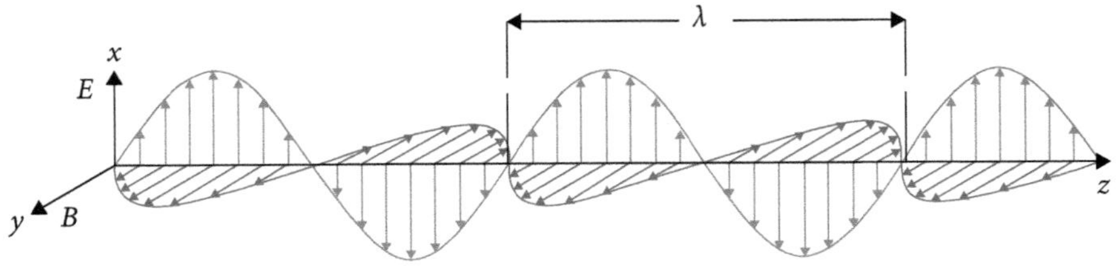

Fig. 15.1 The electromagnetic field is a plane wave, extending to infinity in the x-y plane and propagating in the +z direction. The electric field (shown in the x-direction) is linearly polarized along the x- or y-axis, and the magnetic field is perpendicular to the electric field, as shown.

$\boldsymbol{E}$ and $\boldsymbol{B}$ are given by $\cos(\boldsymbol{k}\cdot\boldsymbol{r}-\omega t)$ which for Fig. 15.1 becomes $\cos(\boldsymbol{kz}-\omega t)$. The pattern moves from left to right as a wave, if it is viewed in time. Here, the natural frequency, ω in radian/sec is related to the frequency, υ, in hertz by $\omega = 2\pi\upsilon$, and the direction of propagation is given by the wave vector, $\boldsymbol{k}$, assumed here to be *positive* and in the z-direction. The magnitude of the k-vector is related to ω by the speed of light in vacuum, namely $\omega = ck$ and $k = \frac{2\pi}{\lambda}$, where λ is the wavelength of the radiation.

The properties of this wave are described by the four equations of Maxwell that are written using basic vector algebra and relate the electric field, $\boldsymbol{E}(\boldsymbol{R},t)$, to the magnetic field, $\boldsymbol{B}(\boldsymbol{R},t)$. The dependence on $(\boldsymbol{R},t)$ is understood and not included in the equations. In charge-free space:

$$\nabla\cdot\boldsymbol{E}=0 \tag{15.1}$$

$$\nabla\cdot\boldsymbol{B}=0 \tag{15.2}$$

$$\nabla\times\boldsymbol{E}+\frac{\partial\boldsymbol{B}}{\partial t}=0 \tag{15.3}$$

$$\nabla\times\boldsymbol{B}-\frac{1}{c^2}\frac{\partial\boldsymbol{E}}{\partial t}=0 \tag{15.4}$$

From here we get the wave equation for the electromagnetic field, starting with taking the curl of 15.3:

$$\nabla\times\nabla\times\boldsymbol{E}+\frac{\partial}{\partial t}\nabla\times\boldsymbol{B}=0 \tag{15.5}$$

Inserting 15.4 we get

$$\nabla\times\nabla\times\boldsymbol{E}+\frac{1}{c^2}\frac{\partial^2\boldsymbol{E}}{\partial t^2}=0 \tag{15.6}$$

Now using the vector identity,

$$\nabla\times\nabla\times\boldsymbol{E}=\nabla(\nabla\cdot\boldsymbol{E})-\nabla^2\boldsymbol{E},\ \text{for Cartesian coordinates} \tag{15.7}$$

and using 15.1, we have

$$\nabla^2 \mathbf{E} - \frac{1}{c^2}\frac{\partial^2 \mathbf{E}}{\partial t^2} = 0 \tag{15.8}$$

This equation is the usual vector form of the **wave equation** for the electromagnetic field, where the components of the wave are in Cartesian coordinates because of the use of Eq. 15.7. So, for the case of Fig. 15.1, for an infinite plane wave, we have simply

$$\frac{\partial^2 E_x}{\partial z^2} - \frac{1}{c^2}\frac{\partial^2 E_x}{\partial t^2} = 0 \tag{15.9}$$

The solution describing the monochromatic field in Fig. 15.1 is, in phasor notation,

$$E_x = \frac{1}{2}E_0 e^{ikz - i\omega_k t} + \frac{1}{2}E_0^* e^{-ikz + i\omega_k t} \tag{15.10}$$

where $\omega_{\mathrm{k}} = ck$ and $c = \frac{1}{\sqrt{\mu_0 \epsilon_0}}$

Here, μ_0 is the magnetic permeability of free space and ϵ_0 is the electric permittivity of free space. Note that for each phasor above,

$$\nabla \cdot \mathbf{E} = \frac{\partial}{\partial z}\mathbf{E}_x = ikE_x \check{\mathbf{z}} \cdot \check{x} = ikE_x \check{\mathbf{z}} \cdot \check{x} = 0 \tag{15.11}$$

because the unit vectors $\check{\mathbf{z}}$ and $\check{x}$ are orthogonal to each other. This shows that Eq. 15.1 is satisfied, but it is also a statement that this is simply a plane wave where the temporal and spatial oscillations in the field are in the plane perpendicular to the direction of propagation and the fields are uniform and infinite in extent in the x-y plane. This is called a **transverse wave,** meaning that it is divergence free.

Everything in this discussion is classical. However, since the field is an observable, the postulates in Chapter 7 provide guidance for developing a quantum treatment. For such a quantum theory of the electromagnetic field, we have to find a canonical coordinate and conjugate momentum and the corresponding Hamiltonian for the field.

We begin by determining the total energy in the field. Starting with the knowledge that the energy density of an electric and magnetic field in charge free space is given by

$$\mathcal{W}_E = \frac{\epsilon_o}{2}|E|^2 \tag{15.12}$$

and

$$\mathcal{W}_B = \frac{1}{2\mu_o}|B|^2 \tag{15.13}$$

for the electric and magnetic components of the field. The total energy becomes

$$W = \frac{1}{2}\int d^3r\left(\epsilon_o |E|^2 + \frac{1}{\mu_o}|B|^2\right) \tag{15.14}$$

where the integral is over all space.

Problem 15.1 *Given that* $\boldsymbol{E} = E_x\check{x}$ *is given by Eq. 15.10 and* $\check{x}$ *is the unit vector, show that* $\boldsymbol{B} = B_y\check{y}$, *and that*

$$B_y = \frac{1}{2c}E_0 e^{ikz-i\omega_k t} + \frac{1}{2c}E_0 e^{-ikz+i\omega_k t}$$

It now remains to find a canonical coordinate and conjugate momentum and then show that we recover the results of Maxwell's equations using Hamilton's equations of motion (Eqs 7.73 and 7.74) for the canonical coordinate and conjugate momentum. It is tempting to imagine that E or B is a canonical coordinate and that the other is the conjugate momentum. The test is to show that when you then make this assignment and apply Hamilton's equations that your results recover what you expect from Maxwell's equations. Making such an assumption about E and B fails that test.

To get to the Hamiltonian, we have to move to a form that involves potentials rather than the fields themselves. This requires backing up and using two more vector identities in the analysis of Eqs 15.1–15.4. Specifically, for any vector field $\boldsymbol{G}$ and scalar field φ we have two well-known vector identities:

$$\nabla \cdot \nabla \times \boldsymbol{G} = 0 \tag{15.15}$$

and

$$\nabla \times \nabla\varphi = 0 \tag{15.16}$$

Applying these, we have from 15.2, $\nabla \cdot \boldsymbol{B} = 0$, then using 15.15, we conclude that

$$\boldsymbol{B} = \nabla \times \boldsymbol{A} \tag{15.17}$$

$\boldsymbol{A}$ is called the **vector potential.**

We now substitute this into Eq. 15.3:

$$\nabla \times \boldsymbol{E} + \frac{\partial \boldsymbol{B}}{\partial t} = \nabla \times \boldsymbol{E} + \frac{\partial}{\partial t}\nabla \times \boldsymbol{A} = \nabla \times \left(\boldsymbol{E} + \frac{\partial}{\partial t}\boldsymbol{A}\right) = 0 \tag{15.18}$$

Now seeing in 15.16 that the curl of the gradient of a scalar field is zero, we set the term in the parentheses equal to the gradient of a scalar function:

$$\boldsymbol{E} + \frac{\partial}{\partial t}\boldsymbol{A} = -\boldsymbol{\nabla}\varphi \tag{15.19}$$

or

$$\boldsymbol{E} = -\frac{\partial}{\partial t}\boldsymbol{A} - \boldsymbol{\nabla}\varphi \tag{15.20}$$

φ is called the **scalar potential**. When using potentials, there is an arbitrariness about the exact form that they can take when solving for their functional form. For example, you recall that in the presence of charge, Eq. 15.1 becomes

$$\boldsymbol{\nabla} \cdot \boldsymbol{E} = \frac{\rho}{\epsilon_o} \tag{15.21}$$

Using Eq. 15.20,

$$\boldsymbol{\nabla} \cdot \left(\frac{\partial}{\partial t} A + \boldsymbol{\nabla} \varphi \right) = \frac{\rho}{\epsilon_o} \tag{15.22}$$

The arbitrariness mentioned above allows us to take

$$\boldsymbol{\nabla} \cdot \boldsymbol{A} = 0 \tag{15.23}$$

Then 15.21 becomes the usual **Poisson's equation**:

$$\boldsymbol{\nabla}^2 \varphi = \frac{\rho}{\epsilon_o} \tag{15.24}$$

which leads to the familiar form for the potential for **Coulomb's law**[2] with the corresponding field being given by

$$\boldsymbol{E} = -\boldsymbol{\nabla} \varphi \tag{15.25}$$

Setting $\boldsymbol{\nabla} \cdot \boldsymbol{A} = 0$ is called the **Coulomb gauge** or the **transverse gauge**. There are also other options. The full implications of this discussion are beyond the scope of this presentation but can be found in most advanced texts on electricity and magnetism.

From vector algebra, the **Helmholtz theorem** shows that an arbitrary vector field $\boldsymbol{E}$ can be written as the sum of a **longitudinal vector component**, $\boldsymbol{E}_\ell$ and a **transverse vector component**, $\boldsymbol{E}_t$, where

$$\nabla \times \boldsymbol{E}_\ell = 0 \tag{15.26}$$

and

$$\nabla \cdot \boldsymbol{E}_t = 0 \tag{15.27}$$

Looking at Eq. 15.20 giving $\boldsymbol{E} = -\frac{\partial}{\partial t}\boldsymbol{A} - \boldsymbol{\nabla}\varphi$, we see that in the transverse gauge, $\boldsymbol{\nabla}\varphi$ gives rise to longitudinal fields arising from Poisson's equation (because $\nabla \times \boldsymbol{\nabla}\varphi = 0$). The vector potential gives rise to the transverse part of the electric field, $\boldsymbol{E}$: i.e.,

$$\boldsymbol{E}_t = -\frac{\partial}{\partial t}\boldsymbol{A} \tag{15.28}$$

Likewise, since $\nabla \cdot \boldsymbol{B} = 0$ (under all conditions), $\boldsymbol{B}$ is also purely transverse. Finally, if we know $\boldsymbol{A}$, we know $\boldsymbol{E}_t$ (Eq. 15.25) and Eq. 15.17 gives $\boldsymbol{B} = \nabla \times \boldsymbol{A}$.

Recall that the goal is to write the total energy in Eq. 15.14 in terms of $\boldsymbol{A}$ to explore whether that form leads to a Hamiltonian. For this, it is helpful to find the functional form of $\boldsymbol{A}$. To find the equation of motion for $\boldsymbol{A}$, we return to Eq. 15.4 and substitute Eqs 15.17 and 15.28:

[2] For a point charge at the origin, the Coulomb potential is simply $\varphi(r) = \frac{q}{4\pi\varepsilon_o r}$.

$$\nabla \times \nabla \times \boldsymbol{A} + \frac{1}{c^2}\frac{\partial^2 \boldsymbol{A}}{\partial t^2} = 0 \tag{15.29}$$

Then recalling that from Eq. 15.7 with the vector components in Cartesian coordinates, $\nabla \times \nabla \times \boldsymbol{A} = \nabla\nabla \cdot \boldsymbol{A} - \nabla^2 \boldsymbol{A}$, and with the transverse gauge condition $\nabla \cdot \boldsymbol{A} = 0$, we find that $\boldsymbol{A}$ satisfies a wave equation identical to Eq. 15.8:

$$\nabla^2 \boldsymbol{A} - \frac{1}{c^2}\frac{\partial^2 \boldsymbol{A}}{\partial t^2} = 0 \tag{15.30}$$

We now consider the solution to Eq. 15.30. Specifically, we are going to work in a cubic *cavity* that will eventually become infinitely large. This will make evaluation of the integrals over the cavity volume well-defined. At present in 15.15, for an infinite plane wave, there is an infinite amount of energy in the field and hence the integral is not defined. We take the volume of the cubic cavity to be $\mathcal{V} = L^3$ where L is the length of one side. With reference to the discussion above, the form of the solution for Eq. 15.30 is given by

$$\boldsymbol{A} = \frac{1}{\sqrt{\mathcal{V}\epsilon_o}} \sum_{\boldsymbol{k},\sigma} \left(\boldsymbol{C}_{\boldsymbol{k},\sigma} e^{i\boldsymbol{k}\cdot\boldsymbol{R} - i\omega_k t} + \boldsymbol{C}^*_{\boldsymbol{k},\sigma} e^{-i\boldsymbol{k}\cdot\boldsymbol{R} + i\omega_k t} \right) \tag{15.31}$$

where $\boldsymbol{k}$ designates the direction of propagation and wavelength $\left(|\boldsymbol{k}| = k = \frac{2\pi}{\lambda}\right)$ and σ indicates which of the two orthogonal polarization states (vertical and horizontal or left and right circularly polarized) and $\omega_k = ck$. The constants out front of the form of $\boldsymbol{A}$ are chosen so that the final result will be easier to interpret. The form in Eq. 15.31 represents a plane wave traveling in all directions with all frequencies and polarizations. We also assume **periodic boundary conditions** meaning that

$$e^{ik_u(u+L)} = e^{ik_u u}; u = x, y, z \tag{15.32}$$

requiring

$$e^{ik_u L} = 1 \text{ or } k_u = \frac{2n\pi}{L}; n = \pm 1, \pm 2, \ldots \tag{15.33}$$

From here, we can go back and re-evaluate the total energy in terms of the vector potential. We begin by determining both E and H, using Eq. 15.31. For the electric field we have from Eq. 15.28,

$$\boldsymbol{E}_t = -\frac{\partial}{\partial t}\boldsymbol{A} = \frac{i}{\sqrt{\mathcal{V}\epsilon_o}} \sum_{\boldsymbol{k},\sigma} \omega_{\boldsymbol{k}} \left(\boldsymbol{C}_{\boldsymbol{k},\sigma} e^{i\boldsymbol{k}\cdot\boldsymbol{R} - i\omega_k t} - \boldsymbol{C}^*_{\boldsymbol{k},\sigma} e^{-i\boldsymbol{k}\cdot\boldsymbol{R} + i\omega_k t} \right) \tag{15.34}$$

and from Eq. 15.17

$$\boldsymbol{B} = \nabla \times \boldsymbol{A} = \frac{i}{\sqrt{\mathcal{V}\epsilon_o}} \sum_{\boldsymbol{k},\sigma} \left(\boldsymbol{k} \times \boldsymbol{C}_{\boldsymbol{k},\sigma} e^{i\boldsymbol{k}\cdot\boldsymbol{R} - i\omega_k t} - \boldsymbol{k} \times \boldsymbol{C}^*_{\boldsymbol{k},\sigma} e^{-i\boldsymbol{k}\cdot\boldsymbol{R} + i\omega_k t} \right) \tag{15.35}$$

These expressions will be inserted in Eq. 15.14 for integration over the cavity volume. There are two important results from math and vector algebra that are important for carrying out the integration. The first is the relation that we have used earlier in a form useful for this work:

$$\int_{Cavity\ volume} d^3 r e^{i(k-k')\cdot R} = \mathcal{V}\delta_{kk'} \tag{15.36}$$

This arises because of the periodic boundary conditions. And the second is the vector identity

$$\begin{aligned}(k \times C_{k,\sigma}) \cdot (k \times C^*_{k,\sigma'}) &= C_{k,\sigma} \cdot C^*_{k,\sigma'} - (k \cdot C_{k,\sigma})(k \cdot C^*_{k,\sigma'}) = C_{k,\sigma} \cdot C^*_{k,\sigma'} \\ &= k^2 |C_{k,\sigma}|^2 \delta_{\sigma\sigma'}\end{aligned} \tag{15.37}$$

where $k \cdot C_{k,\sigma} = 0$ because the field is transverse and hence $C_{k,\sigma}$ lies in the plane perpendicular to the direction of propagation. We now insert 15.34 and 15.35 into 15.14, and use 15.36 and 15.37, integrating over the cavity volume. We find the energy in the electric, W_E, and then the energy in the magnetic field, W_B, and add the results together:[3]

$$\begin{aligned}W_E &= \frac{1}{2}\int d^3 r \epsilon_o |E|^2 = -\frac{1}{2}\frac{1}{V}\int d^3 r \sum_{k,\sigma}\sum_{k',\sigma'} \omega_k \omega_{k'} \left(C_{k,\sigma} e^{ik\cdot R - i\omega_k t} - C^*_{k,\sigma} e^{-k\cdot R + i\omega_k t}\right) \cdot \\ &\quad \left(C_{k',\sigma'} e^{ik\cdot' R - i\omega_{k'} t} - C^*_{k',\sigma'} e^{-ik'\cdot R + i\omega_{k'} t}\right) = \sum_{k,\sigma} \omega_k^2 |C_{k,\sigma}|^2\end{aligned} \tag{15.38}$$

$$\begin{aligned}W_B &= \frac{1}{2}\int d^3 r \frac{1}{\mu_0}|B|^2 = \frac{1}{2}\frac{1}{V\epsilon_o \mu_0}\int d^3 r \sum_{k,\sigma}\sum_{k',\sigma'} kk' \\ &\quad \left(k \times C_{k,\sigma} e^{ik\cdot R - i\omega_k t} - k \times C^*_{k,\sigma} e^{-ik\cdot R + i\omega_k t}\right) \cdot \\ &\quad \left(k' \times C_{k',\sigma'} e^{ik'\cdot R - i\omega_{k'} t} - k' \times C^*_{k',\sigma'} e^{-ik'\cdot R + i\omega_{k'} t}\right) = \sum_{k,\sigma} \omega_k^2 |C_{k,\sigma}|^2\end{aligned} \tag{15.39}$$

Therefore, Eq. 15.14 becomes

$$W = \frac{1}{2}\int_{Volume} d^3 r \left(\epsilon_o |E|^2 + \frac{1}{\mu_o}|B|^2\right) = 2\sum_{k,\sigma} \omega_k^2 |C_{k,\sigma}|^2 \tag{15.40}$$

Equation 15.40 is a statement of the total energy in the cavity in terms of the modal expansion coefficients. We now consider a seemingly arbitrary assignment to $C_{k,\sigma}$ of two new real variables (it is not entirely arbitrary if one thinks back to the operator approach for the nano-vibrator):

$$C_{k,\sigma} = \frac{1}{2}\left(Q_{k,\sigma} + i\frac{P_{k,\sigma}}{\omega_k}\right)\check{\varepsilon}_{k,\sigma} \tag{15.41}$$

where it follows then that

$$C^*_{k,\sigma} = \frac{1}{2}\left(Q_{k,\sigma} - i\frac{P_{k,\sigma}}{\omega_k}\right)\check{\varepsilon}_{k,\sigma} \tag{15.42}$$

[3] In Eqs 15.35 and 15.36, there are terms oscillating at twice the frequency in both the fields. However, they are opposite in sign and cancel.

$\check{\boldsymbol{e}}_{k,\sigma}$ is a unit vector representing the direction of $\boldsymbol{C}_{k,\sigma}$. Then 15.40 becomes

$$W = \sum\nolimits_{\boldsymbol{k},\sigma} \frac{1}{2} \left[P_{\boldsymbol{k},\sigma}^2 + \omega_k^2 Q_{\boldsymbol{k},\sigma}^2\right] \equiv \sum_{\boldsymbol{k},\sigma} H_{\boldsymbol{k},\sigma} \tag{15.43}$$

where

$$H_{\boldsymbol{k},\sigma} = \frac{1}{2} \left[P_{\boldsymbol{k},\sigma}^2 + \omega_k^2 Q_{\boldsymbol{k},\sigma}^2\right] \tag{15.44}$$

Before continuing, we recall that the objective is to find a canonical coordinate and the corresponding conjugate momentum which we can then use to write the total energy in the form of a Hamiltonian. The proof of the correctness is that the resulting canonical coordinate and conjugate momentum will then satisfy Hamilton's equations for the dynamics. Hence the coordinate and momentum are intrinsically time dependent. However, it is evident from Eq. 15.31 that the time dependence is shown explicitly meaning Eq. 15.41 and Eq. 15.42 are time independent. This is remedied by absorbing the time dependence in Eq 15.31 into a new expansion coefficient $\boldsymbol{C}_{\boldsymbol{k},\sigma}(t)$ according to:

$$\boldsymbol{C}_{\boldsymbol{k},\sigma}(t) = \boldsymbol{C}_{\boldsymbol{k},\sigma} e^{-i\omega_k t}; \boldsymbol{C}_{\boldsymbol{k},\sigma}^*(t) = \boldsymbol{C}_{\boldsymbol{k},\sigma}^* e^{-i\omega_k t} \tag{15.45}$$

Except for absorbing the explicit time dependence into the expansion coefficients, none of the above results are affected. We will see below that $H_{\boldsymbol{k},\sigma}$ will emerge as the Hamiltonian for the specific mode, $\boldsymbol{k}, \sigma$.

We use Eq. 15.45 along with Eqs 15.41 and 15.42 to solve for $Q_{\boldsymbol{k},\sigma}(t)$ and $P_{\boldsymbol{k},\sigma}(t)$ in terms of $\boldsymbol{C}_{\boldsymbol{k},\sigma}(t)$ and $\boldsymbol{C}_{\boldsymbol{k},\sigma}^*(t)$:

$$Q_{\boldsymbol{k},\sigma}(t) = \boldsymbol{C}_{\boldsymbol{k},\sigma}(t) \cdot \check{\boldsymbol{e}}_{\boldsymbol{k},\sigma} + \boldsymbol{C}_{\boldsymbol{k},\sigma}^*(t) \cdot \check{\boldsymbol{e}}_{\boldsymbol{k},\sigma} \tag{15.46}$$

$$P_{\boldsymbol{k},\sigma}(t) = -i\omega_k \left[\boldsymbol{C}_{\boldsymbol{k},\sigma}(t) \cdot \check{\boldsymbol{e}}_{\boldsymbol{k},\sigma} - \boldsymbol{C}_{\boldsymbol{k},\sigma}^*(t) \cdot \check{\boldsymbol{e}}_{\boldsymbol{k},\sigma}\right] \tag{15.47}$$

Using Eqs 15.45 and 15.46, we get

$$\frac{d}{dt} Q_{\boldsymbol{k},\sigma}(t) = -i\omega_k \left(\boldsymbol{C}_{\boldsymbol{k},\sigma}(t) \cdot \check{\boldsymbol{e}}_{\boldsymbol{k},\sigma} - \boldsymbol{C}_{\boldsymbol{k},\sigma}^*(t) \cdot \check{\boldsymbol{e}}_{\boldsymbol{k},\sigma}\right) \tag{15.48}$$

Comparing the right-hand side of Eq. 15.48 to the right-hand side of Eq. 15.47, we see that

$$\frac{d}{dt} Q_{\boldsymbol{k},\sigma}(t) = P_{\boldsymbol{k},\sigma}(t) \tag{15.49}$$

Using Eq. 15.47, we further note that

$$\frac{d}{dt} P_{\boldsymbol{k},\sigma}(t) = -\omega_k^2 \left[\boldsymbol{C}_{\boldsymbol{k},\sigma}(t) \cdot \check{\boldsymbol{e}}_{\boldsymbol{k},\sigma} + \boldsymbol{C}_{\boldsymbol{k},\sigma}^*(t) \cdot \check{\boldsymbol{e}}_{\boldsymbol{k},\sigma}\right] \tag{15.50}$$

Comparing the right-hand side of Eq. 15.50 to the right-hand side of Eq. 15.46, we conclude that

$$\frac{d}{dt} P_{\boldsymbol{k},\sigma}(t) = -\omega_k^2 Q_{\boldsymbol{k},\sigma}(t) \tag{15.51}$$

We now show that $Q_{\mathbf{k},\sigma}(t)$ and $P_{\mathbf{k},\sigma}(t)$ correspond to the canonical coordinate and the conjugate momentum for a specific mode described by the Hamiltonian, $H_{\mathbf{k},\sigma}$, in Eq. 15.44. We do this by using Hamilton's equations given in Eqs 7.73. and 7.74:

$$\frac{dq}{dt} = \frac{\partial H}{\partial p} \tag{15.52}$$

and

$$\frac{dp}{dt} = -\frac{\partial H}{\partial q} \tag{15.53}$$

and show that we recover the equations of motion in Eqs 15.49 and 15.51. Assuming then that $H_{\mathbf{k},\sigma}$ in Eq. 15.44 is the Hamiltonian for a specific mode with $Q_{\mathbf{k},\sigma}(t)$ and $P_{\mathbf{k},\sigma}(t)$ corresponding to the canonical coordinate and the conjugate momentum for that mode, we use Eq. 15.52 and immediately recover Eq. 15.49:

$$\frac{\partial H_{\mathbf{k},\sigma}}{\partial P_{\mathbf{k},\sigma}(t)} = \frac{\partial}{\partial P_{\mathbf{k},\sigma}(t)}\frac{1}{2}\left[P^2_{\mathbf{k},\sigma}(t) + \omega_k^2 Q^2_{\mathbf{k},\sigma}(t)\right] = P_{\mathbf{k},\sigma}(t) = \frac{d}{dt}Q_{\mathbf{k},\sigma}(t) \tag{15.54}$$

Using Eq. 15.53 we recover Eq. 15.51:

$$\frac{\partial H_{\mathbf{k},\sigma}}{\partial Q_{\mathbf{k},\sigma}(t)} = \frac{\partial}{\partial Q_{\mathbf{k},\sigma}(t)}\frac{1}{2}\left[P^2_{\mathbf{k},\sigma}(t) + \omega_k^2 Q^2_{\mathbf{k},\sigma}(t)\right] = \omega_k^2 Q_{\mathbf{k},\sigma}(t) = -\frac{d}{dt}P_{\mathbf{k},\sigma}(t) \tag{15.55}$$

Hence, the results show the validity of the assumption that $H_{\mathbf{k},\sigma}$ is the Hamiltonian for a specific mode with $Q_{\mathbf{k},\sigma}(t)$ and $P_{\mathbf{k},\sigma}(t)$ corresponding to the canonical coordinate and the conjugate momentum for that mode.

We have succeeded in showing that by using the vector potential and creating two new (mathematically) real variables, we have arrived at a **Hamiltonian for the electromagnetic field** in Eqs 15.43 and 15.44 in terms of a canonical coordinate, $Q_{\mathbf{k},\sigma}(t)$, and its conjugate momentum, $P_{\mathbf{k},\sigma}(t)$. Importantly, there is no significant meaning to these two variables other than that they are related directly to the expansion coefficients for the vector potential. Furthermore, it must be stressed again that, in spite of the language, *these variables have nothing to do with real space.* A similar kind of situation arose in Ch. 8 in the analysis of the LC-circuit. These variables simply constitute the critical variables for converting the classical formalism into the quantum formalism. This is a result of quantizing a field (not the motion of a massive particle), what matters will be the energy and determining how to work with the fields when, as observables, they become quantum operators.

15.3 Quantizing the Field

The key results from the previous discussion are summarized here so that the implications of the quantization of the field and the path that we have taken are more evident:

1. We introduced the vector potential and showed how it gives rise to the transverse electric and magnetic fields in Coulomb gauge. Working in a cavity with periodic boundary conditions, we then expanded the vector potential in Eq. 15.31 in terms of its solution to the corresponding wave equation:

$$\boldsymbol{A} = \frac{1}{\sqrt{\mathcal{V}\epsilon_o}} \sum_{\boldsymbol{k},\sigma} \left(\boldsymbol{C}_{\boldsymbol{k},\sigma} e^{i\boldsymbol{k}\cdot\boldsymbol{R} - i\omega_k t} + \boldsymbol{C}^*_{\boldsymbol{k},\sigma} e^{-i\boldsymbol{k}\cdot\boldsymbol{R} + i\omega_k t} \right)$$

where $k = \frac{2n\pi}{L}$; $n = \pm 1, \pm 2$. propagation along the z-axis is arbitrary, k is the propagation constant (more generally, a vector) and σ is the polarization.

2. We then showed in Eq. 15.40 that the total energy in the field is given by

$$W = 2 \sum_{\boldsymbol{k},\sigma} \omega_k^2 \left| \boldsymbol{C}_{\boldsymbol{k},\sigma} \right|^2$$

3. We then absorbed the time dependence in the phasors for the vector potential into the expansion coefficients such that $\boldsymbol{C}_{\boldsymbol{k},\sigma}(t) = \boldsymbol{C}_{\boldsymbol{k},\sigma} e^{-i\omega t}$ and $\boldsymbol{C}^*_{\boldsymbol{k},\sigma}(t) = \boldsymbol{C}^*_{\boldsymbol{k},\sigma} e^{i\omega t}$. Then, in Eqs 15.41 and 15.42, we then defined two new variables $Q_{\boldsymbol{k},\sigma}(t)$ and $P_{\boldsymbol{k},\sigma}(t)$ such that

$$\boldsymbol{C}_{\boldsymbol{k},\sigma}(t) = \frac{1}{2} \left(Q_{\boldsymbol{k},\sigma}(t) + i\frac{P_{\boldsymbol{k},\sigma}(t)}{\omega_k} \right) \check{\boldsymbol{\varepsilon}}_{\boldsymbol{k},\sigma} \text{ and } \boldsymbol{C}^*_{\boldsymbol{k},\sigma} = \frac{1}{2} \left(Q_{\boldsymbol{k},\sigma}(t) - i\frac{P_{\boldsymbol{k},\sigma}(t)}{\omega_k} \right) \check{\boldsymbol{\varepsilon}}_{\boldsymbol{k},\sigma}$$

where we then showed that $Q_{\boldsymbol{k},\sigma}$ was a canonical coordinate and $P_{\boldsymbol{k},\sigma}$ was the conjugate momentum satisfying Hamilton's equations. This allowed us to then show that the Hamiltonian for the $\boldsymbol{k}, \sigma$ mode is (Eq. 15.44)

$$H_{\boldsymbol{k},\sigma} = \frac{1}{2} \left[P^2_{\boldsymbol{k},\sigma} + \omega_k^2 Q^2_{\boldsymbol{k},\sigma} \right]$$

and then that

$$H_{Rad} = \sum_{\boldsymbol{k},\sigma} H_{\boldsymbol{k},\sigma} = \sum_{\boldsymbol{k},\sigma} \frac{1}{2} \left[P^2_{\boldsymbol{k},\sigma} + \omega_k^2 Q^2_{\boldsymbol{k},\sigma} \right] \tag{15.56}$$

Referring to the postulates in Chapter 7, the conversion to the quantum form is done by converting $Q_{\boldsymbol{k},\sigma}$ and $P_{\boldsymbol{k},\sigma}$ to operators $\widehat{Q}_{\boldsymbol{k},\sigma}$ and $\hat{P}_{\boldsymbol{k},\sigma}$ and setting their commutator to $i\hbar$, according to Postulate 5 in Chapter 7. The similarity of the Hamiltonian for the $\boldsymbol{k}, \sigma$ mode, $H_{\boldsymbol{k},\sigma}$, to that of the nano-vibrator in Chapter 2 is evident. The Heisenberg formalism of Chapter 8 is now important since it would have no physical meaning to examine eigenfunctions that are depended on some unphysical coordinate, $Q_{\boldsymbol{k},\sigma}$, that has nothing to do with real space.

The commutator for the operators $\widehat{Q}_{\boldsymbol{k},\sigma}$ and $\hat{P}_{\boldsymbol{k}',\sigma'}$ is then given by

$$\left[\widehat{Q}_{\boldsymbol{k},\sigma}, \hat{P}_{\boldsymbol{k}',\sigma'} \right] = i\hbar \delta_{\boldsymbol{k}\boldsymbol{k}'} \delta_{\sigma\sigma'} \tag{15.57}$$

The commutator relationship looks a little different here because we are now dealing with an infinite number of modes identified by $\boldsymbol{k}$ and σ. *Each unique pair of* $\boldsymbol{k}$ and σ *represents a separate Hilbert space.* The entire Hilbert space for the quantized field is the product space of an infinite number of Hilbert spaces. The commutator is zero when $\widehat{Q}_{\boldsymbol{k},\sigma}$ is in the

Hilbert space characterized by mode $\mathbf{k}, \sigma$ and $\widehat{Q}_{k',\sigma'}$ in a separate Hilbert space characterized by mode $\mathbf{k}', \sigma'$.

As we did with the harmonic oscillator in Eqs 8.11 and 8.12, we define two new operators as

$$\hat{a}_{k,\sigma} = \sqrt{\frac{\omega_k}{2\hbar}}\left(\widehat{Q}_{k,\sigma} + i\frac{\hat{P}_{k,\sigma}}{\omega_k}\right) \tag{15.58}$$

and

$$\hat{a}^{\dagger}_{k,\sigma} = \sqrt{\frac{\omega_k}{2\hbar}}\left(\widehat{Q}_{k,\sigma} - i\frac{\hat{P}_{k,\sigma}}{\omega_k}\right) \tag{15.59}$$

It follows then, from Eq. 15.57, that

$$\left[\hat{a}_{k,\sigma}, \hat{a}^{\dagger}_{k',\sigma'}\right] = \delta_{kk'}\delta_{\sigma\sigma'} \tag{15.60}$$

where, as was the case in Chapter 8, $\hat{a}^{\dagger}_{k,\sigma}$ and $\hat{a}_{k,\sigma}$ correspond to the raising and lowering operators except that in the case of the radiation field, they are called the **creation and annihilation operators**, respectively, associated with creating or destroying a quanta of energy. Following the results developed for these operators in Chapter 8, the Hamiltonian for the $\mathbf{k}, \sigma$ mode in Eq. 15.44 becomes

$$H_{k,\sigma} = \hbar\omega_k\left(\hat{a}^{\dagger}_{k,\sigma}\hat{a}_{k,\sigma} + \frac{1}{2}\right) \tag{15.61}$$

and for the entire radiation field, the **Hamiltonian for the quantized electromagnetic field** is

$$\widehat{H}_{Rad} = \sum_{k,\sigma}\hbar\omega_k\left(\hat{a}^{\dagger}_{k,\sigma}\hat{a}_{k,\sigma} + \frac{1}{2}\right) \tag{15.62}$$

The eigenvectors for $H_{k,\sigma}$ are given by

$$|E_{n,k,\sigma}\rangle = |n_{k,\sigma}\rangle \tag{15.63}$$

meaning that there are n quanta of energy $\hbar\omega_k$ in the $\mathbf{k}, \sigma$ mode:

$$H_{k,\sigma}|n_{k,\sigma}\rangle = \hbar\omega_k\left(\hat{a}^{\dagger}_{k,\sigma}\hat{a}_{k,\sigma} + \frac{1}{2}\right)|n_{k,\sigma}\rangle = \hbar\omega_k\left(n_{k,\sigma} + \frac{1}{2}\right)|n_{k,\sigma}\rangle \tag{15.64}$$

$|n_{k,\sigma}\rangle$ is a **number state** or **Fock state**, and $n_{k,\sigma}$ is the **occupation number**. The creation and annihilation operators change the occupation number by 1:

$$\hat{a}^{\dagger}_{k,\sigma}|n_{k,\sigma}\rangle = \sqrt{n_{k,\sigma} + 1}\,|n_{k,\sigma} + 1\rangle \tag{15.65}$$

$$\hat{a}_{k,\sigma}|n_{k,\sigma}\rangle = \sqrt{n_{k,\sigma}}\,|n_{k,\sigma} - 1\rangle \tag{15.66}$$

and the number operator $\hat{n}_{\boldsymbol{k},\sigma} \equiv \hat{a}^{\dagger}_{\boldsymbol{k},\sigma}\hat{a}_{\boldsymbol{k},\sigma}$ where

$$\hat{n}_{\boldsymbol{k},\sigma}|n_{\boldsymbol{k},\sigma}\rangle = n_{\boldsymbol{k},\sigma}|n_{\boldsymbol{k},\sigma}\rangle \tag{15.67}$$

The conventional language has been to call each quanta of energy a **photon**. But it must be clear that the photon is a unit of energy in a specific mode of the field and not a particle, the way we think of a massive particle like an electron.

The radiation eigenvector, $|E_{rad.e.vctr}\rangle$, for the entire radiation field is then a **product state** of an eigenstate of reach mode. Recall that k can take on an infinite number of values each corresponding to a different energy and σ can take on two values corresponding to two orthogonal states of polarization.

$$|E_{rad.e.vctr}\rangle = \prod_{\boldsymbol{k},\sigma}|n_{\boldsymbol{k},\sigma}\rangle = |n_{1,1}\rangle|n_{1,2}\rangle\,|n_{2,1}\rangle|n_{2,2}\rangle\,|n_{3,1}\rangle|n_{3,2}\rangle \dots |n_{s,1}\rangle|n_{s,2}\rangle \dots \tag{15.68}$$

Furthermore, each mode $|n_{\boldsymbol{k},\sigma}\rangle$ with n units of energy in mode $(\boldsymbol{k},\sigma)$ has an infinite number of different states (different values of n) and there are an infinite number of combinations leading to a very large Hilbert space. The eigenenergy is given by

$$\hat{H}_{rad}|E_{rad.e.vctr}\rangle = \left(\sum_{\boldsymbol{k},\sigma}\hbar\omega_k\left(n_{\boldsymbol{k},\sigma}+\frac{1}{2}\right)\right)|E_{rad.e.vctr}\rangle \tag{15.69}$$

Notice that even if $n_{\boldsymbol{k},\sigma} = 0$ for all $\boldsymbol{k},\sigma$, there is an infinite amount of energy in the field. This is called the **vacuum field**:

$$|E_{vacuum}\rangle \equiv |0\rangle = |0_{1,1}\rangle|0_{1,2}\rangle\,|0_{2,1}\rangle|0_{2,2}\rangle\,|0_{3,1}\rangle|0_{3,2}\rangle \dots |0_{s,1}\rangle|0_{s,2}\rangle \dots \tag{15.70}$$

We are now able to determine the operator form for the different physical variables corresponding to the vector potential, $\boldsymbol{A}$, the electric field, $\boldsymbol{E}$, and the magnetic field, $\boldsymbol{B}$. For the vector potential, we recall the expansion in Eq. 15.31 and transform it to an operator (recall that $\boldsymbol{C}^{*}_{\boldsymbol{k},\sigma} \to \hat{\boldsymbol{C}}^{\dagger}_{\boldsymbol{k},\sigma}$):

$$\hat{\boldsymbol{A}} = \frac{1}{\sqrt{\mathcal{V}\epsilon_o}}\sum_{\boldsymbol{k},\sigma}\left(\hat{\boldsymbol{C}}_{\boldsymbol{k},\sigma}e^{i\boldsymbol{k}\cdot\boldsymbol{R}} + \hat{\boldsymbol{C}}^{\dagger}_{\boldsymbol{k},\sigma}e^{-ikz}\right) \tag{15.71}$$

Note that the time dependence is now absent in the Schrödinger picture. Then from Eqs 15.41 and 15.42, we can write the operator form for $\hat{\boldsymbol{C}}_{\boldsymbol{k},\sigma}$ and $\hat{\boldsymbol{C}}^{\dagger}_{\boldsymbol{k},\sigma}$ in terms of the canonical coordinate and conjugate momentum:

$$\begin{aligned}\hat{\boldsymbol{C}}_{\boldsymbol{k},\sigma} &= \tfrac{1}{2}\left(\hat{Q}_{\boldsymbol{k},\sigma} + i\tfrac{\hat{P}_{\boldsymbol{k},\sigma}}{\omega_k}\right)\check{\boldsymbol{\varepsilon}}_{\boldsymbol{k},\sigma}\\ \hat{\boldsymbol{C}}^{\dagger}_{\boldsymbol{k},\sigma} &= \tfrac{1}{2}\left(\hat{Q}_{\boldsymbol{k},\sigma} - i\tfrac{\hat{P}_{\boldsymbol{k},\sigma}}{\omega_k}\right)\check{\boldsymbol{\varepsilon}}_{\boldsymbol{k},\sigma}\end{aligned} \tag{15.72}$$

Combining 15.58 and 15.59 we get

$$\hat{Q}_{\boldsymbol{k},\sigma} = \sqrt{\frac{\hbar}{2\omega_k}}\left(\hat{a}_{\boldsymbol{k},\sigma} + \hat{a}^{\dagger}_{\boldsymbol{k},\sigma}\right) \tag{15.73}$$

and

$$\hat{P}_{k,\sigma} = -i\sqrt{\frac{\hbar\omega_k}{2}}\left(\hat{a}_{k,\sigma} - \hat{a}^\dagger_{k,\sigma}\right) \quad (15.74)$$

we then substitute into Eq. 15.72:

$$\hat{C}_{k,\sigma} = \frac{1}{2}\left(\sqrt{\frac{\hbar}{2\omega_k}}\left(\hat{a}_{k,\sigma} + \hat{a}^\dagger_{k,\sigma}\right) + \sqrt{\frac{\hbar}{2\omega_k}}\left(\hat{a}_{k,\sigma} - \hat{a}^\dagger_{k,\sigma}\right)\right)\check{\epsilon}_{k,\sigma} = \sqrt{\frac{\hbar}{2\omega_k}}\hat{a}_{k,\sigma}\check{\epsilon}_{k,\sigma} \quad (15.75)$$

Likewise, for $\hat{C}^\dagger_{k,\sigma}$ we get

$$\hat{C}^\dagger_{k,\sigma} = \sqrt{\frac{\hbar}{2\omega_k}}\hat{a}^\dagger_{k,\sigma}\check{\epsilon}_{k,\sigma} \quad (15.76)$$

Returning now to Eq. 15.31 and recalling Eq. 15.45, we insert the results in Eqs 15.64 and 15.65 we get for the quantized vector potential,

$$\hat{A} = \sum_{k,\sigma}\sqrt{\frac{\hbar}{2\epsilon_o\omega_k\mathcal{V}}}\left(\hat{a}_{k,\sigma}\check{\epsilon}_{k,\sigma}e^{ik\cdot R} + \hat{a}^\dagger_{k,\sigma}\check{\epsilon}^*_{k,\sigma}e^{-ik\cdot R}\right) = \hat{A}_+ + \hat{A}_- \quad (15.77)$$

where

$$\hat{A}_+ = \sum_{k,\sigma}\mathcal{A}_k\hat{a}_{k,\sigma}\check{\epsilon}_{k,\sigma}e^{ik\cdot R};\ \hat{A}_- = \sum_{k,\sigma}\mathcal{A}_k\hat{a}^\dagger_{k,\sigma}\check{\epsilon}^*_{k,\sigma}e^{-ik\cdot R};\ \mathcal{A}_k = \sqrt{\frac{\hbar}{2\epsilon_o\omega_k\mathcal{V}}}$$

Likewise, for the quantized electric field, we use Eq. 15.34 and get

$$\hat{E}_t = i\sum_{k,\sigma}\mathcal{E}_k\left(\hat{a}_{k,\sigma}\hat{\epsilon}_{k,\sigma}e^{ik\cdot R} - \hat{a}^\dagger_{k,\sigma}\hat{\epsilon}^*_{k,\sigma}e^{-ik\cdot R}\right) = \hat{E}_+ + \hat{E}_- \quad (15.78)$$

where

$$\hat{E}_+ = i\sum_{k,\sigma}\mathcal{E}_k\hat{a}_{k,\sigma}\check{\epsilon}_{k,\sigma}e^{ik\cdot R};\ \hat{E}_- = -i\sum_{k,\sigma}\mathcal{E}_k\hat{a}^\dagger_{k,\sigma}\check{\epsilon}^*_{k,\sigma}e^{-ik\cdot R}\ \text{and}\ \mathcal{E}_k \equiv \sqrt{\frac{\hbar\omega_k}{2\epsilon_o\mathcal{V}}};$$

where the $\pm$ convention corresponds to $e^{\mp i\omega t}$ the positive and negative frequency phaser, respectively, that would be present if these were classical fields rather than Schrödinger operators.

Using Eq. 15.35 we get for the quantized magnetic field (recall that the field is propagating along the z-direction in the cavity).

$$\hat{B} = \nabla\times\hat{A} = i\sum_{k,\sigma}\mathcal{B}_k\left(\hat{a}_{k,\sigma}\,\check{\kappa}\times\check{\epsilon}_{k,\sigma}e^{ik\cdot R} - \hat{a}^\dagger_{k,\sigma}\check{\kappa}\times\check{\epsilon}^*_{k,\sigma}e^{-ik\cdot R}\right) = \hat{B}_+ + \hat{B}_- \quad (15.79)$$

where $\hat{B}_+ = i\sum_{k,\sigma}\mathcal{B}_k\hat{a}_{k,\sigma}\,\check{\kappa}\times\check{\epsilon}_{k,\sigma}e^{ik\cdot R}$; $\hat{B}_- = -i\sum_{k,\sigma}\mathcal{B}_k\hat{a}^\dagger_{k,\sigma}\check{\kappa}\times\check{\epsilon}^*_{k,\sigma}e^{-ik\cdot R}$; and $\mathcal{B}_k \equiv \sqrt{\frac{\hbar\omega_k}{2\epsilon_o c^2\mathcal{V}}}$ and $\check{\kappa} = \frac{k}{k}$ is a unit vector in the direction of propagation.

The results are summarized in Table 15.1:

Table 15.1 Summary of operators and their properties resulting from quantizing the electromagnetic field.

Eigenstate	n-photons (units of energy) in mode $\boldsymbol{k}, \sigma$, with the direction of propagation given $\boldsymbol{k}$ and the polarization, σ	$\lvert n_{\boldsymbol{k},\sigma}\rangle$
Annihilation operator	$\hat{a}_{\boldsymbol{k},\sigma}$	$\hat{a}_{\boldsymbol{k},\sigma}\lvert n_{\boldsymbol{k},\sigma}\rangle = \sqrt{n_{\boldsymbol{k},\sigma}}\lvert n_{\boldsymbol{k},\sigma} - 1\rangle$
Creation operator	$\hat{a}^{\dagger}_{\boldsymbol{k},\sigma}$	$\hat{a}^{\dagger}_{\boldsymbol{k},\sigma}\lvert n_{\boldsymbol{k},\sigma}\rangle = \sqrt{n_{\boldsymbol{k},\sigma} + 1}\lvert n_{\boldsymbol{k},\sigma} + 1\rangle$
Number operator	$\hat{n}_{\boldsymbol{k},\sigma} = \hat{a}^{\dagger}_{\boldsymbol{k},\sigma}\hat{a}_{\boldsymbol{k},\sigma}$	$\hat{n}_{\boldsymbol{k},\sigma}\lvert n_{\boldsymbol{k},\sigma}\rangle = n_{\boldsymbol{k},\sigma}\lvert n_{\boldsymbol{k},\sigma}\rangle$
Commutation relation	$\left[\hat{a}_{\boldsymbol{k},\sigma}, \hat{a}^{\dagger}_{\boldsymbol{k}',\sigma'}\right] = \delta_{\boldsymbol{k}\boldsymbol{k}'}\delta_{\sigma\sigma'}$	
Single mode Hamiltonian	$\hat{H}_{\boldsymbol{k},\sigma} = \hbar\omega_k\left(\hat{a}^{\dagger}_{\boldsymbol{k},\sigma}\hat{a}_{\boldsymbol{k},\sigma} + \frac{1}{2}\right)$	$\hat{H}_{\boldsymbol{k},\sigma}\lvert E_{n,\boldsymbol{k},\sigma}\rangle = \hat{H}_{\boldsymbol{k},\sigma}\lvert n_{\boldsymbol{k},\sigma}\rangle = \hbar\omega_k\left(n_{\boldsymbol{k},\sigma} + \frac{1}{2}\right)\lvert E_{n,\boldsymbol{k},\sigma}\rangle$
Vector potential operator	$\hat{\boldsymbol{A}} = \hat{\boldsymbol{A}}_+ + \hat{\boldsymbol{A}}_-$	$\hat{\boldsymbol{A}}_+ = \sum_{\boldsymbol{k},\sigma}\mathcal{A}_k\hat{a}_{\boldsymbol{k},\sigma}\hat{\boldsymbol{\epsilon}}_{\boldsymbol{k},\sigma}e^{i\boldsymbol{k}\cdot\boldsymbol{R}}$ $\hat{\boldsymbol{A}}_- = [\hat{\boldsymbol{A}}_+]^{\dagger} = \sum_{\boldsymbol{k},\sigma}\mathcal{A}_k\hat{a}^{\dagger}_{\boldsymbol{k},\sigma}\hat{\boldsymbol{\epsilon}}^{*}_{\boldsymbol{k},\sigma}e^{-i\boldsymbol{k}\cdot\boldsymbol{R}}$ $\mathcal{A}_k = \sqrt{\frac{\hbar}{2\epsilon_o\omega_k\mathcal{V}}}$
Transverse electric field operator	$\hat{\boldsymbol{E}}_t = \hat{\boldsymbol{E}}_+ + \hat{\boldsymbol{E}}_-$	$\hat{\boldsymbol{E}}_+ = i\sum_{\boldsymbol{k},\sigma}\mathcal{E}_k\hat{a}_{\boldsymbol{k},\sigma}\hat{\boldsymbol{\epsilon}}_{\boldsymbol{k},\sigma}e^{i\boldsymbol{k}\cdot\boldsymbol{R}}$ $\hat{\boldsymbol{E}}_- = [\hat{\boldsymbol{E}}_+]^{\dagger} = -i\sum_{\boldsymbol{k},\sigma}\mathcal{E}_k\hat{a}^{\dagger}_{\boldsymbol{k},\sigma}\hat{\boldsymbol{\epsilon}}^{*}_{\boldsymbol{k},\sigma}e^{-i\boldsymbol{k}\cdot\boldsymbol{R}}$ $\mathcal{E}_k \equiv \sqrt{\frac{\hbar\omega_k}{2\epsilon_o\mathcal{V}}}$
Magnetic field operator	$\hat{\boldsymbol{B}} = \hat{\boldsymbol{B}}_+ + \hat{\boldsymbol{B}}_-$	$\hat{\boldsymbol{B}}_+ = i\sum_{\boldsymbol{k},\sigma}\mathcal{B}_k\hat{a}_{\boldsymbol{k},\sigma}\,\check{\boldsymbol{\kappa}}\times\check{\boldsymbol{\epsilon}}_{\boldsymbol{k},\sigma}e^{i\boldsymbol{k}\cdot\boldsymbol{R}}$; $\hat{\boldsymbol{B}}_- = [\hat{\boldsymbol{B}}_+]^{\dagger} = -i\sum_{\boldsymbol{k},\sigma}\mathcal{B}_k\hat{a}^{\dagger}_{\boldsymbol{k},\sigma}\check{\boldsymbol{\kappa}} \times \check{\boldsymbol{\epsilon}}^{*}_{\boldsymbol{k},\sigma}e^{-i\boldsymbol{k}\cdot\boldsymbol{R}}$; $\mathcal{B}_k \equiv \sqrt{\frac{\hbar\omega_k}{2\epsilon_o c^2\mathcal{V}}}$

Problem 15.2 *For the modes of the electric field propagating along the z-axis, calculate the expectation value for* $\hat{\boldsymbol{E}}_t$ *and* $\hat{\boldsymbol{E}}_t^{\,2}$ *for the vacuum (no photons in any mode).*

As seen in Problem 15.2, while the average of the vacuum field is 0, the average of the square is not 0. And from the discussion above, the energy density is proportional to the modulus squared of the field. Recall that there is a sum over all modes and since the sum goes to infinity, it says that the energy density in the vacuum is infinite. This result remains an issue in our understanding of the basic physics, but it is not hard to imagine the impact. Consider two perfectly conducting plates, spaced a distance *a* apart that is infinite in extent. The standard theory for guided waves shows that there is a cutoff wavelength determined by the spacing of the plates. For wavelengths longer than that cutoff, there is no propagating solution to Maxwell's equations. This means that between two such plates, there is no vacuum field either. But that field can exist on the outside of those plates. Since there is an energy density there, energy density is the same as force per unit area (pressure) and so it is easy to imagine that there is a force exerted by the vacuum on the plates. This force is referred to as the **Casimir–Polder force**. The exact calculation is a little tricky because the physics involves the difference of two infinities. However, it is easy to get close to the answer by calculating the energy density that is excluded because the fields cannot propagate between two plates: $\frac{\pi^2 \hbar c}{8d^4}$. This is clearly an overestimate, for a number of reasons, including the fact that the TH mode has no cutoff. Much more involved and careful calculations show that this energy density overestimates the force by a factor of 30.[4]

15.4 Spontaneous Emission

Having quantized the field, it is now possible to use the understanding of decay of an excited state coupled to a continuum in Chapter 14 to understand why a quantum system in an excited state will decay to the ground state if that transition can couple to the vacuum field, emitting light in the process. This is **spontaneous emission**. As described above, we quantized the field in a cavity of volume V, resulting in discrete modes. However, in the limit that the box becomes infinite, the modes become the continuum of states that we used in Chapter 14 to examine dissipation. We will use the same math in this section to see the emission of light from an excited quantum system. We rely on the understanding developed in Section 9.3 and start with a two-level system with a non-zero optical dipole moment between levels 1 and 2; i.e., $\boldsymbol{\mu}_{12} \equiv \langle 1|\hat{\boldsymbol{\mu}}|2\rangle = \boldsymbol{\mu}_{21}^* \neq 0$ and $\boldsymbol{\mu}_{11} = \boldsymbol{\mu}_{22} = 0$ and $\hat{\boldsymbol{\mu}} = e\boldsymbol{r}$. We also retain the vector nature of the interaction so that we can correctly account for the behavior in real atoms rather than a simple model. The Hamiltonian for the system in the dipole approximation and Schrödinger picture is then:

$$\widehat{H} = \hbar \begin{bmatrix} 0 & 0 \\ 0 & \omega_{21} \end{bmatrix} - \hat{\boldsymbol{\mu}} \cdot \hat{\boldsymbol{E}} + \hbar \sum_{\boldsymbol{k},\sigma} \omega_k \hat{a}^{\dagger}_{\boldsymbol{k},\sigma} \hat{a}_{\boldsymbol{k},\sigma} = \widehat{H}_{0,atom} + \widehat{V}_{int.} + \widehat{H}_{0,rad.} \tag{15.80}$$

Since any operator, $\hat{A}$, can be expressed as $\hat{A} = \sum_{ij} \langle i|\hat{A}|j\rangle \, |i\rangle\langle j|$, where $\{|i\rangle\}$ forms a complete basis set, we can define new **atomic operators** in terms of the outer products. The language "atomic" is simply to imply the electronic system with discrete energy levels, regardless of whether it is an atom or maybe a molecule or a quantum dot. For a two-state system in the Schrödinger picture, we have

$$\begin{aligned} \hat{\sigma}_{12} &\equiv |1\rangle\langle 2| \\ \hat{\sigma}_{21} &\equiv |2\rangle\langle 1| \\ \hat{\sigma}_{22} &\equiv |2\rangle\langle 2| \\ \hat{\sigma}_{11} &\equiv |1\rangle\langle 1| \end{aligned} \tag{15.81}$$

[4] H. G. B. Casimir and D. Polder *Physical Review* 73, p.360 (1948), Peter W. Milonni and Mei-Li Shih, *Contemporary Physics* 33, 313 (1992).

These operators will be discussed more completely in Chapter 16, where we examine the solutions to their equation of motion in the Heisenberg picture. Their behavior is determined by the orthonormality relations of the basis vectors. For example,

$$\hat{\sigma}_{22}|2\rangle \equiv |2\rangle\langle 2|2\rangle = |2\rangle, \hat{\sigma}_{22}|1\rangle \equiv |2\rangle\langle 2|1\rangle = 0, \hat{\sigma}_{12}|2\rangle \equiv |1\rangle\langle 2|2\rangle = |1\rangle \tag{15.82}$$

and so on. This can obviously be generalized to more complicated systems. The various terms in the Hamiltonian can now be written as

$$\hat{H}_{0,atom} = \hbar \begin{bmatrix} 0 & 0 \\ 0 & \omega_{21} \end{bmatrix} = \hbar\omega_{21}|2\rangle\langle 2| = \hbar\omega_{21}\,\hat{\sigma}_{22} \tag{15.83}$$

$$\hat{V}_{int.} = -\hat{\boldsymbol{\mu}} \cdot \hat{\boldsymbol{E}} = -\left(\boldsymbol{\mu}_{12} \cdot \hat{\boldsymbol{E}}\right)|1\rangle\langle 2| - \left(\boldsymbol{\mu}_{21} \cdot \hat{\boldsymbol{E}}\right)|2\rangle\langle 1| = -\left(\boldsymbol{\mu}_{12} \cdot \hat{\boldsymbol{E}}\right)\hat{\sigma}_{12} - \left(\boldsymbol{\mu}_{21} \cdot \hat{\boldsymbol{E}}\right)\hat{\sigma}_{21} \tag{15.84}$$

where from Eq. 15.78 for the quantized field (the subscript t on E is suppressed):

$$\hat{\boldsymbol{E}} = \hat{\boldsymbol{E}}_+ + \hat{\boldsymbol{E}}_- = i\sum_{\boldsymbol{k},\sigma} \mathcal{E}_k \hat{a}_{\boldsymbol{k},\sigma} \check{\boldsymbol{e}}_{\boldsymbol{k},\sigma} e^{i\boldsymbol{k}\cdot\boldsymbol{R}} - i\sum_{\boldsymbol{k},\sigma} \mathcal{E}_k \hat{a}^{\dagger}_{\boldsymbol{k},\sigma} \check{\boldsymbol{e}}^{*}_{\boldsymbol{k},\sigma} e^{-i\boldsymbol{k}\cdot\boldsymbol{R}} \tag{15.85}$$

and

$$\mathcal{E}_{\mathrm{k}} \equiv \sqrt{\frac{\hbar\omega_{\mathrm{k}}}{2\epsilon_{\mathrm{o}}\mathrm{V}}}.$$

We could move forward in the Schrödinger picture, but to reduce the algebra, we move to the interaction picture, referring to Chapter 9. Using

$$\hat{H}_0 = \hat{H}_{0,atom} + \hat{H}_{0,rad} \tag{15.86}$$

we recall fromChapter 9 for the interaction picture, that we defined an operator $\hat{U}(t) = e^{i\frac{\hat{H}_0 t}{\hbar}}$ then defined the state vector in the interaction picture as $|\psi\rangle_I = \hat{U}(t)|\psi\rangle_S$, where $|\psi\rangle_S$ is the state vector in the Schrödinger picture. The equation of motion for an operator in the interaction picture is

$$\frac{d}{dt}\hat{O}_I(t) = \frac{1}{i\hbar}\left[\hat{O}_I(t), \hat{H}_o\right] \tag{15.87}$$

with the initial condition that $\hat{O}_I(t = 0) = \hat{O}_S$ where $\hat{O}_S$ is the Schrödinger operator. For the operators $\hat{\sigma}_{12}(t)$, $\hat{\sigma}_{21}(t)$, $\hat{a}_{\boldsymbol{k},\sigma}(t)$, and $\hat{a}^{\dagger}_{\boldsymbol{k},\sigma}(t)$, the equation of motion for these operators is (the I subscript is suppressed but the explicit time dependence is shown).

$$\frac{d}{dt}\hat{\sigma}_{12}(t) = -i\omega_{21}\left[\hat{\sigma}_{12}(t), \hat{\sigma}_{22}(t)\right] = -i\omega_{21}\hat{\sigma}_{12}(t) \tag{15.88}$$

$$\frac{d}{dt}\hat{\sigma}_{21}(t) = -i\omega_{21}\left[\hat{\sigma}_{21}(t), \hat{\sigma}_{22}(t)\right] = i\omega_{21}\hat{\sigma}_{21}(t) \tag{15.89}$$

$$\frac{d}{dt}\hat{a}_{\mathbf{k},\sigma}(t) = -i\left[\hat{a}_{\mathbf{k},\sigma}(t), \left(\sum_{k',\sigma'} \omega_{k'}\hat{a}^{\dagger}_{k',\sigma'}\hat{a}_{k',\sigma'}\right)\right] = -i\omega_k\hat{a}_{\mathbf{k},\sigma}(t) \tag{15.90}$$

$$\frac{d}{dt}\hat{a}^{\dagger}_{\mathbf{k},\sigma}(t) = -i\left[\hat{a}^{\dagger}_{\mathbf{k},\sigma}, \left(\sum_{k',\sigma'} \omega_{k'}\hat{a}^{\dagger}_{k',\sigma'}\hat{a}_{k',\sigma'}\right)\right] = i\omega_k\hat{a}_{\mathbf{k},\sigma}(t) \tag{15.91}$$

These equations are not coupled, and so are easily integrated to give the interaction picture form of the operators

$$\hat{\sigma}_{12}(t) = \hat{\sigma}_{12}(0)e^{-i\omega_{21}t} \tag{15.92}$$

$$\hat{\sigma}_{21}(t) = \hat{\sigma}_{21}(0)e^{i\omega_{21}t} \tag{15.93}$$

$$\hat{a}_{\mathbf{k},\sigma}(t) = \hat{a}_{\mathbf{k},\sigma}(0)e^{-i\omega_k t} \tag{15.94}$$

$$\hat{a}^{\dagger}_{\mathbf{k},\sigma}(t) = \hat{a}^{\dagger}_{\mathbf{k},\sigma}(0)e^{i\omega_k t} \tag{15.95}$$

The corresponding Hamiltonian is then written as

$$\hat{V}_{I,int.}(t) = -\left(i\sum_{\mathbf{k},\sigma}\mathcal{E}_{k,\sigma}\hat{a}_{\mathbf{k},\sigma}(0)\left(\boldsymbol{\mu}_{12}\cdot\check{\epsilon}_{\sigma}\right)e^{-i\omega_k t}e^{i\mathbf{k}\cdot\mathbf{R}} - i\sum_{\mathbf{k},\sigma}\mathcal{E}_{k,\sigma}\hat{a}^{\dagger}_{\mathbf{k},\sigma}(0)e^{i\omega_k t}\left(\boldsymbol{\mu}_{12}\cdot\check{\epsilon}^{*}_{\sigma}\right)e^{-i\mathbf{k}\cdot\mathbf{R}}\right)\times$$
$$\hat{\sigma}_{12}(0)e^{-i\omega_{21}t} - \left(i\sum_{\mathbf{k},\sigma}\mathcal{E}_{k,\sigma}\hat{a}_{\mathbf{k},\sigma}(0)\left(\boldsymbol{\mu}_{21}\cdot\check{\epsilon}_{\sigma}\right)e^{-i\omega_k t}e^{i\mathbf{k}\cdot\mathbf{R}} - i\sum_{\mathbf{k},\sigma}\mathcal{E}_{k,\sigma}\hat{a}^{\dagger}_{\mathbf{k},\sigma}(0)e^{i\omega_k t}\left(\boldsymbol{\mu}_{21}\cdot\check{\epsilon}^{*}_{\sigma}\right)e^{-i\mathbf{k}\cdot\mathbf{R}}\right)\times$$
$$\hat{\sigma}_{21}(0)e^{i\omega_{21}t} \tag{15.96}$$

We are now able to easily see how to make the rotating wave approximation (one of the reasons for moving to the interaction picture) by ignoring terms that vary as $e^{\pm i(\omega_{21}+\omega_k)t}$. The interaction Hamiltonian becomes

$$\hat{V}_{I,int.}(t) = -\left(i\sum_{\mathbf{k},\sigma}\mathcal{E}_{k,\sigma}\hat{a}_{\mathbf{k},\sigma}(0)\left(\boldsymbol{\mu}_{21}\cdot\check{\epsilon}_{\mathbf{k},\sigma}\right)e^{-i(\omega_k-\omega_{21})t}e^{i\mathbf{k}\cdot\mathbf{R}}\right)\hat{\sigma}_{21}(0)$$
$$+\left(i\sum_{\mathbf{k},\sigma}\mathcal{E}_{k,\sigma}\hat{a}^{\dagger}_{\mathbf{k},\sigma}(0)e^{i(\omega_k-\omega_{21})t}\left(\boldsymbol{\mu}_{12}\cdot\check{\epsilon}^{*}_{\mathbf{k},\sigma}\right)e^{-i\mathbf{k}\cdot\mathbf{R}}\right)\hat{\sigma}_{12}(0) \tag{15.97}$$

If we now define

$$f_{\mathbf{k},\sigma} = \mathcal{E}_{k,\sigma}\left(\boldsymbol{\mu}_{21}\cdot\check{\epsilon}_{\mathbf{k}\sigma}\right)e^{i\mathbf{k}\cdot\mathbf{R}} = \sqrt{\frac{\hbar\omega_k}{2\epsilon_o\mathcal{V}}}\left(\boldsymbol{\mu}_{21}\cdot\check{\epsilon}_{\sigma}\right)e^{i\mathbf{k}\cdot\mathbf{R}} \tag{15.98}$$

we can rewrite Eq. 15.97:

$$\hat{V}_{int.} = -i\sum_{\mathbf{k},\sigma} f_{\mathbf{k},\sigma}e^{-i(\omega_k-\omega_{21})t}\hat{\sigma}_{21}(0)\hat{a}_{\mathbf{k},\sigma}(0) + i\sum_{\mathbf{k},\sigma} f^{*}_{\mathbf{k},\sigma}e^{i(\omega_k-\omega_{21})t}\hat{\sigma}_{12}(0)\hat{a}^{\dagger}_{\mathbf{k},\sigma}(0)1 \tag{15.99}$$

We can now write the state vector in the interaction representation as

$$|\psi\rangle_I = C_{20}(t)|2\rangle|0\rangle + \sum_{\boldsymbol{k},\sigma} C_{1,\boldsymbol{k},\sigma}(t)|1\rangle|\boldsymbol{k},\sigma\rangle \tag{15.100}$$

where $|0\rangle$ means the vacuum state (no photons in any mode) and $|k,\sigma\rangle$ means one photon in the $\boldsymbol{k},\sigma$ mode and no photons in any other mode. Using Schrödinger's equation in the interaction representation, we have

$$\begin{aligned}\frac{d}{dt}C_{20}|2\,\rangle|0\rangle + \sum_{\boldsymbol{k},\sigma}\frac{d}{dt}C_{1,\boldsymbol{k},\sigma}|1\,\rangle|\boldsymbol{k},\sigma\rangle \\ = \hbar^{-1}\left\{-\sum_{\boldsymbol{k},\sigma} f_{\boldsymbol{k},\sigma}e^{-i(\omega_k-\omega_{21})t}\hat{\sigma}_{21}(0)\hat{a}_{\boldsymbol{k},\sigma}(0) + \sum_{\boldsymbol{k},\sigma} f^*_{\boldsymbol{k},\sigma}e^{i(\omega_k-\omega_{21})t}\hat{\sigma}_{12}(0)\hat{a}^\dagger_{\boldsymbol{k},\sigma}(0)\right\}(C_{20}|2\rangle|0\rangle \\ + \sum_{\boldsymbol{k},\sigma} C_{1,\boldsymbol{k},\sigma}|1\rangle|\boldsymbol{k},\sigma\rangle)\end{aligned} \tag{15.101}$$

We now project onto eigenstates $\langle 2|\langle\, 0|$ and $\langle 1|\langle\, \boldsymbol{k}',\sigma'|$ and then return $\langle \boldsymbol{k}',\sigma'|$ to $\langle \boldsymbol{k},\sigma|$ to get the coupled equations of motion given by

$$\frac{d}{dt}C_{1,\boldsymbol{k},\sigma} = \hbar^{-1} f^*_{\boldsymbol{k},\sigma}e^{i(\omega_k-\omega_{21})t}C_{20} \tag{15.102}$$

$$\frac{d}{dt}C_{20} = -\hbar^{-1}\sum_{k,\sigma} f_{\boldsymbol{k},\sigma}e^{-i(\omega_k-\omega_{21})t}C_{1,\boldsymbol{k},\sigma} \tag{15.103}$$

Following the same approach as used in Chapter 14, we integrate the equation for $C_{1,\boldsymbol{k},\sigma}$ and insert the result into the equation of motion for C_{20} giving

$$\frac{d}{dt}C_{20} = -\hbar^{-2}\sum_{\boldsymbol{k},\sigma}\left|f_{\boldsymbol{k},\sigma}\right|^2\int_0^t dt'^{\,e^{-i(\omega_k-\omega_{21})(t-t')}}\,C_{20}\,(t') \tag{15.104}$$

The integral over t' is the same as in Eq. 14.21. Again, this is an integral equation. In the **Weisskopf–Wigner approximation**, we remove $\boldsymbol{C_{20}}\,(t')$ from the integral and evaluate it at $t' = t$.

$$\frac{d}{dt}C_{20} = -\hbar^{-2}C_{20}(t)\sum_{\boldsymbol{k},\sigma}\left|f_{\boldsymbol{k},\sigma}\right|^2\int_0^t dt'^{\,e^{-i(\omega_k-\omega_{21})(t-t')}} = -\frac{\gamma}{2}C_{20}(t) \tag{15.105}$$

where

$$\begin{aligned}\frac{\gamma}{2} &= \frac{1}{\hbar^2}\sum_{\boldsymbol{k},\sigma}\left|f_{\boldsymbol{k},\sigma}\right|^2\int_0^t dt'^{\,e^{-i(\omega_k-\omega_{21})(t-t')}} \\ &= \frac{1}{\hbar^2}\frac{\mathcal{V}}{(2\pi)^3}\int_0^\infty k^2dk\int_{\Omega_{s.a.}} d\Omega_{s.a.}\sum_\sigma\left|f_{\boldsymbol{k},\sigma}\right|^2\int_0^t dt'\,e^{-i(\omega_k-\omega_{21})(t-t')}\end{aligned} \tag{15.106}$$

where we have used the results of Eq. 11.103 for converting the sum over discrete modes in three dimensions to an integral over k-space and γ is the **spontaneous emission rate (the Einstein A-coefficient)**. From Eq. 14.26, the integral over time becomes

$$\int_0^t dt' e^{-i(\omega_k-\omega_{21})(t-t')} \cong \pi\delta\left(\omega_k-\omega_{21}\right) \tag{15.107}$$

Using this, along with Eq. 15.98, we get

$$\begin{aligned}\frac{\gamma}{2} &= \frac{1}{\hbar^2}\frac{\mathcal{V}}{(2\pi)^3}\int_0^\infty k^2dk\int_{\Omega_{s.a.}} d\Omega_{s.a.}\sum_\sigma \left|f_{k,\sigma}\right|^2\frac{\pi}{c}\delta\left(k-\frac{\omega_{21}}{c}\right)\\ &= \frac{1}{16\pi^2\hbar\epsilon_o}\left(\frac{\omega_{21}}{c}\right)^3\int_{solid\ angle} d\Omega_{s.a.}\sum_\sigma \left|\boldsymbol{\mu}_{21}\cdot\check{\boldsymbol{e}}_\sigma\right|^2\end{aligned} \tag{15.108}$$

To evaluate the integral over the solid angle, we take the z-axis to be in the direction of propagation and because the field is transvers, $\hat{\boldsymbol{e}}_\sigma$ is perpendicular to z. We can then write the integral as

$$\begin{aligned}\int_{solid\ angle} d\Omega_{s.a.}\sum_\sigma \left(\boldsymbol{\mu}_{21}\cdot\check{\boldsymbol{e}}_\sigma\right)^2 &= \left|\mu_{21}\right|^2\int_0^{2\pi} d\varphi\int_0^\pi \sin\theta d\theta\sum_\sigma\left(\check{\boldsymbol{r}}\cdot\check{\boldsymbol{e}}_\sigma\right)^2\\ &= 2\pi\left|\mu_{21}\right|^2\int_0^\pi \sin\theta d\theta\sum_\sigma\left(\check{\boldsymbol{r}}\cdot\check{\boldsymbol{e}}_\sigma\right)^2\end{aligned} \tag{15.109}$$

where $\boldsymbol{\mu}_{21} = er_{21}$ and $\check{\boldsymbol{r}}$ is the unit vector in the direction of the induced dipole. The two polarization components represented by $\check{\boldsymbol{e}}_\sigma$ are orthogonal and because the field is transverse, we have that $\hat{\boldsymbol{e}}_\sigma$ is orthogonal to the direction of propagation. The direction of propagation of the emitted radiation can be in any direction, and then polarization will also be normal to that direction. Our result will be independent of the direction of propagation so we arbitrarily define that direction to be the z-axis (that direction will be different every time the system emits a unit of energy): i.e.:

$$\boldsymbol{k}\cdot\check{\boldsymbol{e}}_\sigma = k\check{\boldsymbol{z}}\cdot\check{\boldsymbol{e}}_\sigma = 0 \tag{15.110}$$

Therefore, the two components of $\check{\boldsymbol{e}}_\sigma$ and $\check{\boldsymbol{z}}$ form a complete orthonormal vector basis for describing three-dimensional space. Hence, the unit vector $\check{\boldsymbol{r}}$ can be written as[5]

$$\check{\boldsymbol{r}} = \frac{r}{r} = \sum_\sigma\left(\check{\boldsymbol{e}}_\sigma\cdot\check{\boldsymbol{r}}\right)\check{\boldsymbol{e}}_\sigma + \left(\check{\boldsymbol{z}}\cdot\check{\boldsymbol{r}}\right)\check{\boldsymbol{z}} \tag{15.111}$$

Since a unit vector has unity magnitude, we take the dot product of 15.111 with itself:

$$1 = \check{\boldsymbol{r}}\cdot\check{\boldsymbol{r}} = \sum_\sigma\left(\check{\boldsymbol{e}}_\sigma\cdot\check{\boldsymbol{r}}\right)^2 + \left(\check{\boldsymbol{z}}\cdot\check{\boldsymbol{r}}\right)^2 \tag{15.112}$$

[5] Stig Stenholm, *Foundations of Laser Spectroscopy*, John Wiley & Sons, New York (1984).

Since $(\hat{z} \cdot \hat{r}) = \cos\vartheta$, we solve Eq. 15.109 for $\sum_\sigma (\check{\epsilon}_\sigma \cdot \check{r})^2$

$$\sum_\sigma (\check{\epsilon}_\sigma \cdot \check{r})^2 = \left(1 - \cos^2\theta\right) \tag{15.113}$$

Substituting into the expression for the integral we have

$$\int_0^\pi \sin\theta d\theta \sum_\sigma (\check{r} \cdot \check{\epsilon}_\sigma)^2 = \int_0^\pi \sin\theta d\theta \left(1 - \cos^2\theta\right) = \frac{4}{3} \tag{15.114}$$

giving for $\frac{\gamma}{2}$:

$$\frac{\gamma}{2} = \frac{1}{16\pi^2\hbar\epsilon_o}\left(\frac{\omega_{21}}{c}\right)^3 \int_{solid\ angle} d\Omega_{s.a.} \sum_\sigma \left|\boldsymbol{\mu}_{21} \cdot \check{\epsilon}_\sigma\right|^2 = \frac{2}{3}\frac{\left|\mu_{21}\right|^2}{4\pi\epsilon_o\hbar}\left(\frac{\omega_{21}}{c}\right)^3 \tag{15.115}$$

Then

$$\frac{dC_{20}(t)}{dt} = -\frac{1}{2}\gamma C_{20}(t) \tag{15.116}$$

$$C_{20}(t) = C_{20}(0)e^{-\frac{1}{2}\gamma t} \tag{15.117}$$

The probability of decaying as a function of t is[6]

$$\left|C_{20}(t)\right|^2 = \left|C_{20}(0)\right|^2 e^{-\gamma t} \tag{15.118}$$

Hence, the probability of occupying an excited state that is dipole coupled to a lower state decays exponentially with the emission of a photon. If the photon is detected, the probability of being in the excited state goes to zero as a measurement was made that collapsed the wave function.

15.5 **Effects of the Quantum Vacuum on Linear Absorption and Dispersion**

From Eq. 15.116, we see the impact of the vacuum field on the probability amplitude of an excited state. From here, we can return to the semiclassical Hamiltonian for many problems, and modify the equations of motion for excited state probability amplitudes. To be clear, we do not have a means to include the effects of the *increase* in the probability amplitude of the lower lying state from the decay of an excited state. By conservation of probability, since $\left|C_{20}(t)\right|^2 +$

[6] The mathematical presentation for spontaneous emission is based on one of several "standard" approaches. However, there remain serious mathematical issues with the solution, even though the author is not aware of any measurements showing a problem with the result. A much more comprehensive and coherent discussion and review is given in Paul R. Berman and George W. Ford, *Spontaneous Decay, Unitarity, and the Weisskopf–Wigner Approximation*, Chapter 5: E. Arimondo, P. R. Berman, and C. C. Lin editors, Advances in Atomic, Molecular, and Optical Physics (Volume 59), Academic Press 2010, p. 175. ISBN: 978-0-12-381021-2.

$|C_{10}(t)|^2 = 1$, then $|C_{10}(t)|^2 = 1 - |C_{20}(0)|^2 e^{-\gamma t}$. However, this is harder to do in the amplitude picture because the decay destroys the coherent quantum superposition. This will be dealt with in the next chapter. However, in the limit of first order perturbation theory, assuming that the system is initially in the ground state, we can calculate the absorption and dispersion spectrum associated with the transition.

To see this, we could return to the form of the Hamiltonian in Chapter 9, Eq. 9.51. However, we exploit the knowledge that we have been building to write the Hamiltonian in a slightly more compact form. We use the form from Eqs 15.83 and 15.84 and write in the rotating wave and two-level approximation:

$$\widehat{H} = \hbar\omega_{21}\,\hat{\sigma}_{22} - \frac{\hbar}{2}\Omega_R e^{-i\omega t}\hat{\sigma}_{21} - \frac{\hbar}{2}\Omega_R^* e^{i\omega t}\hat{\sigma}_{12} \tag{15.119}$$

and $\Omega_R = \frac{\mu_{21}\mathcal{E}_0}{\hbar}$. We have again suppressed the vector nature of the $\boldsymbol{\mu}_{21}\cdot\mathcal{E}_0$, since it only impacts the details of which electronic transitions participate. We are assuming that the applied field $\mathcal{E}_0$ is classical. The impact of the quantized field is the decay of the excited state due to the quantum vacuum.

We write the state vector in the usual way in the interaction picture,

$$|\psi(t)\rangle = C_1(t)|1\rangle + C_2(t)e^{-\omega_{21}t}|2\rangle \tag{15.120}$$

and insert this into Schrödinger's equation and get the equations of motion we developed in Eqs 9.87 and 9.88, namely:

$$\dot{C}_1(t) = -\frac{1}{2i}\Omega_R^* e^{i(\omega-\omega_{21})t}C_2(t) \tag{15.121}$$

$$\dot{C}_2(t) = -\frac{1}{2i}\Omega_R e^{-i(\omega-\omega_{21})t}C_1(t) \tag{15.122}$$

However, we can now include the effect of the vacuum radiation field on the probability amplitude of the excited state since the derivation would not have been impacted by including a classical field applied to the system as in Eq. 15.119. Using Eq. 15.116 and suppressing the "0" in the subscript representing the absence of a photon in the vacuum, we rewrite Eq. 15.122 as

$$\dot{C}_2(t) = -\frac{1}{2}\gamma C_2(t) - \frac{1}{2i}\Omega_R e^{-i(\omega-\omega_{21})t}C_1(t) \tag{15.123}$$

In first order perturbation, we will assume that the system is initially in the ground state, and therefore $C_1(0) = C_1^{(0)} = 1$ and $C_2(0) = C_2^{(0)} = 0$. To first order in the classical field, (Ω_R), we get

$$\dot{C}_2^{(1)}(t) = -\frac{1}{2}\gamma C_2{}^{(1)}(t) - \frac{1}{2i}\Omega_R e^{-i(\omega-\omega_{21})t}C_1^{(0)} \tag{15.124}$$

and after introducing an integrating factor, we get

$$C_2{}^{(1)}(t) = -\frac{1}{2i}\Omega_R e^{-\frac{1}{2}\gamma t}C_1^{(0)}\int_0^t dt'\, e^{\frac{1}{2}\gamma t' - i(\omega-\omega_{21})t'} = -\frac{1}{2i}\Omega_R\frac{e^{-i(\omega-\omega_{21})t} - e^{-\frac{1}{2}\gamma t}}{\frac{1}{2}\gamma - i(\omega-\omega_{21})} \tag{15.125}$$

The $C_1^{(0)}$ in the second term is replaced by its value, 1, in the last term. The decay term on the right side of numerator is a short time response from the sudden turn-on of the field. It vanishes for long times, leaving just the long-time response. Inserting that into the expression for the state vector we get

$$|\psi(t)\rangle = C_1^{(0)}|1\rangle + C_2^{(1)}(t)e^{-\omega_{21}t}|2\rangle = |1\rangle - \frac{1}{2i}\Omega_R \frac{e^{-i\omega t}}{-i(\omega-\omega_{21}) + \frac{1}{2}\gamma}|2\rangle \tag{15.126}$$

We see that spontaneous emission can limit the degree of quantum coherence when the detuning becomes smaller than the decay rate. This can be a potentially serious problem in the design of quantum devices that must be considered in any design.

The absorption and dispersion arise from the induced optical polarization in the material. The polarization, P, is the expectation value of the dipole moment operator, $P = \langle\hat{\mu}\rangle$. This give us

$$P = -\frac{\hat{\mu}_{12}}{2}\Omega_R \frac{(\omega-\omega_{21}) - i\frac{\gamma}{2}}{\left(\frac{\gamma}{2}\right)^2 + (\omega-\omega_{21})^2} e^{-i\omega t} + cc \tag{15.127}$$

This expression could now be inserted into the wave equation for the electromagnetic field. Specifically, if Maxwell's equations (Eqs 15.1–15.4) and the wave equation (Eq. 15.6) were written to include the presence of radiators (see Appendix G), the wave equation would be written as

$$\nabla^2 \mathbf{E} - \frac{1}{c^2}\frac{\partial^2 \mathbf{E}}{\partial t^2} = \frac{1}{\epsilon_0 c^2}\frac{\partial^2 \mathbf{P}}{\partial t^2} \tag{15.128}$$

Had we kept the vector nature of the interaction, $\mathbf{P}$, would be a vector pointing in the direction of the induced dipole. Inserting Eq. 15.127 into 15.128 would then show that the absorption is due to the imaginary part of the phasor amplitude and the dispersion is due to the real part. The functional form of the imaginary part of the amplitude is a **Lorentzian**. The full width at half-maximum is γ which is the decay rate of the state probability.

Problem 15.3 *Use perturbation theory and calculate $P(t)$ when the two-level system above is excited by a pulsed field and the pulse is described by a Dirac delta-function. Specifically, the electric field $\mathcal{E}_0$ in the Hamiltonian in Eq. 15.119 is replaced $\mathcal{E} = \mathcal{E}_0\tau\delta(t)$.*

15.6 **Rabi Oscillations in the Quantum Vacuum: Jaynes–Cummings Hamiltonian**

In the previous section, we saw the effect of the quantum vacuum on a two-level system that has a coupling to the vacuum field. When the atom is in the excited state, it can decay to the lower state by spontaneously emitting into the vacuum. The key physics in the Hamiltonian was the fact that the sum over the modes of the vacuum field became an integral with a continuous density of states as the cavity volume became infinite. Here, we re-examine this problem in case the vacuum is limited to a single mode by a finite size cavity. The modes remain discrete, and there is a single mode in the sense that the other modes are assumed to be far off resonance. It is possible to design a single mode cavity, but that is not necessary to see the impact.

The Hamiltonian in the rotating wave approximation for the case of a single two-level system and a *single mode cavity* at frequency ω_c, called the **Jaynes–Cummings Hamiltonian**, is given by

$$\hat{H} = \omega_{21}\hat{\sigma}_2 - \frac{1}{2}\mu_{21}\hat{E}_+\hat{\sigma}_{21} - \frac{1}{2}\mu_{12}\hat{E}_-\hat{\sigma}_{12} + \hbar\omega_c\hat{a}^\dagger\hat{a} \tag{15.129}$$

where again the constant of $\frac{1}{2}\hbar\omega_c$ is ignored. This is the single mode version of Eq. 15.80, using Eqs 15.83 and 15.84. There is no classical field in this problem. The field is quantized and the vector nature of the coupling between the transition moment and the field is again suppressed since there is only one polarization .The difference between this Hamiltonian and the original Hamiltonian is that there is only one mode for the radiation field, and to emphasize this, we note that with the quantum two-level system at the origin the operators take the form (Eq. 15.78):

$$\mathcal{E} \equiv \sqrt{\frac{\hbar\omega_c}{2\epsilon_o\mathcal{V}}}; \hat{E}_+ = i\mathcal{E}\hat{a}; \hat{E}_- = -i\mathcal{E}\hat{a}^\dagger \tag{15.130}$$

Because of the nature of the Hamiltonian, the state vector can conveniently be written as

$$|\psi\rangle = \sum_{n=0} \left(C_{1;n}(t)e^{-in\omega_c t}|1;n\rangle + C_{2;n-1}(t)e^{-i(\omega_{21}+(n-1)\omega_c)t}|2;n-1\rangle\right) \tag{15.131}$$

The expression is grouped in terms of $|1;n\rangle$ and $|2;n-1\rangle$ because these two states are degenerate on resonance in the absence of any coupling term.

Substituting into Schrödinger'sequation:

$$\begin{aligned}
&\sum_{n=0} \left(\left\{\dot{C}_{1;n}(t) - in\omega_c C_{1;n}(t)\right\} e^{-in\omega_c t}|1;n\rangle\right. \\
&\quad \left. + \left\{\dot{C}_{2;n-1}(t) - i\left(\omega_{21} + (n-1)\,\omega_c\right) C_{2;n-1}(t)\right\} e^{-i(\omega_{21}+(n-1)\omega_c)t}|2;n-1\rangle\right) \\
&= -i\sum_{n=0} \left(\left(n\omega_c - \frac{1}{2\hbar}\mu_{21}\hat{E}_+\hat{\sigma}_{21}\right) C_{1;n}(t)e^{-in\omega_c t}|1;n\rangle\right. \\
&\quad + \left(\omega_{21} + (n-1)\,\omega_c - \frac{1}{2\hbar}\mu_{12}\hat{E}_-\hat{\sigma}_{12}\right) C_{2;n-1}(t)e^{-i(\omega_{21}+(n-1)\omega_c)t}|2;n-1\rangle
\end{aligned} \tag{15.132}$$

Using Eq. 15.130, operating on the electronic and photonic states and canceling terms

$$\begin{aligned}
&\sum_{n=0} \left(\dot{C}_{1;n}(t)e^{-in\omega_c t}|1;n\rangle + \dot{C}_{2;n-1}(t)e^{-i(\omega_{21}+(n-1)\omega_c)t}|2;n-1\rangle\right) \\
&= \sum_{n=0} \left(-\frac{1}{2\hbar}\mu_{21}\left(\mathcal{E}\sqrt{n}\right) C_{1;n}(t)e^{-in\omega_c t}|2;n-1\rangle\right. \\
&\quad \left. + \frac{1}{2\hbar}\mu_{12}\left(\mathcal{E}\sqrt{n}\right) C_{2;n-1}(t)e^{-i(\omega_{21}+(n-1)\omega_c)t}|1;n\rangle\right)
\end{aligned} \tag{15.133}$$

Now project onto $\langle 1, m|$ and replace m with n

$$\dot{C}_{1;n}(t) = \frac{1}{2\hbar}\mu_{12}\mathcal{E}\sqrt{n}C_{2;n-1}(t)e^{-i(\omega_{21}-\omega_c)t} \tag{15.134}$$

and project onto $\langle 2, m-1|$ and replace m with n

$$\dot{C}_{2;n-1}(t) = -\frac{1}{2\hbar}\mu_{21}\mathcal{E}\sqrt{n}C_{1;n}(t)e^{i(\omega_{21}-\omega_c)t} \tag{15.135}$$

Setting

$$\Omega_R = \frac{\mu_{21}\mathcal{E}}{\hbar}; \Omega_R^* = \frac{\mu_{12}\mathcal{E}}{\hbar}; \delta = \omega_{21} - \omega_c; C_{1;n}(t) = \tilde{C}_{1;n}(t)e^{-i\frac{\delta}{2}t}; C_{2;n-1}(t) = \tilde{C}_{2;n-1}(t)e^{i\frac{\delta}{2}t}$$

Then, in the **field interaction picture**

$$\frac{d\tilde{C}_{1;n}}{dt} = +i\frac{\delta}{2}\tilde{C}_{1;n}(t) + \frac{1}{2}\,\Omega_R^*\sqrt{n}\tilde{C}_{2;n-1}(t) \tag{15.136}$$

and

$$\frac{d\tilde{C}_{2;n-1}}{dt} = -i\frac{\delta}{2}\tilde{C}_{2;n-1}(t) - \frac{1}{2}\Omega_R\sqrt{n}\tilde{C}_{1;n}(t) \tag{15.137}$$

or

$$\frac{d}{dt}\begin{bmatrix} \tilde{C}_{1;n} \\ \tilde{C}_{2;n-1} \end{bmatrix} = \begin{bmatrix} +i\frac{\delta}{2} & \frac{1}{2}\Omega_R^*\sqrt{n} \\ -\frac{1}{2}\Omega_R\sqrt{n} & -i\frac{\delta}{2} \end{bmatrix}\begin{bmatrix} \tilde{C}_{1;n} \\ \tilde{C}_{2;n-1} \end{bmatrix} \tag{15.138}$$

Now find the eigenvalues (see Appendix A). Let

$$\begin{bmatrix} C_{1;n} \\ \tilde{C}_{2;n-1} \end{bmatrix} \sim e^{i\varpi t} \tag{15.139}$$

Substitute into 15.138 giving the usual eigenvalue equation. Eigenvalues are found by solving for ϖ

$$\det\begin{bmatrix} +i\frac{\delta}{2} - i\varpi & \frac{1}{2}\Omega_R^*\sqrt{n} \\ -\frac{1}{2}\Omega_R\sqrt{n} & -i\frac{\delta}{2} - i\varpi \end{bmatrix} = 0 \tag{15.140}$$

$$\varpi = \pm\frac{1}{2}\sqrt{\left(\delta^2 + |\Omega_R|^2 n\right)} \tag{15.141}$$

Therefore, the solution has the form

$$\tilde{C}_{2;n-1} = ae^{i\varpi t} + be^{-i\varpi t} \tag{15.142}$$

and

$$\tilde{C}_{1;n}(t) = \frac{2}{\Omega_R\sqrt{n}}\left(+i\left\{\varpi + \frac{\delta}{2}\right\}ae^{i\varpi t} - i\left\{\varpi - \frac{\delta}{2}\right\}e^{-i\varpi t}b\right) \tag{15.143}$$

The unknowns, a and b, are determined by the initial conditions.Assume that at $t = 0$, an atom in the excited state enters the cavity with no photons in the cavity and for simplicity the atomic transition frequency is on resonance with the cavity frequency, so that $\delta = 0$ and therefore $\varpi = \frac{1}{2}|\Omega_R|$. Since the effect is only a phase, we assume Ω_R is real.Then with the initial conditions given as $\tilde{c}_{2;0}\,(t = 0) = 1$ and $\tilde{c}_{1;1}\,(t = 0) = 0$, we get

$$a = b = \frac{1}{2}$$

giving

$$\tilde{c}_{2;0} = \cos\frac{\Omega_R}{2}t \text{ and } \left|\tilde{c}_{2;0}\right|^2 = \frac{1}{2}\left(1 + \cos\Omega_R t\right) \tag{15.144}$$

The system will oscillate at the **vacuum Rabi** frequency with no applied field (i.e., n = 0). The idea of switching between states with a single or even a few photons remains a major technological goal that is becoming increasingly likely as advances in nano-structures and photonic bandgap material develop.[7]

15.7 Summary

This section shows that by converting the usual expression for the total energy in an electromagnetic transverse field to a form based on the vector potential, it is then possible to find the Hamiltonian written in terms of the canonical coordinate and corresponding conjugate momentum for the field. This was the requirement in order to use the last postulate in Chapter 7 regarding the non-commutation of the canonical coordinate and conjugate momentum to quantize the field. Amazingly, the Hamiltonian has exactly the same operator form as the nanovibrator and the LC circuit, i.e., for mode k, σ,

$$H_{\mathbf{k},\sigma} = \hbar\omega_k\left(\hat{a}^{\dagger}_{\mathbf{k},\sigma}\hat{a}_{\mathbf{k},\sigma} + \frac{1}{2}\right)$$

with the eigenstates given by

$$\left|\boldsymbol{E}_{n,\mathbf{k},\sigma}\right\rangle = \left|n_{\mathbf{k},\sigma}\right\rangle$$

[7] See D.A.B. Miller, "Attojoule Optoelectronics for Low-Energy Information Processing and Communications – a Tutorial Review" arXiv:1609.05510 [physics.optics] v3 1 January 2017. Eden Rephaeli, Jung-Tsung Shen, and Shanhui Fan, "Full inversion of a two-level atom with a single-photon pulse in one-dimensional geometries," Phys. Rev. A 82, 033804 (2010).

and

$$\hat{a}^{\dagger}_{k,\sigma}\hat{a}_{k,\sigma}|n_{k,\sigma}\rangle = n_{k,\sigma}|n_{k,\sigma}\rangle$$

The annihilation operator, $\hat{a}_{k,\sigma}$, and the creation operator, $\hat{a}^{\dagger}_{k,\sigma}$, are not Hermitian operators associated with an observer, but they destroy and create single units of energy, $\hbar\omega_k$ sometimes called a photon, from the $\boldsymbol{k}, \sigma$ mode:

$$\hat{a}_{k,\sigma}|n_{k,\sigma}\rangle = \sqrt{n_{k,\sigma}}|n_{k,\sigma} - 1\rangle$$

$$\hat{a}^{\dagger}_{k,\sigma}|n_{k,\sigma}\rangle = \sqrt{n_{k,\sigma} + 1}|n_{k,\sigma} + 1\rangle$$

To many, one of the most dramatic successes of this theory is the correct description of spontaneous emission of radiation from excited quantum states involving charge and a dipole. In the era of nano-technology, however, this theory is becoming increasingly important for the successful design of quantum devices that involve coupling to the radiation field. Often these devices are controlled by a classical electromagnetic field and the primary issue is then spontaneous emission that can destroy the coherence as discussed above and further in subsequent chapters. However, the discussion of the Jaynes–Cummings Hamiltonian shows the possibility of controlling a quantum system with a single photon when the system is engineered to suppress the other modes.

Vocabulary (page) and Important Concepts

- Hamiltonian for the quantized electromagnetic field 257
- number state 257
- Fock state 257
- occupation number 257
- photon 258
- product state 258
- vacuum field 258
- Casimir–Polder Force 261
- spontaneous emission 261
- Atomic operators 261
- Weisskopf-Wigner approximation 264
- spontaneous emission rate 265
- Einstein A-coefficient 265
- Lorentzian 268
- Jaynes–Cummings Hamiltonian 269
- field interaction picture 270
- vacuum Rabi oscillations 271

16 Two-State Systems: The Atomic Operators

16.1 Introduction

The simple problem in engineering of "off and on" represents two states. Whether it represents a bit of information for a 0 and a 1 or the state of a valve, the two states represent stable states of the system and, furthermore, to change from one state to another takes energy. A transitioning from one state to the other requires a given amount of energy, making the relationship between the state of the system and applied energy. Without enough energy nothing happens, and with enough energy it switches, but using more energy does not change anything. This makes the system nonlinear in the driving energy. This was explored in Chapter 9 in the study of Rabi oscillations. In that case we examined a quantum system with many levels (like an atom or an electron trapped in a defect) and then by using the effect of resonance where the frequency of an AC electric field matches the energy difference between the two levels divided by $\hbar$, $\frac{E_{upper}-E_{lower}}{\hbar} > 0$, to restrict the coupling to two states, we made the two-level approximation. In the case of spin ½ systems like the electron, the system is intrinsically just two states. The spin states are usually degenerate unless split by a DC magnetic field, where again they can be switched back and forth by an AC magnetic field.

Because of the importance of the generic two-level system in many potential engineering applications, we examine this problem in the context of the Heisenberg picture. The exercise of creating the appropriate operators is useful, since it can be easily generalized. Furthermore, as we saw in the RLC problem, using operators and the Heisenberg picture facilitated the calculation of problems with time dependent Hamiltonians and enabled including the effects of damping in a relatively straightforward approach, in contrast to the Schrödinger picture. Though the symbols defining the operators use $\hat{\sigma}$, they should not be confused with the Pauli operators sometimes seen in theory of spin, though they are similar. The operators are called the **atomic operators**[1] and are used with any two-level system, not just atoms. They were first introduced in Chapter 15 in Eqs 15.81 and 15.82. In that case, we simply used them to represent the Hamiltonian with operators whose operations corresponded to what we saw in the earlier work and we worked with them in an interaction representation.

Here we describe the complete set of atomic operators and then examine the general problem of dynamics in the Heisenberg representation of these operators, including the effects of the classical and quantum electromagnetic fields. One feature of this approach is that you only

[1] A much more comprehensive presentation with more thoughtful and detailed discussion is to be had in Paul R. Berman and Vladimir S. Malinovsky, *Principle of Laser Spectroscopy and Quantum Optics*, Princeton University Press, Princeton (2011).

Introduction to Quantum Nanotechnology: A Problem Focused Approach. Duncan Steel, Oxford University Press (2021).
 DOI: 10.1093/oso/9780192895073.003.0016

need to know the initial state of the system, since the dynamics is included in the Heisenberg operators. These are particularly powerful equations for understanding the role of quantum fluctuations from the bath (e.g., the vacuum field) in devices. But if you need to understand the final state of the system, sometimes the density matrix (Chapter 18) is more appropriate. In both cases it is possible to include the role of spontaneous emission on the emitting state and the final state. They can both be generalized to an arbitrary number of levels or even a continuum.

16.2 Defining the Atomic Operators

We start by considering the Hamiltonian for the two-level system on the left in Fig. 16.1 and developing the relevant atomic operators. In the problems, you will develop a corresponding set of operators for the three-level Λ system on the right in Fig. 16.1.

We begin with $\hat{H}_0$ for the two-level system. We work exclusively in Dirac notation and, where necessary, representing operators as $\hat{A} = \sum_{ij} \langle i|\hat{A}|j\rangle\, |i\rangle\langle j|$. The only assumptions are (1) the system has two energy levels E_2 and E_1, where for convenience we take $E_2 \geq E_1$; (2) the eigenstates defined by a Hamiltonian, $\hat{H}_0$, for the two levels, are $|2\rangle$ and $|1\rangle$; and (3) there is an interaction potential, $\hat{V}$, in the total Hamiltonian, $\hat{H}$, that couples the two states. In the semiclassical limit, it varies sinusoidally in time. The interaction term could also be a quantized field. Two examples of interest for $\hat{V}$, involving an oscillating magnetic field are a spin ½ electron or a nucleon such as a proton (the latter is key for magnetic resonance imaging) sitting in a static magnetic field. Three electric field examples are a bound state of a positive and negative charge such as in an atom, a defect in a crystal such as the NV center in diamond, or a hole and electron (an exciton) in a semiconductor quantum dot.

For the two-level Hamiltonian, in the absence of the interaction term, we have $\hat{H}_0|2\rangle = E_2|2\rangle$ and $\hat{H}_0|1\rangle = E_1|1\rangle$. Using $\langle i|j\rangle = \delta_{ij}$ we can now write the Hamiltonian as:

$$\hat{H}_0 = \sum_{ij} \langle i|\hat{H}_0|j\rangle |i\rangle\langle j| = E_1|1\rangle\langle 1| + E_2|2\rangle\langle 2| \tag{16.1}$$

We now define two atomic operators called **state operators**:

$$\hat{\sigma}_1 \equiv |1\rangle\langle 1| \tag{16.2}$$

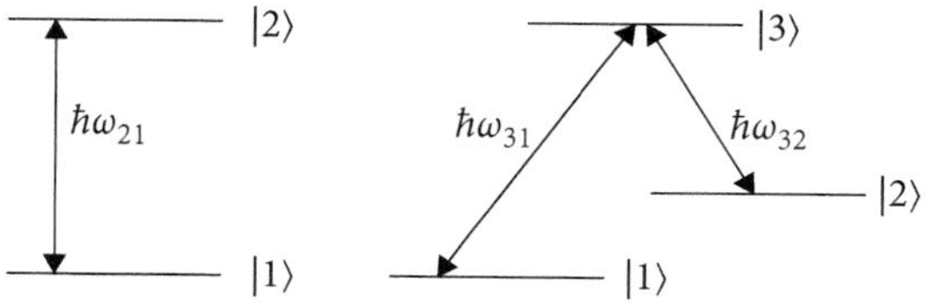

Fig. 16.1 Energy level scheme for a 2-level system (left) and a Λ 3-level system (right). The arrows represent which states are coupled by the interaction term in the Hamiltonian.

$$\hat{\sigma}_2 \equiv |2\rangle\langle 2| \tag{16.3}$$

so that

$$\hat{H}_0 = E_1\hat{\sigma}_1 + E_2\hat{\sigma}_2 \tag{16.4}$$

Then

$$\hat{H}_0|i\rangle = (E_1\hat{\sigma}_1 + E_2\hat{\sigma}_2)|i\rangle = E_i|i\rangle; i = 1,2 \tag{16.5}$$

If we set the zero point for the energy (an arbitrary assignment for such an isolated system) midway between E_2 and E_1 and define $\hbar\omega_{21} = E_2 - E_1$, then

$$\hat{H}_0 = \frac{\hbar\omega_{21}}{2}(\hat{\sigma}_2 - \hat{\sigma}_1) \equiv \frac{1}{2}\hbar\omega_{21}\hat{\sigma}_z \tag{16.6}$$

where we define $\hat{\sigma}_z$ now as[2]

$$\hat{\sigma}_z \equiv |2\rangle\langle 2| - |1\rangle\langle 1| = \hat{\sigma}_2 - \hat{\sigma}_1 \tag{16.7}$$

Had we assumed the lower level $E_1 = 0$, then

$$\hat{H}_0 = \hbar\omega_{21}|2\rangle\langle 2| = \hbar\omega_{21}\hat{\sigma}_2 \tag{16.8}$$

Either would be fine.

We also note that because the basis set is complete, then $\sum_i |i\rangle\langle i| = \hat{I}$ where $\hat{I}$ is the identity operator. This is easily shown by again expanding the operator, this time $\hat{I}$, according to the prescription above $\hat{A} = \sum_{ij}\langle i|\hat{A}|j\rangle|i\rangle\langle j|$. For $\hat{I}$, since by definition $\hat{I}|i\rangle = |i\rangle$ we get:

$$\hat{I} = \sum_{ij=1,2}\langle i|\hat{I}|j\rangle\,|i\rangle\langle j| = |1\rangle\langle 1| + |2\rangle\langle 2| = \hat{\sigma}_1 + \hat{\sigma}_2 \tag{16.9}$$

From the work in Chapter 9 (see for example Eq. 9.79), the interaction term for an atomic-like system with electrons, $\hat{V} = -\hat{\boldsymbol{\mu}} \cdot \boldsymbol{E}(\boldsymbol{R}, t)$ in the Hamiltonian discussed above, is assumed to be 0 along the diagonal because the electromagnetic interaction term has odd parity and it is assumed that parity is a good quantum number for the electronic system. This means the off-diagonal elements are non-zero and $\langle 1|\hat{V}|2\rangle = \langle 2|\hat{V}|1\rangle^*$ because the operator is Hermitian.

In the case of the two-level approximation, we assume that the applied classical field can couple only these two levels, even though there may be an infinite number of states, giving rise to two off-diagonal terms. Recall that the two-level approximation is justified because the other

[2] Typically, the first column is state 1, the second column is state 2, etc. Hence, in the earlier notation using the Pauli spin matrices, there was a minus sign in front of the $\hat{\sigma}_z$ operator when used in a quantum problem. However, for spin, the first column represents spin up, which is a higher energy state, and then spin down for the second column. To be consistent with spin, then, the labeling of the spin operators is that column 1 is state $|2\rangle$ and column 2 is state $|1\rangle$. So no minus sign is needed in front of the $\hat{\sigma}_z$.

coupling terms are much smaller or because the other transitions are far off resonance with the oscillating frequency of the applied field.

To write the interaction term in the Hamiltonian in terms of **two-state operators**, we recall that the dipole moment operator (or transition moment operator), $\hat{\mu} = q\boldsymbol{r} = qr\check{\boldsymbol{r}}$, where $\check{\boldsymbol{r}}$ is the dimensionless unit vector: $\check{\boldsymbol{r}} = \frac{\boldsymbol{r}}{r}$. Using the two basis states and writing $\hat{\mu}$ as a matrix, we get

$$\hat{\mu} = \sum_{ij} \langle i|\hat{\mu}|j\rangle\,|i\rangle\langle j| = \mu_{12}\check{\boldsymbol{r}}_{12}|1\rangle\langle 2| + \mu_{21}\check{\boldsymbol{r}}_{21}|2\rangle\langle 1| = \mu_{12}\check{\boldsymbol{r}}_{12}\hat{\sigma}_{12} + \mu_{21}\check{\boldsymbol{r}}_{21}\hat{\sigma}_{21} \tag{16.10}$$

where $\hat{\sigma}_{12} = |1\rangle\langle 2|$ and $\hat{\sigma}_{21} = |2\rangle\langle 1|$ are the lowering and raising operator, respectively.

The interaction term in the Hamiltonian for a classical field, $\widehat{V} = -\hat{\mu}\cdot\boldsymbol{E}(\boldsymbol{R},t)$ can then be written in the rotating wave approximation as (Chapter 9, Eqs 9.86–9.88)

$$\widehat{V} = -\hat{\mu}\cdot\boldsymbol{E}(\boldsymbol{R},t) = -\frac{(\check{\boldsymbol{r}}_{21}\cdot\check{\varepsilon})}{2}\mu_{21}\mathcal{E}_0(\boldsymbol{R},t)\,e^{-i\omega t}|2\rangle\langle 1| - \frac{(\check{\boldsymbol{r}}_{12}\cdot\hat{\varepsilon})}{2}\mu_{12}\mathcal{E}_0^*(\boldsymbol{R},t)\,e^{i\omega t}|1\rangle\langle 2| \tag{16.11}$$

or

$$\widehat{V} = -\frac{\hbar}{2}\Omega_R(\boldsymbol{R},t)\,e^{-i\omega t}\hat{\sigma}_{21} - \frac{\hbar}{2}\Omega_R^*(\boldsymbol{R},t)\,e^{i\omega t}\hat{\sigma}_{12} \tag{16.12}$$

where the Rabi frequency is again given by $\Omega_R(\boldsymbol{R},t) = \frac{\mu_{21}\mathcal{E}_0(\boldsymbol{R},t)}{\hbar}(\check{\boldsymbol{r}}_{21}\cdot\check{\varepsilon})$ and its complex conjugate $\Omega_R^*(\boldsymbol{R},t) = \frac{\mu_{12}\mathcal{E}_0^*(\boldsymbol{R},t)}{\hbar}(\check{\boldsymbol{r}}_{12}\cdot\check{\varepsilon})$. Then $\mathcal{E}_0(\boldsymbol{R},t)$ is evaluated at the quantum system and, here, usually $\boldsymbol{R} = 0$, and $\check{\varepsilon}$ is a unit vector for the direction of the field polarization. Also, the time dependence in $\mathcal{E}_0(\boldsymbol{R},t)$ is slow compared to the oscillation period, $\frac{2\pi}{\omega}$, and $\Omega_R(\boldsymbol{R},t)$ is small compared to relevant energy level separation, Typically $\mathcal{E}_0$ is constant as a function of time, meaning monochromatic, or it is a short pulse, where the pulse width is long compared to $\frac{2\pi}{\omega}$ but often short compared to any other time scale in the system to create quantum coherence that can exist in the absence of excitation. Sometimes the pulse is square, as we assumed in Chapter 9.

$\hat{\sigma}_{21}$ and $\hat{\sigma}_{12}$ are called the raising and lowing operator because of the following properties:

$$\hat{\sigma}_{21}|1\rangle = |2\rangle\langle 1|1\rangle = |2\rangle \tag{16.13}$$

$$\hat{\sigma}_{21}|2\rangle = |2\rangle\langle 1|2\rangle = 0 \tag{16.14}$$

$$\hat{\sigma}_{12}|2\rangle = |1\rangle\langle 2|2\rangle = |1\rangle \tag{16.15}$$

$$\hat{\sigma}_{12}|1\rangle = |1\rangle\langle 2|1\rangle = 0 \tag{16.16}$$

Hence, these symbols are often replaced, in a two-level system, with **raising and lowering operator** notation:

$$\hat{\sigma}_{21} \equiv \hat{\sigma}_+ \tag{16.17}$$

$$\hat{\sigma}_{12} \equiv \hat{\sigma}_- \tag{16.18}$$

The definition of the new symbols for operators, along with the operators for the quantized radiation field, are summarized in Table 16.1 as well as their *equal time* commutation relations[3] which are immediately derivable from the definition of the operators and the orthonormality of the basis states.

Table 16.1 Summary of atomic and photonic operator definitions and related commutation relations. The subscript on the photonic operators designates the corresponding mode and polarization. The relations hold for the operators in the Schrödinger picture or at the same time in the Heisenberg picture.

$\hat{\sigma}_1 = \|1\rangle\langle 1\| = \hat{\sigma}_-\hat{\sigma}_+$	$[\hat{\sigma}_1, \hat{\sigma}_2] = 0$
$\hat{\sigma}_2 = \|2\rangle\langle 2\| = \hat{\sigma}_+\hat{\sigma}_-$	$[\hat{\sigma}_1, \hat{\sigma}_z] = 0$
$\hat{\sigma}_1\|1\rangle = \|1\rangle, \hat{\sigma}_1\|2\rangle = 0$	$[\hat{\sigma}_2, \hat{\sigma}_z] = 0$
$\hat{\sigma}_2\|2\rangle = \|2\rangle, \hat{\sigma}_2\|1\rangle = 0$	$[\hat{\sigma}_1, \hat{\sigma}_+] = -\hat{\sigma}_+$
$\hat{\sigma}_z = \|2\rangle\langle 2\| - \|1\rangle\langle 1\| = \hat{\sigma}_2 - \hat{\sigma}_1$	$[\hat{\sigma}_2, \hat{\sigma}_+] = \hat{\sigma}_+$
$\hat{\sigma}_- = \|1\rangle\langle 2\|$	$[\hat{\sigma}_1, \hat{\sigma}_-] = \hat{\sigma}_-$
$\hat{\sigma}_-\|2\rangle = \|1\rangle\langle 2\|2\rangle = \|1\rangle,$	$[\hat{\sigma}_2, \hat{\sigma}_-] = -\hat{\sigma}_-$
$\hat{\sigma}_-\|1\rangle = \|1\rangle\langle 2\|1\rangle = 0$	$[\hat{\sigma}_z, \hat{\sigma}_+] = 2\hat{\sigma}_+$
$\hat{\sigma}_+ = \|2\rangle\langle 1\|$	$[\hat{\sigma}_z, \hat{\sigma}_-] = -2\hat{\sigma}_-$
$\hat{\sigma}_+\|1\rangle = \|2\rangle\langle 1\|1\rangle = \|2\rangle$	$[\hat{\sigma}_+, \hat{\sigma}_-] = \hat{\sigma}_z$
$\hat{\sigma}_+\|2\rangle = \|1\rangle\langle 1\|2\rangle = 0$	$\left[\hat{a}_r, \hat{a}_s^\dagger\right] = \delta_{r,s}$
$\hat{I} = \|1\rangle\langle 1\| + \|2\rangle\langle 2\| = \hat{\sigma}_2 - \hat{\sigma}_1$	$[\hat{a}_r, \hat{a}_s] = \left[\hat{a}_r^\dagger, \hat{a}_s^\dagger\right] = 0$
$\hat{a}_r\|n_r\rangle = \sqrt{n_r\rangle}\|n_r - 1\rangle$	$\left[\hat{a}_s^\dagger\hat{a}_s, \hat{a}_s\right] = -\hat{a}_s$
$\hat{a}_r\|n_r\rangle = \sqrt{n_r}\|n_r - 1\rangle$	$\left[\hat{a}_s^\dagger\hat{a}_s, \hat{a}_s^\dagger\right] = \hat{a}_s^\dagger$
$\hat{a}_r^\dagger\|n_r\rangle = \sqrt{n_r + 1}\|n_r + 1\rangle$	
$\hat{a}_r^\dagger\hat{a}_r\|n_r\rangle = n_r\|n_r\rangle$	

The entire two-level semiclassical Hamiltonian (again, semiclassical signifies that the electromagnetic field is classical) with a monochromatic field in the rotating wave approximation can then be written as

$$\widehat{H} = \frac{\hbar\omega_0}{2}(\hat{\sigma}_2 - \hat{\sigma}_1) - \frac{\hbar}{2}\hat{\sigma}_+\Omega_R(\mathbf{R}, t)\, e^{-i\omega t} - \frac{\hbar}{2}\Omega_R^*(\mathbf{R}, t)\, e^{i\omega t}\hat{\sigma}_- \tag{16.19}$$

where we have arbitrarily taken the zero-energy to be midway between the two states.

[3] So far, the operators are time independent as we expect in the Schrödinger picture. However, in the Heisenberg picture or some interaction picture, the operators usually become time dependent. The commutation relations hold only at equal times.

Problem 16.1 *Use the definition of the operators on the left side of Table 16.1 and prove the commutation relations given on the right side of the table for the atomic operators $\hat{\sigma}$. For example in the first case:*

$$\begin{aligned}[\hat{\sigma}_1, \hat{\sigma}_z] &= \hat{\sigma}_1\hat{\sigma}_z - \hat{\sigma}_z\hat{\sigma}_1 = |1\rangle\langle 1|(|2\rangle\langle 2| - |1\rangle\langle 1|) - (|2\rangle\langle 2| - |1\rangle\langle 1|)|1\rangle\langle 1| \\ &= |1\rangle\langle 1|2\rangle\langle 2| - |1\rangle\langle 1|1\rangle\langle 1| - |2\rangle\langle 2|1\rangle\langle 1| + |1\rangle\langle 1|1\rangle\langle 1| \\ &= -|1\rangle\langle 1|1\rangle\langle 1| + |1\rangle\langle 1|1\rangle\langle 1| = -|1\rangle\langle 1| + |1\rangle\langle 1| = 0\end{aligned}$$

because $\langle 1|2\rangle = \langle 1|2\rangle = 0$ and $\langle 1|1\rangle = 1$.

Problem 16.2*. Assuming a three-level system like the Λ three-level system on the right in Fig. 9.6, and assuming that the interaction terms in the Hamiltonian couple states 1 to 3 and states 2 to 3, but that there is no coupling from states 1 to 2, generate the corresponding matrices for the three states and operators $\hat{\sigma}_{13}, \hat{\sigma}_{31}, \hat{\sigma}_{32}, \hat{\sigma}_{23}$ and then write out the full Hamiltonian assuming a classical applied physics.*

16.3 Physical Meaning of the Atomic Operators

The physical meaning of these operators is readily understood by considering the corresponding expectation value for an arbitrary state vector. Hence, for the expectation value, we consider the state vector,

$$|\psi(t)\rangle = C_1 e^{\frac{i\omega_0 t}{2}}|1\rangle + C_2 e^{-\frac{i\omega_0 t}{2}}|2\rangle = C_1|1\rangle + C_2 e^{-i\omega_0 t}|2\rangle \tag{16.20}$$

The C_1 and C_2 may or may not be time dependent depending on whether the total Hamiltonian is time dependent or time independent. We have factored out the phase factor of state $|1\rangle$, $e^{\frac{i\omega_0 t}{2}}$, and suppressed it since it then just represents an overall phase. This is appropriate for the isolated two-level system that we are considering here.

Considering $\langle\hat{\sigma}_1\rangle$, we get:

$$\langle\hat{\sigma}_1\rangle = \langle\psi(t)|\hat{\sigma}_1|\psi(t)\rangle = C_1^* C_1 \tag{16.21}$$

Hence, $\langle\hat{\sigma}_1\rangle$ is the probability of finding the system in state $|1\rangle$. Similarly, $\langle\hat{\sigma}_2\rangle = C_2^* C_2$, is then the probability of being in state $|2\rangle$. Likewise, when the state vector is in a coherent superposition state of the two states, $\langle\hat{\sigma}_{12}\rangle = \langle\hat{\sigma}_{21}\rangle^* = C_1^* C_2$. The effect of the interaction potential is to couple the states together, so for an applied electric field, the induced polarization in the system is given by $\boldsymbol{P} = \langle\psi(t)|\hat{\boldsymbol{\mu}}|\psi(t)\rangle$. To express this in terms of the atomic operators, we use the expression above (Eq. 16.10) to get

$$\begin{aligned}\boldsymbol{P} &= \langle\psi(t)|\hat{\mu}|\psi(t)\rangle = \left(C_1^*\langle 1| + C_2^* e^{i\omega_0 t}\langle 2|\right)(\mu_{12}\check{\boldsymbol{r}}_{12}\hat{\sigma}_- + \mu_{21}\check{\boldsymbol{r}}_{21}\hat{\sigma}_+)\left(C_1|1\rangle + C_2 e^{-i\omega_0 t}|2\rangle\right) \\ &= \mu_{12}\check{\boldsymbol{r}}_{12} C_1^* C_2 e^{-i\omega_0 t} + \mu_{21}\check{\boldsymbol{r}}_{21} C_2^* C_1 e^{i\omega_0 t}\end{aligned} \tag{16.22}$$

Assuming the matrix elements μ_{21} and μ_{12}, which represent coupling between the states due to a field, are non-zero the expectation values of the raising and lowering operators represent the existence of a coherent superposition of the states as evidence by the presence of $C_1^* C_2$ and $C_2^* C_1$ giving rise to an induced polarization.

16.4 Atomic Operators in the Heisenberg Picture

Atomic operators are very useful when working to understand and develop concepts based on dynamic interactions. They can be extended to more levels, but most problems of interest focus on two or three levels. In developing atomic operators, the physics that goes into evaluating a matrix element, like μ_{12}, usually relies on the spatial solution to the time independent Schrödinger equation. However, these are fixed static parameters, like the inductance of a coil. The system becomes active when there are interaction terms in the Hamiltonian and the system is characterized by an initial condition. For this problem, the Heisenberg picture can be a powerful approach to describing the resulting dynamical behavior. Hence, we continue by deriving the appropriate equations of motions.

We refer to the discussion in Chapter 9 on the Heisenberg picture (Eq. 9:35 and earlier discussion). The *time dependent* Heisenberg operator is defined as $\hat{A}(t) = \hat{U}^{\dagger}(t)\hat{A}\hat{U}(t)$ based on the time *independent* Schrödinger operator, $\hat{A}$, and the time evolution operator defined in Eqs 9.10 and 9.11 by the equation $i\hbar\frac{\partial}{\partial t}\hat{U}(t,t_0) = \hat{H}(t)\,\hat{U}(t,t_0)$ and the adjoint equation. The Heisenberg equations of motion for $\hat{A}(t)$ is given in Eq. 9.35 as $\frac{d}{dt}\hat{A}_H(t) = i\hbar^{-1}\left[\hat{H},\hat{A}_H\right]$.

For the balance of this text, we suppress the subscript H on the Heisenberg operators. Heisenberg operators will be represented by their explicit time dependence as $\hat{A}(t)$ and Schrödinger operators as $\hat{A} = \hat{A}\,(t=0)$.

For the two-state operators and using Eq. 16.19 for the Hamiltonian, we then get

$$\begin{aligned}\frac{d}{dt}\hat{\sigma}_+(t) &= i\hbar^{-1}\left[\hat{H},\hat{\sigma}_+(t)\right] = i\frac{\omega_0}{2}\left[\sigma_z(t),\hat{\sigma}_+(t)\right] - i\frac{1}{2}\left[\hat{\sigma}_+(t),\hat{\sigma}_+(t)\right]\Omega_R e^{-i\omega t}\\ &\quad - i\frac{1}{2}\Omega_R^* e^{i\omega t}\left[\hat{\sigma}_-(t),\hat{\sigma}_+(t)\right] = i\omega_0\hat{\sigma}_+(t) + i\frac{1}{2}\Omega_R^* e^{i\omega t}\hat{\sigma}_z(t)\end{aligned} \tag{16.23}$$

$$\begin{aligned}\frac{d}{dt}\hat{\sigma}_-(t) &= i\hbar^{-1}\left[\hat{H},\hat{\sigma}_-(t)\right] = i\frac{\omega_0}{2}\left[\tilde{\sigma}_z,\hat{\sigma}_-(t)\right] - i\frac{1}{2}\Omega_R e^{-i\omega t}\left[\hat{\sigma}_+(t),\hat{\sigma}_-(t)\right]\\ &\quad - i\frac{1}{2}\Omega_R^* e^{i\omega t}\left[\hat{\sigma}_-(t),\hat{\sigma}_-(t)\right] = -i\omega_0\hat{\sigma}_-(t) - i\frac{1}{2}\hat{\sigma}_z(t)\Omega_R e^{-i\omega t}\end{aligned} \tag{16.24}$$

$$\begin{aligned}\frac{d}{dt}\hat{\sigma}_z(t) &= i\hbar^{-1}\left[\hat{H},\hat{\sigma}_z(t)\right] = i\hbar^{-1}\left[-\frac{\hbar}{2}\Omega_R e^{-i\omega t}\hat{\sigma}_+(t) - \frac{\hbar}{2}\Omega_R^* e^{i\omega t}\hat{\sigma}_-,\hat{\sigma}_z(t)\right]\\ &= -\frac{i}{2}\Omega_R e^{-i\omega t}\left[\hat{\sigma}_+(t),\hat{\sigma}_z(t)\right] - \frac{i}{2}\Omega_R^* e^{i\omega t}\left[\hat{\sigma}_-(t),\hat{\sigma}_z(t)\right]\\ &= i\Omega_R e^{-i\omega t}\hat{\sigma}_+(t) - i\Omega_R^* e^{i\omega t}\hat{\sigma}_-(t)\end{aligned} \tag{16.25}$$

$$\begin{aligned}\frac{d}{dt}\hat{\sigma}_1(t) &= i\hbar^{-1}\left[\hat{H},\hat{\sigma}_1(t)\right] = i\hbar^{-1}\left[-\frac{\hbar}{2}\Omega_R e^{-i\omega t}\hat{\sigma}_+(t) - \frac{\hbar}{2}\Omega_R^* e^{i\omega t}\hat{\sigma}_-,\hat{\sigma}_1(t)\right]\\ &= -\frac{i}{2}\Omega_R e^{-i\omega t}\left[\hat{\sigma}_+(t),\hat{\sigma}_1(t)\right] - \frac{i}{2}\Omega_R^* e^{i\omega t}\left[\hat{\sigma}_-(t),\hat{\sigma}_1(t)\right]\\ &= -\frac{i}{2}\Omega_R e^{-i\omega t}\hat{\sigma}_+(t) + \frac{i}{2}\Omega_R^* e^{i\omega t}\hat{\sigma}_-(t)\end{aligned} \tag{16.26}$$

$$\begin{aligned}\frac{d}{dt}\hat{\sigma}_2(t) &= i\hbar^{-1}\left[\hat{H},\hat{\sigma}_2(t)\right] = i\hbar^{-1}\left[-\frac{\hbar}{2}\Omega_R e^{-i\omega t}\hat{\sigma}_+(t) - \frac{\hbar}{2}\Omega_R^* e^{i\omega t}\hat{\sigma}_-,\hat{\sigma}_2(t)\right]\\ &= -\frac{i}{2}\Omega_R e^{-i\omega t}\left[\hat{\sigma}_+(t),\hat{\sigma}_2(t)\right] - \frac{i}{2}\Omega_R^* e^{i\omega t}\left[\hat{\sigma}_-(t),\hat{\sigma}_2(t)\right]\\ &= \frac{i}{2}\Omega_R e^{-i\omega t}\hat{\sigma}_+(t)(t) - \frac{i}{2}\Omega_R^* e^{i\omega t}\hat{\sigma}_-(t)\end{aligned} \tag{16.27}$$

The above equations are summarized in Table 16.2.

Since $\hat{\sigma}_z(t) = \hat{\sigma}_2(t) - \hat{\sigma}_1(t)$, including the equation for $\hat{\sigma}_z(t)$ is unnecessary. However, $\hat{\sigma}_+(t)$ and $\hat{\sigma}_-(t)$ are written more compactly in terms of $\hat{\sigma}_z(t)$ and for a closed system, using $\hat{\sigma}_z(t)$ is simpler than using $\hat{\sigma}_2(t)$ and $\hat{\sigma}_1(t)$ as we see below.

16.5 **Exact Solution for the Atomic Operators for a Monochromatic Field**

To facilitate the solution to these equations, we move to the **field interaction picture** as we did in Chapter 9 (Eqs 9.97 and 9.98). We define two new operators to eliminate the time dependence due to the classical driving field associated with $e^{\pm i\omega t}$. Specifically,

$$\hat{\sigma}_+(t) = \tilde{\sigma}_+(t)e^{i\omega t} \tag{16.28}$$

$$\hat{\sigma}_-(t) = \tilde{\sigma}_-(t)e^{-i\omega t} \tag{16.29}$$

where $\tilde{\sigma}_+(t)$ and $\tilde{\sigma}_-(t)$ remain Heisenberg operators, but without the fast oscillation associated with ω. Substituting this in to Eqs 16.23 and 16.24 and noting that

Table 16.2 Summary of Heisenberg equations of motion for a two-level system in the presence of a classical electromagnetic field in the rotating wave approximation.

$\frac{d}{dt}\hat{\sigma}_+(t) = i\omega_0\hat{\sigma}_+(t) + i\frac{1}{2}\Omega_R^* e^{i\omega t}\hat{\sigma}_z(t)$
$\frac{d}{dt}\hat{\sigma}_-(t) = -i\omega_0\hat{\sigma}_-(t) - i\frac{1}{2}\Omega_R e^{-i\omega t}\hat{\sigma}_z(t)$
$\frac{d}{dt}\hat{\sigma}_z(t) = i\Omega_R e^{-i\omega t}\hat{\sigma}_+(t) - i\Omega_R^* e^{i\omega t}\hat{\sigma}_-(t)$
$\frac{d}{dt}\hat{\sigma}_1(t) = -\frac{i}{2}\Omega_R e^{-i\omega t}\hat{\sigma}_+(t) + \frac{i}{2}\Omega_R^* e^{i\omega t}\hat{\sigma}_-(t)$
$\frac{d}{dt}\hat{\sigma}_2(t) = \frac{i}{2}\Omega_R e^{-i\omega t}\hat{\sigma}_+(t) - \frac{i}{2}\Omega_R^* e^{i\omega t}\hat{\sigma}_-(t)$

$$\frac{d}{dt}\hat{\sigma}_+(t) = e^{i\omega t}\left(\frac{d}{dt}\tilde{\sigma}_+(t) + i\omega\tilde{\sigma}_+(t)\right) \tag{16.30}$$

$$\frac{d}{dt}\hat{\sigma}_-(t) = e^{-i\omega t}\left(\frac{d}{dt}\tilde{\sigma}_-(t) - i\omega\tilde{\sigma}_-(t)\right) \tag{16.31}$$

we collect terms, setting $\delta = \omega - \omega_0$ to get:

$$\frac{d}{dt}\tilde{\sigma}_+(t) = -i\delta\tilde{\sigma}_+(t) + i\frac{1}{2}\Omega_R^*\hat{\sigma}_z(t) \tag{16.32}$$

$$\frac{d}{dt}\tilde{\sigma}_-(t) = i\delta\tilde{\sigma}_-(t) - i\frac{1}{2}\Omega_R\hat{\sigma}_z(t) \tag{16.33}$$

$$\frac{d}{dt}\hat{\sigma}_z(t) = i\Omega_R\hat{\sigma}_+(t) - i\Omega_R^*\hat{\sigma}_-(t) \tag{16.34}$$

$$\frac{d}{dt}\hat{\sigma}_1(t) = -\frac{i}{2}\Omega_R\tilde{\sigma}_+(t) + \frac{i}{2}\Omega_R^*\tilde{\sigma}_-(t) \tag{16.35}$$

$$\frac{d}{dt}\hat{\sigma}_2(t) = \frac{i}{2}\Omega_R\tilde{\sigma}_+(t) - \frac{i}{2}\Omega_R^*\tilde{\sigma}_-(t) \tag{16.36}$$

This is a **closed quantum system**, meaning that $\hat{\sigma}_2(t) + \hat{\sigma}_1(t) = \widehat{U}^\dagger(t)|1\rangle\langle 1|\widehat{U}(t) + \widehat{U}^\dagger(t)|2\rangle\langle 2|\widehat{U}(t) = \widehat{U}^\dagger(t)(|1\rangle\langle 1| + |2\rangle\langle 2|)\widehat{U}(t) = \widehat{U}^\dagger(t)\,\hat{I}\,\widehat{U}(t) = \hat{I}$, where because of completeness, $|1\rangle\langle 1| + |2\rangle\langle 2| = \hat{I}$, and the last equality follows because $\widehat{U}(t)$ is unitary, meaning $\widehat{U}^\dagger(t) = \widehat{U}^{-1}(t)$. We then work with the three coupled operator equations for $\hat{\sigma}_+(t)$, $\hat{\sigma}_-(t)$, and $\hat{\sigma}_z(t)$. If information on $\hat{\sigma}_1(t)$ or $\hat{\sigma}_2(t)$ is needed, we note that $\hat{\sigma}_1(t) = \frac{1}{2}\left(\hat{I} - \hat{\sigma}_z(t)\right)$ and $\hat{\sigma}_2(t) = \frac{1}{2}\left(\hat{I} + \hat{\sigma}_z(t)\right)$.

The equations of motion for the atomic operators in the field interaction picture can now be written in matrix form:

$$-i\frac{d}{dt}\begin{bmatrix}\tilde{\sigma}_+(t)\\ \tilde{\sigma}_-(t)\\ \hat{\sigma}_z(t)\end{bmatrix} = \begin{bmatrix}-\delta & 0 & \frac{1}{2}\Omega_R^*\\ 0 & \delta & -\frac{1}{2}\Omega_R\\ \Omega_R & -\Omega_R^* & 0\end{bmatrix}\begin{bmatrix}\tilde{\sigma}_+(t)\\ \tilde{\sigma}_-(t)\\ \hat{\sigma}_z(t)\end{bmatrix} \tag{16.37}$$

The square matrix is now time independent, making the solution straightforward. Assuming a solution of the form

$$\begin{bmatrix}\tilde{\sigma}_+(t)\\ \tilde{\sigma}_-(t)\\ \hat{\sigma}_z(t)\end{bmatrix} = \begin{bmatrix}\tilde{\sigma}'_+\\ \tilde{\sigma}'_-\\ \hat{\sigma}'_z\end{bmatrix}e^{i\nu t} \tag{16.38}$$

we get an eigenvalue equation:

$$\begin{bmatrix}-\delta-\nu & 0 & \frac{1}{2}\Omega_R^*\\ 0 & \delta-\nu & -\frac{1}{2}\Omega_R\\ \Omega_R & -\Omega_R^* & -\nu\end{bmatrix}\begin{bmatrix}\tilde{\sigma}'_+\\ \tilde{\sigma}'_-\\ \hat{\sigma}'_z\end{bmatrix} = 0 \tag{16.39}$$

giving for the characteristic equation (i.e., the equation for ν for the case that the determinant of the above matrix is zero):

$$(\delta+\nu)\left[(\delta-\nu)(-\nu)-\frac{1}{2}[\Omega_R]^2\right]+\frac{1}{2}[\Omega_R]^2(\delta-\nu)=0 \tag{16.40}$$

After rewriting and canceling terms we get

$$-\nu\left(\delta^2-\nu^2\right)-\nu[\Omega_R]^2=0 \tag{16.41}$$

One of the roots is zero. The other two roots are given by the solution to

$$\left(\delta^2-\nu^2\right)+|\Omega_R|^2=0 \tag{16.42}$$

or

$$\nu=\pm\Omega_{GR}\equiv\pm\sqrt{\delta^2+|\Omega_R|^2} \tag{16.43}$$

Ω_{GR} is the **generalized Rabi frequency**, as we defined in Chapter 9. The original equation of motion is cubic so there are three solutions: $\nu=\pm\Omega_{GR}, 0$. To find the Heisenberg form $\tilde{\sigma}_+(t)$, we write the operators as

$$\tilde{\sigma}_+(t)=Ae^{-i\Omega_{GR}t}+Be^{+i\Omega_{GR}t}+C \tag{16.44}$$

$$\tilde{\sigma}_-(t)=\tilde{\sigma}_+^\dagger(t)=A^\dagger e^{i\Omega_{GR}t}+B^\dagger e^{-i\Omega_{GR}t}+C^\dagger \tag{16.45}$$

$$\hat{\sigma}_z(t)=Fe^{-i\Omega_{GR}t}+Ge^{+i\Omega_{GR}t}+H \tag{16.46}$$

where the capital letters are *unknown time independent operators*. Recall that at $t=0$, the Heisenberg form of the operators is equal to the Schrödinger form. Hence,

$$A+B+C=\hat{\sigma}_+ \tag{16.47}$$

$$A^\dagger+B^\dagger+C^\dagger=\hat{\sigma}_- \tag{16.48}$$

$$F+G+H=\tilde{\sigma}_z \tag{16.49}$$

The atomic operators in Eqs 16.47–16.49 are in the Schrödinger picture and are time independent. Then from Eq. 16.23, we see that the equation of motion for $\tilde{\sigma}_+(t)$ involves $\hat{\sigma}_z(t)$. Set $\delta=0$ and assuming Ω_R is real to make it simpler and inserting Eqs 16.44–16.46 into the equations of motion, we find

$$\hat{\sigma}_z(t)=\hat{\sigma}_z\cos\Omega_R t+i\left(\hat{\sigma}_+-\hat{\sigma}_-\right)\sin\Omega_R t \tag{16.50}$$

$$\tilde{\sigma}_+(t)=\hat{\sigma}_+\cos\Omega_R t+i\frac{\hat{\sigma}_z}{2}\sin\Omega_R t;\quad \tilde{\sigma}_-(t)=\hat{\sigma}_-\cos\Omega_R t-i\frac{\hat{\sigma}_z}{2}\sin\Omega_R t \tag{16.51}$$

Problem 16.3 *Using the details above, find the solutions for $\hat{\sigma}_z(t)$, $\hat{\sigma}_+(t)$, and $\hat{\sigma}_-(t)$ given in Eqs 16.50 and 16.51*

Physically from the discussion of Rabi oscillations in Chapter 9, we expect similar behavior here. The problem is identically the same, but the method of solution is quite different. In fact, while we clearly see evidence of oscillation at the Rabi frequency, the oscillations are associated with the amplitudes of the operators $\hat{\sigma}_z$ and $\tilde{\sigma}_+$. In Chapter 9 we found the time dependence of the probability amplitude of the wave function. Here, we can now directly calculate $\langle \hat{\sigma}_z \rangle (t)$. We assume, as before, that at $t = 0$, $|\psi(t=0)\rangle = |1\rangle$. Then

$$\langle \hat{\sigma}_z \rangle (t) = \langle 1|\hat{\sigma}_z(t)|1\rangle = -\cos|\Omega_R|t \tag{16.52}$$

Problem 16.4 *Find $\hat{\sigma}_2(t)$ given at $t = 0$, $|\psi(t=0)\rangle = |1\rangle$*

As in the case of Chapter 9, we have assumed a monochromatic field, the field can be turned on and off at some point in time, and the solution remains valid during that time.

16.6 **Operator Equations of Motion Including the Vacuum Field**

We now consider the effect of spontaneous emission in the system. We start with the multimode Hamiltonian with the quantized field from Chapter 15, Eq. 15.80, where we have already incorporated the dipole and rotating wave approximation and take the position of the quantum system to be at the origin ($\boldsymbol{R} = 0$).

$$\hat{H} = \frac{\hbar\omega_{21}}{2}\hat{\sigma}_z - \sum_j \left(\boldsymbol{\mu}_{21}\cdot\hat{\boldsymbol{E}}_{j+}\hat{\sigma}_+ + \boldsymbol{\mu}_{12}\cdot\hat{\boldsymbol{E}}_{j-}\hat{\sigma}_-\right) + \sum_j \hbar\omega_j\hat{a}_j^\dagger\hat{a}_j \tag{16.53}$$

$$\hat{\boldsymbol{E}}_{j+} = i\mathcal{E}_j\check{\varepsilon}_j\hat{a}_j;\ \hat{\boldsymbol{E}}_{j-} = -i\mathcal{E}_j\check{\varepsilon}_j\hat{a}_j^\dagger \tag{16.54}$$

$$\mathcal{E}_j \equiv \sqrt{\frac{\hbar\omega_j}{2\epsilon_o V}} \tag{16.55}$$

We have combined the sum over $\boldsymbol{k}, \sigma$ into single subscript, j. The classical field terms of the previous section (e.g., Eq. 16.19) are not included because the results of this section are not impacted by the presence of the classical fields. However, the results of this section have an important impact on the equations of motions for the operators, and the classical fields can be introduced in the usual way at the end of this calculation if desired.

By defining yet an additional symbol, we can rewrite this Hamiltonian in a slightly more compact manner:

$$\hat{H} = \frac{\hbar\omega_{21}}{2}(\hat{\sigma}_2 - \hat{\sigma}_1) + \hbar\sum_j\left(g_j\hat{\sigma}_+\hat{a}_j + g_j^*\hat{a}_j^\dagger\hat{\sigma}_-\right) + \sum_j \hbar\omega_j\hat{a}_j^\dagger\hat{a}_j \tag{16.56}$$

$$g_j = -i\frac{\mathcal{E}_j\,\mu_{\mathbf{21}}}{\hbar}\check{\boldsymbol{r}}_{\mathbf{21}}\cdot\check{\varepsilon}_j;\quad g_j^* = i\frac{\mathcal{E}_j\,\mu_{\mathbf{12}}}{\hbar}\check{\boldsymbol{r}}_{\mathbf{12}}\cdot\check{\varepsilon}_j \tag{16.57}$$

We now write out the equations of motion for the relevant Heisenberg two-level operators. The system remains closed, so we will find that decay of the excited state will lead to an increase in the probability of the ground state. This means that the time evolution of $\hat{\sigma}_2$ will be correlated with the time evolution of $\hat{\sigma}_1$. So, we must solve equations for $\hat{\sigma}_1, \hat{\sigma}_2. \hat{\sigma}_+, \hat{\sigma}_-, \hat{a}$, and $\hat{a}^\dagger$. But since $\hat{\sigma}_1 + \hat{\sigma}_2 = \hat{I}$ and $\hat{\sigma}_- = \hat{\sigma}_+^\dagger$, we only need to consider the equations of motion for $\hat{\sigma}_+, \hat{\sigma}_1$, and $\hat{a}$. We will return to this issue at the end of this discussion.

Using Table 16.1 and the Hamiltonian in Eq. 16.56, we get

$$\frac{d}{dt}\hat{\sigma}_+(t) = i\hbar^{-1}\left[\widehat{H}, \hat{\sigma}_+(t)\right] = i\frac{\omega_0}{2}\left[(\hat{\sigma}_2(t) - \hat{\sigma}_1(t)), \hat{\sigma}_+(t)\right] + i\left(\sum_{k,\sigma} g_j^* \hat{a}_{k,\sigma}^\dagger(t)\right)\left[\hat{\sigma}_-(t), \hat{\sigma}_+(t)\right]$$
$$= i\omega_0\hat{\sigma}_+(t) - i\sum_j g_j^* \hat{a}_j^\dagger(t)\left(\hat{\sigma}_2(t) - \hat{\sigma}_1(t)\right) \tag{16.58}$$

$$\frac{d}{dt}\hat{\sigma}_1(t) = i\hbar^{-1}\left[\widehat{H}, \hat{\sigma}_1(t)\right] = i\sum_j\left(g_j\hat{\sigma}_+(t)\hat{a}_j(t) - g_j^*\hat{a}_j^\dagger(t)\hat{\sigma}_-(t)\right) \tag{16.59}$$

$$\frac{d}{dt}\hat{\sigma}_2(t) = i\hbar^{-1}\left[\widehat{H}, \hat{\sigma}_2(t)\right] = -i\sum_j\left(g_j\hat{\sigma}_+(t)\hat{a}_j(t) - g_j^*\hat{a}_j^\dagger(t)\hat{\sigma}_-(t)\right) \tag{16.60}$$

$$\frac{d}{dt}\hat{a}_j(t) = i\hbar^{-1}\left[\widehat{H}, \hat{a}_j(t)\right] = -i\omega_j\hat{a}_j(t) - ig_j^*\hat{\sigma}_-(t) \tag{16.61}$$

Integrating the equations for $\hat{a}_j(t)$ with the integrating factor $e^{i\omega_j t}$ we get

$$\hat{a}_j(t) = \hat{a}_j e^{-i\omega_j t} - ig_j^*\int_0^t dt'\, e^{-i\omega_j(t-t')}\hat{\sigma}_-(t') \tag{16.62}$$

It follows that

$$\hat{a}_j^\dagger(t) = \hat{a}_j^\dagger e^{i\omega_j t} + ig_j\int_0^t dt'\, e^{i\omega_j(t-t')}\hat{\sigma}_+(t') \tag{16.63}$$

Recall that, without the explicit time dependence, $\hat{a}_j^\dagger$ and $\hat{a}_j$ are the time independent Schrödinger operators. We substitute both of these expressions into Eq. 16.59:

$$\frac{d}{dt}\hat{\sigma}_1(t) = i\sum_j g_j\hat{\sigma}_+(t)\hat{a}_j e^{-i\omega_j t} - i\sum_j g_j^*\hat{a}_j^\dagger e^{i\omega_j t}\hat{\sigma}_-(t)$$
$$+\sum_j |g_j|^2\left[\hat{\sigma}_+(t)\int_0^t dt'\, e^{-i\omega_j(t-t')}\hat{\sigma}_-(t') + \int_0^t dt'\, e^{i\omega_j(t-t')}\hat{\sigma}_+(t')\,\hat{\sigma}_-(t)\right] \tag{16.64}$$

At this point, the form in the integrands is similar to the integrand in Eq. 15.104. However, the assumption that $\hat{\sigma}_-(t')$ changes slowly and thus can be removed from the integral is not valid here. This is because, even in the absence of coupling to the vacuum, $\hat{\sigma}_-(t) \sim e^{-i\omega_0 t}$. This is evident from Eq. 16.24 by setting $\Omega_R = 0$ and integrating, giving $\hat{\sigma}_-(t) = \hat{\sigma}_-(0)e^{-i\omega_0 t}$. Similarly, $\hat{\sigma}_+(t) \sim \hat{\sigma}_+(0)e^{i\omega_0 t}$. To deal with this, we write the operator in the form[4]

$$\hat{\sigma}_-(t) \equiv \left(\hat{\sigma}_-(t)e^{i\omega_0 t}\right)e^{-i\omega_0 t} \text{ and } \hat{\sigma}_+(t) \equiv \left(\hat{\sigma}_+(t)e^{-i\omega_0 t}\right)e^{i\omega_0 t} \tag{16.65}$$

Hence for the first integral in 16.64,

$$\int_0^t dt' e^{-i\omega_j(t-t')}\hat{\sigma}_-(t') = \int_0^t dt' e^{-i\omega_j(t-t')}e^{-i\omega_0 t'}\left\{\hat{\sigma}_-(t')\,e^{+i\omega_0 t'}\right\} \tag{16.66}$$

Now the term $\left\{\hat{\sigma}_-(t')\,e^{+i\omega_0 t'}\right\}$ is evolving slowly in time and can be evaluated at $t' = t$ and removed from the integral. We could have avoided this discussion had we worked in an interaction picture. This is again the application of the **Weisskopf–Wigner** approximation.

$$\int_0^t dt' e^{-i\omega_j(t-t')}\hat{\sigma}_-(t') = \left\{\hat{\sigma}_-(t)e^{i\omega_0 t}\right\}\int_0^t dt' e^{-i\omega_j(t-t')}e^{-i\omega_0 t'} = \hat{\sigma}_-(t)\int_0^t dt' e^{-i(\omega_j-\omega_0)(t-t')} \tag{16.67}$$

Similarly,

$$\int_0^t dt' e^{i\omega_j(t-t')}\hat{\sigma}_+(t') = \hat{\sigma}_+(t)\int_0^t dt' e^{i(\omega_j-\omega_0)(t-t')} \tag{16.68}$$

We then write Eq. 16.64 as

$$\frac{d}{dt}\hat{\sigma}_1(t) = i\sum_j g_j\hat{\sigma}_+(t)\hat{a}_j e^{-i\omega_j t} - i\sum_j g_j^*\hat{a}_j^\dagger e^{i\omega_j t}\hat{\sigma}_-(t) + 2\hat{\sigma}_2(t)\sum_j |g_j|^2\pi\delta\left(\omega_j - \omega_0\right) \tag{16.69}$$

where we have used the fact that $\hat{\sigma}_2(t) = \hat{\sigma}_+(t)\hat{\sigma}_-(t)$ and Eq. 14.26 to evaluate the integrals over time, giving the Dirac delta-function.

We now evaluate the sum in the last term of Eq. 16.69. Namely, following the discussion in Eq. 15.106, the sum over modes of the cavity becomes an integral over $k = \frac{\omega}{c}$ as the volume goes to infinity:[5]

$$\frac{\gamma}{2} = \sum_j |g_j|^2\pi\delta\left(\omega_j - \omega_0\right) = \frac{\mathcal{V}}{(2\pi)^3}\int_0^\infty k^2 dk\int_{\Omega_{s.a.}} d\Omega_{s.a.}|g_k|^2\frac{\pi}{c}\delta\left(k - \frac{\omega_0}{c}\right) = \frac{2}{3}\frac{|\mu_{21}|^2}{4\pi\epsilon_o\hbar}\left(\frac{\omega_{21}}{c}\right)^3 \tag{16.70}$$

Eq. 16.69 becomes

$$\frac{d}{dt}\hat{\sigma}_1(t) = \gamma\hat{\sigma}_2 + i\sum_j g_j\,e^{-i\omega_j t}\hat{\sigma}_+(t)\hat{a}_j - i\sum_j g_j^*\,e^{+i\omega_j t}\hat{a}_j^\dagger\hat{\sigma}_-(t) \tag{16.71}$$

[4] See Stig Stenholm, Foundations of Laser Spectroscopy, John Wiley & Sons, New York (1984) for a discussion on Weisskopf–Wigner theory.

[5] The symbol $\int_{\Omega_{s.a.}} d\Omega_{s.a.} \equiv \int_0^{2\pi} d\phi \int_0^{\pi} \sin\Theta\, d\Theta$.

This shows that the upper state decays to the lower state, as expected. This is a challenging problem in the Schrödinger picture. A similar calculation for the equation of motion for $\hat{\sigma}_2$ shows a decay term $-\gamma\hat{\sigma}_2$, as summarized below. Terms like $\hat{\sigma}_+(t)\,\hat{a}_j$ do not commute unless they are evaluated at the same time. Hence, since $\hat{a}_j^\dagger$ is a time independent Schrödinger operator, it only commutes with $\hat{\sigma}_-(t)$ at $t = 0$. The two sums represent fluctuations, and are often ignored except when the focus is photon statistics or other effects of quantum fluctuations.

We now consider the case again of an applied classical field but including the presence of decay, while ignoring fluctuations. The equations of motion are

$$\frac{d}{dt}\hat{\sigma}_+(t) = \left(i\omega_0 - \frac{\gamma}{2}\right)\hat{\sigma}_+(t) + i\frac{1}{2}\Omega_R^* e^{i\omega t}\hat{\sigma}_z(t) \tag{16.72}$$

$$\frac{d}{dt}\hat{\sigma}_-(t) = -\left(i\omega_0 + \frac{\gamma}{2}\right)\hat{\sigma}_-(t) - i\frac{1}{2}\Omega_R e^{-i\omega t}\hat{\sigma}_z(t) \tag{16.73}$$

$$\frac{d}{dt}\hat{\sigma}_1(t) = +\gamma\hat{\sigma}_2(t) - \frac{i}{2}\Omega_R e^{-i\omega t}\hat{\sigma}_+(t) + \frac{i}{2}\Omega_R^* e^{i\omega t}\hat{\sigma}_-(t) \tag{16.74}$$

$$\frac{d}{dt}\hat{\sigma}_2(t) = -\gamma\hat{\sigma}_2(t) + \frac{i}{2}\Omega_R e^{-i\omega t}\hat{\sigma}_+(t) - \frac{i}{2}\Omega_R^* e^{i\omega t}\hat{\sigma}_-(t) \tag{16.75}$$

Or, in the field interaction picture (Eqs 16.28 and 16.29),

$$\frac{d}{dt}\tilde{\sigma}_+(t) = -\left(i\delta + \frac{\gamma}{2}\right)\tilde{\sigma}_+(t) + i\frac{1}{2}\Omega_R^*\hat{\sigma}_z(t) \tag{16.76}$$

$$\frac{d}{dt}\tilde{\sigma}_-(t) = \left(i\delta - \frac{\gamma}{2}\right)\tilde{\sigma}_-(t) - i\frac{1}{2}\Omega_R\hat{\sigma}_z(t) \tag{16.77}$$

$$\frac{d}{dt}\hat{\sigma}_1(t) = +\gamma\hat{\sigma}_2(t) - \frac{i}{2}\Omega_R\tilde{\sigma}_+(t) + \frac{i}{2}\Omega_R^*\tilde{\sigma}_-(t) \tag{16.78}$$

$$\frac{d}{dt}\hat{\sigma}_2(t) = -\gamma\hat{\sigma}_2(t) + \frac{i}{2}\Omega_R e^{-i\omega t}\tilde{\sigma}_+(t) - \frac{i}{2}\Omega_R^*\tilde{\sigma}_-(t) \tag{16.79}$$

It is important to note that when the fluctuation terms are not included, the equal time commutation relations are not satisfied.

Problem 16.5 *Using first order perturbation theory, find the induced polarization by a monochromatic field:* $P(t) = \langle\hat{\mu}\rangle(t)$*, where the form of the dipole momentum operator using atomic operators is given in Eq. 15.82. Assume that at* $t = 0$*, the system is in the ground state. In perturbation theory,* $\hat{\sigma}_z(t) = \hat{\sigma}_z$ *at zero order.*

Show that for long times, you recover the result in Eq. 15.127.

$$P = \frac{\hat{\mu}_{12}}{2}\Omega_R\frac{(\omega - \omega_{21}) - i\frac{\gamma}{2}}{\left(\frac{\gamma}{2}\right)^2 + (\omega - \omega_{21})^2}e^{-i\omega t} + cc$$

Problem 16.6 *Repeat problem 16.5 but replace* $\Omega_R e^{-i\omega t}$ *with* $\Omega_R \tau \delta(t) e^{-i\omega t}$. *Show that the polarization oscillates at the frequency* ω_{21} *and decay exponentially as* $\frac{\gamma}{2}$. *The effect is called* ***free induction decay****, after it was first observed from the nuclear spins.*

16.7 Summary

The atomic operators along with the field operators provide a very important ability to calculate numerous quantities of importance to quantum technology. This capability is the subject of more advanced books, but this section introduces the Heisenberg formulation. The problem of a quantum system coupled to a classical electromagnetic field while undergoing spontaneous emission is a nonlinear system of equations. However, there are many cases that can be evaluated analytically that have been very important for recent work. If one is interested only in dynamical problems with an applied classical field, then these equations also provide a common analytical approach where the effect of decay is fully included. The advantage of these equations is that calculation of measurable quantities depends only on the initial state of the system. However, if the final state of the system is needed, then the density matrix formulation of the final chapter of this book (Chapter 18) is often useful.

Vocabulary (page) and Important Concepts

17 Quantum Electromagnetics

17.1 Introduction

Quantum electromagnetics or quantum optics is a major sub-field of quantum mechanics and quantum electrodynamics. With the advent of ideas like quantum information, quantum secure communications, quantum computing, and quantum lidar/radar, the field of quantum optics has become profoundly important to advancing technology. Even before the field of quantum optics exploded in interest for new applications, the fundamental statistical properties of optical fields limits the signal to noise in fiber optical communications. To be clear, while the field has the word "optics" in the title, the bandwidth is extending to much lower frequencies as detector technology improves and now even to higher frequencies.

When a standard source of coherent radiation such as a laser or microwave oscillator is used to drive a quantum system, there is usually no advantage to using a quantized electromagnetic field to characterize that source and the interactions with a quantum system. However, if the system is using a single- or few-photon source or if signal detection is based on photon counting or any number of other single to few photon based technologies, then it is likely important to work with the quantized field for the detected emission. The intent of this chapter is to introduce some of the basic ideas in the field.

Typically, especially at high frequencies as in the telecom band (1.5 microns) or higher, the radiation is detected by detecting the power, where the power per unit area in the vacuum is intensity, given by $I = \frac{\epsilon_0 c}{2}|E|^2$. Recall that the quantized field operator is given in Eq. 15.78 as:

$$\hat{\boldsymbol{E}}_t = i\sum_{\boldsymbol{k},\sigma} \mathcal{E}_k \left(\hat{a}_{\boldsymbol{k},\sigma}\hat{\boldsymbol{\epsilon}}_{\boldsymbol{k},\sigma}e^{ik\cdot R} - \hat{a}^{\dagger}_{\boldsymbol{k},\sigma}\hat{\epsilon}^{*}_{k,\sigma}e^{-ik\cdot R}\right) = \hat{\boldsymbol{E}}_+ + \hat{\boldsymbol{E}}_- \tag{17.1}$$

where $\hat{\boldsymbol{E}}_+ = i\sum_{\boldsymbol{k},\sigma} \mathcal{E}_k\ \hat{a}_{\boldsymbol{k},\sigma}\hat{\boldsymbol{\epsilon}}_{\boldsymbol{k},\sigma}e^{ik\cdot R}$; $\hat{\boldsymbol{E}}_- = -i\sum_{\boldsymbol{k},\sigma} \mathcal{E}_k\hat{a}^{\dagger}_{\boldsymbol{k},\sigma}\hat{\boldsymbol{\epsilon}}^{*}_{\boldsymbol{k},\sigma}e^{-ik\cdot R}$, and $\mathcal{E}_k \equiv \sqrt{\frac{\hbar\omega_k}{2\epsilon_o V}}$. The intensity is then proportional to $\langle\hat{\boldsymbol{E}}_-\hat{\boldsymbol{E}}_+\rangle$.

Since introducing the quantized field, the focus has been on primarily examining the effects of the vacuum on transitions. In this chapter, we discuss some of the basic features of the quantized field and the impact on optical design. Not only has quantum optics impacted quantum computing, it has also been used to demonstrate secure quantum communications and is being examined for use in quantum lidar (light detection and ranging) and quantum radar (radio detection and ranging) where in all the cases, the use of quantum entangled photons leads to new capabilities.

Introduction to Quantum Nanotechnology: A Problem Focused Approach. Duncan Steel, Oxford University Press (2021).
 DOI: 10.1093/oso/9780192895073.003.0017

In the following, we examine a number of different aspects of quantum optics to give a sense of the possibilities and the differences that can emerge when comparing classical quantum systems.

17.2 Number State Representation

From Chapter 15, recall that the field is quantized by defining the electromagnetic modes of the field in a cavity, determining the total energy in each mode, and then finding the appropriate canonical coordinate and conjugate momentum operators in terms of the expansion coefficients of the field. Following the postulates, the operators satisfied the commutation relation, and from there we were able to find the eigenstates of the Hamiltonian. Specifically, for a single mode radiation field from Eq. 15.61, we have

$$\hat{H}_{Rad} = \hbar\omega\left(\hat{a}^\dagger\hat{a} + \frac{1}{2}\right) \tag{17.2}$$

The form of Hamiltonian is analogous to the nano-vibrator or quantum LC circuit problem, as discussed in Chapter 8. The eigenstates are the number states, $|n\rangle$, where n is a positive integer and represents the quanta of energy given by $\hbar\omega$:

$$\hat{H}_{Rad}|n\rangle = \hbar\omega\left(\hat{a}^\dagger\hat{a} + \frac{1}{2}\right)|n\rangle = \hbar\omega\left(n + \frac{1}{2}\right)|n\rangle \tag{17.3}$$

The eigenstates, $|n\rangle$, are called **number states** or **Fock states**. Recall that $\hat{a}^\dagger$ and $\hat{a}$ are the creation and annihilation operators (corresponding to the raising and lowering operator for the nano-vibrator) with the properties

$$\left[\hat{a}, \hat{a}^\dagger\right] = 1 \tag{17.4}$$

$$\hat{a}^\dagger|n\rangle = \sqrt{n+1}\,|n+1\rangle \tag{17.5}$$

$$\hat{a}|n\rangle = \sqrt{n}\,|n-1\rangle \tag{17.6}$$

The number state, $|n\rangle$, is also an eigenstate of the number operator, $\hat{n} = \hat{a}^\dagger\hat{a}$, namely:

$$\hat{n}|n\rangle = n|n\rangle \tag{17.7}$$

Fock states are quantum states and are not produced in, say, a laser or regular LED. However, if we assume that we can produce these states, then an immediate question is to evaluate the noise in such a signal. We examined similar questions associated with measurement earlier in these notes, by using the definition of the quantum variance. We will do the same here. Of course, in comparing to any experiment, it is important to consider the noise in the detector itself. This is a separate subject. However, for many decades we have had detectors that are sensitive enough to see a single quanta of energy, albeit with efficiencies of order a few percent.

Today, new technology is enabling single photon detection with efficiencies in excess of 80% with virtually no noise. So, we will assume a perfect photon counting (i.e., number) detector.[1]

Then for a radiation field to be in a single Fock state, $|n\rangle$, the noise is given by the quantum variance,

$$(\Delta n^2) = \left\langle (\hat{n} - \langle \hat{n} \rangle)^2 \right\rangle = \langle \hat{n}^2 \rangle - \langle \hat{n} \rangle^2 = \langle n|\hat{n}^2|n\rangle - \langle n\,|\,\hat{n}\,|\,n\rangle^2 = 0 \tag{17.8}$$

Remarkably, there is no noise for a system in a Fock state. Photon counting does not imply a Fock state. If we arrange a standard laser to produce a sufficiently slow photon arrival rate at a detector, we can count those photons, also. However, you will see below that the standard deviation (square root of the variance) goes like $\sqrt{n}$ because the output is close to that of a coherent state.

Problem 17.1 *Find the photon counting noise for a radiation field in a superposition of two number states of the same mode:* $|\psi\rangle = \alpha|n\rangle + \beta|m\rangle$, *recalling that the number states in the same mode are orthonormal. Show*

$$\Delta\hat{n}^2 = |\alpha|^2 \left(1 - |\alpha|^2\right)(n - m)^2 \tag{17.9}$$

Problem 17.2 *Find the photon counting noise for a radiation field in a product state of two distinct modes:* $|\psi\rangle = |n_i\rangle|m_j\rangle = |n_i m_j\rangle$. *Show* $\Delta\hat{n}^2 = \langle n_i m_j|\left(\hat{n}_i + \hat{n}_j\right)^2|n_i m_j\rangle - \langle n_i m_j|\hat{n}_i + \hat{n}_j|n_i m_j{}^2\rangle = n_i^2 + 2n_i m_j + m_j^2 - \left(n_i + m_j\right)^2 = 0.$

More sophisticated detectors may detect just the field. The details of how to do this are beyond the scope of this book, but the question remains as to what the noise is in this case. We refer back to Eq. 17.1 to see that, for a single mode field and ignoring the vector nature of the coupling, the field is a linear combination of the raising and lowering operators at $x = 0$:

$$\hat{E} = i\mathcal{E}\left(\hat{a} - \hat{a}^\dagger\right) \text{ for } x = 0 \text{ and } \mathcal{E} = \sqrt{\frac{\hbar\omega}{2\epsilon_o V}} \tag{17.10}$$

Problem 17.3 *Show that for a Fock state,* $|n\rangle$, *The variance of the field is*

$$\langle \Delta\hat{E}^2 \rangle = \mathcal{E}^2 (2n + 1) \tag{17.11}$$

Notice that even for the vacuum $|n = 0\rangle$, there are fluctuations in the field. These are referred to as **vacuum field fluctuations** and are a direct consequence of the non-commutation of the operators.

Problem 17.4 *Measurements of energy may be characterized by the photon number operator or possibly by the corresponding measurements of* $\hat{E}^2$, *which is proportional to the intensity. Show that the variance for* $\hat{E}^2$ *for a system in a Fock state is given by*

[1] While assuming that the number operator is an appropriate representation for a number calculation involving photon counting detectors, most detectors work on the same general principle of the field inducing a polarization or other physical process such as being absorbed and transforming the energy in that electromagnetic mode to an electron or hole that is then electrically detected. The physics of how a detector transforms a quantum system into a classical signal remains a subject of continued discussion.

$$\langle \Delta \hat{E}^4 \rangle = \langle \hat{E}^4 \rangle - \langle \hat{E}^2 \rangle^2 = \mathcal{E}^4 2\left(n^2 + n + 1\right) \tag{17.12}$$

Note: There are 16 terms in $\langle \hat{E}^4 \rangle$ *that have to be evaluated from the expression for* $\langle \hat{E}^4 \rangle = \langle (\hat{a} - \hat{a}^\dagger)^4 \rangle$. *However, this is greatly simplified by first recognizing that since the bra and ket are the same, only terms that have two annihilation operators and two creation operators like* $\hat{a}\hat{a}^\dagger\hat{a}\hat{a}^\dagger$ *will contribute. The next step is then to put each of these combinations in* ***normal order****, meaning that all the creation operators are on the left and annihilation operators on the right, so that the operator written like* $\hat{a}\hat{a}^\dagger$ *is written as* $\hat{a}\hat{a}^\dagger = 1 + \hat{a}^\dagger\hat{a}$. *This often facilitates calculations since then* $\hat{a}^\dagger$ *operating on the bra (to the left) operates as* $\hat{a}$ *on the ket to the right. So, for example, using the commutator* $\left[\hat{a}, \hat{a}^\dagger\right] = 1$ *to replace* $\hat{a}\hat{a}^\dagger$ *with* $\left(\hat{a}^\dagger\hat{a} + 1\right)$ *we get:*

$$\begin{aligned} \hat{a}\hat{a}^\dagger\hat{a}\hat{a}^\dagger &= \left(\hat{a}^\dagger\hat{a} + 1\right)\left(\hat{a}^\dagger\hat{a} + 1\right) = \hat{a}^\dagger\hat{a}\hat{a}^\dagger\hat{a} + 2\hat{a}^\dagger a + 1 = \hat{a}^\dagger\left(\hat{a}^\dagger\hat{a} + 1\right)\hat{a} + 2\hat{a}^\dagger a + 1 \\ &= \hat{a}^\dagger\hat{a}^\dagger\hat{a}\hat{a} + 3\hat{a}^\dagger a + 1 \end{aligned} \tag{17.13}$$

In measurements, the finite variance gives rise to a spread in the results for multiple measurements of the system, following reinitialization. The spread in measurements is given by the quantum standard deviation, which is the square root of the variance. Hence, in Eq. 17.12 for example, the standard deviation of the square of the field (intensity) is given by

$$\sqrt{\langle \Delta \hat{E}^4 \rangle} = \mathcal{E}^2\sqrt{2\left(n^2 + n + 1\right)} \tag{17.14}$$

17.3 The Coherent State

In Chapter 7, the non-commutation relation of the operators associated with the canonical coordinate and conjugate momentum led directly to a fundamental uncertainty associated with the measurement of these quantities. The uncertainty is reflected in the Heisenberg uncertainty relation. It was found that there existed a state that could minimize this uncertainty to the fundamental limit set by the Heisenberg relation, which we called the minimum uncertainty state (Eq. 7.97) that was the solution to a first order differential equation in terms of the coordinate and momentum operator (Eq. 7.95). We saw then in Problem 8.8 that the operator that gave rise to the minimum uncertainty state in Chapter 7 is easily related to the annihilation operator, $\hat{a}$. There we explored the properties of the *eigenfunctions* of $\hat{a}$ called the **coherent state**. We consider the same problem here, but now for the quantized radiation field.

This state has many properties similar to a classical monochromatic field. The coherent state of the radiation field, $|\alpha\rangle$, is defined by the eigenstate of the annihilation operator,

$$\hat{a}|\alpha\rangle = \alpha|\alpha\rangle \tag{17.15}$$

As we did in Problem 8.8, the form of the coherent state is found by expanding in terms of the number states, $|n\rangle$:

$$|\alpha\rangle = \sum_n c_n|n\rangle \tag{17.16}$$

The normalized result from Problem 8.8 is

$$|\alpha\rangle = \sum_{n=0} e^{-\frac{|\alpha|^2}{2}} \frac{\alpha^n}{\sqrt{n!}} |n\rangle \tag{17.17}$$

Problem 17.5 *For*

$$|\psi_{rad}(t=0)\rangle = |\alpha\rangle$$

show that

$$|\psi_{rad}(t)\rangle = e^{-i\frac{\omega}{2}t} \sum_{n=0} e^{-\frac{|\alpha|^2}{2}} e^{-i\omega n t} \frac{\alpha^n}{\sqrt{n!}} |n\rangle \tag{17.18}$$

The coherent state is remarkably important because this corresponds, for example, to the output of an ideal laser. Classically, the output of an ideal laser is a pure monochromatic sinewave. To see this in the coherent state, we consider the expectation value of the classical field. Recall that the operator for single mode the transverse electric field from Eq. 17.10 at position z is (ignoring the vector nature of the field)

$$\hat{E} = i\mathcal{E}\left(\hat{a}e^{ikz} - \hat{a}^{\dagger}e^{-ikz}\right) \tag{17.19}$$

Problem 17.6 Using $|\psi_{rad}(t)\rangle$ in Eq. 17.18,

a) Show that, for a coherent state with $\alpha = |\alpha| e^{i\varphi}$,

$$\langle \hat{E}(t)\rangle = \langle \hat{E}(t)\rangle = -2\mathcal{E}|\alpha|\sin(kz - \omega t + \varphi) \tag{17.20}$$

b) Repeat problem (a) using the Heisenberg picture and recover the same result, beginning with showing that in the Heisenberg picture,

$$\hat{a}(t) = \hat{a}e^{-i\omega t} \tag{17.21}$$

$$\hat{a}^{\dagger}(t) = \hat{a}^{\dagger}e^{i\omega t} \tag{17.22}$$

Hence the coherent state results in a monochromatic field that propagates like a simple plane wave with an amplitude given by $2\mathcal{E}|\alpha|$. *This is quite in contrast to the result if the system were in a Fock state, where we saw above that* $\langle \hat{E}\rangle = 0$.

Problem 17.7 *Show that* $\langle n\rangle$, $\langle n^2\rangle$, *and* $\langle \Delta n^2\rangle$ *for a coherent state are given by:*

$$\langle n\rangle = |\alpha|^2 \equiv \bar{n} \tag{17.23}$$

$$\langle n^2\rangle = |\alpha|^4 + |\alpha|^2 \tag{17.24}$$

$$\langle \Delta n^2\rangle = |\alpha|^2 \tag{17.25}$$

Problem 17.8 *a) Find the probability of measuring n photons in a coherent state, and show that it is a Poisson distribution given by*

$$P(n) = |\langle n|\alpha\rangle|^2 = e^{-|\alpha|^2}\frac{\alpha^{2n}}{n!} \tag{17.26}$$

b) Show the variance $\langle \Delta\hat{E}^2(t)\rangle$ *is*

$$\begin{aligned}\langle \Delta\hat{E}^2(t)\rangle &= \mathcal{E}^2\left(4|\alpha|^2\sin^2(kz-\omega t+\varphi)+1\right)-\mathcal{E}^2|\alpha|^2\\ &= \mathcal{E}^2\left(4|\alpha|^2\sin^2(kz-\omega t+\varphi)+1-|\alpha|^2\right)\end{aligned} \tag{17.27}$$

17.4 Quantum Beam Splitter: Quantum Interference

Boundary Conditions and Energy Conversation Requirements

One of the more important elements in electromagnetics is a power splitter that takes incident power and divides it into two different optical paths. It has different names, depending on the operating frequency, but from the infrared to the near ultraviolet, the element is usually called a **beam splitter**. This idea works well with our description of the quantized radiation field, because it assumes the electromagnetic mode is a plane wave with a well-defined k-vector, meaning a well-defined direction of propagation. The quantum description of the beam splitter is not only important for correctly describing the behavior of quantized fields, it is also important because the quantum properties of the beam splitter are exploited in numerous applications of this element for realizing important quantum devices.

Rather than go into the details of Maxwell's equations and boundary condition, we use energy conservation to establish the fundamental behavior of two classical beam-splitters. The details of implying boundary conditions simply give quantitative details which are not relevant to the fundamental results here. Just as with the quantization of the field applied to the classical electromagnetic modes of the problem (a cavity), the classical analysis of mode-behavior for the beam splitter applies to the quantum system. The classical analysis of the modes in a cavity served the same purpose in the original discussion of quantization of the field in Chapter 15.

The basic beam splitter, a four-port device, is shown in two standard configurations in Fig. 17.1. Since we are talking about a solution to the wave equation, we are dealing with fields. The physical origin of reflection at the interface is the subject of solutions to Maxwell's equations. However, assuming a reflection and transmission at the interface and no loss, we can establish the key fundamental relationships that are preserved, independent of the details, based simply on energy conservation.

Considering first the field E_a entering the a-port (left). We see that after the beam splitter the reflected and transmitted fields are given by $E_d = r_{ad}E_a$ and $E_c = t_{ac}E_a$ where r_{ad} is the reflectance relating the incident field, E_a, to the reflected field, E_d, and t_{ac} is the transmittance

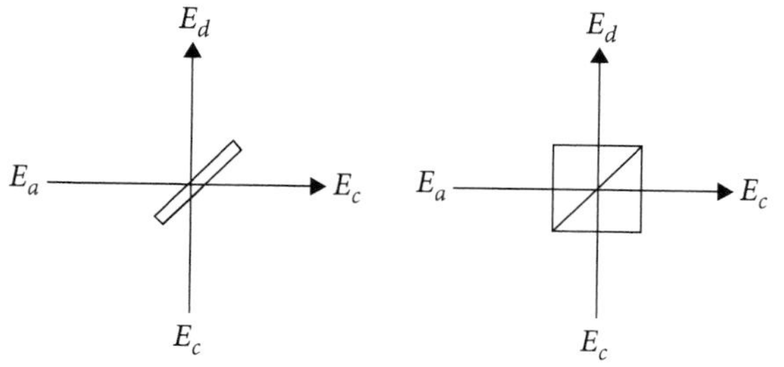

Fig. 17.1 A dielectric beam splitter. The left shows a typical asymmetric beam splitter made of high index material (compared to the vacuum). We assume there is only a reflection at one interface, the second interface has a coating to suppressed the reflection. The right symmetric beam splitter is made from two prisms where reflection occurs at the interface because of a small gap and partial total internal reflection.

relating the incoming field E_a to the transmitted field, E_c, respectively.[2] In general, r_{ad} and t_{ac} are complex numbers, implying the possible existence of a phase change associated with the reflected and transmitted fields. The power per unit area (intensity) in the vacuum is $I = \frac{\epsilon_0 c}{2}|E|^2$. Since energy is conserved, the power exiting the beam splitter must equal the power entering. Hence, we require:

$$\frac{\epsilon_0 c}{2}|r_{ad}E_a|^2 + \frac{\epsilon_0 c}{2}|t_{ac}E_a|^2 = \frac{\epsilon_0 c}{2}|E_a|^2 \tag{17.28}$$

or, by cancelling the terms in common, a simpler and more intuitive expression is obtained:

$$|r_{ad}|^2 + |t_{ac}|^2 = 1 \tag{17.29}$$

Likewise, for the case of power entering from the b-port, energy conservation gives

$$|t_{bd}|^2 + |r_{bc}|^2 = 1 \tag{17.30}$$

If power is entering both a- and b-ports, then the total power in must equal the total power out. Neglecting factors common to each term, we get

$$|E_a|^2 + |E_b|^2 = |E_c|^2 + |E_d|^2 \tag{17.31}$$

Now writing $|E_c|^2$ and $|E_d|^2$ in terms of E_a and E_b we start with:

$$E_c = t_{ac}E_a + r_{bc}E_b \tag{17.32}$$

$$E_d = r_{ad}E_a + t_{bd}E_b \tag{17.33}$$

[2] Recall that the reflectance, r_{ij}, and transmittance, t_{ij} refer to the complex field amplitude, the reflection $R_{ij} = |r_{ij}|^2$, and transmission, $T_{ij} = |t_{ij}|^2$, coefficients refer to the power (or the square of the modulus of the complex field).

and then using these to rewrite $|E_c|^2 + |E_d|^2$ for the equality in Eq. 17.31, we get

$$|E_a|^2 + |E_b|^2 = |t_{ac}E_a + r_{bc}E_b|^2 + |r_{ad}E_a + t_{bd}E_b|^2 = \left(t_{ac}t_{ac}^* + r_{ad}r_{ad}^*\right)E_aE_a^* \\ + \left(t_{ac}r_{bc}^* + r_{ad}t_{bd}^*\right)E_aE_b^* + \left(r_{bc}t_{ac}^* + t_{bd}r_{ad}^*\right)E_bE_a^* + \left(r_{bc}r_{bc}^* + t_{bd}t_{bd}^*\right)E_bE_b^* \tag{17.34}$$

For the first and last terms on the right side of the equation, we get

$$\left(t_{ac}t_{ac}^* + r_{ad}r_{ad}^*\right)E_aE_a^* + \left(r_{bc}r_{bc}^* + t_{bd}t_{bd}^*\right)E_bE_b^* = \left(|t_{ac}|^2 + |r_{ad}|^2\right)|E_a|^2 \\ + \left(|r_{bc}|^2 + |t_{bd}|^2\right)|E_b|^2 = |E_a|^2 + |E_b|^2 \tag{17.35}$$

which is equal to the left-hand side of the equation. This establishes the equality and also proves that the middle terms must each be zero,

$$\left(t_{ac}r_{bc}^* + r_{ad}t_{bd}^*\right) = 0 \\ \left(r_{bc}t_{ac}^* + t_{bd}r_{ad}^*\right) = 0 \tag{17.36}$$

otherwise, energy would not be conserved. The requirement that these terms be zero establishes a phase relationship between the complex reflectance and transmittance coefficients, which must be satisfied. Writing each term (Eq. 17.36) in phasor form for the top equation (the results hold for the bottom equation because it is just the complex conjugate of the top equation) gives

$$|t_{ac}|e^{i\phi_{ac}}|r_{bc}|e^{-i\phi_{bc}} + |r_{ad}|e^{i\phi_{ad}}|t_{bd}|e^{-i\phi_{bd}} = 0 \tag{17.37}$$

or

$$|t_{ac}||r_{bc}| + |r_{ad}||t_{bd}|e^{i(\phi_{ad}-\phi_{ac}-\phi_{bd}+\phi_{bc})} = 0 \tag{17.38}$$

Because the absolute values of the prefactors are real and positive, this is satisfied if

$$(\phi_{ad} - \phi_{ac} - \phi_{bd} + \phi_{bc}) = \pm\pi \tag{17.39}$$

which then shows that

$$|t_{ac}||r_{bc}| - |r_{ad}||t_{bd}| = 0 \tag{17.40}$$

From here, we can compare the ratios of reflectivity and transmission:

$$\frac{|r_{bc}|^2}{|t_{bd}|^2} = \frac{|r_{ad}|^2}{|t_{ac}|^2} \tag{17.41}$$

Then, using Eqs 17.29 and 17.30,

$$\frac{1 - |t_{bd}|^2}{|t_{bd}|^2} = \frac{1 - |t_{ac}|^2}{|t_{ac}|^2} \tag{17.42}$$

and we find that

$$|t_{ac}| = |t_{bd}| = t \tag{17.43}$$

and

$$|r_{ad}| = |r_{bc}| = r \tag{17.44}$$

With these constraints, we can construct a transfer matrix from the input ports to the output ports:

$$\begin{bmatrix} E_c \\ E_d \end{bmatrix} = \begin{bmatrix} te^{i\phi_{ac}} & re^{i\phi_{bc}} \\ re^{i\phi_{ad}} & te^{i\phi_{bd}} \end{bmatrix} \begin{bmatrix} E_a \\ E_b \end{bmatrix} \tag{17.45a}$$

It is also possible to find the input fields based on the output fields:

$$\begin{bmatrix} E_a \\ E_b \end{bmatrix} = \begin{bmatrix} te^{-i\phi_{ac}} & re^{-i\phi_{ad}} \\ re^{-i\phi_{bc}} & te^{-i\phi_{bd}} \end{bmatrix} \begin{bmatrix} E_c \\ E_d \end{bmatrix} \tag{17.45b}$$

Note the transfer matrix, M is unitary, meaning $M^{\dagger} = M^{-1}$ and therefor

$$MM^{\dagger} = \begin{bmatrix} te^{i\phi_{ac}} & re^{i\phi_{bc}} \\ re^{i\phi_{ad}} & te^{i\phi_{bd}} \end{bmatrix} \begin{bmatrix} te^{-i\phi_{ac}} & re^{-i\phi_{ad}} \\ re^{-i\phi_{bc}} & te^{-i\phi_{bd}} \end{bmatrix} = \begin{bmatrix} 1 & 0 \\ 0 & 1 \end{bmatrix} \tag{17.46}$$

We note here that for the symmetric beam splitter on the right in Fig. 17.1, the behavior of the fields is the same, regardless of which direction they enter and exit. From the phase relationship given in Eq. 17.39, this means that $\phi_{ad} - \phi_{ac} = \phi_{bc} - \phi_{bd} = \pm\frac{\pi}{2}$. In this case, it is common to take $\phi_{ac} = \phi_{bd} = 0$ and $\phi_{ad} = \phi_{bc} = \pm\frac{\pi}{2}$.

In the case of the asymmetric beam splitter on the left of Fig. 17.1, it is known from the Fresnel equations that there is no sign difference on transmission from either direction.[3] However, the sign of the reflection coefficient is set by difference in the index of refraction. For $n_1 - n_2 < 0$ where n_1 is the index of refraction for the medium with the incident field and n_2 is the medium on the other side of the boundary, then the sign is negative. For $n_1 - n_2 > 0$; the sign is positive. Hence, for the left case in Fig. 17.1 for field E_a in the vacuum, $n_1 = 1$ and $n_2 > 1$, so the reflection phase $\phi_{ad} = \pm\pi$ giving for the reflectance $-r$ and for E_b, the sign is positive. So to be clear, $re^{i\phi_{ad}} = -r$, $re^{i\phi_{bc}} = +r$, and $te^{i\phi_{bd}} = te^{i\phi_{ac}} = t$.

Qualitatively, the results below are not sensitive to the above details.

Quantized Field for the Beam Splitter

If we now replace these fields with the field operators, we can find the relationship between the annihilation and creation operators at ports a and b to the annihilation and creation operators at ports c and d. We keep the discussion general to start with, leaving the phases arbitrary, subject to Eqs 17.38 and 17.39. Since the relationships between the operators is known, we work just with the annihilation operators,

$$\begin{aligned} \hat{a}_c &= te^{i\phi_{ac}}\hat{a}_a + re^{i\phi_{bc}}\hat{a}_b \\ \hat{a}_d &= re^{i\phi_{ad}}\hat{a}_a + te^{i\phi_{bd}}\hat{a}_b \end{aligned} \tag{17.47}$$

[3] Max Born and Emil Wolfe, *Principles of Optics*, Pergamon Press Ltd. London (1959).

or, expressing the input modes in terms of the output modes,

$$\begin{aligned}\hat{a}_a &= te^{-i\phi_{ac}}\hat{a}_c + re^{-i\phi_{ad}}\hat{a}_d \\ \hat{a}_b &= re^{-i\phi_{bc}}\hat{a}_c + te^{-i\phi_{bd}}\hat{a}_d\end{aligned} \tag{17.48}$$

Recalling that taking the adjoint of an equation includes taking the complex conjugate of all c-numbers (complex numbers), we find the raising operators for the output modes:

$$\begin{aligned}\hat{a}_c^\dagger &= te^{-i\phi_{ac}}\hat{a}_a^\dagger + re^{-i\phi_{bc}}\hat{a}_b^\dagger \\ \hat{a}_d^\dagger &= re^{-i\phi_{ad}}\hat{a}_a^\dagger + te^{-i\phi_{bd}}\hat{a}_b^\dagger\end{aligned} \tag{17.49}$$

or the input modes in terms of the output modes:

$$\begin{aligned}\hat{a}_a^\dagger &= te^{i\phi_{ac}}\hat{a}_c^\dagger + re^{i\phi_{ad}}\hat{a}_d^\dagger \\ \hat{a}_b^\dagger &= re^{i\phi_{bc}}\hat{a}_c^\dagger + te^{i\phi_{bd}}\hat{a}_d^\dagger\end{aligned} \tag{17.50}$$

We know that before the beam splitter, the operators satisfy the commutation relation $\left[\hat{a}_i, \hat{a}_j^\dagger\right] = \delta_{ij}$. We evaluate $\left[\hat{a}_c, \hat{a}_c^\dagger\right]$ to find:

$$\begin{aligned}\left[\hat{a}_c, \hat{a}_c^\dagger\right] &= \left[te^{i\phi_{ac}}\hat{a}_a, te^{-i\phi_{ac}}\hat{a}_a^\dagger\right] + \left[re^{i\phi_{bc}}\hat{a}_b, re^{-i\phi_{bc}}\hat{a}_b^\dagger\right] \\ &= t^2\left[\hat{a}_a, \hat{a}_a^\dagger\right] + r^2\left[\hat{a}_b, \hat{a}_b^\dagger\right] = r^2 + t^2 = 1\end{aligned} \tag{17.51}$$

This is a primary requirement in this problem to be consistent with the postulates in Chapter 7. Similarly $\left[\hat{a}_d, \hat{a}_d^\dagger\right] = 1$. In the case of $\left[\hat{a}_c, \hat{a}_d^\dagger\right]$, we get

$$\begin{aligned}\left[\hat{a}_c, \hat{a}_d^\dagger\right] &= \left[te^{i\phi_{ac}}\hat{a}_a + re^{i\phi_{bc}}\hat{a}_b, re^{-i\phi_{ad}}\hat{a}_a^\dagger + te^{-i\phi_{bd}}\hat{a}_b^\dagger\right] = te^{i\phi_{ac}}\, re^{-i\phi_{ad}} + re^{i\phi_{bc}}te^{-i\phi_{bd}} \\ &= tre^{i\phi_{ac}-i\phi_{ad}}\left(1 + e^{-i\phi_{ac}+i\phi_{ad}+i\phi_{bc}-i\phi_{bd}}\right) = 0\end{aligned} \tag{17.52}$$

where we used the fact that $(\phi_{ad} - \phi_{ac} - \phi_{bd} + \phi_{bc}) = \pm\pi$. The above results show that, as required, the commutation relations hold for all modes.

We note that it might have been assumed that, as in the classical case, if there is no field present in port b, then the output would only be in terms of port a. So, in Eq. 17.49, for example, there would have been no port b term. In this case, the required commutation relations would not have been preserved after the beam splitter. Moreover, in the case of the quantized field, *even if there is no input in port b, the vacuum field is still present.* As we show below, this leads to noise in the measurement of the output.

With the above information, we are now able to examine the effect of a beam splitter on the quantized field. To make the details of the calculations that follow more straightforward but with no qualitative impact on the final results, we assume the asymmetric form of the beam splitter in Fig. 17.1, and use the above results to assume: $re^{i\phi_{ad}} = -r$, $re^{i\phi_{bc}} = r$, and $te^{i\phi_{bd}} = te^{i\phi_{ac}} = t$. We also we assume a 50/50 beam splitter meaning $t^2 = r^2 = \frac{1}{2}$. The equations that are needed are summarized in Table 17.1.

Table 17.1 Input and output operators for the asymmetric 50/50 beam splitter on the left of Fig. 17.1.

$\hat{a}_c = (t\hat{a}_a + r\hat{a}_b)$	$\hat{a}_a = (t\hat{a}_c - r\hat{a}_d)$
$\hat{a}_d = (-r\hat{a}_a + t\hat{a}_b)$	$\hat{a}_b = (r\hat{a}_c + t\hat{a}_d)$
$\hat{a}_c^\dagger = \left(t\hat{a}_a^\dagger + r\hat{a}_b^\dagger\right)$	$\hat{a}_a^\dagger = \left(t\hat{a}_c^\dagger - r\hat{a}_d^\dagger\right)$
$\hat{a}_d^\dagger = \left(-r\hat{a}_a^\dagger + t\hat{a}_b^\dagger\right)$	$\hat{a}_b^\dagger = \left(r\hat{a}_c^\dagger + t\hat{a}_d^\dagger\right)$

Effect of the Beam Splitter on the Noise of an n-Photon Fock State and Coherent State Entering Through One Port

The question of interest here is to understand the effect of the beam splitter on the noise in the measurement: specifically, using photon counting in port c (or d), what is the noise? If we are measuring say port c, then we must calculate $\langle \Delta\hat{n}_c^2 \rangle$. However, port c has contributions from ports a and b. Hence, referring to Table 17.1,

$$\hat{n}_c = \hat{a}_c^\dagger \hat{a}_c = \left(t\hat{a}_a^\dagger + r\hat{a}_b^\dagger\right)(t\hat{a}_a + r\hat{a}_b) = \left(t^2\hat{a}_a^\dagger\hat{a}_a + tr\hat{a}_a^\dagger\hat{a}_b + tr\hat{a}_b^\dagger\hat{a}_a + r^2\hat{a}_b^\dagger\hat{a}_b\right) \tag{17.53}$$

We must evaluate $\langle \Delta\hat{n}_c{}^2 \rangle = \langle \hat{n}_c{}^2 \rangle - \langle \hat{n}_c \rangle^2$. We first write out the expression for $\hat{n}_c^2$ in terms of the operators for ports a and b and then put them in normal order. The expression for $\hat{n}_c$ above is already in normal order. In Problem 17.9 you will examine $\langle \Delta\hat{n}_c^2 \rangle$ for a Fock state entering in port a and the vacuum at port b. That is to say, the input state is $|n_a 0_b\rangle$. If there was no vacuum included in port b, then the result for $\langle \Delta\hat{n}_c^2 \rangle$ would be 0, as expected for a Fock state. However, this is unphysical because it implies that somehow the photons "sort themselves out" so as to give exactly $t^2 n_a$ for the output. In problem 17.9, you find that there is noise and that $\langle \Delta\hat{n}_c^2 \rangle = t^2 r^2 n_a$. This term arises from the presence of the vacuum in port b as evidenced by the r^2 that represents reflection of port b input into port c. The fluctuations (standard deviation) scale as $\sqrt{n_a}$, like a Poissonian.

Problem 17.9 *Evaluate the quantum variance,* $\langle \Delta\hat{n}_c^2 \rangle$*, assuming an input state* $|n_a 0_b\rangle$.

Problem 17.10 *Evaluate the quantum variance, assuming that a coherent state enters through port c, the input state for the systems being* $|\alpha_a 0_b\rangle$.

Two-Photon Interference at a Beam Splitter: Hong-Ou-Mandel Interferometer

The **Hong-Ou-Mandel (HOM) interometer** is an especially important device for confirming the identical nature of the emission from a single photon emitter. Consider a one-photon Fock state entering through port a and another one-photon Fock state entering through port b. The initial state is then given by $\hat{a}_a^\dagger \hat{a}_b^\dagger |0\rangle = |1_a 1_b\rangle$. In Problem 17.11, you will show that for a 50/50 beam splitter, the unnormalized final state is given by $\frac{1}{2}$ $(|2_c 0_d\rangle - |0_c 2_d\rangle)$. The important physics here is that these two photons are identical, but in two different modes (a and b), while the beam splitter transforms both of them into the same mode. This result, first discovered by Hong, Ou, and Mandel, is a central tool in demonstrating that a quantum emitter is creating identical states. The system is also used in measurement-based entanglement.

Problem 17.11 *Show that, for the input state of* $|1_a 1_b\rangle$, *the unnormalized output state is given by* $\frac{1}{2}\left(|2_c 0_d\rangle - |0_c 2_d\rangle\right)$.

17.5 Resonant Rayleigh Scattering: A Single Quantum Emitter

Single quantum emitters are a key component in various quantum information applications, including for secure communications and sensors. The understanding of the mechanism by which radiation is emitted was developed in Chapter 15 when we discussed how the vacuum radiation field gives rise to spontaneous emission from an excited state. A standard approach to create a single photon emitter is to excite the system using a **π-pulse** of resonant radiation. Recall from Chapter 9 that a π-pulse is a half a cycle of a Rabi oscillation, when $|\Omega_R|t = \pi$ seen in Fig. 9.4, so that the entire probability which is initially unity for the ground state is now unity for the excited state. A single photon will be emitted with exponentially decreasing probability according to the decay rate of the state. Hence, we know with exponential probability when the photon is emitted.

If the excitation radiation is monochromatic, weak, tuned on or near resonance and on continuously (called CW for continuous wave), then the source will emit single photons but at unknown times. This kind of emission can also be important and is called **resonant Rayleigh scattering**. As you will see, the idea that the system absorbs light and then spontaneously decays, emitting a photon, is incorrect and leads to common errors.

Since the excitation field is a classical source, but emission arises due to coupling with the vacuum, we start with a Hamiltonian including both kinds of interaction. Hence, we start with the Hamiltonian in 16.56 and add to it the interactions terms with the classical field from Eq. 15.119:

$$\widehat{H} = \frac{\hbar\omega_{21}}{2}(\hat{\sigma}_2 - \hat{\sigma}_1) - \frac{\hbar}{2}\Omega_R e^{-i\omega t}\hat{\sigma}_{21} - \frac{\hbar}{2}\Omega_R^* e^{i\omega t}\hat{\sigma}_{12}$$
$$+ \hbar\sum_j \left(g_j\hat{\sigma}_+\hat{a}_j + g_j^*\hat{a}_j^\dagger\hat{\sigma}_-\right) + \sum_j \hbar\omega_k\hat{a}_j^\dagger\hat{a}_j \tag{17.54}$$

where $g_j = -i\frac{\mathcal{E}_j\mu_{21}}{\hbar}\hat{r}_{21}\cdot\check{\boldsymbol{\varepsilon}}_j$; $g_j^* = i\frac{\mathcal{E}_j^*\mu_{12}}{\hbar}\check{r}_{12}\cdot\check{\boldsymbol{\varepsilon}}_j$, $\mathcal{E}_j \equiv \sqrt{\frac{\hbar\omega_J}{2\epsilon_o V}}$ and $\check{\mathbf{r}}_{ij}$ is a matrix element of the unit vector. The $\hat{\sigma}_i$ operators are defined in Chapter 16 (see, e.g., Table 16.1) and $\Omega_R = \frac{\mu_{21}\mathcal{E}_0}{\hbar}(\check{\mathbf{r}}_{21}\cdot\check{\varepsilon})$, $\Omega_R^* = \frac{\mu_{12}\mathcal{E}_0^*}{\hbar}(\check{\mathbf{r}}_{12}\cdot\check{\varepsilon})$.

In the earlier work on spontaneous emission we simply integrated over all the states of the radiation field, assuming the system was prepared in an excited state. Now we will start the calculation as we did when we started linear absorption, with the system in the ground state $|1,0\rangle$ but now we note there are no photons in the vacuum. The system will be excited to state $|2,0\rangle$, again with no photons in the vacuum using a classical monochromatic source, and then in the process of coupling to the vacuum field, it will return to the ground state with one photon in mode $\boldsymbol{k}\sigma$, $|1, \boldsymbol{k}\sigma\rangle$ where $\boldsymbol{k}$ represents the $\boldsymbol{k}$-vector (direction of propagation with $|\boldsymbol{k}|\rangle = k = \frac{\omega}{c}$) with σ polarization. We assume weak excitation, use perturbation theory, and modify the equation of motion to include spontaneous emission. Intuitively, we might expect the emission spectrum for scattered light to mirror the absorption spectrum. That would make sense if the idea was that the system absorbed a photon and then emitted a photon. The results will be quite different.

We adapt Eq. 15.123:

$$\dot{C}_{2,0}(t) = -\frac{1}{2}\gamma C_{2,0}(t) - \frac{1}{2i}\Omega_R e^{-i(\omega-\omega_{21})t} C_{1,0}(t) \tag{17.55}$$

where we have included the decay of the excited state due to spontaneous emission. In perturbation theory, we will assume $C_{1,0}(0) = 1$, (no photons in the k σ mode) as we did for linear absorption, though there we did not keep track of scattered photons. To second order in perturbation theory, we could consider using Eq. 15.121 to see the effect of driving the system back to the ground state. However, that is not the interest here, nor is it necessary to calculate all possible second order terms. We need to calculate only the term of interest, which is $C_{1,k\sigma}(t)$, one photon in the k σ mode, in the perturbation limit, when $\Omega_R \ll \gamma$. Here, the equation of motion is given by

$$\dot{C}_{1,k\sigma}(t) = -ig^*_{k\sigma} e^{i(\omega_k-\omega_{21})} C_{2,0}(t) \tag{17.56}$$

Problem 17.12 *Starting with Eq. 15.119, and the Hamiltonian in 17.54, derive Eq. 17.56 by ignoring the coupling by the classical field from states 2 to 1. This assumption is justified since the measurement is on the scattered field, not the ground state of the emitter.*

Integrating Eq. 17.55 we get, Eq. 15.125,

$$C_{2,0}(t) = -\frac{1}{2i}\Omega_R \frac{e^{-i(\omega-\omega_{21})t} - e^{-\frac{1}{2}\gamma t}}{\frac{1}{2}\gamma - i(\omega - \omega_{21})} \tag{17.57}$$

As before, we ignore the transient response since we are interested in long time scales. Inserting the result into the equation of motion for $\dot{C}_{1,k\sigma}(t)$ (Eq. 17.56) and integrating

$$\begin{aligned} C_{1,k\sigma}(t) &= -ig^*_{k\sigma}\left(-\frac{1}{2i}\Omega_R\right)\frac{1}{\frac{\gamma}{2} - i(\omega - \omega_{21})}\int_0^t dt' e^{-i(\omega-\omega_k)t'} \\ &= \left(\frac{ig^*_{k\sigma}\Omega_R}{2(\omega-\omega_k)}\right)\frac{e^{-i(\omega-\omega_k)t} - 1}{\frac{\gamma}{2} - i(\omega-\omega_{21})} = \left(\frac{-g^*_{k\sigma}\Omega_R}{2(\omega-\omega_k)}\right)\frac{2\sin\frac{(\omega-\omega_k)}{2}t}{\frac{\gamma}{2} - i(\omega-\omega_{21})} e^{-i\frac{(\omega-\omega_k)}{2}t} \end{aligned} \tag{17.58}$$

Using Appendix C for functions that behave like a Dirac delta-function[4] gives for the probability,

[4] $\delta(\omega - \omega_k) = 2\lim_{t\to\infty}\frac{\sin^2\frac{(\omega-\omega_k)t}{2}}{(\omega-\omega_k)^2 t}$: note the discussion of Eq. 14.29 to understand the physical requirement that justifies assuming t "effectively" goes to infinity while in reality it can be quite short.

$$|C_{1,k\sigma}(t)|^2 = \frac{|g_{k\sigma}^* \Omega_R|^2}{\left(\frac{\gamma}{2}\right)^2 + (\omega - \omega_{21})^2} \frac{sin^2 \frac{(\omega-\omega_k)}{2} t}{(\omega - \omega_k)^2} \underset{lm\ t\to\infty}{\longrightarrow} \frac{2|g_{k\sigma}^* \Omega_R|^2 t}{\left(\frac{\gamma}{2}\right)^2 + (\omega - \omega_{21})^2} \delta(\omega - \omega_k) \quad (17.59)$$

The probability increases linearly with t and hence the scattering rate is given by dividing by t. Assuming that the monochromatic field has some small (compared to $\frac{\gamma}{2}$) but non-zero distribution of frequencies, integration over ω_k then shows that the frequency spectrum of the scattered photon is exactly the frequency spectrum of the classical incident monochromatic field centered at the excitation frequency ω. If the problem had been one of absorption followed by spontaneous emission, the spectrum would be centered at ω_{21}. Typical numbers for γ commonly range anywhere from ~MHz to THz. The output bandwidth of modern monochromatic lasers is typically <10 kHz to MHz. So in most cases, the linewidth of the scattered light is then the laser linewidth. This is a very narrowband compared to the transition linewidth, an important feature for application of **single photon emitters**.

We refer to the system as a single photon emitter because it can only emit one photon at a time. In the laboratory, it is important to prove this. Thinking physically, as the system is excited and the probability amplitude of the upper state increases, the system emits a photon at time t. The probability of emitting a second photon at time t is zero. A function that describes the conditional probability is $g_2(\tau)$, and is called the **second order correlation function** in intensity (or fourth order in the field). It means that if a photon is detected at $t = 0$, what is the probability of detecting a photon at time $t = \tau$. For a single photon $g_2(\tau = 0) = 0$. For $\tau > 0$, $g_2(\tau > 0)$ rises to an asymptotic value. This effect is called **anti-bunching**, meaning that the chance of finding two photons close in time is vanishingly small as $\tau \to 0$. The calculation of g_2 is beyond the scope of this book, but a contemporary textbook discussion is presented by Berman and Malinovsky.[5]

17.6 Creating Quantum Entangled States Between a Photon and an Electron

In this problem, we now consider spontaneous emission from an excited state to one of two lower-lying (ground) states of the kind shown in Fig. 17.2, known as a three-level Λ-system. We assume the two transitions emit linearly orthogonally polarized light. The arrows representing the spontaneously emitted radiation field are going in opposite directions for convenience. There is no correlation between the direction of the corresponding k-vectors. The system is assumed to be initialized in $|3, 0\rangle$. This is normally done by first applying a long pulse from say state $|2\rangle$ to state $|3\rangle$ which, via spontaneous emission, eventually transfers the population in state $|2\rangle$ to state $|1\rangle$ and then applying a π-pulse from state $|1\rangle$ to state $|3\rangle$ to put the probability of state $|3\rangle$ to unity.

[5] Paul R. Berman and Vladimir S. Malinovsky, *Principles of Spectroscopy and Quantum Optics*, Princeton (2011).

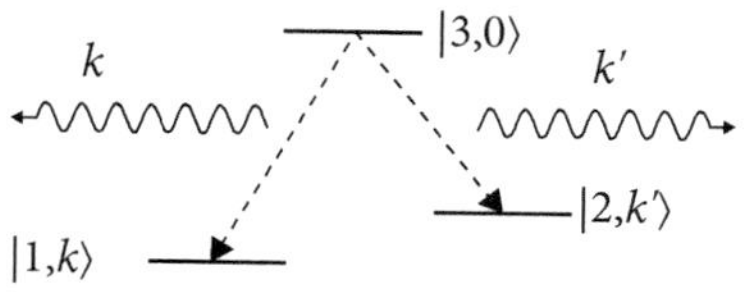

Fig. 17.2 A three level Λ system, initially in the excited state, $|3,0\rangle$ with no photons in any mode, decays to one of two ground states creating a coherent superposition of the and $|1,k\rangle$ the $|2,k'\rangle$ state.

The Hamiltonian for spontaneous emission in this system is then

$$\hat{H} = \hbar\left(\omega_3\hat{\sigma}_3 - \omega_1\hat{\sigma}_1\right) + \hbar\left(\omega_3\hat{\sigma}_3 - \omega_2\hat{\sigma}_2\right) + \hbar\sum_{k\sigma,k'\sigma'}\left(g^*_{1,k\sigma}\hat{a}^\dagger_{k\sigma}\hat{\sigma}_{31} + g^*_{2,k'\sigma'}\hat{a}^\dagger_{k'\sigma'}\hat{\sigma}_{32}\right)$$
$$+\hbar\sum_{k\sigma}\omega_k\hat{a}^\dagger_{k\sigma}\hat{a}_{k\sigma} \tag{17.60}$$

The state vector is then

$$|\psi\rangle = \sum_{k\sigma} c_{1,k\sigma}|1,k\sigma\rangle e^{-i(\omega_1+\omega_k)t} + \sum_{k'\sigma'} c_{2,k'\sigma'}|2,k'\sigma'\rangle e^{-i(\omega_2+\omega_{k'})t} + c_{3,0}|3,0\rangle e^{-i\omega_3 t} \tag{17.61}$$

After substituting into the time dependent Schrödinger equation and projecting onto each state as we did above in the study of Rayleigh scattering (Eq. 17.56), we get

$$\frac{dc_{1,k\sigma}}{dt} = -ig^*_{1,k\sigma}e^{-i(\omega_{31}-\omega_k)t}c_{3,0} \tag{17.62}$$

$$\frac{dc_{2,k'\sigma'}}{dt} = -ig^*_{2,k'\sigma'}e^{-i\left(\omega_{32}-\omega_{k'}\right)t}c_{3,0} \tag{17.63}$$

We assume that the system is initially in state $c_{3,0}$ as described above. The equation of motion is then

$$\frac{dc_{3,0}}{dt} = -\frac{1}{2}\left(\gamma_{13}+\gamma_{23}\right)c_{3,0} \tag{17.64}$$

Since the initial condition is that $|\psi\rangle = |3,0\rangle$, i. e., $c_{3,0}\,(t=0) = 1$ we have $c_{3,0}(t) = e^{-\frac{\gamma_3}{2}t}$ where $\gamma_3 = \gamma_{13} + \gamma_{23}$. We can then integrate $c_{i,j}$:

$$c_{1,k}(t) = -ig^*_{1,k\sigma}\frac{e^{-i(\omega_{31}-\omega_k)t-\frac{\gamma_3}{2}t}-1}{i\left(\omega_{31}-\omega_k\right)+\frac{\gamma_3}{2}} \tag{17.65}$$

$$c_{2,k'}(t) = -ig^*_{2,k'\sigma'}\frac{e^{-i\left(\omega_{32}-\omega_{k'}\right)t-\frac{\gamma_3}{2}t}-1}{i\left(\omega_{32}-\omega_{k'}\right)+\frac{\gamma_3}{2}} \tag{17.66}$$

At long times, we then have for the state vector,

$$(|\psi\rangle = \sum_k ig_{1,k\sigma} \frac{e^{-i(\omega_1+\omega_k)t}}{i(\omega_{31}-\omega_k)+\frac{\gamma_3}{2}} |1, k\sigma\rangle + \sum_{k'} ig_{2,k'\sigma'} \frac{e^{-i(\omega_2+\omega_{k'})t}}{i(\omega_{32}-\omega_{k'})+\frac{\gamma_3}{2}} |2, k'\sigma'\rangle \quad (17.67)$$

The result is a non-factorizable state in a linear superposition of $|1, \boldsymbol{k}\rangle$ and $|2, \boldsymbol{k}'\rangle$ where, because they are orthogonally polarized and centered at different frequencies, the degrees of freedom that are entangled are the two electron quantum states (often spin up and spin down) and the two degrees of polarization or two different frequencies. The emitted photon in this case is a called a **flying qubit**, because it carries information about the two low-lying states.

As a side note to this section, note that, unlike the Rayleigh scattering spectrum in the previous section, this emission is directly from spontaneous emission and is centered at the emission peak and has the expected Lorentzian profile.

17.7 Engineering the Quantum Vacuum

The language of the quantum vacuum was introduced in Chapter 15 and applies to the quantized electromagnetic multimode spectrum in a cavity with all allowed modes empty, i.e., no photons. Following the quantization of the electromagnetic field in Chapter 15, it was used to show that a quantum system capable of coupling to the electromagnetic field would spontaneously decay from an excited state with no photons in the radiation field to a lower state with one photon spontaneously emitted into the radiation field. We then examined the Jaynes–Cummings Hamiltonian that had only a single mode with what amounts to an ideal cavity. The excited state then did not decay but oscillated at the vacuum Rabi frequency.

Practically, the decay is linked to the performance of modern devices such as light emitting diodes, lasers, etc. However, with the capability of today's technology, it is now possible to control this decay. The spontaneous emission rate can be enhanced or suppressed. Depending on how the results are exploited, it is possible to improve device performance, reduce energy consumption, reduce noise, and alter the quantum coherence. It is even possible to control the direction of emission so that, compared to spontaneous emission which is into 4π, in principle it can be limited to one mode.

Not unexpectedly, these ideas can become relatively complex and detailed as more possibilities are considered. However, the basic idea is seen by considering a quantum emitter in a cavity with mirrors (Fig. 17.3). The goal here is to understand the underlying quantum behavior, learn how to do these calculations, do this in a system of practical interest, and avoid some of the complexities associated with simultaneously working with Maxwell's equations beyond what was done in earlier chapters. Earlier, the field was quantized in the cavity but there were no mirrors in the cavity to feedback energy. The cavity worked to established conditions on the mode. Here, we assume that the cavity is bounded by mirrors. This is the case of a weak cavity compared to Jaynes–Cummings, where there is only one mode. The physics we observe is called the **Purcell effect**.

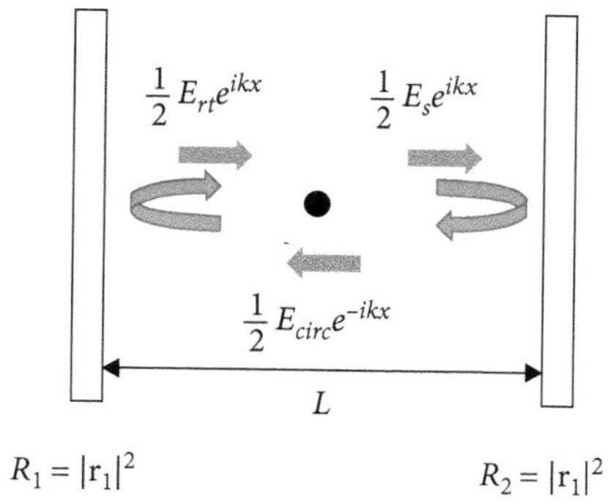

Fig. 17.3 Intracavity field strength, Eq. 17.63, as a function of mode number

The quantum emitter has two levels and is confined in a cavity formed by two reflecting mirrors, shown in Fig. 17.3. The mirrors have a reflectivity of R_1 and R_2, respectively, where $R_i = |r_i|^2$ and r_i is the reflectance. Such a cavity is called a **resonator** or a **Fabry–Pérot etalon.**

We first have to determine the strength of the field inside the cavity, assuming the emitter generates a field. That field propagates to the mirrors and then most of it is reflected back to add, constructively or destructively, to the source field. The cavity is small enough that retardation effects are ignored. We assume a CW source to determine the field enhancement and then normalize it for a single quantum of energy (i.e., one photon).

Assume that the electromagnetic field is a simple linearly polarized plane wave and the quantum emitter is located at x. The propagating field at x arising from the two-level system (called the source) is given by

$$E = \frac{1}{2}E_S e^{\pm ikx} + c.c. \tag{17.68}$$

where c.c. means complex conjugate and the plus (minus) sign is for propagating to the right (left). Then $k = \frac{2\pi}{\lambda}$, $\lambda = \frac{c}{\upsilon}$ and υ is the frequency of the field in hertz. We take E_S to be real and linearly polarized. This is an approximation since the emitter radiates as an electric dipole. In the resonator, the field that is propagating to the right is then reflected by the mirror on the right and propagates to the left where it is reflected by the other mirror and then propagates back to the right. At the source, the round trip has resulted in the field accumulating a phase given by 2φ where

$$2\varphi = k2L = \frac{\omega}{c}2L \tag{17.69}$$

where as usual $\omega = 2\pi\upsilon$ is the natural frequency of the field in radians. For the purposes of illustration, we assume that the cavity is the vacuum and the index of refraction is 1.

In steady state, the source is emitting and there is a steady state field in the cavity described as the circulating field given at the position x by $\frac{1}{2}E_{circ}e^{\pm ikx}$. If we relate the fields propagating in the positive direction at the position x, we get in steady state

$$E_{circ} = E_s + E_{rt} = E_s + r_1 r_2 e^{2i\varphi}E_{circ} \tag{17.70}$$

where $E_{rt} = r_1 r_2 e^{2i\varphi} E_{circ}$. Solving for E_{circ} we then get[6]

$$E_{circ} = \frac{E_s}{1 - r_1 r_2 e^{2i\varphi}} \tag{17.71}$$

The ratio of the circulating intensity to the source intensity (recall that the electromagnetic field intensity in vacuum is $\frac{\epsilon_0 c}{2}|E|^2$) is given by

$$f(\varphi) = \left|\frac{E_{circ}}{E_s}\right|^2 = \frac{1}{1 - 2\sqrt{R_1 R_2} cos\, 2\varphi + R_1 R_2} = \frac{1}{1 - 2\sqrt{R_1 R_2}\,(1 - 2\,sin^2\,\varphi) + R_1 R_2}$$
$$= \frac{1}{\left(1 - \sqrt{R_1 R_2}\right)^2 + 4\sqrt{R_1 R_2}\, sin^2\,\varphi} \tag{17.72}$$

where $\varphi = kL = 2\pi\frac{L}{\lambda} = 2\pi\frac{L}{c}\,\nu = \frac{L}{c}\omega$. The function $f(\varphi)$ in Eq. 17.72 is called the **Airy distribution**. Notice that when

$$\varphi = m\pi, \quad m = 0, \pm 1, \pm 2, \ldots \tag{17.73}$$

corresponding to $m\lambda = 2L$, then the denominator is a minimum and the circulating field intensity is a maximum, given by

$$f(\varphi = m\pi) = \left|\frac{E_{circ}}{E_s}\right|^2 = \frac{1}{\left(1 - \sqrt{R_1 R_2}\right)^2} \tag{17.74}$$

The resonance or mode condition can be rewritten so it is linear in ν

$$m = \frac{2L}{c}\,\nu \tag{17.75}$$

and the separation between successive values of m is the **free spectral range** given for the frequency as

$$\Delta\nu_{fsr} = \frac{c}{2L} \tag{17.76}$$

This is shown in Fig. 17.4.

There is an infinite series of resonances going to higher frequencies (shorter wavelengths). Since the bandwidth of interest to us is given by the inverse of the spontaneous emission rate which, by design, tends to be small compared to the free spectral range, we work with a cavity resonance at or near the frequency of the quantum emitter. Define the cavity resonance wavelength for the etalon, λ_{cav} for a specific m as

$$\lambda_{cav} = \frac{c}{\nu_{cav}} = \frac{2L}{m} \tag{17.77}$$

[6] See discussion by Nur Ismail, Cristine Calil Kores, Dimitri Geskus, and Markus Pollnau, "Fabry-Pérot resonator: spectral line shapes, generic and related Airy distributions, linewidths, finesses, and performance at low or frequency-dependent reflectivity" *Optics Express* 24, 16366 (2016).

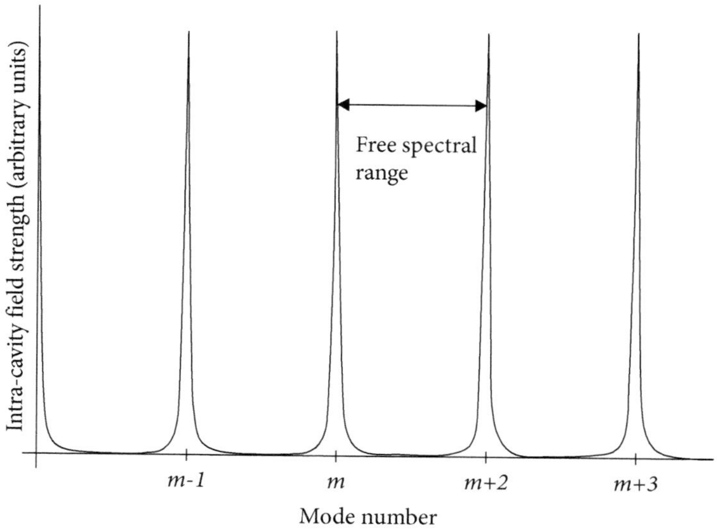

Fig. 17.4 Intra-cavity field strength as a function of the mode number showing resonances at integer numbers, separated by the free spectral range.

with $\lambda_{source} \sim \lambda_{cav}$. Often, the approach tunes either the source or the cavity so that the two wavelengths are identical (on resonance). For the source frequency slightly away from the cavity resonance frequency we can write for the phase for frequency, $\nu = \nu_{cav} + \delta\nu$, near the cavity resonance, ν_{cav}, as

$$\varphi = 2\pi \frac{L}{c} \nu = 2\pi \frac{L}{c} (\nu_{cav} + \delta\nu) = \left(m\pi + 2\pi \frac{L}{c} \delta\nu \right) \tag{17.78}$$

where $\delta\nu$ is the detuning from the cavity resonance, $\delta\nu \equiv \nu - \nu_{cav}$. Then

$$\sin\varphi = \sin\left(m\pi + 2\pi \frac{L}{c} \delta\nu \right) = \sin\left(\frac{2\pi L}{c} \delta\nu \right) \tag{17.79}$$

In the case of interest here, we will work with the quantum system on a cavity resonance, and assume $\frac{\delta\nu}{\nu_{cav}} \ll 1$.

Hence, we use the small argument expansion for the sine (see Appendix B) and set

$$\sin\varphi = \varphi = 2\pi \frac{L}{c} \delta\nu \tag{17.80}$$

The Airy function then becomes

$$f(\varphi) \equiv f(\delta\nu) = \frac{1}{\left(1 - \sqrt{R_1 R_2}\right)^2 + R_1 R_2 \left(\frac{2\pi L}{c} \delta\nu \right)^2} \tag{17.81}$$

Or converting to radians per second,

$$f(\delta\omega) = \frac{1}{\left(1-\sqrt{R_1R_2}\right)^2 + R_1R_2\left(\frac{L}{c}\delta\omega\right)^2} = \frac{1}{\left(\sqrt{R_1R_2}\frac{L}{c}\right)^2}\frac{1}{\left(\frac{1-\sqrt{R_1R_2}}{\frac{L}{c}\sqrt{R_1R_2}}\right)^2 + (\omega-\omega_{cav})^2} \qquad (17.82)$$

This functional form is a **Lorentzian**. The spectral response is shown in Fig. 17.5.

We set the first term in the Lorentzian denominator to γ and $\delta = \omega - \omega_{cav}$ so that

$$f(\delta) = \frac{1}{\left(\sqrt{R_1R_2}\frac{L}{c}\right)^2}\frac{1}{\gamma^2+\delta^2} \qquad (17.83)$$

The full width at half maximum (FWHM) is given by

$$FWHM(radians) = 2\,\gamma = 2\frac{1-\sqrt{R_1R_2}}{\frac{L}{c}\sqrt{R_1R_2}} \qquad (17.84)$$

The above problem is appropriate when the frequency of the source is well-defined, meaning relatively monochromatic (the bandwidth is much less than the FSR). However, if there had been energy stored in the cavity at $t = 0$, we would have proceeded by examining the temporal response. Specifically, with a given amount of field intensity, $I_s\,(t=0) = \frac{\epsilon_0 c}{2}|E_S|^2$, initially stored in the cavity, the intensity of the field strength in the cavity after one round trip is $R_1 R_2\, I_s\,(t = 0)$. After n round trips, there is $(R_1R_2)^n\, I_s\,(t = 0)$. Physically, the time for n round trips is $t = nt_{rt}$ where the round trip time is $t_{rt} = \frac{2L}{c}$. To represent this as a rate, we set:

$$I_s(R_1R_2)^{\frac{t}{t_{rt}}} = I_s e^{-\frac{t}{\tau_{cav}}} \qquad (17.85)$$

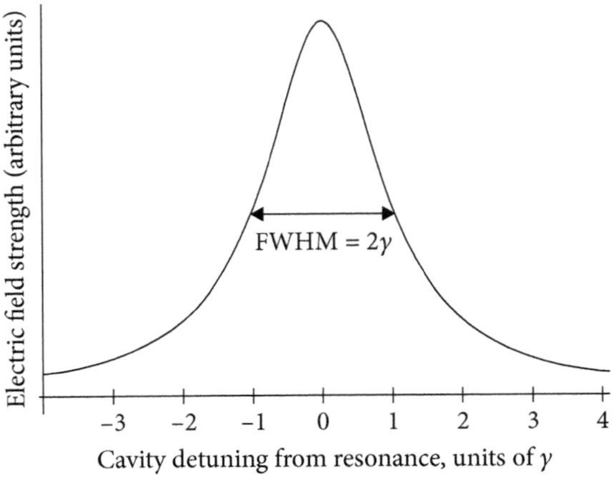

Fig. 17.5 The electric field strength plotted as a function frequency (Eq.18.82) centered at a cavity resonance (Fig. 17.4).

and solve for τ_{cav} where τ_{cav} is the **cavity lifetime**:

$$-\frac{ln\ R_1R_2}{t_{rt}} = \frac{1}{\tau_{cav}} > 0 \tag{17.86}$$

To compare this to the earlier result, we assume again that R_1R_2 approaches 1. Then

$$\begin{aligned}\frac{1}{\tau_{cav}} &= -\frac{ln\ R_1R_2}{t_{rt}} = -\frac{c}{2L}\ln R_1R_2 = -\frac{c}{L}\ln\sqrt{R_1R_2} = -\frac{c}{L}\ln\left[1-\left(1-\sqrt{R_1R_2}\right)\right]\\ &\cong \frac{c}{L}\left(1-\sqrt{R_1R_2}\right)\end{aligned} \tag{17.87}$$

Since in the limit

$$R_1R_2 \to 1, 1-\sqrt{R_1R_2}$$

is a small number and therefore we can use the expansion for the ln-function of $\ln(1-\epsilon) \cong -\epsilon$ for small ϵ (see Appendix B). Comparing this to the expression for γ from the Airy function, in the limit $R_1R_2 \to 1$

$$\gamma = \frac{1-\sqrt{R_1R_2}}{\frac{L}{c}\sqrt{R_1R_2}} \cong \frac{c}{L}\left(1-\sqrt{R_1R_2}\right) \tag{17.88}$$

We see

$$\frac{1}{\tau_{cav}} = \gamma \tag{17.89}$$

This is expected since the frequency response of the Airy function must be related to the Fourier transform of the temporal function, i. e., the *Fourier transform of an exponential decay is a Lorentzian.*

In studies of applications of the resonators to various devices, there is an important parameter used to characterize the resonator: the **cavity Q**, where Q is called the **quality factor**. As in LC circuits and other resonator effects, the Q is defined in terms of the resonant frequency and the full width at half maximum of the resonance:

$$Q = \frac{\omega_{cav}}{2\gamma} \tag{17.90}$$

Physically, it corresponds to the ratio of the maximum energy stored in the cavity in one round trip time to the energy lost in the time corresponding to one radian of the cycle. This corresponds to the cavity resonance frequency, ω_{cav}, divided by the FWHM of the resonance, 2γ. In terms of Q we can rewrite the Airy function as

$$f(\omega) = \frac{1}{\left(\sqrt{R_1R_2}\frac{L}{c}\right)^2}\frac{1}{\left(\frac{\omega_{cav}}{2Q}\right)^2 + (\omega-\omega_{cav})^2} \tag{17.91}$$

The narrower the linewidth, i. e., the smaller γ corresponding to higher the mirror reflectivities, the stronger the circulating field. This is associated with a high-Q-cavity. Another measure of a cavity's performance is the **cavity finesse**, defined as the ratio of the free spectral range to the cavity linewidth:

$$\mathcal{F}_{cav} = \frac{\Delta\omega_{FSR}}{2\gamma} = \frac{\pi c}{2\gamma L} \tag{17.92}$$

As with Q, a high finesse is also associated with high circulating power.

To examine the impact on **spontaneous emission** in a cavity compared to in free space as we did in Chapter 15, we need to recall that the modes in the cavity were taken to be normalized. This is to ensure that one unit of energy in the field corresponds to one unit of energy in the cavity. To preserve this, we need to normalize the expression for $f(\omega - \omega_{cav})$. The result for the normalized distribution of modes is

$$f_N(\omega - \omega_{cav}) = \frac{\omega_{cav}}{2\pi Q} \frac{1}{\left(\frac{\omega_{cav}}{2Q}\right)^2 + (\omega - \omega_{cav})^2} \tag{17.93}$$

The lower limit on the normalization integral, used to determine the factor before the Lorentzian in Eq. 17.93, is 0 because $\omega \geq 0$. However, to make the result simpler to evaluate, we can extend the lower limit $-\infty$ in the limit where $Q >> 1$, which is the case of physical interest here since the contribution for unphysical negative frequency components is small.

The amplitude of the vacuum field for mode k is $\varepsilon_k \equiv \sqrt{\frac{\hbar\omega_k}{2\varepsilon_o V}}$. In the cavity, this is now modified by the normalized Lorentzian to be

$$\varepsilon_{k,cav} \equiv \sqrt{\frac{\hbar\omega_k}{2\varepsilon_o V}}\sqrt{f_N(\omega)} \tag{17.94}$$

Anticipating the results below, a quantum radiator on resonance in a high Q cavity will experience a strong circulating field associated with its emitted field if the cavity is tuned to resonance, meaning stronger coupling to the field. We would expect that the spontaneous emission rate would increase. Similarly, when the radiator is far off resonance in a high Q cavity the coupling will be extremely weak. Hence, we would expect the spontaneous emission to be reduced.

To understand how we can engineer the coupling of the emitter with the vacuum, we return to the Hamiltonian in Eq. 16.53 without the applied classical field:

$$\hat{H} = \omega_{21}\hat{\sigma}_2 - \frac{1}{2}\mu_{21} \cdot \hat{E}_+\hat{\sigma}_{21} - \frac{1}{2}\mu_{12} \cdot \hat{E}_-\hat{\sigma}_{12} + \hbar\sum_k \omega_k\hat{a}_k^\dagger\hat{a}_k \tag{17.95}$$

and $\hat{E}_+ = i\sum_k \varepsilon_{k,cav}\ \hat{a}_k\hat{e}_k e^{i\boldsymbol{k}\cdot\boldsymbol{r}}$; and $\hat{E}_- = -i\sum_k \varepsilon_{k,cav}\ \hat{a}_k^\dagger\hat{e}_k e^{-i\boldsymbol{k}\cdot\boldsymbol{r}}$. To simplify the notation, the symbol k now represents the direction of propagation and polarization. The cavity performance is assumed to be independent of polarization. In the problem of interest, we assume that radiation from the emitter that is captured by the etalon defines the modes in the sum that are described by the above analysis. Hence the remaining modes are those that are free to radiate in other directions, such as parallel to the etalon mirrors. At this point in the calculation, we simply note that point about the sum. A more accurate analysis would need to account for the more complex dipole radiation field emitted when the mirror separation is small. Additional simplifications will be discussed at the end.

The two-level emitter is given by the state vector,

$$|\psi\rangle = C_{20}\,|2\rangle\,|0\rangle + \sum_k C_{1_s} e^{-i(\omega_k-\omega_{21})t}\,|1\rangle\,|k\rangle \tag{17.96}$$

We then follow the same steps as presented earlier (see steps leading to Eq. 15.108) to get to the equation of motion for excited state probability amplitude, except in this case, the calculation is done in a one-dimensional system for the radiation field meaning that $\sum_k \to \int d\omega_k d\Omega_{SA}$.

$$\begin{aligned}\frac{dC_{20}(t)}{dt} &= -\,C_{20}(t)\sum_k \frac{e^2\epsilon_{k,cav}^2}{\hbar^2}\langle k|\langle 1|\,\boldsymbol{r}\cdot\hat{\epsilon}_k^*\,|2\rangle\langle 2|\,\boldsymbol{r}\cdot\hat{\epsilon}_k\,|1\rangle\,|k\rangle\,\pi\delta\left(\omega_k-\omega_{21}\right)\\ &= -\,C_{20}(t)\frac{\pi\left|\mu_{12}\right|^2}{\hbar^2}\eta_{\Omega_{SA}}\int d\omega_k\frac{\hbar\omega_k}{2\epsilon_o v}f_N\left(\omega_k\right)\delta\left(\omega_k-\omega_{21}\right)\\ &= -\,C_{20}(t)\frac{\pi\left|\mu_{12}\right|^2}{\hbar}\eta_{\Omega_{SA}}\frac{\omega_{21}}{2\epsilon_o v}\frac{\omega_{cav}}{2\pi Q}\frac{1}{\left(\frac{\omega_{cav}}{2Q}\right)^2+\left(\omega_{21}-\omega_{cav}\right)^2}\xrightarrow[\omega_{21}=\omega_{cav}]{}-C_{20}(t)2\pi\eta_{\Omega_{SA}}\frac{\left|\mu_{12}\right|^2}{4\pi\epsilon_o\hbar}\frac{2Q}{v}\end{aligned} \tag{17.97}$$

The factor $\eta_{\Omega_{SA}}$ represents the integration over the solid angle (including the angular dependence of the emission) which in free space was $\frac{8\pi}{3}$. The physicals results are clear.[7] In free space, the spontaneous emission decay rate was given in Eq. 15.115 as

$$\gamma_{free\ space} = \frac{4}{3}\frac{\left|\mu_{21}\right|^2}{4\pi\epsilon_o\hbar}\left(\frac{\omega_{21}}{c}\right)^3 \tag{17.98}$$

and for the cavity

$$\gamma_{cavity} = 4\pi\eta_{\Omega_{SA}}\frac{\left|\mu_{12}\right|^2}{4\pi\epsilon_o\hbar}\frac{2Q}{\mathcal{V}} \tag{17.99}$$

The factor of two in Eq. 17.99 compared to 17.97 is because γ_{cavity} reports on the decay of the probability, not the probability amplitude. The ratio of the transition probability per unit time of the emitter in the cavity to the free space result is

$$3\pi\eta_{\Omega_{SA}}\frac{2Q}{\mathcal{V}}\left(\frac{\lambda_{21}}{2\pi}\right)^3 \tag{17.100}$$

[7] Oliver Benson, Chapter 12 from "Lecture on Quantum Optics" by Oliver Benson, Humboldt-University, Berlin. https://www.physik.hu-berlin.de/de/nano/lehre/copy_of_quantenoptik09, Chapter 12.

This shows the spontaneous emission rate on resonance in a cavity, increases as the quality factor divided by the mode volume (**Purcell effect**). Far off resonance the rate decreases as $\frac{3}{16\pi^2}\eta\Omega_{SA}\frac{1}{Q}\left(\frac{\lambda_{21}}{\mathcal{V}}\right)^3$. Hence by using a cavity with a volume close to the wavelength $\mathcal{V} \sim \lambda^3$, it is possible to engineer the decay rate to optimize performance, depending on the objectives of the device. For example, extending the lifetime could extend memory time for information stored, while shortening the lifetime could improve brightness and timing of a single quantum emitter. More complete analysis shows that this also allows the engineer to improve directionality of the emission and therefore improve the collection efficiency. High Q cavities in the near IR (telecom) and visible can have Q' s in excess of 10^6 to closer to a billion, comparable to superconducting systems in the microwave.

17.8 **Summary**

Once the electromagnetic field is quantized, a completely new set of opportunities and interesting questions arises. To be clear, quantizing the field from say a laser does not typically provide any new insight into possible devices or fundamental physics. Hence, this fields remain "classical" in its properties. However, single photon emitters like a single atom or quantum dot generate a highly non-classical field and for this, the field must be quantized. The vacuum itself is a quantum "object" and, as seen, its effects are significant, causing ideal stationary eigenstates to decay. The vacuum field can be engineered with care, to control and channel the decay in ways that turn a liability into an asset, such as being able to collect all the light from a quantum emitter rather than just the usual fraction determined by the solid angle of the collection optics.

Vocabulary (page) and Important Concepts

- Purcell effect 304, 311
- resonator 305
- Fabry-Pérot etalon 305
- Airy distribution 306
- free spectral range 306
- Lorentzian 308
- cavity lifetime 309
- quality factor 309
- cavity Q 309

18 The Density Matrix: Bloch Equations

18.1 Introduction

The last topic in this text is an introduction to another very powerful formalism that provides a means to deal with time dependent problems that can be very challenging in the Schrödinger picture. The approach is based on developing a new operator, called the **density matrix** operator. We approach development of this operator by starting with Schrödinger's equation. In this approach we will see how to study the behavior of a given quantum system and simultaneously incorporate the interaction of the system with the bath. The **bath** was a concept first introduced in Chapter 14 and is a term meant to include the quantum vacuum, thermal fluctuations, and other physical behaviors that may alter the state of the system but are not directly coupled to operators of interest.

18.2 The Density Matrix Operator

We start with Schrödinger's equation:

$$i\hbar\frac{d}{dt}\,|\psi(t)\rangle = \hat{H}\,|\psi(t)\rangle = \left(\hat{H}_0 + \hat{V}(t)\right)|\psi(t)\rangle \tag{18.1}$$

Without loss of generality, but to reduce the number of symbols and math, we work with just a two-level system where the basis states are eigenstates of the Hamiltonian $\hat{H}_0$. The final result, however, will be quite general.

In the Schrödinger picture, the state vector is given by:

$$|\psi(t)\rangle = c_1(t)|1\rangle + c_2(t)|2\rangle \tag{18.2}$$

where

$$\langle 1|\psi(t)\rangle = c_1(t) \tag{18.3}$$

$$\langle 2|\psi(t)\rangle = c_2(t) \tag{18.4}$$

For the expectation value of an operator, we get

$$\begin{aligned}\langle \hat{A}\rangle(t) &= \langle \psi(t)|\hat{A}|\psi(t)\rangle = c_1^*c_1\langle 1|\hat{A}|1\rangle + c_1^*c_2\langle 1|\hat{A}|2\rangle + c_2^*c_1\langle 2|\hat{A}|1\rangle + c_2^*c_2\langle 2|\hat{A}|2\rangle \\ &= c_1^*c_1\hat{A}_{11} + c_1^*c_2\hat{A}_{12} + c_2^*c_1\hat{A}_{21} + c_2^*c_2\hat{A}_{22}\end{aligned} \tag{18.5}$$

Introduction to Quantum Nanotechnology: A Problem Focused Approach. Duncan Steel, Oxford University Press (2021).
 DOI: 10.1093/oso/9780192895073.003.0018

where

$$\hat{A}_{ij} = \langle i|\hat{A}|j\rangle \tag{18.6}$$

Inserting expressions in Eqs 18.3 and 18.4 for probability amplitudes,

$$\begin{aligned}\langle\hat{A}\rangle(t) &= \langle\psi(t)|1\rangle\langle 1|\psi(t)\rangle\hat{A}_{11} + \langle\psi(t)|1\rangle\langle 2|\psi(t)\rangle\hat{A}_{12} + \langle\psi(t)|2\rangle\langle 1|\psi(t)\rangle\hat{A}_{21} \\ &\quad + \langle\psi(t)|2\rangle\langle 2|\psi(t)\rangle\hat{A}_{22} = \langle 1|\psi(t)\rangle\langle\psi(t)|1\rangle\hat{A}_{11} + \langle 2|\psi(t)\rangle\langle\psi(t)|1\rangle\hat{A}_{12} \\ &\quad + \langle 1|\psi(t)\rangle\langle\psi(t)|2\rangle\hat{A}_{21} + \langle 2|\psi(t)\rangle\langle\psi(t)|2\rangle\hat{A}_{22}\end{aligned} \tag{18.7}$$

The prefactor for every matrix element is of the form

$$\langle\psi(t)|j\rangle\langle i|\psi(t)\rangle = \langle i|\psi(t)\rangle\langle\psi(t)|j\rangle \tag{18.8}$$

The outer product of the state vector with itself, $|\psi(t)\rangle\langle\psi(t)|$, is defined as the **density matrix operator**:

$$\hat{\rho}(t) = |\psi(t)\rangle\langle\psi(t)| \tag{18.9}$$

This is a square matrix with matrix elements:

$$\rho_{ij}(t) = \langle i|\psi(t)\rangle\langle\psi(t)|j\rangle = c_i(t)c_j^*(t) \tag{18.10}$$

We then see for $\langle A\rangle(t)$:

$$\begin{aligned}\langle\hat{A}\rangle(t) &= \rho_{11}(t)A_{11} + \rho_{21}(t)A_{12} + \rho_{12}(t)A_{21} + \rho_{22}(t)A_{22} = \sum_{n,m} A_{nm}\rho_{mn}(t) \\ &= Tr\hat{A}\hat{\rho}(t) = Tr\hat{\rho}(t)\hat{A}\end{aligned} \tag{18.11}$$

The density matrix operator in Eq. 18.9 is usually in the Schrödinger picture because it is time dependent.

As expected from the definition of the expectation value, the expectation value is determined by bilinear products of the probability amplitudes in the state vector or wave function. The density matrix includes all possible bilinear combinations. If we use the matrix to develop an explicit form for the operator, we get

$$\hat{\rho}(t) = \sum_{ij}\langle i|\psi(t)\rangle\langle\psi(t)|j\rangle|i\rangle\langle j| = \sum_{ij}\langle i|\hat{\rho}(t)|j\rangle|i\rangle\langle j| = \begin{bmatrix} c_1(t)c_1^*(t) & c_1(t)c_2^*(t) \\ c_2(t)c_1^*(t) & c_2(t)c_2^*(t)\end{bmatrix} \tag{18.12}$$

and

$$\hat{A} = \sum_{ij}\langle i|\hat{A}|j\rangle|i\rangle\langle j| = \begin{bmatrix}\hat{A}_{11} & \hat{A}_{12} \\ \hat{A}_{21} & \hat{A}_{22}\end{bmatrix} \tag{18.13}$$

Hence, calculating these matrix elements of the density matrix provides immediate insight into the underlying physical processes associated with a given measurement.

Physically, the diagonal terms $\rho_{11} = c_1 c_1^*$ and $\rho_{22} = c_2 c_2^*$ correspond to the probability of being in the state 1 or 2. A non-zero off-diagonal term like $\rho_{ij} = c_i c_j^*$ describes the level of coherence associated with a superposition between states i and j. Operators associated with observables that report on coherence must have the corresponding non-zero matrix element. Moreover, while the density matrix does not provide the state vector resulting from some dynamics, it does provide the dynamics of the resulting bilinear products of probability amplitudes which are directly associated with some measurement. Sometimes, this is easier to understand than seeing the dynamics of the operator in the Heisenberg picture, as done in Chapter 16.

To derive the equation of motion for the density matrix operator, we use Schrödinger's equation for the ket and bra:

$$i\hbar\frac{d}{dt}|\psi(t)\rangle = \widehat{H}|\psi(t)\rangle \tag{18.14}$$

$$-i\hbar\frac{d}{dt}\langle\psi(t)| = \langle\psi(t)|\widehat{H} \tag{18.15}$$

Then we use these equations to evaluate $i\hbar\frac{d}{dt}\rho$:

$$\begin{aligned} i\hbar\frac{d}{dt}\hat{\rho} &= i\hbar\frac{d}{dt}\,|\psi(t)\rangle\langle\psi(t)| = i\hbar\left(\frac{d}{dt}\,|\psi(t)\rangle\right)\langle\psi(t)| + i\hbar\,|\psi(t)\rangle\left(\frac{d}{dt}\,\langle\psi(t)|\right) \\ &= H|\psi(t)\langle\psi(t)| - |\psi(t)\rangle\langle\psi(t)|\,H = \left[\widehat{H},\hat{\rho}\right] \end{aligned} \tag{18.16}$$

Resulting in the **equation of motion for the density matrix**:

$$i\hbar\frac{d}{dt}\hat{\rho} = \left[\widehat{H},\hat{\rho}\right] \tag{18.17}$$

When the equation of motion is combined with

$$\left\langle\hat{A}\right\rangle(t) = Tr\hat{A}\hat{\rho} \tag{18.18}$$

we have a complete description of the system. The equation of motion for dynamics is equivalent to Schrödinger's equation, but will be seen shortly to be more powerful than Schrödinger's equation.

Continuing to work with the two-state system in Eq. 18.2, with a classical monochromatic electromagnetic field interaction, where $\widehat{H} = \widehat{H}_0 + \widehat{V}(t)$, we use the Hamiltonian in Eq. 16.19 in the rotating wave approximation:

$$\widehat{H} = \frac{\hbar\omega_{21}}{2}\hat{\sigma}_z - \hbar\frac{1}{2}\Omega_R\left(\mathbf{R},t\right)e^{-i\omega t}\hat{\sigma}_+ - \hbar\frac{1}{2}\Omega_R^*\left(\mathbf{R},t\right)e^{i\omega t}\hat{\sigma}_- \tag{18.19}$$

corresponding to the energy level scheme in Fig. 18.1 and where $\widehat{H}_0 = \frac{\hbar\omega_0}{2}\hat{\sigma}_z$ and $\Omega_R\left(\mathbf{R},t\right) = \frac{\mu_{21}\check{r}_{21}\cdot\mathbf{E}(\mathbf{R},t)}{\hbar}$. Recall that $\mu_{21} = \langle 2|qr|1\rangle$ where q is the charge of the particle. If q is for an electron, $q = -e$.

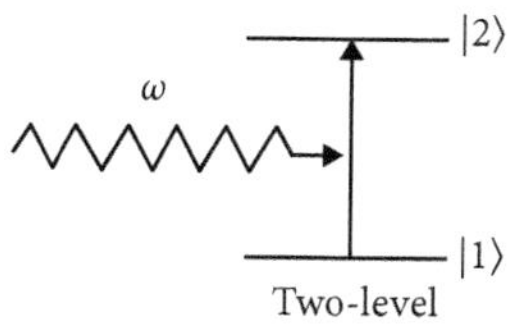

Fig. 18.1 Two-level systems with a monochromatic field incident near resonance at frequency ω. The energy separation of the two states is $\hbar\omega_{21}$.

We now use Eq. 18.17 to work out the equations of motion for the different matrix elements:[1]

$$i\frac{d}{dt}\rho_{11} = -\frac{1}{2}\Omega_R^* e^{i\omega t}\rho_{21} + \rho_{12}\frac{1}{2}\Omega_R e^{-i\omega t} \tag{18.20}$$

$$i\frac{d}{dt}\rho_{22} = -\frac{1}{2}\Omega_R e^{-i\omega t}\rho_{12} + \rho_{21}\frac{1}{2}\Omega_R^* e^{i\omega t} \tag{18.21}$$

$$i\frac{d}{dt}\rho_{21} = \omega_{21}\rho_{21} - \frac{1}{2}\Omega_R e^{-i\omega t}(\rho_{11} - \rho_{22}) \tag{18.22}$$

$$i\frac{d}{dt}\rho_{12} = -\omega_{21}\rho_{12} + \frac{1}{2}\Omega_R^* e^{i\omega t}(\rho_{11} - \rho_{22}) \tag{18.23}$$

These equations are called the **optical Bloch equations** or sometimes just the **Bloch equations**, since they were first developed to describe the dynamics of nuclear spin in an oscillating magnetic field. If they are combined with Maxwell's equations, they are the **Maxwell–Bloch equations**. In Section 18.3, we include relaxation, making these equations very powerful. It is worth noting, however, that in the absence of relaxations, it is often easier to find the probability amplitudes for Eq. 18.10 by solving for the amplitudes directly from Schrödinger's equation.

From Eq. 18.10, we see that $\rho_{ij} = \rho_{ji}^*$, so that it is not necessary to solve a separate equation for ρ_{12}. If we assume that at $t = 0$, $\rho_{11} = 1$ and $\rho_{22} = \rho_{21} = 0$, we get to first order in the field using time dependent perturbation theory (Eq. 11.80):

$$\int_0^t dt' \frac{d}{dt'}\left(e^{i\omega_{21}t'}\rho_{21}(t')\right) = +i\frac{1}{2}\Omega_R \int_0^t dt' e^{-i(\omega-\omega_{21})t'} \tag{18.24}$$

$$\rho_{21}^{(1)}(t) = -\frac{1}{2}\Omega_R \frac{e^{-i\omega t} - e^{-i\omega_{21}t}}{(\omega - \omega_{21})} \tag{18.25}$$

Using Eq. 18.11, the poplarization is given by $\mathrm{P} = \mathrm{q}\langle \boldsymbol{r}\rangle$ or

$$\boldsymbol{P} = q\langle \boldsymbol{r}\rangle = q\boldsymbol{r}_{12}\rho_{21}^{(1)}(t) + q\boldsymbol{r}_{21}\rho_{12}^{(1)}(t) = q\boldsymbol{r}_{12}\rho_{21}^{(1)}(t) + cc \tag{18.26}$$

Of course, this is the same expression we got in Chapter 9. We have succeeded in perhaps reducing the algebra a little bit with this approach, but the information is the same.

[1] In the course of determining the equation of motion for an arbitrary matrix element, ρ_{mn}, we must determine the corresponding matrix element of the commutator given by $\left[\hat{\rho}, \hat{V}\right]_{mn}$. Operators are like square matrices with the same order, so $\left[\hat{\rho}, \hat{V}\right]_{mn} = \sum_j V_{mj}\rho_{jn} - \rho_{mj}V_{jn}$. The sum over j also follows by an application of completeness when evaluating $\langle m|V\rho|n\rangle = \sum_j \langle m|V|j\rangle\langle j|\rho|n\rangle$.

18.3 The Density Matrix Equations Including Relaxation

We now consider a more complicated problem. We consider the case of a quantum system described by a Hamiltonian, $\hat{H}_S$, with eigenstates $|S_j\rangle$ that describes the system of interest to us, for example a nano-vibrator or an electron with spin in a magnetic field. In general, $\hat{H}_S = \hat{H}_0 + \hat{V}$, where $\hat{H}_0|S_j\rangle = E_j|S_j\rangle$.

We then consider a second Hamiltonian, $\hat{H}_B$, that represents the bath that includes the vacuum radiation field, phonons, or anything else that might interact with the system independent of what manipulations are taking place in $\hat{H}_S$. We then also assume that there is some kind of interaction between the bath and the system of interest, represent by $\hat{V}_{SB}$ similar to the problem discussed in Chapters 14, 15, and 16. The entire Hamiltonian is then

$$\hat{H} = \hat{H}_S + \hat{H}_B + \hat{V}_{SB} \tag{18.27}$$

The Hilbert spaces of $\hat{H}_S$ and $\hat{H}_B$ are separate and hence the eigenstates of the entire Hamiltonian are a product state of the eigenstates of the two Hilbert spaces. The eigenstates of $\hat{H}$ given by $|H\rangle = |S_j\rangle|B_\alpha\rangle$ span the Hilbert space defined by $|\hat{H}_0\rangle \otimes |\hat{H}_B\rangle$ (the notation meaning a product of two spaces). It is important to understand that operators associated with the observable in the quantum system of interest do not operate on states in the bath. In general then, a state vector describes the entire Hilbert space:

$$|\psi\rangle = \sum_{j,\alpha} C_{j\alpha}|S_j\rangle|B_\alpha\rangle \tag{18.28}$$

We now use the above definition of the density matrix operator and create a density matrix operator for the product space of the two Hilbert sub-spaces:

$$\rho = |\psi\rangle\langle\psi| = \sum_{j,\alpha,k,\beta} C_{j\alpha}C^*_{k\beta}|S_j\rangle|B_\alpha\rangle\langle B_\beta|\langle S_k| \tag{18.29}$$

Using this result, we consider the expectation value of an operator defined only in the Hilbert sub-space of $\hat{H}_S$, represented by the operator $\hat{A}_S$:

$$\begin{aligned}\langle\hat{A}\rangle_S = Tr\rho\hat{A}_S &= \sum_{j,\alpha,j',\alpha'} \langle B_{\alpha'}|\langle S_{j'}|C_{j'\alpha'}C^*_{j\alpha}|S_j\rangle|B_\alpha\rangle\langle B_\alpha|\langle S_j|\hat{A}_S|S_{j'}\rangle|B_{\alpha'}\rangle \\ &= \sum_{j,j',\alpha} C_{j'\alpha}C^*_{j\alpha}\langle S_j|\hat{A}_S|S_{j'}\rangle\end{aligned} \tag{18.30}$$

We see from here, that we can define a **reduced density matrix** by returning to Eq. 18.28, forming the density matrix operator and tracing over the bath:

$$\rho_S = Tr_{Bath}\rho = \sum_\alpha \langle B_\alpha|\psi\rangle\langle\psi|B_\alpha\rangle = \sum_{j,j'\alpha} C_{j'\alpha}C^*_{j\alpha}|S_j\rangle\langle S_{j'}| \tag{18.31}$$

From Eq. 18.29, we see that

$$\langle\hat{A}\rangle_S = Tr\rho_S\hat{A}_S \tag{18.32}$$

It is often possible to start with the equations of motion for the product space and then trace over the bath to get a set of equations for ρ_S. For example, working with the quantized radiation field from Chapter 15 formalism, you can show that for the reduced matrix[2] that (the subscript S is now suppressed) $\dot{\rho}_{22} = -\gamma_{sp}\rho_{22}$, $\dot{\rho}_{11} = \gamma_{sp}\rho_{22}$, $\dot{\rho}_{21} = \frac{1}{2}\gamma_{sp}\rho_{21}$, $\dot{\rho}_{12} = \frac{1}{2}\gamma_{sp}\rho_{12}$. Furthermore, if there are random phase fluctuation between the two states, it increases the decay rate of the coherence, by an amount γ_{deph}, giving a total decay rate for the coherences of $\gamma = \frac{1}{2}\gamma_{sp} + \gamma_{deph}$.

18.4 Solving the Reduced Density Matrix for a Two-Level System in the Presence of a Near Resonant Classical Electromagnetic Field

These results lead to an important new set of equations, but for the *reduced* density matrix:

$$\frac{d}{dt}\rho_{11} = i\frac{1}{2}\Omega_R^*(\mathbf{R},t)\,e^{i\omega t}\rho_{21} - i\rho_{12}\frac{1}{2}\Omega_R(\mathbf{R},t)\,e^{-i\omega t} + \gamma_{sp}\rho_{22} \tag{18.33}$$

$$\frac{d}{dt}\rho_{22} = i\frac{1}{2}\Omega_R(\mathbf{R},t)\,e^{-i\omega t}\rho_{12} - i\rho_{21}\frac{1}{2}\Omega_R^*(\mathbf{R},t)\,e^{i\omega t} - \gamma_{sp}\rho_{22} \tag{18.34}$$

$$\frac{d}{dt}\rho_{21} = -i\omega_{21}\rho_{21} + i\frac{1}{2}\Omega_R(\mathbf{R},t)\,e^{-i\omega t}(\rho_{11}-\rho_{22}) - \gamma\rho_{21} \tag{18.35}$$

$$\frac{d}{dt}\rho_{12} = i\omega_{21}\rho_{12} - i\frac{1}{2}\Omega_R^*(\mathbf{R},t)\,e^{i\omega t}(\rho_{11}-\rho_{22}) - \gamma\rho_{12} \tag{18.36}$$

subject to the condition that

$$\rho_{22}(t) + \rho_{11}(t) = 1 \tag{18.37}$$

These equations are identical to the Bloch equations but now also include the effects of decay and decoherence.

Assuming an applied monochromatic classical field, we change to the **field interaction representation** to eliminate the rapidly varying terms. For this, we let $\rho_{12}(t) = \tilde{\rho}_{12}(t)e^{i\omega t}$ and $\rho_{21}(t) = \tilde{\rho}_{21}(t)e^{-i\omega t}$. This eliminates the explicit time dependence. Setting $\delta = \omega - \omega_{21}$:

$$\frac{d}{dt}\rho_{11} = i\frac{1}{2}\Omega_R^*(\mathbf{R},t)\,\tilde{\rho}_{21} - i\frac{1}{2}\Omega_R(\mathbf{R},t)\,\tilde{\rho}_{12}(t) + \gamma_{sp}\rho_{22} \tag{18.38}$$

$$\frac{d}{dt}\rho_{22} = i\frac{1}{2}\Omega_R(\mathbf{R},t)\,\tilde{\rho}_{12}(t) - i\frac{1}{2}\Omega_R^*(\mathbf{R},t)\,\tilde{\rho}_{21} - \gamma_{sp}\rho_{22} \tag{18.39}$$

$$\frac{d}{dt}\tilde{\rho}_{21} = (i\delta - \gamma)\,\tilde{\rho}_{21} + i\frac{1}{2}\Omega_R(\mathbf{R},t)\,(\rho_{11}-\rho_{22}) \tag{18.40}$$

$$\frac{d}{dt}\tilde{\rho}_{12} = -i(\delta+\gamma)\,\rho_{12} - i\frac{1}{2}\Omega_R^*(\mathbf{R},t)\,(\rho_{11}-\rho_{22}) \tag{18.41}$$

[2] There are many books that do this. Two modern discussions are in Stenholm *Foundations of Laser Spectroscopy* and Berman and Malinofsky *Principles of Laser Spectroscopy and Quantum Optics.*

which can be written in matrix form where none of the elements of the matrix are explicitly time dependent:

$$\frac{d}{dt}\begin{bmatrix}\rho_{11}\\ \rho_{22}\\ \tilde{\rho}_{21}\\ \tilde{\rho}_{12}\end{bmatrix}=\begin{bmatrix}0 & \gamma_{sp} & i\frac{1}{2}\Omega_R^*(\mathbf{R},t) & -i\frac{1}{2}\Omega_R(\mathbf{R},t)\\ 0 & -\gamma_{sp} & -i\frac{1}{2}\Omega_R^*(\mathbf{R},t) & i\frac{1}{2}\Omega_R(\mathbf{R},t)\\ i\frac{1}{2}\Omega_R(\mathbf{R},t) & -i\frac{1}{2}\Omega_R(\mathbf{R},t) & i\delta-\gamma & 0\\ -i\frac{1}{2}\Omega_R^*(\mathbf{R},t) & i\frac{1}{2}\Omega_R^*(\mathbf{R},t) & 0 & -i\delta-\gamma\end{bmatrix}\begin{bmatrix}\rho_{11}\\ \rho_{22}\\ \tilde{\rho}_{21}\\ \tilde{\rho}_{12}\end{bmatrix} \tag{18.42}$$

An exact solution to this problem is straightforward when $\gamma=\gamma_{sp}=0$ and $\Omega_R(\mathbf{R},t)$ is *independent* of time. Exact solutions are usually only required if the intent is to control the state of the system, such as in a switch. Often, for measurements to characterize the general properties of the system, time dependent perturbation theory is adequate. In this case, we follow the approach in Chapter 11, where the classical electromagnetic field is the perturbation. We expand the density matrix is powers of λ, where λ is a number between 0 and 1:

$$\hat{\rho}(t)=\lambda^0\hat{\rho}^{(0)}(t)+\lambda^1\hat{\rho}^{(1)}(t)+\lambda^2\hat{\rho}^{(2)}(t)+\cdots \tag{18.43}$$

and the Hamiltonian is given by

$$\widehat{H}=\widehat{H}_0+\lambda\widehat{V}(t) \tag{18.44}$$

We get then that the nth order correction:

$$i\hbar\frac{d\hat{\rho}^{(n)}(t)}{dt}=\left[H_0,\hat{\rho}^{(n)}(t)\right]+\left[V(t),\hat{\rho}^{(n-1)}(t)\right]-i\hbar\left.\frac{d\hat{\rho}^{(n)}}{dt}\right|_{relaxation} \tag{18.45}$$

We now consider a few examples of solutions using perturbation theory. Using the **field interaction picture**, it is easy to see the response to first order including decay of the optically induced coherence. Assuming that the system is initially in the ground state ($\rho_{11}(t=0)=1$) and $\tilde{\rho}_{21}(t=0)=\tilde{\rho}_{21}(t=0)=\rho_{22}(t=0)=0$, we get:

$$\begin{aligned}\tilde{\rho}_{21}(t)&=i\Omega_R e^{(i\delta-\gamma)t}\int_0^t e^{-(i\delta-\gamma)t'}dt'=\left(-i\frac{1}{2}\Omega_R\right)\frac{1-e^{(i\delta-\gamma)t}}{(i\delta-\gamma)}\\ &=\lim_{t\to\infty}\left(-i\frac{1}{2}\Omega_R\right)\frac{1}{(i\delta-\gamma)}=\frac{1}{2}\Omega_R\frac{-\delta+i\gamma}{\delta^2+\gamma^2}\end{aligned} \tag{18.46}$$

We see that the singularity at $\delta=0$ present in our earlier discussion (see for example Eq. 11.91) is now gone. Recall again that the polarization for electromagnetic wave equation is given by $Tr\hat{\mu}\hat{\rho}$, where the imaginary part is the **absorption** and the real part is **dispersion** (see Appendix F). We see that the absorption line shape is described by a **Lorentzian**, where the full width at half-maximum (FWHM) (in hertz) is $\frac{\gamma}{\pi}$. The dispersive part is the dielectric constant (index of refraction). If $\gamma_{deph}=0$, the linewidth is broadened still by spontaneous emission, and the FWHM (in hertz) is $\frac{\gamma_{sp}}{2\pi}$. This is called the **radiative lifetime limit**.

Problem: 18.1 *Using Eq. 18.34 and first order perturbation theory in the field, find the* $\rho_{21}(t)$ *for a short pulse of resonant electromagnetic radiation. We assume that the pulse width,* τ*, is sufficiently short that* $\tau \ll \frac{\gamma_{sp}}{2}, \gamma, \Delta = \omega - \omega_0$*. It is easiest then to use a delta-function pulse as an approximation for a square pulse:* $E = E_0 e^{-i\omega t}\tau\delta(t)$ *so that* $V_{21}(t) = -e\hbar\frac{\mathbf{r}_{21}\cdot\mathbf{E}_0}{\hbar}\tau\frac{1}{2}\delta(t)e^{-i\omega t} = -\hbar\frac{1}{2}\Omega_R\tau\delta(t)e^{-i\omega t}$*. The Rabi frequency,* Ω_R*, is now defined as* $\frac{\mathbf{r}_{21}\cdot\mathbf{E}_0}{\hbar}$ *and* $\Omega_R\tau$ *is the* ***pulse area.*** *Referring to Eqs 9.84 and 9.85,* $\Omega_R\tau$ *determines the fraction of the probability amplitude lost in one state and gained in the other. Starting in the ground state,* $\tau = \frac{\pi}{\Omega_R}$ *thus, giving* τ *for a specified* Ω_R*, we get a 100% inversion for a* π*-pulse. In this calculation, however, you assume that* $\Omega_R\tau \ll \pi$*, so that perturbation theory is valid. Show that the polarization is given by*

$$P = \langle\mu\rangle = \mu_{12}(t)\rho_{21}(t) + cc = i\mu_{12}\frac{1}{2}\Omega_R\tau e^{-(i\omega_{21}+\gamma)t} + cc$$

This is the same problem we did in Problem 16.6. This means that a quantum system will radiate at its resonance frequency after being excited by a shortpulse. The radiation will decay at 2γ *with the intensity (goes like the square modulus of the field). This is called* ***free induction decay*** *(FID) after its first observation in nuclear spin systems. In using the density matrix, we calculate the quantum coherence,* $\hat{\rho}_{21}(t)$*, to first order. The decay is seen as a manifestation of the loss of quantum coherence. In the operator approach, we calculate the expectation value of a Heisenberg operator* $\hat{\sigma}_{\pm}(t)$ *using the initial state.*

Problem 18.2 *Radio frequency (RF) radiation is used in magnetic resonance imaging to probe the nuclear spin of hydrogen in water. In Problem 10.6, you considered this problem in the amplitude picture for a* ***proton****. The Hamiltonian for a spin ½* ***electron*** *is given by*

$$\hat{H} = -\boldsymbol{\mu}_S \cdot B_0\hat{\mathbf{z}} - \boldsymbol{\mu}_S \cdot B_x\hat{\mathbf{x}}\cos\omega t = \hbar\frac{geB_0}{4m_p}\begin{bmatrix}1 & 0\\ 0 & -1\end{bmatrix} + \hbar\frac{geB_x}{4m_p}\begin{bmatrix}0 & 1\\ 1 & 0\end{bmatrix}\cos\omega t$$

$$= \hbar\frac{\omega_0}{2}\begin{bmatrix}1 & 0\\ 0 & -1\end{bmatrix} + \hbar\Omega\begin{bmatrix}0 & 1\\ 1 & 0\end{bmatrix}\cos\omega t$$

where $\omega_0 = g\frac{eB_0}{2m_p}$ *and* $\Omega = g\frac{eB}{4m_p}$*. Using this result, find the equations of motion for the density matrix.*

Problem 18.3 *Use second order PT and the results in Problem 18.1 to find* $\rho_{22}(t)$*, following excitation with a weak delta-function pulse.*

18.5 **Rate Equation Approximation**

Rate equations are common in many areas of physics, chemistry, and engineering, even when dealing with quantum behavior. Lasers and LEDs are common examples. Often, they are presented phenomenologically, based on rate constants. Those that deal with quantum behavior can be developed rigorously from the density matrix equations.

We work with Eqs 18.38–18.41, which are the density matrix equations for a monochromatic oscillating classical field in the rotating wave approximation coupled to a two-level system and work in the field interaction picture. We integrate Eq. 18.40 to get

$$\tilde{\rho}_{21}(t) = i\int_0^t dt' e^{(i(\omega-\omega_{21})-\gamma)(t-t')}\frac{1}{2}\Omega_R(t')(\rho_{11}(t') - \rho_{22}(t')) \tag{18.47}$$

where we introduced the integrating factor $e^{(-i(\omega-\omega_0)+\gamma)t}$ and suppressed the dependence on $\boldsymbol{R}$. In the **rate equation approximation**, we assume the time rate of change of the quantities $\Omega_{21}(t)$, $\rho_{11}(t)$, and $\rho_{22}(t)$ is slow compared to the decay rate, γ. In this way, we can remove these terms from the integral and evaluate them at $t' = t$, giving

$$\begin{aligned}
\tilde{\rho}_{21}(t) &= i\frac{1}{2}\Omega_R(t)(\rho_{11}(t) - \rho_{22}(t))\int_0^t dt' e^{(i(\omega-\omega_{21})-\gamma)(t-t')}\\
\tilde{\rho}_{21}(t) &= -i\frac{1}{2}\Omega_R(t)(\rho_{11}(t) - \rho_{22}(t))\frac{1}{i(\omega-\omega_{21})-\gamma}
\end{aligned} \tag{18.48}$$

This is the rate equation approximation. It is valid when the detuning $|\omega - \omega_{21}|$ or dephasing rate, γ, is larger than both the excited state decay rate, γ_2, and the Rabi frequency, $\Omega_R(t)$.

To be consistent with the approximation, we must assume $t \gg_{\gamma_2^{-1}}$ and hence the decaying term is suppressed. In this way, we can now substitute this expression along with the complex conjugate for $\tilde{\rho}_{12}(t)$ into the equations of motion for the upper and lower states using Eqs 18.38 and 18.39:

$$\begin{aligned}
\frac{d}{dt}\rho_{11} &= \frac{1}{2}\Omega_R^*(t)\frac{1}{2}\Omega_R(t)(\rho_{11}(t) - \rho_{22}(t))\frac{1}{i(\omega-\omega_{21})-\gamma}\\
&\quad + \frac{1}{2}\Omega_R(t)\frac{1}{2}\Omega_R^*(t)(\rho_{11}(t) - \rho_{22}(t))\frac{1}{-i(\omega-\omega_{21})-\gamma} + \gamma_{sp}\rho_{22}\\
&= +\gamma_{sp}\rho_{22} + \left(\frac{|\Omega_R(t)|}{2}\right)^2(\rho_{11}(t) - \rho_{22}(t))\frac{-2\gamma}{(\omega-\omega_{21})^2+\gamma^2}
\end{aligned} \tag{18.49}$$

$$\begin{aligned}
\frac{d}{dt}\rho_{22} &= -\frac{1}{2}\Omega_R(t)\frac{1}{2}\Omega_R^*(t)(\rho_{11}(t) - \rho_{22}(t))\frac{1}{-i(\omega-\omega_{21})-\gamma}\\
&\quad - \frac{1}{2}\Omega_R^*(t)\frac{1}{2}\Omega_R(t)(\rho_{11}(t) - \rho_{22}(t))\frac{1}{i(\omega-\omega_{21})-\gamma} - \gamma_{sp}\rho_{22}\\
&= -\gamma_{sp}\rho_{22} + \left(\frac{|\Omega_R(t)|}{2}\right)^2(\rho_{11}(t) - \rho_{22}(t))\frac{2\gamma}{(\omega-\omega_{21})^2+\gamma^2}
\end{aligned} \tag{18.50}$$

If we define

$$\mathcal{L}(\omega) = \frac{2\gamma}{(\omega-\omega_{21})^2+\gamma^2} \tag{18.51}$$

and recall that the intensity coupled to the system is given by $\Omega_R(\boldsymbol{R}, t) = \frac{\mu_{21}\check{\boldsymbol{r}}_{21}\cdot \boldsymbol{E}(\boldsymbol{R},t)}{\hbar}$, then

$$I(R,t) = \frac{\varepsilon_0 c}{2}\left|\check{\boldsymbol{r}}_{21}\cdot E(R,t)\right|^2 \tag{18.52}$$

The equations of motion simplify to

$$\frac{d}{dt}\rho_{11} = +\gamma_{sp}\rho_{22} - \left(\frac{|\mu_{21}|}{\hbar}\right)^2 \frac{1}{2\varepsilon_0 c}\mathcal{L}(\omega)\, I(R,t)\,(\rho_{11}(t)-\rho_{22}(t)) \tag{18.53}$$

$$\frac{d}{dt}\rho_{22} = -\gamma_{sp}\rho_{22} + \left(\frac{|\mu_{21}|}{\hbar}\right)^2 \frac{1}{2\varepsilon_0 c}\mathcal{L}(\omega)\, I(R,t)\,(\rho_{11}(t)-\rho_{22}(t)) \tag{18.54}$$

Assuming that $I(R,t)$ is independent of time, the steady state solutions take the time derivatives on the left to be zero, and the population terms are independent of time. The system is closed, meaning $\rho_{11}(t)+\rho_{22}(t)=1$. Hence, we only need one of the above equations. If we are interested in calculating either absorption or the dispersion associated with the two-level system, we are interested then in calculating $(\rho_{11}(t)-\rho_{22}(t))$ which we then substitute back into the expression for ρ_{21} to calculate the polarization for Maxwell's equations. Using the condition that the system is closed, we can write ρ_{22} as $\rho_{22}=\frac{1}{2}\left[1-(\rho_{11}(t)-\rho_{22}(t))\right]$, substituting into the Eq. 18.54 with the time derivative equal to zero becomes

$$-\gamma_{sp}\frac{1}{2}\left[1-(\rho_{11}(t)-\rho_{22}(t))\right] + \left(\frac{|\mu_{21}|}{\hbar}\right)^2 \frac{1}{2\varepsilon_0 c}\mathcal{L}(\omega)\, I(R,t)\,(\rho_{11}(t)-\rho_{22}(t)) = 0$$

$$\rho_{11}(t)-\rho_{22}(t) = \frac{(\omega-\omega_{21})^2+\gamma^2}{\left[(\omega-\omega_{21})^2+\gamma^2\left(1+\frac{I}{I_{sat}}\right)\right]} \tag{18.55}$$

where the saturation intensity is defined to be

$$I_{sat} = \frac{\varepsilon_0 c\gamma\gamma_{sp}}{2}\left(\frac{\hbar}{|\mu_{21}|}\right)^2 \tag{18.56}$$

Substituting 18.55 into 18.48 gives for $\tilde{\rho}_{21}(t)$

$$\begin{aligned}\tilde{\rho}_{21}(t) &= -i\frac{1}{2}\Omega_R(t)\,(\rho_{11}(t)-\rho_{22}(t))\,\frac{1}{i(\omega-\omega_{21})-\gamma}\\ &= \frac{1}{2}\Omega_R(t)\frac{-(\omega-\omega_{21})-i\gamma}{\left[(\omega-\omega_{21})^2+\gamma^2\left(1+\frac{I}{I_{sat}}\right)\right]}\end{aligned} \tag{18.57}$$

Recalling now that the polarization that goes into the wave equation for the electromagnetic field is $P = Tr\mu\rho$, we see that the imaginary part of Eq. 18.57 gives rise to absorption, and the line shape is a Lorentzian that is broadened as the intensity approaches the saturation intensity. The real part contributes to the dispersion of the system. The broadening of the linewidth with intensity, as seen in the denominator, is a result when $|\Omega_R| > \gamma_{sp}$

18.5 Three-Level System: Emerging Importance in Quantum Technology

The focus in much of this text has been on applications to various technologies using two-level systems. Using semiconductor structures, for example, it is possible to arrange two-level systems to emit in the telecom band (1.55 micron fiber optics) for long-distance applications. However, for some applications it might be useful to work with two electron spin states where even with a relatively large magnetic field, the splitting is in the microwave region. In these cases, adding a third level at higher energy could then allow a higher frequency field to drive the system as we shall see below. Adding a third level has also been shown, in Chapter 17, to enable generation of entangled states of photons.

Figure 18.2 shows the generic three-level systems. The two on the left—old and the Λ three level systems—have emerged as important for quantum photonics. A slightly more complicated version of the cascade-up three-level can produce entangled photons as it cascades radiatively, emitting a photon in going from $|3\rangle$ to $|2\rangle$ and then another photon from $|2\rangle$ to $|1\rangle$. Λ systems have been used to enable manipulation of spin states of electrons or holes, where their spin orientation is associated with levels $|1\rangle$ and $|2\rangle$, and both are optically accessible through level $|3\rangle$. The V-three level system is an important subsystem in structures like quantum gates.

As an example, we will just consider the simple problem of creating a coherent superposition state between states $|1\rangle$ and $|3\rangle$ in both the cascade-up three-level and the Λ three-level systems.

Assuming two monochromatic fields at frequency ω_1 and ω_2 resonant with the corresponding transition shown in Fig. 18.2 and assuming that the field does not interact significantly with the other transition, the corresponding Hamiltonians for these three systems are then given by

$$\hat{H}_{cascade} = \sum_j \hbar\omega_j |j\rangle\langle j| - \hbar\frac{1}{2}\Omega_{R_1}(\mathbf{R},t)\,e^{-i\omega_1 t}\hat{\sigma}_{+_{21}} - \hbar\frac{1}{2}\Omega^*_{R_1}(\mathbf{R},t)\,e^{i\omega_1 t}\hat{\sigma}_{-_{12}}$$
$$- \hbar\frac{1}{2}\Omega_{R_2}(\mathbf{R},t)\,e^{-i\omega_2 t}\hat{\sigma}_{+_{32}} - \hbar\frac{1}{2}\Omega^*_{R_2}(\mathbf{R},t)\,e^{i\omega_2 t}\hat{\sigma}_{-_{23}} \tag{18.58}$$

$$\hat{H}_{\Lambda} = \sum_j \hbar\omega_j |j\rangle\langle j| - \hbar\frac{1}{2}\Omega_{R_1}(\mathbf{R},t)\,e^{-i\omega_1 t}\hat{\sigma}_{+_{31}} - \hbar\frac{1}{2}\Omega^*_{R_1}(\mathbf{R},t)\,e^{i\omega_1 t}\hat{\sigma}_{-_{13}}$$
$$- \hbar\frac{1}{2}\Omega_{R_2}(\mathbf{R},t)\,e^{-i\omega_2 t}\hat{\sigma}_{+_{32}} - \hbar\frac{1}{2}\Omega^*_{R_2}(\mathbf{R},t)\,e^{i\omega_2 t}\hat{\sigma}_{-_{23}} \tag{18.59}$$

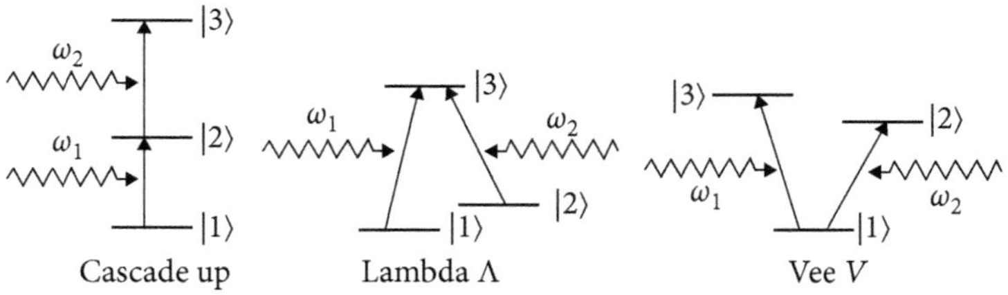

Fig. 18.2 Three-level systems that can be electromagnetically manipulated. The straight arrows represent dipole allowed transitions being driven by monochromatic fields at frequency ω_1 and ω_2.

$$\hat{H}_V = \sum_j \hbar\omega_j |j\rangle\langle j| - \hbar\frac{1}{2}\Omega_{R_1}(\mathbf{R},t)\,e^{-i\omega_1 t}\hat{\sigma}_{+_{31}} - \hbar\frac{1}{2}\Omega^*_{R_1}(\mathbf{R},t)\,e^{i\omega_1 t}\hat{\sigma}_{-_{13}}$$
$$- \hbar\frac{1}{2}\Omega_{R_2}(\mathbf{R},t)\,e^{-i\omega_2 t}\hat{\sigma}_{+_{21}} - \hbar\frac{1}{2}\Omega^*_{R_2}(\mathbf{R},t)\,e^{i\omega_2 t}\hat{\sigma}_{-_{12}} \tag{18.60}$$

The general equation of motion for the density matrix operator with $\hat{H} = \hat{H}_0 + \hat{V}(t)$ is

$$i\hbar\frac{d\hat{\rho}(t)}{dt} = [\hat{H}_0, \hat{\rho}(t)] + [\hat{V}(t), \hat{\rho}(t)] - i\hbar\left.\frac{d\hat{\rho}}{dt}\right|_{relaxation} \tag{18.61}$$

where the last term is a place holder for including relaxation along the lines as done above for the two-level system.

We consider the **cascade-up three-level system** as an example of how to use time dependent perturbation theory to reveal the important physics. The Hamiltonian for the cascade-up three-level system is given in Eq. 18.58. There are two monochromatic fields at frequency ω_1 and ω_2. We assume that the fields are constant and turn on at $t = 0$. We note that, while the fields are monochromatic (like a source that has been on for a long time and then the shutter opens or a switch closes), the sudden turning on results in a distribution of frequencies observed in the Fourier transform at short times. All of this is included automatically in the analysis without additional calculations. The interest here is to simply understand how resonant excitation can lead to a coherent superposition state in a more complex system. Specifically, we want to create a **two-photon coherence** between states $|1\rangle$ and $|3\rangle$. It is called a two-photon coherence because it takes two resonant fields.

The evolution of the coherent superposition between these two states is described by the density matrix element $\rho_{31} = \rho^*_{13}$. The equation of motion at n^{th} order is given above in Eq. 18.45. We assume at $t = 0$ that the probability of being in state $|1\rangle$ is unity meaning $\rho^0_{11} = 1$ and all the other density matrix elements are 0.

The details of the potential are in Eq. 18.58, but we know that the only non-zero terms for V_{3j} and V_{j1} are V_{32} and V_{21}, respectively. Hence, to get a first order contribution to ρ_{31} would mean that $\rho^{(0)}_{21}$ and/or $\rho^{(0)}_{32}$ have to be non-zero at zero order. However, we have stated that these are indeed zero at zero order at $t = 0$. Hence, the lowest possible order for a non-zero $\rho^{(n)}_{31}$ is $n = 2$. This requires then that $\rho^{(1)}_{21}$ and/or $\rho^{(1)}_{32}$ must be non-zero. To determine if this is possible, we then consider the equations of motion for these two matrix elements. For $\rho^{(1)}_{21}$

$$i\hbar\frac{d\rho^{(1)}_{21}(t)}{dt} = \hbar\omega_{21}\rho^{(1)}_{21} + \sum_j\left(V_{2j}(t)\rho^{(0)}_{j1} - \rho^{(0)}_{2j}V_{j1}(t)\right) - i\hbar\left.\frac{d\rho^{(1)}_{21}}{dt}\right|_{relaxation}$$
$$= \hbar\left(\omega_{21} - i\gamma_{21}\right)\rho^{(1)}_{21} + V_{21}(t)\rho^{(0)}_{11} \tag{18.61}$$

We see that the only term in the sum that contributes is $V_{21}(t)\rho^{(0)}_{11}$, because the other terms for $j = 1, 2,$ or 3 are all zero.

We now examine the equation of motion for $\rho^{(1)}_{32}$

$$i\hbar\frac{d\rho^{(1)}_{32}(t)}{dt} = \hbar\omega_{32}\rho^{(1)}_{32} + \sum_j\left(V_{3j}(t)\rho^{(0)}_{j2} - \rho^{(0)}_{3j}V_{j2}(t)\right) - i\hbar\left.\frac{d\rho^{(1)}_{32}}{dt}\right|_{relaxation} \tag{18.62}$$

We see immediately that $\rho_{32}^{(1)} = 0$, because there are no terms in the sum that are non-zero, since $\rho_{j2}^{(0)} = \rho_{3j}^{(0)} = 0$.

For $\rho_{21}^{(1)}(t)$ and inserting the matrix element for V_{21} from the Hamiltonian we get the result shown in Eq. 18.46 except that $\omega \to \omega_1$, namely

$$\frac{d\rho_{21}^{(1)}(t)}{dt} = -\left(i\omega_{21} + \gamma_{21}\right)\rho_{21}^{(1)} + i\frac{\Omega_{R_1}(\mathbf{R})}{2}e^{-i\omega_1 t}\rho_{11}^{(0)} \tag{18.63}$$

$$\rho_{21}^{(1)}(t) = -i\lim_{\gamma_{21}t\gg 1}\left(\frac{\Omega_{R_1}(\mathbf{R})}{2}\right)\rho_{11}^{(0)}\frac{e^{-i\omega_1 t}}{i\left(\omega_1 - \omega_{21}\right) - \gamma_{21}} \tag{18.64}$$

We take the long time limit to reduce the amount of algebra while preserving the key physics. The equation of motion for $\rho_{31}^{(2)}(t)$ now follows from Eq. 18.45, where we find a non-zero first order term for the right-hand side of the equation leading to a second order contribution for $\rho_{31}^{(2)}(t)$:

$$i\hbar\frac{d\rho_{31}^{(2)}(t)}{dt} = \hbar\left(\omega_{31} - i\gamma_{31}\right)\rho_{31}^{(2)} + V_{32}(t)\rho_{21}^{(1)} \tag{18.65}$$

We insert $V_{32}(t) = -\hbar\frac{1}{2}\Omega_{R_2}(\mathbf{R})e^{-i\omega_2 t}$ and cancel terms to get

$$\frac{d\rho_{31}^{(2)}(t)}{dt} = -\left(i\omega_{31} + \gamma_{31}\right)\rho_{31}^{(2)} + \left(\frac{\Omega_{R_2}(\mathbf{R})}{2}\right)\left(\frac{\Omega_{R_1}(\mathbf{R})}{2}\right)\rho_{11}^{(0)}\frac{e^{-i(\omega_1+\omega_2)t}}{i\left(\omega_1 - \omega_{21}\right) - \gamma_{21}} \tag{18.66}$$

and integrate giving

$$\begin{aligned}\rho_{31}^{(2)}(t) &= -\rho_{11}^{(0)}\left(\frac{\Omega_{R_2}(\mathbf{R})}{2}\right)\left(\frac{\Omega_{R_1}(\mathbf{R})}{2}\right)e^{-(i\omega_{31}+\gamma_{31})t}\int_0^t dt'\frac{e^{[-i(\omega_1+\omega_2-\omega_{31})+\gamma_{31}]t'}}{-i\left(\omega_1-\omega_{21}\right)+\gamma_{21}} \\ &= \lim_{\gamma_{31}t\gg 1} -\rho_{11}^{(0)}\left(\frac{\Omega_{R_2}(\mathbf{R})}{2}\right)\left(\frac{\Omega_{R_1}(\mathbf{R})}{2}\right)\frac{e^{-i(\omega_1+\omega_2)t}}{\left[-i\left(\omega_1-\omega_{21}\right)+\gamma_{21}\right]\left[-i\left(\omega_1+\omega_2-\omega_{31}\right)+\gamma_{31}\right]}\end{aligned} \tag{18.67}$$

The same result could have been had more quickly by finding the equivalent steady state response.[3]

[3] Note that while we did the integrals allowing for more complex behavior, we could recover these results with less algebra by moving to the field interaction picture and solving for the steady state response. In that case, we would have

$$\rho_{21}^{(1)}(t) = \tilde{\rho}_{21}^{(1)}e^{-i\omega_1 t}$$

$$\rho_{31}^{(2)}(t) = \tilde{\rho}_{31}^{(2)}e^{-i(\omega_1+\omega_2)t}$$

Insert these into Eqs 18.63 and 18.65, and we get

$$\tilde{\rho}_{21}^{(1)} = \rho_{11}^{(0)}\left(\frac{\Omega_{R_1}(\mathbf{R})}{2}\right)\frac{-i}{i\left(\omega_1-\omega_{21}\right)-\gamma_{21}}$$

$$\tilde{\rho}_{31}^{(2)} = -\rho_{11}^{(0)}\left(\frac{\Omega_{R_2}(\mathbf{R})}{2}\right)\left(\frac{\Omega_{R_1}(\mathbf{R})}{2}\right)\frac{1}{\left[-i\left(\omega_1-\omega_{21}\right)+\gamma_{21}\right]\left[-i\left(\omega_1+\omega_2-\omega_{31}\right)+\gamma_{31}\right]}$$

The result shows the strong effect of resonance: the physical effect of driving the system with frequencies such that $\omega_1 = \omega_{21}$ (a single-photon resonance) and $\omega_1 + \omega_2 = \omega_{31}$ (a two-photon resonance) leads to the maximum value of the two-photon coherence ρ_{31}. An important point is that it is possible to create arbitrary coherent superposition states even between states differing by a significant energy gap, by using coherent excitation with intermediates states.

While we took the approach of starting by examining the behavior of the density matrix term of interest and then worked backwards to find what terms and at what order would give rise to the lowest order term for desired matrix element, it is now evident that a simpler way is possible. Starting with the zeroth order state, $\rho_{11}^{(0)}$, we see that, by studying which terms in the potential $\hat{V}(\mathbf{R}, t)$ have non-zero matrix elements, we can readily get $\rho_{11}^{(0)} \xrightarrow{\Omega_{R_1}} \rho_{21}^{(1)} \xrightarrow{\Omega_{R_2}} \rho_{31}^{(2)}$. Such a diagram is called a **perturbation sequence**. Finally, this kind of problem could also have been approached with a little less algebra, by moving to the field interaction picture where $\rho_{21} = \tilde{\rho}_{21}(t)e^{-i\omega_1 t}, \rho_{32} = \tilde{\rho}_{32}(t)e^{-i\omega_2 t}, \rho_{31} = \tilde{\rho}_{31}(t)e^{-i(\omega_1+\omega_2)t}$, etc. and where in steady state (long time response as above), $\tilde{\rho}_{jk}(t) = \tilde{\rho}_{jk}$.

The **lambda three-level system** has also played a central role in many quantum systems because the two low-lying states $|1\rangle$ and $|2\rangle$ can correspond to states with coherence times of order milliseconds or longer. In some cases, they actually correspond to electron spin states that are oriented parallel and antiparallel to an applied magnetic field. The **vee three-level system** has played an important role in experiments that actually involve four states where the arrangement of the states forms the four corners of diamond in some pictures. This happens when two two-level quantum systems are sufficiently close so that there is coupling between them, such as in van der Waals (dipole-dipole) coupling.

Problem 18.4 *Solve the density matrix to second order in the field interaction picture to find the long time behavior of the coherence between states $|1\rangle$ and $|2\rangle$ in the Λ-three-level system in the center of Fig. 18.2 Specifically, show that assuming that everything is in state $|1\rangle$ at $t = 0$, then $\rho_{11}^{(0)} \xrightarrow{\Omega_{R_1}} \rho_{31}^{(1)} \xrightarrow{\Omega_{R_2}^*} \rho_{21}^{(2)}$ leads to*

$$\rho_{21}^{(2)}(t) = \rho_{11}^{(0)} \frac{\Omega_{R_1}(\mathbf{R})}{2} \frac{\Omega_{R_2}^*(\mathbf{R})}{2} \frac{1}{-i(\omega_1 - \omega_{31}) + \gamma_{31}} \frac{e^{-i(\omega_1-\omega_2)t}}{-i(\omega_1 - \omega_2 - \omega_{21}) + \gamma_{21}}$$

*When created this way, the coherence is a **two-photon Raman coherence** or just **Raman coherence**.*

Problem 18.5 *Repeat Problem 18.4 but assume that $\omega_1 = \omega_2$ and the field is described by a short pulse given by $= \frac{1}{2}\mathbf{E}(\mathbf{R})\tau\delta(t)e^{-i\omega_1 t} + cc.$*

18.7 **Summary**

For few-level systems, these equations apply also to the study and use of electronic and nuclear spin in a DC magnetic field in the z-direction and an oscillating magnetic field usually in the x-y plane. There is very little contribution from spontaneous emission at these low frequencies, but other physics leads to relaxation. In addition, often the discussion of these systems is in terms of the three components of the magnetization rather the density matrix. However, the equations are the same. The same basic physics permeates all these areas of quantum behavior. Indeed, Rabi oscillations were first demonstrated using nuclear spin systems. The basic theory of magnetic resonance imaging follows from these equations. In addition, with a more detailed theoretical description of relaxation, it is possible to then use these equations to see how to ameliorate the effects of decoherence. An example is a three-pulse Carr–Purcell spin or photon echo to cancel the effects of spectral wandering and/or inhomogeneous broadening.

Vocabulary (page) and Important Concepts

- bath 314
- density matrix 314
- density matrix operator 314
- equation of motion for the density matrix 316
- optical Bloch equations 317
- Bloch equations 317
- Maxwell–Bloch equations 317
- field interaction picture 320
- absorption 320
- dispersion 320
- Lorentzian 320
- radiative lifetime limit 320
- pulse area 321
- free induction decay 321
- rate equation approximation 321, 322
- three-level systems 325
- cascade up three-level system 327
- lambda three-level system 327
- vee three-level system 327
- two-photon coherence 327
- perturbation sequence 327

Appendix A Review of Essential Math

Here we review the math that is used in the text.

A.1 Differential Equations

Differential equations for physical systems establish a relationship between differentials. A differential such as $\frac{df(t)}{dt}$ or $\frac{df(x)}{dx}$ represents the change in f for a small change in the independent variable for space (x) or time (t). A second order differential, like say $\frac{d^2f(t)}{dt^2}$, is then the change of $\frac{df(t)}{dt}$ for a small change in t. From simple mechanics, Newton's law has the form $\frac{d^2x(t)}{dt^2} = \frac{F(x)}{m}$ where $F(x)$ is the force, m is the mass, and now $x(t)$ (the position) is the dependent variable and t for time is the independent variable. So $\frac{dx(t)}{dt} \equiv \dot{x}$ is then the change in x for a small change in t and is the velocity. And $\frac{d^2x(t)}{dt^2} \equiv \ddot{x}$ is the change in $\frac{dx(t)}{dt}$ for a small change in t and is the acceleration. Newton's law then says that the acceleration, $\frac{d^2x(t)}{dt^2}$, is $\frac{F(x)}{m}$ where F is the force and m is the mass.

We are interested in ***linear* differential equations**, meaning that terms such as $\frac{df(x)}{dx}, \frac{d^2f(x)}{dx^2}, \frac{df(t)}{dt}, \frac{d^2f(t)}{dt^2}$ may appear, but not terms that involve higher powers of these terms such as $\left(\frac{df}{dx}\right)^2$. The order of the differential equation is given by the number of times a given operator such as $\frac{d}{dx}$ *or* $\frac{d}{dt}$ operates on the function $f(x)$ or $f(t)$, respectively. So[1]

$$\frac{df(x)}{dx} + f(x) = 0 \tag{A.1}$$

is **first order**, while

$$\frac{d^2f(x)}{dx^2} + a\frac{df(x)}{dx} + cf(x) = 0 \tag{A.2}$$

is **second order**.

When all the operators and the unknown function are on the left-hand side and the right-hand side is zero such as

$$\frac{df(x)}{dx} + f(x) = 0 \tag{A.3}$$

[1] Note that in this discussion, there is just one independent variable for the sake of simplicity. That variable is usually either space or time (x or t, respectively). Rather than give each discussion in what follows in terms of say first time and then space, we will just arbitrarily pick x or t, assuming then that the reader can easily substitute the variable interest for their own problem. Language such as the word "gradient" for $\frac{df(x)}{dx}$ or "rate" for $\frac{df(t)}{dt}$ should be changed, as required.

or

$$\frac{d^2f(x)}{dx^2} + a\frac{df(x)}{dx} + cf(x) = 0 \tag{A.4}$$

then the equations are ***homogeneous***. When there is a function on the right such as

$$\frac{df(x)}{dx} + f(x) = g(x) \tag{A.5}$$

or

$$\frac{d^2f(t)}{dt^2} + a\frac{df(t)}{dt} + cf(t) = \sin\omega t \tag{A.6}$$

the equations are ***inhomogeneous***.

First order linear homogeneous differential equations have one solution, second order linear differential equations have two solutions, etc. An inhomogeneous differential equation has the **particular solution**, $f_p(t)$ or $f_p(x)$, associated with the function on the right-hand side, and also with the solution or solutions to the homogeneous equation. So for Eq. A.6, the total solution is $f(t) = A_1f_1(t) + A_2f_2(t) + f_p(t)$. In the case of the a time dependent problem, A_1 and A_2 are determined by the initial conditions, two in the case of a second order differential equation.

Examples of Solutions to Homogeneous Differential Equations

1. The solution to a first order homogeneous equation of the form

$$\frac{df(x)}{dx} + bf(x) = 0 \tag{A.7}$$

can be found by rearranging and integrating:

$$\frac{df(x)}{f(x)} = -bdx \tag{A.8}$$

$$\int \frac{df(x)}{f(x)} = -b\int dx \tag{A.9}$$

$$\ln f(x) = -bx + a \tag{A.10}$$

$$f(x) = Ae^{-bx} \tag{A.11}$$

where $A = e^a$ is a constant. to be determined by the boundary conditions.

Note on boundary and initial conditions: In solving a differential equation, since the equation represents a relationship between different derivatives of a function and the function, the solution for the function is going to depend on the starting value of that function when the independent variable (say x or t) is zero. So, in the above example, A is going to be determine by the value of $f(x)$ when $x = 0$. That is called a **boundary condition**. If the problem involved time instead of space, like $f(t)$, you would need to know $f(t = 0)$ which is then an **initial condition**. If a differential equation is nth order, meaning the highest order derivative is $\frac{d^nf(t)}{dt^n}$. or $\frac{d^nf(x)}{dx^n}$, then there must be n initial or boundary conditions. Think of Newton's second law, which is second order in time, then the position $x(t)$ and velocity $v(t) = \frac{dx(t)}{dt}$ must be specified at $t = 0$.

2. The solution to a second order homogeneous equation of the form

$$\frac{d^2f(t)}{dt^2} + a\frac{df(t)}{dt} + bf(t) = 0 \tag{A.12}$$

can be found by assuming a solution of the form

$$f(t) = e^{\gamma t} \tag{A.13}$$

The same approach used in case 1 above would work, but the above approach is now a little simpler. Substituting this form, taking the derivatives and dividing through by $e^{\gamma x}$ results in a quadratic equation given by

$$\gamma^2 + a\gamma + b = 0 \tag{A.14}$$

With the solution given by the quadratic formula

$$\gamma_\pm = \frac{1}{2}\left(-a \pm \sqrt{a^2 - 4b}\right) \tag{A.15}$$

Note that when $a = 0$ and b is real and > 0, then

$$\gamma_\pm = \pm i\frac{1}{2}\sqrt{b} \tag{A.16}$$

where

$$i = \sqrt{-1} \tag{A.17}$$

The two solutions to the homogeneous equation are

$$f_{h1} = A_+ e^{\gamma_+ t} \tag{A.18}$$

and

$$f_{h2} = A_- e^{\gamma_- t} \tag{A.19}$$

The complete solution is

$$f_h(t) = A_+ e^{\gamma_+ t} + A_- e^{\gamma_- t} \tag{A.20}$$

where the constants are determined by the initial conditions (boundary conditions if the independent variable is space).

In case

$$a^2 - 4b = 0 \tag{A.21}$$

then

$$\gamma_+ = \gamma_- = \frac{\gamma - a}{2} \tag{A.22}$$

and the general solution is then

$$f_h(t) = A_1 e^{\gamma t} + A_2 t e^{\gamma t} \tag{A.23}$$

Solutions to most common and physically relevant differential equations are tabulated, meaning they can be found in published tables.[2]

Examples of Solutions to Inhomogeneous Differential Equations

The solution to an inhomogeneous differential equation is called the *particular solution* as indicated earlier.

1. To solve an inhomogeneous first order differential equation of the form

$$\frac{df(x)}{dx} + bf(x) = g(x) \tag{A.24}$$

we introduce an **integrating factor:**

$$e^{bx} \tag{A.25}$$

Then we can recover the above inhomogeneous equation by noting that

$$\frac{d}{dx}\left(e^{bx}f(x)\right) = e^{bx}\left(\frac{df(x)}{dx} + bf(x)\right) = e^{bx}g(x) \tag{A.26}$$

We can integrate this equation:

$$\int dx \frac{d}{dx}\left(e^{bx}f(x)\right) = \int dx\, e^{bx} g(x) \tag{A.27}$$

$$e^{bx}f(x) = \int dx'\, e^{bx'} g(x') \tag{A.28}$$

where x' has been substituted on the right to distinguish between the x that is in the integral and the x on the left that is not in the integral, so that

$$f_p(x) = \int dx'\, e^{-b(x-x')} g(x') \tag{A.29}$$

The subscript p denotes the particular solution to the inhomogeneous equation.

As indicated earlier, the complete solution to an inhomogeneous first order differential equation is then the sum of the homogeneous solution and the inhomogeneous solution:

$$f(x) = f_h(x) + f_p(x) = Ae^{-bx} + e^{-bx}\int dx'\, e^{bx'} g(x') \tag{A.30}$$

and A is determined for the boundary condition for $f(x)$ (not just for the homogeneous part).

[2] For this and many more helpful mathematical relationships, see for example Murray R. Spiegel, Seymour Lipschutz and John Liu, *Schaum's Outline of Mathematical Handbook of Formulas and Tables*, 4th Edition, McGraw Hill (2013). Also, Milton Abramowitz and Irene A. Stegun, Handbook of Mathematical Functions, National Bureau of Standards Applied Mathematics Series 55, US Government Printing Office (1964).

2. To solve a second order inhomogeneous differential equation, the complete solution is of the form

$$f(x) = f_h(x) + f_p(x) \tag{A.31}$$

where the two constants in the solution $f_h(x)$ are again determined for the boundary conditions for $f(x)$. The steps for generally finding the **particular solution**, $f_p(x)$ are beyond the current discussion, but this is often done with a Green's function approach.[3]

A.2 Partial Differential Equations (PDE)—Method of Separation of Variables

Much of the analysis and models in this text are limited to one dimension in order to keep the math simple and because in today's technology, a one-dimensional system has technological advantages.

A one-dimensional differential equation of the form:

$$\frac{d^2 f(x)}{dx^2} + bf(x) = 0 \tag{A.32}$$

may be generalized to two or three dimensions when appropriate. For example, the equation for electromagnetic waves including light and radio is often a three-dimensional problem. Like mechanical vibrations of a three-dimensional object, quantum systems like atoms, and thermal transport for heat management, many devices require a solution to the equivalent three-dimensional equations. We can deal with this by replacing the differential operator with the appropriate ∇ operator where, in three dimensions in Cartesian coordinates, for example:

$$\nabla^2 f(x, y, z) + bf(x, y, z) = 0 \tag{A.33}$$

where

$$\nabla^2 = \frac{\partial^2}{\partial x^2} + \frac{\partial^2}{\partial y^2} + \frac{\partial^2}{\partial z^2} \tag{A.34}$$

This equation is solved by the **method of separation of variables** and, since the three coordinates are independent of each other, it must be that this equation holds only when

$$f(x, y, z) = f_x(x) f_y(y) f_z(z) \tag{A.35}$$

If we substitute this into the partial differential equation and then divide by $f(x, y, z)$ we get

$$\frac{1}{f_x(x)}\frac{\partial^2 f_x(x)}{\partial x^2} + \frac{1}{f_y(y)}\frac{\partial^2 f_y(y)}{\partial y^2} + \frac{1}{f_z(z)}\frac{\partial^2 f_z(z)}{\partial z^2} + b = 0 \tag{A.36}$$

Since this has to hold for all values of x, y, and z, it means that

$$\frac{1}{f_x(x)}\frac{\partial^2 f_x(x)}{\partial x^2} = a_x \tag{A.37}$$

[3] George B Arfken, Hans J. Weber and Frank E. Harris Mathematical Methods for Physicists, 7th ed., Elsevier, Amsterdam (2013).

$$\frac{1}{f_y(y)}\frac{\partial^2 f_y(y)}{\partial y^2} = a_y \tag{A.38}$$

$$\frac{1}{f_z(z)}\frac{\partial^2 f_z(z)}{\partial z^2} = a_z \tag{A.39}$$

where

$$a_x + a_y + a_z = -b \tag{A.40}$$

A.3 Eigenvalue Problems

In general, if $\hat{A}$ is an operator like $\frac{d^2}{dx^2}$ or ∇^2, then the equation

$$\hat{A}u_a(x) = au_a(x) \tag{A.41}$$

where a is a constant, is called an **eigenvalue equation**. Then $u_a(x)$ is the eigenfunction and a is the **eigenvalue**. So, if $\hat{A} = \frac{d^2}{dx^2}$, then

$$\frac{d^2}{dx^2}u_a(x) = -au_a(x) \tag{A.42}$$

The solution is

$$u_a(x) = e^{\pm i\sqrt{a}x} \tag{A.43}$$

If there are no constraints like boundary conditions, then a can take on any positive value., and $\frac{d^2}{dx^2}$ is said to have a *continuous* ***spectrum of eigenvalues.*** Sometimes, in this case, the eigenfunction is written as

$$u_a(x) \equiv u(a, x) \tag{A.44}$$

If there are constraints, such as that the function must satisfy **periodic boundary conditions**, e.g.,

$$e^{\pm i\sqrt{a}x} = e^{\pm i\sqrt{a}(x+L)} \tag{A.45}$$

then this is only satisfied for specific values of a. Namely,

$$e^{\pm i\sqrt{a}L} = 1$$

Limiting the values of a to

$$\sqrt{a}L = 2m\pi \quad \text{where } m \text{ is a positive integer} \tag{A.46}$$

or

$$a = \left(\frac{2m\pi}{L}\right)^2 \tag{A.47}$$

In this case, the spectrum of eigenvalues is discrete and we write $u_a(x)$ as

$$u_a(x) \equiv u_m(x) \tag{A.48}$$

For either a discrete or continuous spectrum of eigenvalues and assuming that the operators are **Hermitian**, meaning an operator $\hat{P}$ is Hermitian if $\int_0^\infty dx\, g^*(x)\hat{P}f(x) = \int_0^\infty dx\big(\hat{P}g(x)\big)^* f(x)$ for an arbitrary and well behaved $g(x)$ and $f(x)$, there are an infinite number of eigenfunctions and *the set of all eigenfunctions for each case forms a complete set*. If the boundary conditions allow $-\infty < x < \infty$, then any $f(x)$ can be expanded in terms of the eigenfunctions. In the case that the spectrum of eigenvalues is discrete, it means that $f(x)$ can be written as:

$$f(x) = \sum_{m=0}^{\infty} c_m u_m(x) \tag{A.49}$$

where c_m is called an expansion coefficient. This is a countably infinite series. This is very similar to the ideas learned in the study of the Fourier series. If the spectrum of eigenvalues is continuous, we could then expand $f(x)$ in terms of an uncountably infinite series,

$$f(x) = \int_0^\infty da\, c(a) u(a, x) \tag{A.50}$$

where again $c(a)$ is an expansion coefficient.

It can be shown that, for the operators of interest in these notes and that are associated with an observable (i.e., the operators are Hermitian), the eigenfunctions are orthogonal and can be normalized (i.e., they are orthonormal), mathematically meaning that, for eigenfunctions with discrete eigenvalues,

$$\int_{-\infty}^{\infty} dx\, u_n^*(x) u_m(x) \equiv (u_n(x)|u_m(x)) = \delta_{nm} \equiv \begin{cases} 1 & n = m \\ 0 & n \neq m \end{cases} \tag{A.51}$$

where δ_{nm} is a Kronecker delta. A shorthand notation has been introduced to simplify the writing and calculations and looks similar to Dirac notation though it is definitely not Dirac notation, which describes eigenvectors in a Hilbert space. For eigenfunctions with a continuous spectrum of eigenvalues, the orthonormality is given by

$$\int_{-\infty}^{\infty} dx\, u^*(a', x)\, u(a, x) \equiv (u(a', x)\,|u(a, x)) = \delta(a' - a) \tag{A.52}$$

where $\delta(a' - a)$ is the **Dirac delta-function** with the property that $\int_{-\infty}^{\infty} dx f(x)\delta(x - x_0) = f(x_0)$, discussed below.

We can use these results to find the expansion coefficient above. For the case of a discrete spectrum of eigenvalues,

$$f(x) = \sum_{m=0}^{\infty} c_m u_m(x) \tag{A.53}$$

We then multiply both sides by $u_n^*(x)$ and integrate over x:

$$\int_{-\infty}^{\infty} dx\, u_n^*(x) f(x) = \int_{-\infty}^{\infty} dx \sum_{m=0}^{\infty} c_m u_n^*(x) u_m(x) = \sum_{m=0}^{\infty} c_m \int_{-\infty}^{\infty} dx\, u_n^*(x) u_m(x)$$
$$= \sum\nolimits_{m=0}^{\infty} c_m \delta_{nm} = c_n \tag{A.54}$$

Therefore, with n reverting now to the symbol m:

$$c_m = \int_{-\infty}^{\infty} dx\, u_m^*(x) f(x) \tag{A.55}$$

For the case of a continuous spectrum of eigenvalues,

$$f(x) = \int_{0}^{\infty} da\, c(a) u(a, x) \tag{A.56}$$

We then multiply both sides by $u^*(a', x)$ and again integrate over x:

$$\int_{-\infty}^{\infty} dx\, u^*(a', x) f(x) = \int_{-\infty}^{\infty} dx \int_{0}^{\infty} da\, c(a) u^*(a', x)\, u(a, x)$$
$$= \int_{0}^{\infty} da\, c(a) \int_{-\infty}^{\infty} dx\, u^*(a', x)\, u(a, x) \int_{0}^{\infty} da\, c(a)\, \delta(a' - a) = c(a') \tag{A.57}$$

Therefore, after a' reverting back to the symbol a,

$$c(a) = \int_{-\infty}^{\infty} dx\, u^*(a, x) f(x) \tag{A.58}$$

A.4 Complex Numbers and Euler's Theorem

For real numbers x, and y, a complex number z is written as

$$z = x + iy \tag{A.59}$$

where

$$i = \sqrt{-1} \tag{A.60}$$

Euler's theorem says that

$$z = x + iy = R \exp(i\theta) = R\cos\theta + iR\sin\theta \tag{A.61}$$

where

$$R = \sqrt{x^2 + y^2} \text{ and } \tan\theta = \frac{y}{x} \tag{A.62}$$

Complex numbers are represented in the complex plane as shown in Fig. A.1

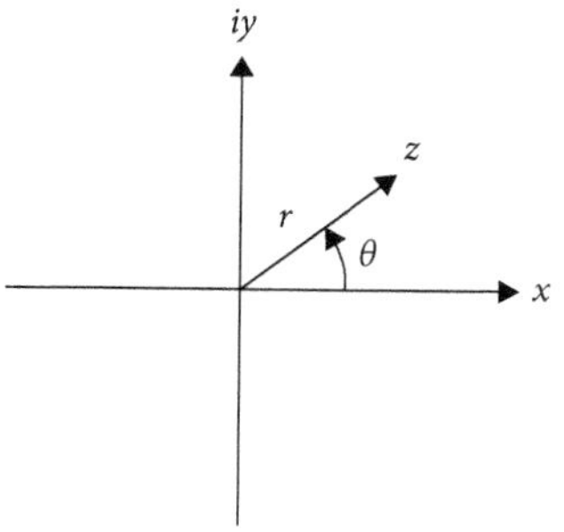

Fig. A.1 Complex numbers are often represented in the complex plane.

A.5 Dirac Delta-Function

The Dirac delta-function has physical meaning only under an integral and defined by the property that

$$\int_{-\infty}^{\infty} dx\, f(x)\delta(x) = f(0) \tag{A.63}$$

Or, by changing variables, it is easy to show that

$$\int_{-\infty}^{\infty} dx\, f(x)\delta(x - x_0) = f(x_0) \tag{A.64}$$

Likewise,

$$\int_{-\infty}^{\infty} dx\, f(x)\delta(ax) = \frac{1}{|a|} f(0);\, |a| > 0 \tag{A.65}$$

From the Fourier theorem it can be shown that

$$\int_{-\infty}^{\infty} dk\, e^{-ik(x-x')} = 2\pi\delta\,(x - x') \tag{A.66}$$

Proof: It follows from Fourier theory that if

$$\mathcal{F}(k) = \frac{1}{\sqrt{2\pi}} \int_{-\infty}^{\infty} dx\, e^{ikx}\, f(x) \tag{A.67}$$

and

$$f(x) = \frac{1}{\sqrt{2\pi}} \int_{-\infty}^{\infty} dk\, e^{-ikx}\, \mathcal{F}(k) \tag{A.68}$$

Then the Fourier integral theorem says that

$$f(x) = \frac{1}{2\pi} \int_{-\infty}^{\infty} dk\, e^{-ikx} \int_{-\infty}^{\infty} dx'\, e^{ikx'}\, f(x') \tag{A.69}$$

Reorganizing, we get

$$f(x) = \int_{-\infty}^{\infty} dx' \left(\frac{1}{2\pi} \int_{-\infty}^{\infty} dk\, e^{-ik(x-x')} \right) f(x') \tag{A.70}$$

This is only true if

$$\frac{1}{2\pi} \int_{-\infty}^{\infty} dk\, e^{-ik(x-x')} = \delta\,(x - x') \tag{A.71}$$

Another form for the Dirac delta-function that is convenient to use is

$$\delta\,(k - k_0) = \lim_{a\to\infty} \frac{1}{a\sqrt{\pi}} e^{-\left(\frac{k-k_0}{a}\right)^2} \tag{A.72}$$

A.6 Gaussian Integral

The functional form $e^{-\alpha x^2}$ or $e^{-\alpha x^2+bx}$ is called a Gaussian. The second form is equivalent to displaced Gaussian since the exponent can be rewritten as $\alpha x^2 - bx = a\left(x^2 - \frac{b}{a} + \frac{b^2}{4a^2}\right) - \frac{b^2}{4a} = a\left(x - \frac{b}{2a}\right)^2 - \frac{b^2}{4a}$

$$\int_{-\infty}^{\infty} dx\, e^{-\alpha x^2} = \sqrt{\frac{\pi}{a}};\quad \text{Re } a > 0 \tag{A.73}$$

$$\int_{-\infty}^{\infty} dx\, e^{-\alpha x^2+bx} = \sqrt{\frac{\pi}{a}} e^{\frac{b^2}{4a}};\quad \text{Re } a > 0 \tag{A.74}$$

Note that more complex integrals involve a Gaussian form, such as $\int_{-\infty}^{\infty} dx\, xe^{-\alpha x^2+bx}$ or $\int_{-\infty}^{\infty} dx\, x^2 e^{-\alpha x^2+bx}$, which can be evaluated by **differentiating under the integral**. For example: $\int_{-\infty}^{\infty} dx\, xe^{-\alpha x^2+bx} = \frac{d}{db}\int_{-\infty}^{\infty} dx\, e^{-\alpha x^2+bx} = \frac{b}{2a}\sqrt{\frac{\pi}{a}} e^{\frac{b^2}{4a}}$ and $\int_{-\infty}^{\infty} dx\, x^2 e^{-\alpha x^2} = -\frac{d}{da}\int_{-\infty}^{\infty} dx\, e^{-\alpha x^2+bx} = \sqrt{\frac{\pi}{a^3}}$.

A.7 Linear Algebra: Matrices, Determinants, Permanents, and the Eigenvector

Multiplication

Matrices are a two-dimensional array of numbers: $n - rows \times m - column$. A square matrix, such as frequently encountered in quantum problems, is one where the number of columns and rows are the same. An $n \times n$ matrix $\hat{A}$ is written as:

$$\hat{A} = \begin{bmatrix} a_{11} & \cdots & a_{1n} \\ \vdots & \ddots & \vdots \\ a_{n1} & \cdots & a_{nn} \end{bmatrix} \tag{A.75}$$

The subscripts refer to the rows and columns, respectively. A square $n \times n$ matrix has **order** $n \times n$ which is sometimes just n. Two matrices, $\hat{A}$ and $\hat{B}$ of order $j \times k$ and $m \times n$ can be multiplied together if $k = m$ to give a resulting matrix that is order $j \times n$. **Matrix multiplication** is given by

$$\hat{A} * \hat{B} = \begin{bmatrix} a_{11} & \cdots & a_{1m} \\ \vdots & \ddots & \vdots \\ a_{j1} & \cdots & a_{jm} \end{bmatrix} \begin{bmatrix} b_{11} & \cdots & b_{1n} \\ \vdots & \ddots & \vdots \\ b_{m1} & \cdots & b_{mn} \end{bmatrix} = \begin{bmatrix} \sum_{i=1}^{m} a_{1i} b_{i1} & \cdots & \sum_{i=1}^{m} a_{1i} b_{in} \\ \vdots & \ddots & \vdots \\ \sum_{i=1}^{m} a_{ji} b_{i1} & \cdots & \sum_{i=1}^{m} a_{ji} b_{in} \end{bmatrix} \tag{A.76}$$

Three examples:

$$\begin{bmatrix} 1 & 4 \\ 3 & 5 \end{bmatrix} \begin{bmatrix} 1 & 6 \\ 2 & 3 \end{bmatrix} = \begin{bmatrix} 1*1+4*2 & 1*6+4*3 \\ 3*1+5*2 & 3*6+5*3 \end{bmatrix} = \begin{bmatrix} 9 & 18 \\ 13 & 33 \end{bmatrix} \tag{A.77a}$$

$$\begin{bmatrix} 1 & 2 \end{bmatrix} \begin{bmatrix} 1 & 6 \\ 2 & 3 \end{bmatrix} = \begin{bmatrix} 1*1+2*2 & 1*6+2*3 \end{bmatrix} = \begin{bmatrix} 5 & 12 \end{bmatrix} \tag{A.77b}$$

$$\begin{bmatrix} 1 & 6 \\ 2 & 3 \end{bmatrix} \begin{bmatrix} 4 \\ 5 \end{bmatrix} = \begin{bmatrix} 1*4+6*5 \\ 2*4+3*5 \end{bmatrix} = \begin{bmatrix} 34 \\ 23 \end{bmatrix} \tag{A.77c}$$

Note that when a scalar multiplies a matrix, it multiples every term in that matrix:

$$3 * \begin{bmatrix} 1 & 6 \\ 2 & 3 \end{bmatrix} = \begin{bmatrix} 3 & 18 \\ 6 & 9 \end{bmatrix} \tag{A.77d}$$

Determinants

The **determinant** of a square matrix is a scalar and is given by selecting any row or column in the matrix, and for each element in that row or column, a_{ij}, multiply it by $(-1)^{i+j}$ and the determinant of the matrix formed by removing the row i and column j. As an example, we evaluate the determinant of the square matrix $\hat{A}$ by removing the top row:

$$\det \hat{A} = \begin{vmatrix} a_{11} & \cdots & a_{1n} \\ \vdots & \ddots & \vdots \\ a_{n1} & \cdots & a_{nn} \end{vmatrix} = a_{11} \begin{vmatrix} a_{21} & \cdots & a_{2n} \\ \vdots & \ddots & \vdots \\ a_{n2} & \cdots & a_{nn} \end{vmatrix} - a_{21} \begin{vmatrix} a_{11} & \cdots & a_{2n} \\ \vdots & \ddots & \vdots \\ a_{n2} & \cdots & a_{nn} \end{vmatrix} + \cdots \tag{A.78}$$

where the determinant associated with the multiplying factor out front, a_{j1}, is missing the first column and j^{th} row of the original determinant. The process is continued with each resulting matrix until the final result is a scalar. This can be written more succinctly as:

$$\det \hat{A} = \sum_{i \text{ or } j}^{n} (-1)^{i+j} a_{ij} M_{ij} \tag{A.79}$$

M_{ij} is the **first minor** of the a_{ij} element and is computed by forming the sub-matrix of $\hat{A}$ by removing row i and column j and calculating the determinant. Then $(-1)^{i+j} M_{ij}$ is called the **cofactor** of the a_{ij} element.

For the matrix:

$$\hat{A} = \begin{bmatrix} 3 & 1 & 2 \\ 4 & 2 & 3 \\ 2 & 5 & 1 \end{bmatrix} \tag{A.80}$$

the minor of $\hat{A}_{12}$ (note that $\hat{A}_{12} = a_{12} = 1$, the second entry from the upper left) is

$$\hat{M}_{12} = \det \begin{bmatrix} 4 & 3 \\ 2 & 1 \end{bmatrix} \tag{A.81}$$

So for the determinant of $\hat{A}$:

$$\det\begin{vmatrix}3 & 1 & 2\\ 4 & 2 & 3\\ 2 & 5 & 1\end{vmatrix} = 3\begin{vmatrix}2 & 3\\ 5 & 1\end{vmatrix} - 4\begin{vmatrix}1 & 2\\ 5 & 1\end{vmatrix} + 2\begin{vmatrix}1 & 2\\ 2 & 3\end{vmatrix}$$
$$= 3\,(2-15) - 4\,(1-10) + 2\,(3-4) = -5. \tag{A.82}$$

The **rank** of a square matrix is the order of the largest sub-matrix with linearly independent rows (columns). This corresponds to the largest **sub-matrix** with a non-zero determinant. In the case above with the matrix of order three, corresponding to the determinant with value −45, since the determinant is non-zero, the rank and order are both three.

The determinant is used to preserve exchange symmetry for fermions.

Permanents

The **permanent** of a square matrix is identical to the determinant except that there is no $(-1)^{i+j}$ factor. So in the case of Eq. A.82, switch the minus sign in front of the 4 in the first equality to a plus sign: i.e., $-4\begin{vmatrix}1 & 2\\ 2 & 5\end{vmatrix} \rightarrow +4\begin{vmatrix}1 & 2\\ 2 & 5\end{vmatrix}$.

The permanent is used to preserve the exchange symmetry for bosons.

Adjoint, Hermiticity, and Unitarity

The **adjoint** of a matrix $\hat{A}$ is the **complex-transpose** of the original matrix. With the matrix $\hat{A}$ above then the corresponding adjoint, $\hat{A}^{\dagger}$, is given by

$$\hat{A}^{\dagger} = \begin{bmatrix} a_{11}^* & \cdots & a_{n1}^* \\ \vdots & \ddots & \vdots \\ a_{1n}^* & \cdots & a_{nn}^* \end{bmatrix} \tag{A.83}$$

If $\hat{A} = \hat{A}^{\dagger}$, then the matrix is **Hermitian**. If $\hat{A}^{\dagger} = \hat{A}^{-1}$, then the matrix is **unitary**.

Vectors: Inner and Outer Products and Dirac Notation

A matrix $\hat{V}$ with a single column is an $n \times 1$ **column vector**:

$$\hat{V} = \begin{bmatrix} v_1 \\ \vdots \\ v_n \end{bmatrix} \tag{A.84}$$

and the complex transpose is a $1 \times n$ **row vector**:

$$\hat{V}^{\dagger} = \begin{bmatrix} v_1^* & \cdots & v_n^* \end{bmatrix} \tag{A.85}$$

In **Dirac notation**, the **ket** can be represented as a column vector. So for a three-dimensional **Hilbert space**,

$$|V\rangle = \begin{bmatrix} v_1 \\ v_2 \\ v_3 \end{bmatrix} \tag{A.86}$$

The bra is the corresponding complex transpose (row vector) of the **ket**. The **inner product** is similar to a dot product and produces a **scalar** (a matrix of order unity) that can be complex.

$$\langle W|V\rangle = \begin{bmatrix} w_1^* & w_2^* & w_3^* \end{bmatrix} \begin{bmatrix} v_1 \\ v_2 \\ v_3 \end{bmatrix} = w_1^* v_1 + w_2^* v_2 + w_3^* v_3 \tag{A.87}$$

The **outer product** of two vectors of the same size is a square matrix:

$$|V\rangle\langle W| = \begin{bmatrix} v_1 \\ v_2 \\ v_3 \end{bmatrix} \begin{bmatrix} w_1^* & w_2^* & w_3^* \end{bmatrix} = \begin{bmatrix} v_1 w_1^* & v_1 w_2^* & v_1 w_3^* \\ v_2 w_1^* & v_2 w_2^* & v_2 w_3^* \\ v_3 w_1^* & v_3 w_2^* & v_3 w_3^* \end{bmatrix} \tag{A.88}$$

An arbitrary matrix can then be written out as

$$\hat{A} = \sum_{ij} \langle i|\hat{A}|j\rangle\, |i\rangle \langle j| \tag{A.89}$$

where $|i\rangle$ $(\langle j|)$ are unit vectors with 0's in all the positions except position i (j), where there is a 1.

$$\langle i|\hat{A}|j\rangle \equiv a_{ij} = \begin{bmatrix} 0_1 & \cdots & 1_i & \cdots & 0_n \end{bmatrix} \begin{bmatrix} a_{11} & \cdots & a_{1n} \\ \vdots & \ddots & \vdots \\ a_{n1} & \cdots & a_{nn} \end{bmatrix} \begin{bmatrix} 0_1 \\ \vdots \\ 1_j \\ \vdots \\ 0_n \end{bmatrix} \tag{A.90}$$

Example for an operator of order two:

$$\langle 1|\hat{A}|2\rangle = \begin{bmatrix} 1 & 0 \end{bmatrix} \begin{bmatrix} a_{11} & a_{12} \\ a_{21} & a_{22} \end{bmatrix} \begin{bmatrix} 0 \\ 1 \end{bmatrix} = \begin{bmatrix} a_{11} & a_{12} \end{bmatrix} \begin{bmatrix} 0 \\ 1 \end{bmatrix} = a_{12} \tag{A.91}$$

For consistency in notation, we note that

$$\hat{A}_{ij} = \langle i|\hat{A}|j\rangle = a_{ij} \tag{A.92}$$

The Identity Matrix and the Inverse Matrix

The **identity matrix** has zero's in all the positions except along the diagonal where it has ones. For example for a matrix of arbitrary order,

$$\hat{I} = \begin{bmatrix} 1 & 0 & 0 \\ 0 & 1 & 0 \\ 0 & 0 & \ddots \end{bmatrix} \tag{A.93}$$

Therefore, for any matrix $\hat{A}$ of order n,

$$\hat{A} * \hat{I} = \hat{I} * \hat{A} = \hat{A} \tag{A.94}$$

where $\hat{I}$ is the identity matrix of order n.

If $\hat{A}$ is a matrix with a non-singular determinant, then there exists an **inverse matrix** $\hat{A}^{-1}$ such that

$$\hat{A} * \hat{A}^{-1} = \hat{A}^{-1} * \hat{A} = \hat{I} \tag{A.95}$$

To find the inverse of a matrix, we use **Cramer's rule**:

$$\hat{A}^{-1} = \frac{1}{|\hat{A}|}\hat{C}^T \tag{A.96}$$

where

$$|\hat{A}| \equiv \det \hat{A} \tag{A.97}$$

and $\hat{C}^T$ is the *transpose* of the matrix of **cofactors** (see A.79 and discussion) where the cofactor of the a_{ij} element is again $(-1)^{i+j}M_{ij}$, as discussed above. So for say a 4th order matrix:

$$\hat{C}_{24} \equiv (-1)^{2+4}\widehat{M}_{24} = (-1)^{2+4}\det\begin{bmatrix} 1 & 4 & 7 & \cancel{5} \\ \cancel{3} & \cancel{0} & \cancel{5} & \cancel{3} \\ -1 & 9 & 2 & \cancel{7} \\ 2 & -4 & 8 & \cancel{1} \end{bmatrix} = (-1)^{2+4}\begin{vmatrix} 1 & 4 & 7 \\ -1 & 9 & 2 \\ 2 & -4 & 8 \end{vmatrix}$$
$$= 1\begin{bmatrix} 9 & 2 \\ -4 & 8 \end{bmatrix} - 4\begin{bmatrix} -1 & 2 \\ 2 & 8 \end{bmatrix} + 7\begin{bmatrix} -1 & 9 \\ 2 & -4 \end{bmatrix} = 80 + 48 - 98 = 30 \tag{A.98}$$

Finally, with $\hat{C}$ being given by

$$\hat{C} = \begin{bmatrix} c_{11} & \cdots & c_{1n} \\ \vdots & \ddots & \vdots \\ c_{n1} & \cdots & c_{nn} \end{bmatrix} \tag{A.99}$$

$$\hat{A}^{-1} = \frac{1}{|\hat{A}|}\hat{C}^T = \begin{bmatrix} \frac{c_{11}}{|\hat{A}|} & \cdots & \frac{c_{n1}}{|\hat{A}|} \\ \vdots & \ddots & \vdots \\ \frac{c_{1n}}{|\hat{A}|} & \cdots & \frac{c_{nn}}{|\hat{A}|} \end{bmatrix} \tag{A.100}$$

The factor $\frac{1}{|\hat{A}|}$ was taken inside to emphasize that when a number multiplies a matrix, it multiplies every element of that matrix (A.77d).

Cramer's Rule for Solving *n* Linear Inhomogeneous Algebraic Equations in *n* Unknowns

Suppose we have n linear equations in n unknowns:

$$\begin{aligned} a_{11}x_1 + a_{12}x_2 + \cdots + a_{1n}x_n &= b_1 \\ a_{21}x_1 + a_{22}x_2 + \cdots + a_{2n}x_n &= b_2 \\ &\vdots \\ a_{n1}x_1 + a_{n2}x_2 + \cdots + a_{nn}x_n &= b_n \end{aligned} \tag{A.101}$$

These can be written in matrix form by putting the coefficients in a matrix and converting the unknowns, x_i and the right-hand side into vectors. Specifically,

$$\begin{bmatrix} a_{11} & a_{12} & \dots & a_{1n} \\ a_{21} & a_{22} & \dots & a_{2n} \\ \vdots & \vdots & \vdots & \vdots \\ a_{n1} & a_{n2} & \dots & a_{nn} \end{bmatrix} \begin{bmatrix} x_1 \\ x_2 \\ \vdots \\ x_n \end{bmatrix} = \begin{bmatrix} b_1 \\ b_2 \\ \vdots \\ b_n \end{bmatrix} \quad \text{(A.102)}$$

Or written more compactly:

$$\hat{A}\hat{X} = \hat{B} \quad \text{(A.103)}$$

If one or more of the $b_i's \neq 0$, the set of algebraic equations is **inhomogeneous**. In this case, one can derive **Cramer's rule**, which says that the value of ith entry of $\hat{X}$ (a column vector) is:

$$x_i = \frac{\det \hat{A}_i}{\det \hat{A}}; i = 1, \dots, n \quad \text{(A.104)}$$

where the matrix $\hat{A}_i$ is formed by taking the matrix $\hat{A}$ above and replacing the ith column with the vector $\hat{B}$.

For example

$$\begin{bmatrix} 2 & 3 \\ 1 & 4 \end{bmatrix} \begin{bmatrix} x_1 \\ x_2 \end{bmatrix} = \begin{bmatrix} -1 \\ 3 \end{bmatrix} \quad \text{(A.105)}$$

$$\det A = \det \begin{bmatrix} 2 & 3 \\ 1 & 4 \end{bmatrix} = 5 \quad \text{(A.106)}$$

$$x_1 = \frac{\det \hat{A}_1}{\det \hat{A}} = \frac{\det \begin{bmatrix} -1 & 3 \\ 3 & 4 \end{bmatrix}}{\det \begin{bmatrix} 2 & 3 \\ 1 & 4 \end{bmatrix}} = \frac{-13}{5} \quad \text{(A.107)}$$

$$x_2 = \frac{\det \hat{A}_2}{\det \hat{A}} = \frac{\det \begin{bmatrix} 2 & -1 \\ 1 & 3 \end{bmatrix}}{\det \begin{bmatrix} 2 & 3 \\ 1 & 4 \end{bmatrix}} = \frac{7}{5} \quad \text{(A.108)}$$

The Eigenvector Problem

In the above case, a special case is had if

$$\begin{bmatrix} b_1 \\ b_2 \\ \vdots \\ b_n \end{bmatrix} = \lambda \begin{bmatrix} x_1 \\ x_2 \\ \vdots \\ x_n \end{bmatrix} \quad \text{(A.109)}$$

In this case,

$$\hat{A}\hat{X} = \lambda\hat{X} = \lambda\hat{I}\hat{X} \quad \text{(A.110)}$$

This is called an **eigenvector** or **eigenvalue** equation. Then $\hat{X}$ is the eigenvector and λ is the eigenvalue.

Then

$$(\hat{A} - \lambda\hat{I})\,\hat{X} = \begin{bmatrix} 0_1 \\ 0_2 \\ \vdots \\ 0_n \end{bmatrix} = 0 \tag{A.111}$$

We rewrite this as

$$\begin{bmatrix} a_{11} - \lambda & a_{12} & \cdots & a_{1n} \\ a_{21} & a_{22} - \lambda & \cdots & a_{2n} \\ \vdots & \vdots & \vdots & \vdots \\ a_{n1} & a_{n2} & \cdots & a_{nn} - \lambda \end{bmatrix} \begin{bmatrix} x_1 \\ x_2 \\ \vdots \\ x_n \end{bmatrix} = 0 \tag{A.112}$$

This is again a set of n equations in n unknowns but now the right-hand side is 0 and so the equations are a set of **homogeneous** algebraic equations. *In order for a solution to exist, the determinant of coefficients must be 0*:

$$\begin{vmatrix} a_{11} - \lambda & a_{12} & \cdots & a_{1n} \\ a_{21} & a_{22} - \lambda & \cdots & a_{2n} \\ \vdots & \vdots & \vdots & \vdots \\ a_{n1} & a_{n2} & \cdots & a_{nn} - \lambda \end{vmatrix} = 0 \tag{A.113}$$

Solving this equation will yield n different values for λ and requires solving an nth order polynomial. We consider the case for two unknowns:

$$\hat{A} = \begin{bmatrix} a & b \\ c & d \end{bmatrix} \tag{A.114}$$

$$\hat{X} = \begin{bmatrix} x_1 \\ x_2 \end{bmatrix} \tag{A.115}$$

and

$$\begin{bmatrix} a - \lambda & b \\ c & d - \lambda \end{bmatrix} \begin{bmatrix} x_1 \\ x_2 \end{bmatrix} = 0 \tag{A.116}$$

We require that

$$\begin{vmatrix} a - \lambda & b \\ c & d - \lambda \end{vmatrix} = 0 \tag{A.117}$$

or

$$(a - \lambda)(d - \lambda) - cb = 0 \tag{A.118}$$

and after rearranging

$$\lambda^2 - (a + d)\lambda + ad - cb = 0 \tag{A.119}$$

Giving as expected two different eigenvalues,

$$\lambda_{\pm} = \frac{1}{2}\left((a+d) \pm \sqrt{(a+d)^2 - 4(ad - cb)}\right) \quad \text{(A.120)}$$

The complete solution only gives one unknown, say x_2, in terms of x_1 for each of two eigenvalues. So, using the equation from above:

$$\begin{bmatrix} a-\lambda_{\pm} & b \\ c & d-\lambda_{\pm} \end{bmatrix}\begin{bmatrix} x_1 \\ x_2 \end{bmatrix} = 0 \quad \text{(A.121)}$$

$$(a-\lambda_{\pm})x_1 + bx_2 = 0 \quad \text{(A.122)}$$

or

$$x_2 = x_1 \frac{(\lambda_{\pm} - a)}{b} \quad \text{(A.123)}$$

$$\widehat{X}_{\lambda_{\pm}} = \begin{bmatrix} x_1 \\ x_2 \end{bmatrix} = x_1 \begin{bmatrix} 1 \\ \frac{(\lambda_{\pm}-a)}{b} \end{bmatrix} \quad \text{(A.124)}$$

Assuming that the eigenvalues are not degenerate (i.e., $\lambda_+ \neq \lambda_-$), each eigenvector is different. You can also show that they are orthogonal; i.e., $\widehat{X}^*_{\lambda_+} \cdot \widehat{X}_{\lambda_-} = 0$. In quantum systems using Dirac notation, it would be

$$\widehat{X}_{\lambda_{\pm}} \rightarrow |\lambda_{\pm}\rangle = x_1 \begin{bmatrix} 1 \\ \frac{(\lambda_{\pm}-a)}{b} \end{bmatrix} \quad \text{(A.125)}$$

and x_1 would be determined then by normalization:

$$x_1 = \frac{1}{\sqrt{1 + \left(\frac{(\lambda_{\pm}-a)}{b}\right)^2}} \quad \text{(A.126)}$$

$$\langle \lambda_i | \lambda_j \rangle = \delta_{ij}, i,j = +/- \quad \text{(A.127)}$$

We note now that we have found that

$$\hat{A}\widehat{X}_{\lambda_{\pm}} = \lambda_{\pm}\widehat{X}_{\lambda_{\pm}} \quad \text{(A.128)}$$

If we converted

$$\hat{A} \rightarrow \hat{A}' = \begin{bmatrix} \lambda_+ & 0 \\ 0 & \lambda_- \end{bmatrix} \quad \text{(A.129)}$$

then it is easy to show that

$$\begin{bmatrix} \lambda_+ & 0 \\ 0 & \lambda_- \end{bmatrix}\widehat{X}'_{\lambda_{\pm}} = \lambda_{\pm}\widehat{X}'_{\lambda_{\pm}} \quad \text{(A.130)}$$

where the prime represents the eigenstates for this form of the matrix; the corresponding eigenvectors are

$$\hat{X}'_{\lambda_+} = \begin{bmatrix}1\\0\end{bmatrix} \text{ and } \hat{X}'_{\lambda_-} = \begin{bmatrix}0\\1\end{bmatrix} \tag{A.131}$$

So, we say the matrix $\hat{A}$ has been **diagonalized**.

Matrix Diagonalization

In the above example of solving the eigenvalue problem we found two eigenvectors of the original matrix and then saw how replacing $\hat{A}$ with the diagonal matrix, $\hat{A}'$, the new eigenvectors were the unit vectors *in that basis*, meaning in terms of the eigenvectors of $\hat{A}'$. It is important to know how to mathematically convert from $\hat{A}$ to $\hat{A}'$. For an arbitrary non-singular matrix of order n, $\hat{A}$, we assume that we have found the eigenvalues λ_i and the corresponding normalized eigenvectors $\hat{X}_{\lambda_i}$ such that

$$\hat{A}\hat{X}_{\lambda_i} = \lambda_i \hat{X}_{\lambda_i} \tag{A.132}$$

Consider a square matrix where the columns are the eigenvectors,

$$\hat{S} = \left[\hat{X}_{\lambda_1} \;\; \hat{X}_{\lambda_2} \;\; \cdots \;\; \hat{X}_{\lambda_n}\right] \tag{A.133}$$

Then

$$\hat{A}\hat{S} = \left[\lambda_1\hat{X}_{\lambda_1} \;\; \lambda_2\hat{X}_{\lambda_2} \;\; \cdots \;\; \lambda_n\hat{X}_{\lambda_n}\right] \tag{A.134}$$

We now form the adjoint of $\hat{S}$, which you recall is the **complex transpose** of $\hat{S}$:

$$\hat{S}^{\dagger} = \hat{S}^{*T} = \begin{bmatrix}\hat{X}^{*T}_{\lambda_1}\\ \hat{X}^{*T}_{\lambda_2}\\ \vdots \\ \hat{X}^{*T}_{\lambda_n}\end{bmatrix} \tag{A.135}$$

Then

$$\hat{S}^{*T}\hat{A}\hat{S} = \hat{S}^{\dagger}\hat{A}\hat{S} = \begin{bmatrix}\hat{X}^{*T}_{\lambda_1}\\ \hat{X}^{*T}_{\lambda_2}\\ \vdots \\ \hat{X}^{*T}_{\lambda_n}\end{bmatrix}\left[\lambda_1\hat{X}_{\lambda_1} \;\; \lambda_2\hat{X}_{\lambda_2} \;\; \cdots \;\; \lambda_n\hat{X}_{\lambda_n}\right]$$

$$= \begin{bmatrix}\hat{X}^{*T}_{\lambda_1}\lambda_1\hat{X}_{\lambda_1} & \hat{X}^{*T}_{\lambda_1}\lambda_2\hat{X}_{\lambda_2} & \cdots & \hat{X}^{*}_{\lambda_1}\lambda_n\hat{X}_{\lambda_n}\\ \hat{X}^{*T}_{\lambda_2}\lambda_1\hat{X}_{\lambda_1} & \hat{X}^{*T}_{\lambda_2}\lambda_2\hat{X}_{\lambda_2} & \cdots & \hat{X}^{*T}_{\lambda_2}\lambda_n\hat{X}_{\lambda_n}\\ \vdots & \vdots & \vdots & \vdots \\ \hat{X}^{*T}_{\lambda_n}\lambda_1\hat{X}_{\lambda_1} & \hat{X}^{*T}_{\lambda_n}\lambda_2\hat{X}_{\lambda_2} & \cdots & \hat{X}^{*T}_{\lambda_n}\lambda_n\hat{X}_{\lambda_n}\end{bmatrix} = \begin{bmatrix}\lambda_1 & 0 & \cdots & 0\\ 0 & \lambda_2 & \cdots & 0\\ \vdots & \vdots & \vdots & \vdots\\ 0 & 0 & \cdots & \lambda_n\end{bmatrix} \tag{A.136}$$

Where, by orthonormality

$$\hat{X}^{*T}_{\lambda_i}\lambda_k\hat{X}_{\lambda_k} = 0 \text{ for } i \neq k \tag{A.137}$$

and

$$\hat{X}^{*T}_{\lambda_j}\lambda_j\hat{X}_{\lambda_j} = \lambda_j \tag{A.138}$$

Unitary Transformation

Consider an operator (e.g., a matrix), $\hat{A}$. Then let $\hat{U}$ be a unitary operator (A.83 and discussion, meaning $\hat{U}^{\dagger} = \hat{U}^{-1}$) of the same order as $\hat{A}$ in the same Hilbert space. Then a new operator, $\hat{B} = \hat{U}\hat{A}\hat{U}^{\dagger}$, is the result of a **unitary transformation** of $\hat{A}$. For example, let $\hat{A}\hat{X} = \lambda\hat{X}$. Since $\hat{U}^{\dagger}\hat{U} = \hat{U}\hat{U}^{\dagger} = \hat{I}$, then inserting $\hat{U}^{\dagger}\hat{U}$ between $\hat{A}$ and $\hat{X}$ and then multiplying both sides by $\hat{U}$, we get

$$\hat{U}\hat{A}\hat{U}^{\dagger}\hat{U}\hat{X} = \lambda\hat{U}\hat{X} \tag{A.139}$$

Hence, since $\hat{X}$ is an eigenvector of $\hat{A}$ with eigenvalue λ, then $\hat{U}\hat{X}$ is an eigenvector of $\hat{B} = \hat{U}\hat{A}\hat{U}^{\dagger}$ with the same eigenvalue. The same is also true for

$$\hat{U}^{\dagger}\hat{A}\hat{U}\hat{U}^{\dagger}\hat{X} = \lambda\hat{U}^{\dagger}\hat{X} \tag{A.140}$$

The transformation above from a matrix $\hat{A}$ to the matrix $\hat{S}^{\dagger}\hat{A}\hat{S}$, which is now diagonal, is also a unitary transformation. It is easy to show that the magnitude of both eigenvectors is the same.

Vocabulary (page) and Important Concepts

- cofactor 339
- Permament 340
- Hermitian 340
- Unitary 340
- Adjoint 340
- Row vector 340
- Column vector 340
- Inner product 341
- Outer product 341
- Identity matrix 341
- Inverse matrix 342
- Cramer's Rule 342
- Inhomogeneous algebraic equations 343
- Eigenvalue 344
- Eigenvector 344
- Matrix diagonalization 346
- Unitary transformation 347

Appendix B Power Series for Important Functions

$$\frac{1}{1-x} = 1 + x + x^2 + \cdots \quad -1 < x < 1$$

$$\frac{1}{\sqrt{1+x}} = 1 - \frac{x}{2} + \frac{3x^2}{8}$$

$$\sqrt{1+x} = 1 + \frac{x}{2} - \frac{x^2}{8} + \cdots$$

$$\cos x = 1 - \frac{x^2}{2!} + \frac{x^4}{4!} - \cdots$$

$$\tan x = x + \frac{x^3}{3} + \frac{2x^5}{15} + \cdots$$

$$\sin x = x - \frac{x^3}{3!} + \frac{x^5}{5!}$$

$$e^x = 1 + x + \frac{x^2}{2!} + \cdots$$

$$\ln(1+x) = x - \frac{x^2}{2} + \frac{x^3}{3} - \frac{x^4}{4} + \cdots \quad -1 < x < 1$$

Taylor series expansion of $f(x)$ about a:

$$f(x) = f(a) + f'(a)(x-a) + \frac{1}{2!}f''(a)(x-a)^2 + \cdots$$

Appendix C Properties and Representations of the Dirac Delta-Function

Definition:

$$\int_{-\infty}^{\infty} dx\, f(x)\delta\,(x - x_0) = f(x_0)$$

Identities:

$$\delta(ax) = \frac{1}{|a|}\delta(x)$$

Representations:

$$\delta\,(k - k_0) = \frac{1}{2\pi}\int_{-\infty}^{\infty} dx\; e^{i(k-k_0)x}$$

$$\delta\,(k - k_0) = \frac{1}{\sqrt{\pi}}\lim_{a\to+0}\frac{1}{a}e^{-\left(\frac{k-k_0}{a}\right)^2}$$

$$\delta\,(k - k_0) = \frac{1}{\pi}\int_{0}^{\infty} dx\cos x\,(k - k_0)$$

$$\delta\,(k - k_0) = \frac{1}{\pi}\lim_{a\to\infty}\frac{\sin a\,(k - k_0)}{(k - k_0)}$$

$$\delta\,(k - k_0) = \frac{1}{\pi}\lim_{a\to 0}\frac{\mathrm{a}}{(k - k_0)^2 + a^2}$$

Appendix D Vector Calculus and Vector Identities

Cartesian coordinates:

$$\nabla f = \frac{\partial f}{\partial x}\hat{x} + \frac{\partial f}{\partial y}\hat{y} + \frac{\partial f}{\partial z}\hat{z}$$

$$\nabla \cdot \boldsymbol{A} = \frac{\partial A_x}{\partial x} + \frac{\partial A_y}{\partial y} + \frac{\partial A_z}{\partial z}$$

$$\nabla^2 f = \frac{\partial^2 f}{\partial x^2} + \frac{\partial^2 f}{\partial y^2} + \frac{\partial^2 f}{\partial z^2}$$

Cylindrical coordinates:

$$\nabla f = \frac{\partial f}{\partial \rho}\hat{\rho} + \frac{1}{\rho}\frac{\partial f}{\partial \varphi}\hat{\varphi} + \frac{\partial f}{\partial z}\hat{z}$$

$$\nabla \cdot \boldsymbol{A} = \frac{1}{\rho}\frac{\partial \rho A_\rho}{\partial \rho} + \frac{1}{\rho}\frac{\partial A_\varphi}{\partial \varphi} + \frac{\partial A_z}{\partial z}$$

$$\nabla^2 f = \frac{1}{\rho}\frac{\partial}{\partial \rho}\rho\frac{\partial f}{\partial \rho} + \frac{1}{\rho^2}\frac{\partial^2 f}{\partial \varphi^2} + \frac{\partial^2 f}{\partial z^2}$$

Spherical coordinates:

$$\nabla f = \frac{\partial f}{\partial r}\hat{r} + \frac{1}{r}\frac{\partial f}{\partial \theta}\hat{\theta} + \frac{1}{r\sin\theta}\frac{\partial f}{\partial \varphi}\hat{\varphi}$$

$$\nabla \cdot \boldsymbol{A} = \frac{1}{r^2}\frac{\partial r^2 A_r}{\partial r} + \frac{1}{r\sin\theta}\frac{\partial\left(\sin\theta A_\theta\right)}{\partial \theta} + \frac{1}{r\sin\theta}\frac{\partial A_\varphi}{\partial \varphi}$$

$$\begin{aligned}\nabla^2 f &= \frac{1}{r^2}\frac{\partial}{\partial r}r^2\frac{\partial f}{\partial r} + \frac{1}{r^2\sin\theta}\frac{\partial}{\partial \theta}\sin\theta\frac{\partial f}{\partial \theta} + \frac{1}{r^2\sin^2\theta}\frac{\partial^2 f}{\partial \varphi^2}\\ &= \frac{1}{r}\frac{\partial^2}{\partial r^2}rf + \frac{1}{r^2\sin\theta}\frac{\partial}{\partial \theta}\sin\theta\frac{\partial f}{\partial \theta} + \frac{1}{r^2\sin^2\theta}\frac{\partial^2 f}{\partial \varphi^2}\end{aligned}$$

Divergence theorem:

$$\int_{Volume} dv\, \nabla \cdot \boldsymbol{A}(\boldsymbol{x}) = \int_{Surface} ds\, \hat{\boldsymbol{n}} \cdot \boldsymbol{A}(\boldsymbol{x})$$

Stokes' theorem:

$$\int_{Volume} d\boldsymbol{\ell} \cdot \boldsymbol{A}(\boldsymbol{x}) = \int_{Surface} ds\, \hat{\boldsymbol{n}} \cdot \nabla \times \boldsymbol{A}(\boldsymbol{x})$$

Appendix E The Electromagnetic Hamiltonian and the Goeppert Mayer Transformation

The correct form for the Hamiltonian operator for a charged particle in an electromagnetic field that includes fields like the Coulomb term is given in terms of the vector potential and the longitudinal electric field potential (see Chapter 15 for discussion of potentials in electromagnetics)

$$\hat{H} = \frac{1}{2m}\left(\hat{\boldsymbol{p}} - q\hat{\boldsymbol{A}}\right)^2 + V_{longitudinal} \tag{E.1}$$

in the transverse (Coulomb) gauge, $\boldsymbol{\nabla} \cdot \boldsymbol{A} = 0$. In the dipole approximation, we evaluate the fields at the position $\boldsymbol{R}$, and $\boldsymbol{r}$ is the distance from $\boldsymbol{R}$ to the electron. In this case, the vector potential and electric field have the form:

$$\hat{\boldsymbol{A}}(\boldsymbol{R} + \boldsymbol{r}, t) \approx \hat{\boldsymbol{A}}(\boldsymbol{R}, t) \tag{E.2}$$

$$\hat{\boldsymbol{E}}(\boldsymbol{R}, t) = -\frac{\partial}{\partial t}\hat{\boldsymbol{A}}(\boldsymbol{R}, t) \tag{E.3}$$

We start with Schrödinger's equation:

$$\hat{H}\left|\psi\right\rangle = i\hbar\frac{\partial}{\partial t}\left|\psi\right\rangle \tag{E.4}$$

Using the above form for the Hamiltonian does not result in the usual kind of interaction considered in the previous chapters of the form $V = -\boldsymbol{\mu} \cdot \boldsymbol{E}$, where $\boldsymbol{\mu} = q\boldsymbol{r}$ ($q = -e$ for an electron). However, it is possible to make a unitary transformation called the Goeppert Mayer (after Marie Goeppert Mayer, Nobel Laureate 1963) transformation that leads to the more familiar Hamiltonian. Let

$$\left|\psi(t)\right\rangle = e^{i\frac{q}{\hbar}\boldsymbol{A}(\boldsymbol{R},t)\cdot\boldsymbol{r}}\left|\phi(t)\right\rangle \tag{E.5}$$

Then from the left-hand side of E.4 and using the fact that

$$\hat{\boldsymbol{p}}e^{i\frac{q}{\hbar}\boldsymbol{A}(\boldsymbol{R},t)\cdot\boldsymbol{r}}\left|\phi(t)\right\rangle = -i\hbar\frac{\partial}{\partial \boldsymbol{r}}e^{i\frac{q}{\hbar}\boldsymbol{A}(\boldsymbol{R},t)\cdot\boldsymbol{r}}\left|\phi(t)\right\rangle = e^{i\frac{q}{\hbar}\boldsymbol{A}(\boldsymbol{R},t)\cdot\boldsymbol{r}}\left(q\boldsymbol{A}(\boldsymbol{R}, t) + \boldsymbol{p}\right)\left|\phi(t)\right\rangle \tag{E.6}$$

we have

$$\left[\frac{1}{2m}\left(\hat{p}^2 - 2q\hat{A}(\boldsymbol{R},t)\cdot\hat{p} + q^2\hat{A}^2(\boldsymbol{R},t)\right) + V_{longitudinal}\right] e^{i\frac{q}{\hbar}A(\boldsymbol{R},t)\cdot \boldsymbol{r}}\,|\phi(t)\rangle$$

$$= \left[\frac{e^{i\frac{q}{\hbar}A(R)\cdot \boldsymbol{r}}}{2m}\left(q^2\boldsymbol{A}^2(\boldsymbol{R},t) + 2q\boldsymbol{A}(\boldsymbol{R},t)\cdot\hat{p} + \hat{p}^2 - 2q\hat{A}(\boldsymbol{R},t)\cdot \boldsymbol{p} - 2q^2\boldsymbol{A}^2(\boldsymbol{R},t) + q^2\hat{A}^2(\boldsymbol{R},t)\right)\right.$$

$$\left. + V_{longitudinal}\right]|\phi(t)\rangle = e^{i\frac{q}{\hbar}A(\boldsymbol{R},t)\cdot \boldsymbol{r}}\left[\frac{1}{2m}\hat{p}^2 + V_{longitudinal}\right]|\phi(t)\rangle \tag{E.7}$$

On the right-hand side we have

$$i\hbar\frac{\partial}{\partial t}e^{i\frac{q}{\hbar}A(R,t)\cdot \boldsymbol{r}}\,|\phi(t)\rangle = i\hbar e^{i\frac{q}{\hbar}A(R,t)\cdot \boldsymbol{r}}\left(i\frac{q}{\hbar}\left(\frac{\partial}{\partial t}\boldsymbol{A}(R,t)\right)\cdot \boldsymbol{r} + \frac{\partial}{\partial t}\right)|\phi(t)\rangle$$

$$= e^{i\frac{q}{\hbar}A(R,t)\cdot \boldsymbol{r}}\left(q\boldsymbol{r}\cdot \boldsymbol{E}(R,t) + i\hbar\frac{\partial}{\partial t}\right)|\phi(t)\rangle \tag{E.8}$$

where from Eq. E.3, we have $\boldsymbol{E}(R.t) = -\frac{\partial}{\partial t}\boldsymbol{A}(R,t)$. Equating the result in E.7 (the left-hand side of Eq. E.4) to the result in E.8 (the right-hand side of Eq. E.4), we get, after rearranging terms and cancelling the phase factor from both sides,

$$\left(\frac{\hat{p}^2}{2m} + V_{longitudinal} - \boldsymbol{\mu}\cdot \boldsymbol{E}(R,t)\right)|\phi(t)\rangle = i\hbar\frac{\partial}{\partial t}|\phi(t)\rangle \tag{E.9}$$

where $\boldsymbol{\mu} = q\boldsymbol{r}$. Since the first two terms on the left-hand side above correspond to the Hamiltonian for the charged particle in the absence of an applied electromagnetic field, we set

$$\hat{H}_0 = \frac{\hat{p}^2}{2m} + V_{longitudinal} \tag{E.10}$$

The entire Hamiltonian for the problem becomes the one used in the preceding chapters:

$$\hat{H} = H_0 - \boldsymbol{\mu}\cdot \boldsymbol{E}(R) \tag{E.11}$$

The original state vector is given by Eq. E.4, hence after calculating $|\phi(t)\rangle$, we must then calculate $|\psi(t)\rangle = e^{i\frac{q}{\hbar}A(R)\cdot \boldsymbol{r}}\,|\phi(t)\rangle$. While this is strictly true, in most cases of interest, calculating $\langle\hat{O}\rangle$ for some operator of interest, $\hat{O}$, such as the polarization $\boldsymbol{\mu} = q\boldsymbol{r}$, we find that $\left[\hat{O}, e^{i\frac{q}{\hbar}A(R)\cdot \boldsymbol{r}}\right] = 0$, hence we can work with just $|\phi(t)\rangle$.

Appendix F Maxwell's Equations in Media, the Wave Equation, and Coupling of a Field to a Two-Level System

In general, Maxwell's equations describe the behavior of the electromagnetic field. Of interest here is to show how the observables that were calculated in the text relate directly to Maxwell's equations. For this discussion, we assume that the field generated is classical though the source is an ensemble of quantum systems. The charge and the electric dipole related to the displacement of the charge are central to Maxwell's equations. The electron spin is another quantum system that also couples directly to Maxwell's equations through the magnetic field. The discussion is nearly parallel.

We start by writing down Maxwell's equations in their most general form:

$$\nabla \times \boldsymbol{E} + \frac{\partial \boldsymbol{B}}{\partial t} = 0 \tag{F.1}$$

$$\nabla \times \boldsymbol{H} - \frac{\partial \boldsymbol{D}}{\partial t} = \boldsymbol{J} \tag{F.2}$$

$$\nabla \cdot \boldsymbol{B} = 0 \tag{F.3}$$

$$\nabla \cdot \boldsymbol{D} = \rho \tag{F.4}$$

In the above, $\boldsymbol{J}$ is the current density and ρ is the charge density. Charge is conserved by the continuity relation given by:

$$\nabla \cdot \boldsymbol{J} + \frac{\partial \rho}{\partial t} = 0 \tag{F.5}$$

The constitutive relationship between $\boldsymbol{D}$ and $\boldsymbol{E}$ is

$$\boldsymbol{D} = \varepsilon_0 \boldsymbol{E} + \boldsymbol{P} \tag{F.6}$$

and between $\boldsymbol{B}$ and $\boldsymbol{H}$ is

$$\boldsymbol{B} = \mu_0 \boldsymbol{H} + \mu_0 \boldsymbol{M} \tag{F.7}$$

where $\boldsymbol{P}$ is polarization per unit volume (corresponding to charge displacement), $\boldsymbol{M}$ is the magnetization per unit volume (corresponding to the magnetic field resulting from extrinsic and intrinsic angular momentum), $\varepsilon_0 \cong 8.85 \times 10^{-12}$ farads/meter the permittivity of free space, and $\mu_0 \cong 1.2566 \times 10^{-6}$ henries/meter is the permeability of free space.

For a single charge, the source terms become

$$\rho(\boldsymbol{r}, t) = q\left|\psi(\boldsymbol{r}, t)\right|^2 \tag{F.8}$$

The quantum current is

$$\boldsymbol{J} = -\frac{i\hbar}{2m}\left(\psi^*(\boldsymbol{r},t)\,\nabla\psi(\boldsymbol{r},t) - \psi(\boldsymbol{r},t)\,\nabla\psi^*(\boldsymbol{r},t)\right) \tag{F.9}$$

In bulk media, however, with N being the number of quantum systems per unit volume, the polarization is

$$\boldsymbol{P}(t) = qN\langle\psi(\boldsymbol{r},t)\,|\boldsymbol{r}|\psi(\boldsymbol{r},t)\rangle \tag{F.10}$$

and the magnetization is

$$\boldsymbol{M} = \frac{q}{2m}N\langle\psi(\boldsymbol{r},t)\,|\hat{\boldsymbol{J}}|\psi(\boldsymbol{r},t)\rangle \tag{F.11}$$

The angular momentum, $\hat{\boldsymbol{J}}$ in Eq. F.11, is intended to be general here, meaning that if the magnetic moment is due to intrinsic spin, it would be $\boldsymbol{S}$ and there would be a corresponding g-factor).

Here we are interested in the electron described by a two-level Hamiltonian. Hence, we focus just on $\boldsymbol{P}(t) = qN\langle\psi(\boldsymbol{r},t)\,|\boldsymbol{r}|\psi(\boldsymbol{r},t)\rangle$ and set the other source terms to 0. Maxwell's equations then become

$$\nabla\times\boldsymbol{E} + \frac{\partial\boldsymbol{B}}{\partial t} = 0 \tag{F.12}$$

$$\nabla\times\boldsymbol{B} - \frac{1}{c^2}\frac{\partial\boldsymbol{E}}{\partial t} = \mu_0\frac{\partial\boldsymbol{P}}{\partial t} \tag{F.13}$$

$$\nabla\cdot\boldsymbol{B} = 0 \tag{F.14}$$

$$\nabla\cdot\boldsymbol{E} = 0 \tag{F.15}$$

Combining with the curl equation, we get

$$\nabla\times\nabla\times\boldsymbol{E} + \frac{1}{c^2}\frac{\partial^2\boldsymbol{E}}{\partial t^2} = -\mu_0\frac{\partial^2\boldsymbol{P}}{\partial t^2} \tag{F.16}$$

Assuming that the vector components are in the Cartesian coordinate system, then $\nabla\times\nabla\times\boldsymbol{E} = \nabla\nabla\cdot\boldsymbol{E} - \nabla^2\boldsymbol{E} = -\nabla^2\boldsymbol{E}$ since $\nabla\cdot\boldsymbol{E}$. Substituting this result, we get a wave equation,

$$\nabla^2\boldsymbol{E} - \frac{1}{c^2}\frac{\partial^2\boldsymbol{E}}{\partial t^2} = \mu_0\frac{\partial^2\boldsymbol{P}}{\partial t^2} \tag{F.17}$$

with

$$\boldsymbol{P} = N\langle\boldsymbol{\mu}\rangle = \varepsilon_0\chi\boldsymbol{E} \tag{F.19}$$

and

$$\langle\boldsymbol{\mu}\rangle = Tr\boldsymbol{\mu}\hat{\rho} = (\boldsymbol{\mu}_{12}\hat{\rho}_{21} + \boldsymbol{\mu}_{21}\hat{\rho}_{12}) \tag{F.20}$$

where $\hat{\rho}$ is the density matrix operator in Chapter 18 and χ is the electric susceptibility.

To see the implication of the field-induced polarization and ignoring the vector nature of the field as well as the transient response, we have from Chapter 18 that $\hat{\rho}_{21} \sim E(z)e^{ikz-i\omega t}$ and $\hat{\rho}_{12} \sim E^*(z)e^{-ikz+i\omega t}$. If we set $P_+ = \mathrm{N}\mu_{12}\hat{\rho}_{21}$ and $P_- = \mathrm{N}\mu_{21}\hat{\rho}_{12}$, then we can write P_+ as

$$P_+ = \varepsilon_0 \left(\chi_R + i\chi_I\right) E(z) e^{ikz-i\omega t} \tag{F.21}$$

If the field amplitude, $E(z)$, varies slowly on the scale length of a wavelength, then we can ignore terms in the wave equation that go like $\frac{\partial^2}{\partial z^2}E(z)$ (there is no dependence in the field amplitude on x or y for a transverse plane wave). We can then write the wave equation (Eq. F.17) in the form

$$\left(-k^2 + \frac{\omega^2}{c^2}\right) E(z) + ik\frac{\partial}{\partial z}E(z) = -\frac{\omega^2}{c^2}\left(\chi_R + i\chi_I\right) E(z) \tag{F.22}$$

Setting imaginary and real parts equal to each other, we get two equations:

$$\left(-k^2 + \frac{\omega^2}{c^2}\right) = -\frac{\omega^2}{c^2}\chi_R \tag{F.23}$$

and

$$k\frac{\partial}{\partial z}E(z) = -\frac{\omega^2}{c^2}\chi_I E(z) \tag{F.24}$$

where the first equation represents the linear dispersion relation with

$$k^2 = \frac{\omega^2}{c^2}\left(1 + \chi_R\right) \tag{F.25}$$

$$k = n\frac{\omega}{c} \tag{F.26}$$

where $n^2 = (1 + \chi_R)$ is the index of refraction (the ratio of the speed of light in vacuum to the speed of light in the medium)

The second equation can be solved as

$$E(z) = E\left(z = 0\right) e^{-\frac{\alpha}{2}z} \tag{F.27}$$

where the absorption coefficient (gain if it is negative) is

$$\alpha = \frac{2\omega^2}{kc^2}\chi_I = \frac{2\omega}{n\,c}\chi_I \tag{F.28}$$

To see how to relate the **absorption cross section** to fundamental parameters in the density matrix, we start with Eq. F.20 and use the density matrix from Chapter 18 using first order perturbation theory. From the density matrix,

$$i\dot{\rho}_{21} = \frac{1}{\hbar}[H, \rho]_{21} - i\left(\frac{d\rho_{21}}{d1}\right)_{decoherence} = \left(\omega_0 - i\gamma\right)\rho_{21} - \frac{\mu_{21}\tilde{E}e^{-i\omega t}}{2\hbar}\left(\rho_{11} - \rho_{22}\right) \tag{F.29}$$

Take for ρ_{21} in the field interaction picture:

$$\rho_{21} = \tilde{\rho}_{21} e^{-i\omega t} \tag{F.30}$$

Then

$$\omega \tilde{\rho}_{21} = \omega_0 \tilde{\rho}_{21} - i\gamma \tilde{\rho}_{21} - \frac{\mu_{21}\tilde{E}}{2\hbar}(\rho_{11} - \rho_{22}) \tag{F.31}$$

$$\tilde{\rho}_{21} = -\frac{\mu_{21}\tilde{E}}{2\hbar\left[(\omega - \omega_0) + i\gamma\right]}(\rho_{11} - \rho_{22}) \tag{F.32}$$

At low power, $\rho_{11} - \rho_{22} = 1$.

$$\tilde{\rho}_{21} = -\frac{\mu_{21}\tilde{E}}{2\hbar\left[(\omega - \omega_0) + i\gamma\right]} \tag{F.33}$$

and so (to first order in E, linear response)

$$\tilde{P}_{+} = -N\frac{\mu_{12}\mu_{21}\tilde{E}}{2\hbar\left[(\omega - \omega_0) + i\gamma\right]} \tag{F.34}$$

From this,

$$\alpha = \frac{2\omega^2}{c^2 k}\chi_I = \frac{2\omega}{c}\chi_I \equiv N\sigma \tag{F.35}$$

where σ is the cross section for absorption and is defined by this relationship. Since this is linear theory, we use the phasor $e^{-i\omega t}$ as used in the Maxwell equation work and the quantum work, and we have for the prefactor to the phasor,

$$\tilde{P}_{+} = \epsilon_0\left(\chi_R + i\chi_I\right)N\tilde{E} = \mu_{12}\tilde{\rho}_{21} = -\frac{\mu_{12}\mu_{21}\tilde{E}}{2\hbar\left[(\omega - \omega_0) + i\gamma\right]}N = -\frac{\mu_{12}\mu_{21}\tilde{E}\left[(\omega - \omega_0) - i\gamma\right]}{2\hbar\left[(\omega - \omega_0)^2 + \gamma^2\right]}N \tag{F.36}$$

$$\chi_I = \frac{\mu_{12}\mu_{21}\gamma}{2\hbar\epsilon_0\left[(\omega - \omega_0)^2 + \gamma^2\right]}N = \frac{\mu_{12}\mu_{21}\gamma}{2\hbar\epsilon_0\left[(\omega - \omega_0)^2 + \gamma^2\right]}N = \frac{\mu_{12}\mu_{21}}{2\hbar\epsilon_0\gamma}N\mathcal{L}(\Delta) \tag{F.37}$$

where $\mathcal{L}(\Delta) = \frac{\gamma^2}{\left[(\omega - \omega_0)^2 + \gamma^2\right]}$. Finally, from the form for α above,

$$\sigma = 2\frac{\omega}{c}\frac{\chi_I}{N} = 2\frac{\mu_{12}\mu_{21}\omega}{2\hbar\epsilon_0 c\gamma}\mathcal{L}(\Delta) = \alpha_{FS}\frac{4\pi r_{12}r_{21}\omega}{\gamma}\mathcal{L}(\Delta) \tag{F.38}$$

where the fine structure constant is

$$\alpha_{FS} = \frac{e^2}{4\pi\epsilon_0\hbar c} \tag{F.39}$$

Note that there is a correction factor of order unity in Eq. F.38 associated with the radial matrix elements in the last expression discussed in Appendix G. The fine structure constant is dimensionless and is approximately $\alpha_{FS} \sim \frac{1}{137}$. The dependence on the fine structure constant is frequently cited as a hall-mark of optical interactions and it reflects the intrinsically weak interaction between light and charged particles. However, at resonance,

$$\sigma_0 = \alpha_{FS} \frac{4\pi r_{12} r_{21} \omega}{\gamma} \tag{F.40}$$

In the absence of pure dephasing,

$$\gamma = \frac{1}{2}\Gamma_{sp.em.} = \frac{1}{2}A \tag{F.41}$$

where A is the Einstein A-coefficient, or the inverse radiative lifetime.

$$A = \frac{4}{3}\frac{e^2 r_{12} r_{21} \omega^3}{4\pi\epsilon_0 \hbar c^3} = \frac{4}{3}\frac{\alpha_{FS} r_{12} r_{21} \omega^3}{c^2} \tag{F.42}$$

Substituting into the cross section at resonance,

$$\sigma_0 = \frac{6\pi c^2}{\omega^2} = \frac{6\pi}{k^2} = \frac{3}{2\pi}\lambda^2 \tag{F.43}$$

There are many important results here for real device studies; however, an important piece of fundamental physics that impacts technology is the following.

In quantum electrodynamics *a famous result is that electromagnetic radiation interacts with charged particles only weakly, resulting from the fine structure constant.* This is misleading when the transition is a resonance and the transition is lifetime broadened. Just for comparison, for the optical wavelength, $\sigma_0 \sim 10^{-12}$ square meters, but for scattering from a free electron (no resonance, Thompson scattering) $\sigma \sim 10^{-28}$ square meters.

Appendix G Wigner–Eckart Theorem and Irreducible Tensors

The discussions in the text have shown that it is important to understand the impact of symmetry in quantum systems. There are many symmetries that are important to the field, beyond spatial symmetry, but their impact in quantum lies beyond the scope of this presentation. However, spherical symmetry has been seen to be particularly important as evidence not only in the solution for the hydrogen atom in Chapter 6 but also was seen as being fundamental to the concept of angular momentum in Chapter 10. Since atoms are building blocks of all matter it is not surprising that when it is time to get the numbers for a specific problem, one is often having to evaluate matrix elements of operators involving spherically symmetric basis states. The development of this problem and the proofs of fundamental results is the focus of entire books heavily cited for their pioneering work. Fortunately, understanding that level of detail is not required in order to exploit and implement the results. Historically, the results would send the engineer or physicist to tables where numerous results were tabulated, but standard symbolic manipulation software like Mathematica has made this anachronistic. Here, we will present some of the important results for calculating matrix elements associated with angular momentum, without proof for the most part, and explain how to use them.[1]

We begin by identifying the irreducible tensor of rank k as having $2k + 1$ components, q, represented by $T_q^{(k)}$. You recall in Chapter 10 that we introduced the rotation operator and you found the rotation matrix for the spin ½ system. There are rotation matrices for any arbitrary j, $D^j_{m\,m'}$ for rotation from the primed coordinate system to the unprimed coordinates. An operator $T\left(jm'\right)$ that transforms under rotation according to $T(jm) = \sum_{m'} D^j_{m\,m'} T\left(jm'\right)$ is defined to be irreducible. An alternate definition is from Racah and is given by the commutation relations:

$$\left[\hat{J}_\pm,\ T_q^{(k)}\right] = \sqrt{(k \mp q)\,(k \pm q + 1)}\,T_{q\pm1}^{(k)} \tag{G.1}$$

$$\left[\hat{J}_z, T_q^{(k)}\right] = qT_q^{(k)} \tag{G.2}$$

where

$$\hat{J}_\pm = \hat{J}_x \pm i\hat{J}_y \tag{G.3}$$

Examples of a scalar irreducible operator are $\hat{J}^2$, $\hat{L}^2$, and $\hat{S}^2$. Notice that, in the definition, $\hat{J}_\pm$ and $\hat{J}_z$ are place holders for the operator of the corresponding Hilbert space. The spherical harmonics, Y_{lm}, are each an irreducible tensor operator of rank $k = l$ and component $q = m$.

Any vector $\boldsymbol{A} = A_x\check{x} + A_y\check{y} + A_z\check{z}$ can be written in irreducible form by writing the different components as $|\boldsymbol{A}|$ multiplied by the appropriate combination of spherical harmonics. The resulting

[1] I.I. Sobelman *Introduction to the Theory of Atomic Spectra*, Pergamon Press, 1972 (Oxford) presents many more useful results that are essential for many calculations. See also K.T. Hecht, *Quantum Mechanics*, Springer-Verlag, New York.

vector components are called spherical vectors. They are given by

$$\begin{aligned} A_{-1} &= +\tfrac{A_x - iA_y}{\sqrt{2}} = |\mathbf{A}|\sqrt{\tfrac{4\pi}{3}}\,Y_{1-1} = |\mathbf{A}|C^{(1)}_{-1} \\ A_0 &= A_z = |\mathbf{A}|\sqrt{\tfrac{4\pi}{3}}\,Y_{1\ 0} = |\mathbf{A}|C^{(1)}_{0} \\ A_{+1} &= -\tfrac{A_x + iA_y}{\sqrt{2}} = |\mathbf{A}|\sqrt{\tfrac{4\pi}{3}}\,Y_{1\ 1} = |\mathbf{A}|C^{(1)}_{1} \end{aligned} \tag{G.4}$$

where

$$C^{(l)}_m = \sqrt{\frac{4\pi}{2l+1}}\,Y_{l\ m} \tag{G.5}$$

is the **Racah tensor**. The corresponding vector is then written out completely as:

$$\mathbf{A} = A_x\check{x} + A_y\check{y} + A_z\check{z} = -A_{-1}\check{e}_1 + A_0\check{e}_0 - A_1\check{e}_{-1} \tag{G.6}$$

where $\check{e}_i$ is the corresponding spherical unit vector given by

$$\begin{aligned} \check{e}_{-1} &= \tfrac{\check{x} - i\check{y}}{\sqrt{2}} \\ \check{e}_0 &= \check{z} \\ \check{e}_1 &= -\tfrac{\check{x} + i\check{y}}{\sqrt{2}} \end{aligned} \tag{G.7}$$

For the inner product of two vectors, such as $\boldsymbol{A} \cdot \boldsymbol{B}$ then you can show that this is generalized for two irreducible tensors of the same rank as

$$T^{(k)} \cdot U^{(k)} = \sum_q (-1)^q T^{(k)}_{-q} U^{(k)}_q \tag{G.8}$$

Now thinking back to various problems in dynamics that were considered starting in Chapter 9, it is common to have an interaction such as $\hat{\mu} \cdot \boldsymbol{E}$ where $\hat{\mu} = q\boldsymbol{r}$ for the electric dipole operator or $\hat{\boldsymbol{\mu}}_{\boldsymbol{B}} \cdot \boldsymbol{B}$ where $\hat{\boldsymbol{\mu}}_{\boldsymbol{B}} = \beta g \frac{\hat{\mathbf{S}}}{\hbar}$ for the magnetic dipole operator. Both $\boldsymbol{r}$ and $\hat{\mathbf{S}}$ can now be written in terms of irreducible tenor operators. We consider the case for the electron dipole moment interacting with an electric field:

$$\hat{\boldsymbol{\mu}} \cdot \mathbf{E} = -e\sum_q (-1)^q rC^{(1)}_{-q} E^{(k)}_q \tag{G.9}$$

The matrix element that must be found for a spherically symmetric system is $\langle nlm|\hat{\boldsymbol{\mu}}|n'l'm'\rangle$. But the component of $\hat{\boldsymbol{\mu}}$ can be written as the product of the scalar r and a Racah tensor, C^1_q. Hence, that portion of the matrix element that depends on the angular part is separate from the part that depends on the radial part. Moreover, the radial component clearly has no dependence on the magnetic quantum number, m. The matrix element is evaluated with the **Wigner–Eckart theorem**:

$$\langle nlm|\hat{\boldsymbol{\mu}}|n'l'm'\rangle = C(l,1,m,q|l'm')\frac{\langle nl||rC^{(1)}||n'l'\rangle}{\sqrt{2l+1}} = (-1)^{l-m}\langle nl||rC^{(1)}||n'l'\rangle \begin{pmatrix} l & k & l' \\ -m & q & m' \end{pmatrix} \tag{G.10}$$

or more generally

$$\langle \alpha jm | T_q^{(k)} | \alpha' j' m' \rangle = (-1)^{j-m} \langle \alpha j || T^{(k)} ||, \alpha', j' \rangle \begin{pmatrix} j & k & j' \\ -m & q & m' \end{pmatrix} \tag{G.11}$$

where $C(l, l', m, m'|1m)$ is the Clebsch–Gordan coefficient from Chapter 10, the symbol $\langle \alpha j || T^{(k)} || \alpha' j' \rangle$ is called the reduced matrix element, and $\begin{pmatrix} j & k & j' \\ -m & q & m' \end{pmatrix}$ is related to the Clebsch –Gordan coefficient, but is called the Wigner 3j symbol:

$$\begin{pmatrix} j & k & j' \\ m & q & -m' \end{pmatrix} = (-1)^{-j+k-m'} \frac{C(j, k, m, q|j'm')}{\sqrt{2j'+1}} \tag{G.12}$$

For the Wigner 3j symbol to be non-zero, we require $-m + q + m = 0$ and the triangle condition $\Delta(j, k, j')$ meaning $|j - k| \leq j' \leq j + k$. The Wigner 3j is introduced in this discussion because it has more readily exploited symmetries and summation properties that are important in some calculations. Two important relations for 3j symbols are the effects of interchanging columns:

$$\begin{pmatrix} j_1 & j_2 & j \\ m_1 & m_2 & m \end{pmatrix} = \begin{pmatrix} j_2 & j & j_1 \\ m_2 & m & m_1 \end{pmatrix} = \begin{pmatrix} j & j_1 & j_2 \\ m & m_1 & m_2 \end{pmatrix} = (-1)^{j_1+j_2+j} \begin{pmatrix} j_2 & j_1 & j \\ m_2 & m_1 & m \end{pmatrix}$$
$$= (-1)^{j_1+j_2+j} \begin{pmatrix} j_1 & j & j_2 \\ m_1 & m & m_2 \end{pmatrix} = (-1)^{j_1+j_2+j} \begin{pmatrix} j & j_2 & j_1 \\ m & m_2 & m_1 \end{pmatrix}$$

and

$$\begin{pmatrix} j_1 & j_2 & j \\ m_1 & m_2 & m \end{pmatrix} = (-1)^{j_1+j_2+j} \begin{pmatrix} j_1 & j_2 & j \\ -m_1 & -m_2 & -m \end{pmatrix}$$

Before continuing with evaluating G.10, some additional discussion is needed as well as some examples. The Wigner–Eckart theorem is often initially confusing because there is no definition of the reduced matrix element. Rather, it represents a number, to be determined, that relates the matrix element to the 3j symbol. What is important is that the number depends on the details of the overlap between the eigenfunctions corresponding to the bra and ket and the operator itself. That number does not change as the magnetic quantum numbers change or the component of q changes. In the case of Eq. G.10 for the electric dipole moment operator, the number does not change for any of the polarizations or changes in the magnetic substates. The number itself is defined from Eq. G.11 as

$$\langle \alpha j || T^{(k)} || \alpha' j' \rangle = (-1)^{j-m} \frac{\langle \alpha jm | T_q^{(k)} | \alpha' j' m' \rangle}{\begin{pmatrix} j & k & j' \\ -m & q & m' \end{pmatrix}} \tag{G.13}$$

Of course, this is circular. However, the importance is that once the number is determined, it becomes a known number that you record and proceed to use G.11, in order to evaluate the desired matrix elements.

A simple example is the reduced matrix element for $\hat{\mathbf{S}}$ for a spin ½ system. The components of the corresponding irreducible tensor are $S_q^{(1)}$. To find $\langle \alpha s || S^{(1)} ||, \alpha', s' \rangle$, we pick an easy form of $\langle \alpha sm | S_q^{(1)} | \alpha' s' m' \rangle$. For this, take $s = \frac{1}{2}$; $m = \frac{1}{2}$; and $q = 0$. Recall from G.4 that $S_0^{(1)} = S_z$. Then

$$\left\langle \alpha \frac{1}{2} \frac{1}{2} \left| S_q^{(1)} \right| \alpha' s' m' \right\rangle = \left\langle \alpha \frac{1}{2} \frac{1}{2} \left| S_z \right| \alpha' s' m' \right\rangle = m' \hbar \delta_{\alpha\alpha'} \delta_{ss'} \delta_{\frac{1}{2} m'} \tag{G.14}$$

From tabulated values next evaluate the 3j symbol $s = s' = \frac{1}{2}$ and $m = m' = \frac{1}{2}$

$$\begin{pmatrix} \frac{1}{2} & 1 & \frac{1}{2} \\ -\frac{1}{2} & 0 & \frac{1}{2} \end{pmatrix} = \frac{1}{3}$$

Given G. 14, we require $\alpha = \alpha'; s = s' = \frac{1}{2}$, This gives us $\langle \alpha\frac{1}{2}||S^{(1)}||\alpha'\frac{1}{2}\rangle = \dfrac{m\langle \alpha\frac{1}{2}\frac{1}{2}\,|S_q^{(1)}|\alpha' s' m'\rangle}{\begin{pmatrix} j & k & j' \\ -m & q & m' \end{pmatrix}} = \frac{3}{2}\hbar\delta_{\alpha\alpha'}$. Then for an arbitrary matrix element and vector component

$$\left\langle \alpha\frac{1}{2}m\middle|S_q^{(1)}\middle|\alpha'\,\frac{1}{2}m'\right\rangle = (-1)^{\frac{1}{2}-m}\left\langle \alpha\frac{1}{2}\middle\|S^{(1)}\middle\|\,\alpha'\frac{1}{2}\right\rangle\begin{pmatrix} \frac{1}{2} & 1 & \frac{1}{2} \\ -m & q & m' \end{pmatrix} \tag{G.15}$$

Consider now the problem of determining the matrix elements of the electron dipole moment operator in a spherically symmetric system, given in Eq. G.11. We examine the matrix element for the q component of the operator, corresponding to the $-q$ component of the field:

$$\langle nlm|rC_q^{(1)}|n'l'm'\rangle = (-1)^{l-m}\langle nl||rC^{(1)}||n'l'\rangle\begin{pmatrix} l & k & l' \\ -m & q & m' \end{pmatrix} \tag{G.16}$$

To evaluate $\langle nl||rC^{(1)}||n'l'\rangle$, we must evaluate a matrix element of our choice. For this, take $n, l, m = n, l, 0$ and $q = 0$. Therefore $C_0^{(1)} = \sqrt{\frac{4\pi}{3}}Y_{1,0}$. To go further, we must now replace the Dirac representation of the matrix element, $\langle nlm|rC_q^{(1)}|n'l'm'\rangle$ with the coordinate representation (recall the notes have frequently mentioned that to get numbers, it is often necessary to use the spatial solutions to the time independent Schrödinger equation). In this case

$$\begin{aligned}
&\langle nlm|rC_q^{(1)}|n'l'm'\rangle \\
&= \int_0^\infty r^2 dr \int_0^{2\pi} d\phi \int_0^\pi \sin\theta d\theta R_{nl}^*(r)Y_{lm}^*(\theta,\phi)\left(r\sqrt{\frac{4\pi}{3}}Y_{1,0}\right)R_{n'l'}(r)Y_{l'm'}(\theta,\phi) \\
&= \sqrt{\frac{4\pi}{3}}\int_0^\infty r^3 R_{nl}^*(r)R_{nl}(r)dr\int_0^{2\pi} d\phi\int_0^\pi \sin\theta d\theta R_{nl}^*(r)Y_{lm}^*(\theta,\phi)\left[Y_{1,0}(\theta,\phi)\right]R_{nl}(r)Y_{l'm'}(\theta,\phi)
\end{aligned} \tag{G.17}$$

The radial integral is defined by the symbol

$$\langle nl||r||\,n'l'\rangle = \int_0^\infty r^3 R_{nl}^*(r)R_{n'l'}(r)dr \tag{G.18}$$

To get the angular part of the integral, we use a tabulated integral over three spherical harmonics given as[2]

$$\begin{aligned}
&\int_0^{2\pi} d\phi\int_0^\pi \sin\theta d\theta Y_{lm}^*(\theta,\phi)\left[Y_{k,q}(\theta,\phi)\right]Y_{l'm'}(\theta,\phi) \\
&= (-1)^1\sqrt{\frac{(2l+1)(2k+1)(2l'+1)}{4\pi}}\begin{pmatrix} l & k & l' \\ 0 & 0 & 0 \end{pmatrix}\begin{pmatrix} l & k & l' \\ -m & q & m' \end{pmatrix}
\end{aligned} \tag{G.19}$$

[2] I.I. Sobelman, ibid.

Referring now to the angular part of the integral in G.17, we take $k = 1$. Then referring to the tabulated listing of 3 j symbols $\begin{pmatrix} l & 1 & l' \\ 0 & 0 & 0 \end{pmatrix} = 0$ if $l + l' + 1$ is an odd number. This means that the change in l called $\Delta l \neq 0$, this is often the most important conclusion of this exercise. However, to complete the problem, we note that if $l + l' + 1 \equiv 2g$ where g is an integer, then

$$\begin{pmatrix} l & 1 & l' \\ 0 & 0 & 0 \end{pmatrix} = (-1)^g \sqrt{\frac{(2g-2l)!\,(2g-2)!\,(2g-2l')!}{(2g+1)!}}\,\frac{g!}{(g-l)!\,(g-1)!\,(g-l')!} \tag{G.20}$$

For an S to P transition, $g = 1$ and $\begin{pmatrix} 0 & 1 & 1 \\ 0 & 0 & 0 \end{pmatrix} = -\sqrt{\frac{1}{3}}$. Note that the **selection rule** (meaning the requirement for the matrix element to be non-zero) for the orbital angular momentum, we have seen that l must change, but we also have the triangle condition $\Delta(l, 1, l')$. This means that $|l-1| \leq l' \leq l+1$, which means that $l = 0$ or ± 1. But since $\Delta l \neq 0$, the selection rule is $\Delta l = \pm 1$. Since a photon carries one unit ($\hbar$) of angular momentum, this rule is sometimes cited as evidence for conservation of angular momentum in photonic transitions.

The selection rule on the field polarization is determined by requirement on the change in the magnetic substate quantum number, m, for a given polarization determined by q. The condition for a non-zero matrix element is that $-m + q + m' = 0$. Therefore $m' = m - q$. For $q = -1, 0, 1$, $m' = m+1, m, m-1$ or $\Delta m = 0, \pm 1$.

Based on this analysis, we can now evaluate the angular part of the integral in Eq. G.17:

$$\int_0^{2\pi} d\phi \int_0^{\pi} \sin\,\theta d\theta R^*_{nl}(r) Y^*_{lm}(\theta, \phi)\left[Y_{1,0}(\theta, \phi)\right] R_{nl}(r) Y_{l'm'}(\theta, \phi)$$
$$= -\sqrt{\frac{(2l+1)(2+1)(2l \pm 2 + 1)}{4\pi}} \begin{pmatrix} l & 1 & l \pm 1 \\ 0 & 0 & 0 \end{pmatrix} \begin{pmatrix} l & 1 & l \pm 1 \\ -m & q & m' \end{pmatrix} \tag{G.21}$$

For the most common transition of S to P, we get

$$\sqrt{\frac{3}{4\pi}} \begin{pmatrix} 0 & 1 & 1 \\ 0 & q & m' \end{pmatrix} = (-1)^{-1-q}\sqrt{\frac{1}{4\pi}} \text{ with } q = -m \text{ and } m = 0, \pm 1 \text{ for the P-state.}$$

Finally, for the matrix element, we get

$$\langle nlm | rC_q^{(1)} | n'l'm' \rangle = (-1)^{-1-q} \sqrt{\frac{1}{4\pi}} \langle nl \| r \| n'l' \rangle \tag{G.22}$$

There is no constraint on n. Assuming that the change in n is small so that the wave functions for each state spatially overlap and that the transition involves starting in the ground state, the magnitude of the matrix element is on the order of the Bohr radius for the system.

Another important result that is not hard to derive and that we used in developing the Landé g-factor in Chapter 11 is:

$$\langle \alpha jm | A_q^{(1)} | \alpha' jm' \rangle = \frac{\langle \alpha jm | \mathbf{A} \cdot \mathbf{J} | \alpha' jm' \rangle}{j(j+1)\hbar^2} \langle \alpha jm | J_q^{(1)} | \alpha' jm' \rangle \delta_{\alpha\alpha'} \tag{G.23}$$

Index

H

I

J

K

L

M

N

O

P

Q

R

S